Coalitions and alliances
in humans and other animals

Coalitions and alliances in humans and other animals

Edited by

ALEXANDER H. HARCOURT
Department of Zoology, Cambridge University, Cambridge, UK
and
Department of Anthropology, University of California, Davis, USA

and

FRANS B. M. DE WAAL
Wisconsin Regional Primate Research Center, University of Wisconsin-Madison, USA
and
Yerkes Regional Primate Center/Psychology Department, Emory University, Atlanta, USA

OXFORD NEW YORK TOKYO
OXFORD UNIVERSITY PRESS
1992

Oxford University Press, Walton Street, Oxford OX2 6DP
Oxford New York Toronto
Delhi Bombay Calcutta Madras Karachi
Petaling Jaya Singapore Hong Kong Tokyo
Nairobi Dar es Salaam Cape Town
Melbourne Auckland
and associated companies in
Berlin Ibadan

Oxford is a trade mark of Oxford University Press

Published in the United States
by Oxford University Press, New York

A catalogue record for this book is
available from the British Library

Library of Congress Cataloging in Publication Data
Coalitions and alliances in humans and other animals/edited by A. H. Harcourt and Frans B. M. de Waal.
Includes bibliographical references and index.
1. Primates—Behavior. 2. Cooperativeness. 3. Competition (Psychology) 4. Psychology. Comparative. 5. Social behavior in animals. 6. Sociobiology.
I. Harcourt, A. H. (Alexander H.) II. Waal, F. B. M. de (Frans B. M.), 1948–
QL737.P9U78 1992 599.8′0451—dc20 91–4285
ISBN 0–19–854273–9 (hbk.)

Typeset by Joshua Associates Ltd, Oxford
Printed and bound in Great Britain by
Bookcraft (Bath) Ltd
Midsomer Norton, Avon.

Preface

The well-known human proclivity to form coalitions and alliances undoubtedly has a long evolutionary history. We share the characteristic with our closest animal relatives—the monkeys and the apes—as well as with a number of other species. Coalitions and alliances are defined in both ethology and the social sciences as a joining of forces of several parties in order to gain an advantage over another party. Coalitions are therefore of a competitive character. Since this is one of the most conspicuous and ubiquitous types of cooperation observed in primate societies, we are faced with the profound paradox (some would say irony) that cooperative tendencies are probably an evolutionary offshoot of competitive and aggressive tendencies.

Given the tremendous increase in knowledge about animal coalitions over the past two decades, and the absence of any serious attempts to take stock of the progress, each of the editors independently judged the time to be ripe for a volume that would bring together some of the most important studies. Instead of pursuing our separate plans, we followed the spirit of our subject and decided on a cooperative endeavour.

The purpose of the book is to integrate different approaches and disciplines, ranging from experimental psychology to evolutionary biology, and to include a variety of species and taxa. As we wanted each chapter to stand on its own, we allowed some overlap of material between chapters. In addition, since research on coalitions and alliances is still very much in development, we realized from the outset that there would be clashing theoretical views. Instead of trying to suppress or iron out such 'inconsistencies' within the volume, we thought it more interesting to point them out to the readers through editorials and cross-referencing. The editorial contributions form an introduction to each part of the book. The result is a book that looks at coalition formation from a number of quite different angles, and brings together authors from such different fields that many have been unaware of each other's work. In biology, taxonomic specialization is quite common, and exchanges between biologists and social scientists are notoriously limited.

One source of misunderstandings between social scientists and biologists is the level of analysis. Evolutionary biologists try to understand how a particular behaviour came into existence. In order to do so, they develop theories about a behaviour's impact on reproductive success. Provided a behaviour is partially under genetic control (no behaviour is under complete genetic control) and increases the survival and reproduction of the performer or its close relatives, the behaviour will spread through the population. This is known as the *ultimate* explanation of the behaviour, which needs to be sharply distinguished from its *proximate* explanation. The latter concerns the direct experiences, stimuli, and situations that evoke the behaviour. For example, the proximate cause of a coalition between two male primates may be the sight of a sexually receptive female near another male. If they manage to defeat the third male, steal the female, and mate with her, then the act of cooperation might affect their reproductive success and—although the primates themselves are probably totally unaware of this connection—this may be the ultimate reason for the behaviour.

Social scientists deal almost exclusively with the proximate level, whereas biologists deal with both levels and are trained to keep them apart and view them as complementary. It is important to realize that when some of the ethologists in this volume discuss the biological basis of cooperative strategies in animals and humans they are not arguing that these strategies are completely or even largely genetically determined. Hereditary influences by no means exclude the role of learning and experience; several other authors discuss the same animal behaviour in proximate terms.

Throughout this volume the following terminology will be employed. *Cooperation* is defined as an acting together of two or more individuals such that at least one amongst them stands to gain benefits unavailable through solitary action (this definition leaves in the middle the exact cost/benefit balance per individual party, i.e. all parties may benefit, or one party may benefit from costly behaviour by another). A *coalition* is defined as a one-time cooperative action by at least two individuals or units against at least one other individual or unit. Usually a coalition results from a response by outsiders to an ongoing confrontation between two parties. Dependent on their role in the interaction, the parties are called the *intervener* or *supporter* (the individual entering the original conflict by supporting one party or the other); the *recipient* (the party receiving the intervener's support), and the *target* (the party against whom the coalition is formed). The term *alliance*, finally, is reserved for long-term cooperative relationships, that is, partnerships that form coalitions on a regular basis.

We appreciate the authors' enthusiasm for this project, and their responsiveness to our long lists of critiques, questions, and suggestions. Several authors helped referee one another's papers, and we thank them for their assistance. We also wish to express our gratitude to a number of outside referees, Patricia J. Turner, Randy M. Siverson, Barbara B. Smuts, and Kelly J. Stewart. The last is also thanked for editorial help.

Madison, WI, USA F. B. M. de W.
Davis, CA, USA A. H. H.
May 1991

Contents

Contributors

Irwin S. Bernstein, Department of Psychology, University of Georgia, Athens, GA 30602, USA

Christopher Boehm, Department of Anthropology, Northern Kentucky University, Highland Heights, KY 41076, USA

Robert Boyd, Department of Anthropology, University of California, Los Angeles, CA 90024, USA

Bernard Chapais, Département d'Anthropologie, Université de Montréal, C.P 6128, Succ. A, Montréal, H3C 3J7, Canada

Richard C. Connor, Department of Biology, University of Michigan, Ann Arbor, MI 48109, USA. *Current address*: Woods Hole Oceanographic Institution, Woods Hole, MA 02543, USA

S. B. Datta, MRC Unit on the Development and Integration of Behaviour, University of Cambridge, Madingley, Cambridge CB3 8AA, UK

Carolyn L. Ehardt, Department of Anthropology, University of Georgia, Athens, GA 30602, USA

Vincent S. E. Falger, Department of International Relations, University of Utrecht, Janskerhof 3, 3512 BK Utrecht, The Netherlands

Laurence G. Frank, Department of Psychology, University of California, Berkeley, CA 97420, USA

Stephen E. Glickman, Department of Psychology, University of California, Berkeley, CA 94720, USA

Karl Grammer, Forschungsstelle fur Humanethologie in der Max-Planck-Gesellschaft, D-W8138 Andechs, Germany

A. H. Harcourt, Department of Anthropology, University of California, Davis, California 95616, USA

Jan A. R. A. M. van Hooff, Ethologie en Socio-oecologie Groep, Universiteit Utrecht, Postbus 80.086, 3508 TB, Utrecht, The Netherlands

J. A. Johnson, Department of Zoology, University of Edinburgh, West Mains Road, Edinburgh EH9 3JT, UK

Geoffrey Keppel, Department of Psychology, University of California, Berkeley, CA 94720, USA

P. C. Lee, Department of Biological Anthropology, University of Cambridge, Downing Street, Cambridge CB2 3DZ, UK

Ronald Noë, Ethologie und Wildforschung, Universität Zürich, Switzerland; Ethologie en Socio-oecologie, Universiteit Utrecht, The Netherlands. *Current address*: Max Planck Institut für Verhaltensphysiologie, Abteilung Wickler, Seewiesen, WD-8130 Starnberg, Germany

Jacob M. Rabbie, Institute of Social Psychology, University of Utrecht, Postbus 80140, 3508 TC, Utrecht, The Netherlands

Andrew F. Richards, Department of Biology, University of Michigan, Ann Arbor, MI 48109, USA

Carel P. van Schaik, Department of Biological Anthropology and Anatomy, Duke University, The Wheeler Building, 3750-B Erwin Road, Durham, NC 27705, USA

Joan B. Silk, Department of Anthropology, University of California, Los Angeles, CA 90024, USA; California Primate Research Center, University of California, Davis, CA 95616, USA

Rachel A. Smolker, Psychology Department, University of Michigan, Ann Arbor, MI 48109, USA

Frans B. M. de Waal, Yerkes Regional Primate Center and Psychology Department, Emory University, Atlanta, GA 30322, USA

Katya B. Woodmansee, Department of Psychology, University of California, Berkeley, CA 94720, USA

Cynthia J. Zabel, Department of Psychology, University of California, Berkeley, CA 94720, USA. *Current address*: Pacific SW Forestry Station, Redwood Scientific Lab., 1700 Bay View Dr., Arcata, CA 95521, USA

1

Coalitions and alliances: a history of ethological research

FRANS B. M. DE WAAL AND A. H. HARCOURT

COALITIONS AND OTHER FORMS OF COOPERATION

It is not hard to understand why the study of cooperative behaviour in animals has introduced biologists to cost/benefit analysis and to the language of economics. Behaviour that assists the survival and reproduction of the recipient while diminishing the survival and reproduction of the performer poses a serious dilemma for evolutionary explanations. How could such behaviour ever have evolved without at least some hidden or indirect benefit?

Interest in this problem dates back to Darwin (1859) and Kropotkin (1902), but it is only relatively recently that satisfactory solutions have been proposed. One is known as 'kin selection'; that is, the genes for a particular cooperative behaviour may spread if this behaviour promotes the survival and reproduction of kin. From a genetic perspective it does not matter whether the genes are multiplied through the performer's own reproduction or through that of its relatives (Hamilton 1964). The other explanation is known as 'reciprocal altruism'; that is, behaviour that is costly in the short run may nevertheless benefit the performer in the long run provided that recipients tend to show the same or other valuable services in return (Trivers 1971).

There exists an immense array of costly cooperative behaviour in animals, such as warning calls by birds that allow conspecifics to escape a predator's claws yet attract attention to the caller itself; sterile castes in the social insects that do little else than serve food to the larvae of their queen, or sacrifice themselves in defence of the colony; food sharing in carnivores from vampire bats to wild dogs; support of an injured conspecific close to the surface by dolphins in order to prevent drowning; helpers at the nest that enable a breeding pair of jays to fill more hungry mouths and thus raise more offspring than otherwise possible, etc. (Norris and Schilt 1988; Trivers 1985; Wilson 1975).

However, by no means all forms of cooperation involve important costs to the performer. If, for example, the killing of a particular prey cannot be achieved by a single member of a carnivorous species, should we call the killing of this type of prey by a pack of predators an exchange of altruism among the pack members? Inasmuch as the benefits of this cooperation are immediate and apply to all hunters, the cooperation seems rather selfish. Rothstein and Pierotti (1988) review the entire range of cooperative schemes from so—called by-product beneficence to reciprocal altruism and simultaneous cooperation. Even within the rather limited category of cooperative activities considered in the present volume, important variation in the cost/benefit balance can be recognized.

In ethology, a coalition is defined as cooperation in an aggressive or competitive context. This ranges from acts of support that are potentially extremely costly to the performer (e.g. a mother monkey taking a physical risk while defending her offspring against a dominant male) to those that involve little risk and great benefits (e.g. the infliction of injuries upon a rival which is being subdued by a dominant party). Whether altruistic or opportunistic in nature, these acts share one characteristic: the interests of the cooperating parties are served at the expense of the interests of a third party. It is this well-coordinated 'us' against 'them' character that sets coalition formation apart from other cooperative interactions among conspecifics.

Students of animal behaviour have borrowed the term 'coalition' from other disciplines. In the social sciences the term refers to a number of individuals or social units cooperating in order to obtain any kind of advantage over other individuals or units (including the advantage of making a decision for the entire group). The keywords are cooperation and competition: a coalition is a social tool to defeat others. Competition may also occur within the coalition itself: coalition members do not automatically share the advantages of their joint efforts equally.

While additional criteria, such as exclusiveness of membership, may be adopted, the above criteria seem to be the principal ones. Shaw (1971) summarizes a number of definitions when stating that a coalition occurs when (1) three or more individuals are involved, (2) two or more act as a unit against at least one other, and (3) the joint action potentially produces a result superior to any result possible by individual action. Similarly, Gamson (1961) views a coalition as the joint use of resources to determine the outcome of a decision, and Boissevain (1971) as a temporary cooperation among distinct parties for a limited purpose. Distinctiveness of the parties is an important criterion to distinguish coalitions from coordinated actions by mobs or groups without any clear differentiation among their members in terms of contributions and pay-offs. As Boissevain (1971, p. 470) notes about coalition partners: 'their individual identity within the group is not replaced by a group identity, nor is their individual commitment replaced by a uniform set of rights and obligations'.

With regard to animal behaviour, definitions including terms such as 'purpose' and 'decision' pose a problem. Animal behaviour may be perfectly intentional, but this is not easily substantiated in the field. In addition, there are alternative explanations that often appear adequate. For example, a team of individuals may benefit from concerted action without having known this beforehand. They may simply have learned that their rivals are less likely to fight back and more likely to retreat if they act together. This effect may have reinforced their cooperation. In other words, the supportive relationship does function to improve their position vis-à-vis conspecifics, but as a result of conditioning rather than foresight. It seems judicious, therefore, to avoid terms such as 'purpose' and 'decision' in the definition of animal coalitions, and to emphasize effects and functions instead of intentions and motivations.

If we define a coalition as the joining of forces by two or more parties during a conflict of interest with other parties, the label fits most mixed-motive cooperative interactions described for animals in the present volume. Still, one may question its application to a mother monkey protecting her offspring against the aggression of its peers, or to females jointly chasing off a male attacker. Social scientists may not classify this kind of defensive solidarity as a coalition because it gives the impression of occurring on the spur of the moment rather than as a deliberate strategy to gain an advantage or resource. Also, in these instances other individuals are by no means excluded from participation. Ethologists do not make these distinctions *a priori*. We apply the term coalition to all types of cooperation during aggressive or competitive encounters within the social group, often extending the concept to intragroup cooperation during intergroup encounters.

The interaction pattern most commonly observed is that of an ongoing aggressive encounter in which outsiders intervene and take sides by supporting one party or the other. This may result in two or more individuals chasing or subduing a third. The primatological literature employs terms such as 'agonistic aid', 'agonistic intervention', 'fight interference', 'aiding behaviour', 'support choice', etcetera, to indicate the basic pattern of individual A assisting B against C (Fig. 1.1). Throughout the literature, two main types of interventions are distinguished depending on whether it is the aggressive party or the recipient of aggression that is supported (Fig. 1.2).

While the term coalition covers these two interaction patterns, we reserve the term 'alliance' to denote enduring cooperative relationships. So, an alliance manifests itself in repeated coalition formation between two individuals, but not every coalition needs to reflect an alliance. In principle, coalitions can be formed on a case-by-case basis without long-term commitments.

Fig. 1.1. Four chimpanzees temporarily united against one. It started with an attack by a young adult male on a female. The victim received support from an adult male and female, and the three of them (background) now chase the original attacker (right). A juvenile male (center, foreground) takes advantage of the uneven situation by threatening the fleeing target of the coalition. From de Waal (1982).

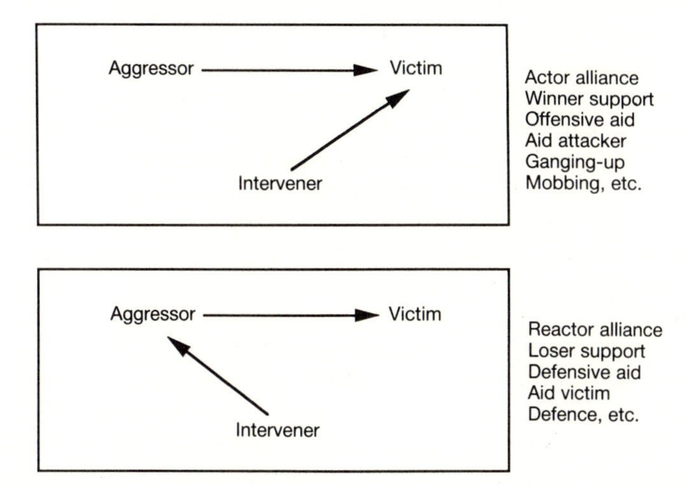

Fig. 1.2. Coalitions manifest themselves by interference by a third party in an ongoing aggressive confrontation. Most studies distinguish between interventions in favour of aggressors or victims. Some common labels for the two types of interaction are provided alongside the interaction diagrams in which arrows indicate the direction of aggressive behaviour among the three parties.

PIONEERING STUDIES IN PRIMATOLOGY

Because coalitions are such a conspicuous part of the social life of well-known primate species, it is not surprising that ethological studies of coalition formation were initiated by primatologists. If we view the dominance hierarchy as the vertical component of social organization, the network of affiliative and kinship ties can be viewed as the horizontal component. In many species these two components exist side by side without much interplay. The remarkable social complexity of human and non-human primates is brought about by their capacity to (1) alter competitive outcomes and dominance positions through collaboration, and especially (2) establish social bonds for this very reason. Alliance formation links the vertical and horizontal components of social organization by making an individual's dominance position dependent on its place in the affiliative network. Consequently, this network becomes an arena of dominance-related strategy.

The first studies demonstrating that a monkey's rank position may depend on its affiliative relationship appeared in the 1950s in Japanese. English translations of these reports reached Western primatologists with Imanishi and Altmann (1965). This series of studies revealed the relevance of genealogical relationships and introduced useful non-hierarchical concepts such as 'central' and 'peripheral' social positions. Patiently, for years on end, Japanese primatologists had studied the macaques (*Macaca fuscata*) native to their country. The great advantage of their longitudinal approach was the knowledge of kin relationships. They knew who was the mother, grandmother, brother, and sister of all immature and many adult group members. This knowledge considerably enhanced their insight into group processes.

One of these primatologists, Kawai (1958), carried out simple field experiments, such as throwing a sweet potato between two juvenile monkeys and recording the outcome of the competition. The monkey taking the food or winning the ensuing aggressive encounter was regarded as dominant. Kawai conducted many such tests and found that some dominance relationships depended on the distance of the two youngsters to their respective mothers. For instance, juvenile A dominated its peer B if their mothers were far away, yet the dominance relationship was reversed if their mothers were nearby. Such reversals occurred if B's mother dominated A's mother. Thus, the offspring of a high-ranking mother benefits from her presence. Kawai termed the dominance relationship between the juveniles in their mothers' absence the *basic rank*, and in their mothers' presence the *dependent rank*.

Dependent rank is of special interest as it is a triadic phenomenon: i.e. although it manifests itself at the level of dyadic (two-animal) interaction, it can be explained only by the relationships of the two parties with third individuals and the inter-relationships among these third ones. Triadic phenomena reflect the integration of dyadic social relationships at a higher level. This integration is likely to stem from repeated interactions involving three or more individuals. That is, the influence of the mere presence of one individual on the relationship between two others is probably caused by previous interferences by this individual in interactions between these others. Dependent rank may be caused, for example, by a mother's inclination to defend her offspring against individuals that she is able to dominate. After years of such assistance, the offspring's position will become so well established that maternal intervention is rarely necessary. The result is the so-called matrilineal hierarchy in which daughters occupy stable ranks adjacent to but below their mothers (Kawamura 1958; Sade 1967).

At the same time that these ideas were being developed, other primatologists began observing and documenting multi-party fights. These interactions are not easy to study. Aggressive interactions among monkeys and apes are extremely complicated. Everything happens fast, many events occur simultaneously, and the participants constantly change position. It takes great experience and familiarity with species-typical communication to recognize the structure of such polyadic confrontations. A pioneer study was carried out in the Zurich Zoo. Kummer observed how female baboons (*Papio hamadryas*) try to enlist support of an adult male against a female rival. The aggressor manoeuvres herself between the male and her opponent, presenting her hindquarters to him while loudly threatening her. This behaviour became known as *gesicherte Drohung*, or protected threat (Kummer 1957, 1967). This was the first indication that primates do more than merely join each other in aggressive encounters; they seem to actively recruit support from others.

Kawai and Kummer may be considered the first explorers of coalitions in non-human primates. A third classical paper was Hall and DeVore's (1965) study of wild savannah baboons (*Papio cynocephalus*). The rank of an adult male baboon seems to depend both on individual fighting ability and mutual cooperation. Hall and DeVore observed how an entire baboon troop was controlled jointly by three adult males, which formed the so-called central hierarchy. Individually, without support from the others, none of the central males had much clout. They needed each other to make a common front against individually more powerful rivals. The case is described of a male named Kula who, individually, was the most dominant animal of the troop. When the central males Dano, Pua, and Kovu were together, however, the odds were strongly against Kula. 'The fact that Pua and Kovu would support Dano when he was challenged, however, meant that Dano was almost always in control and that Kula could only assert himself briefly' (Hall and DeVore 1965, p. 62; Fig. 1.3). If we adopt Kawai's (1958) terminology, one male

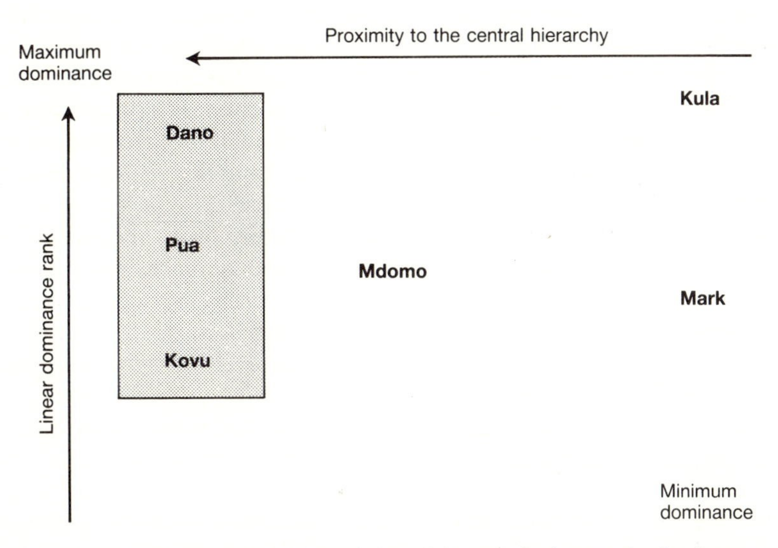

Fig. 1.3. The dominance positions of six adult male baboons. In the linear rank order, based on individual fighting abilities, Kula occupies the top rank. But three other males have formed a coalition—the central hierarchy—which gives Dano the most powerful position. After Hall and DeVore (1965).

(Dano) had a high dependent rank, whereas another (Kula) had a high basic rank. The difference is explained by coalition formation.

The pay-offs to the central coalition may have consisted mainly of access to sexually receptive females. Insofar as this increased access translates into reproductive success, it is of great relevance in relation to evolutionary explanations. Hall and DeVore (1965) counted 53 complete copulations with estrous females by the six adult males represented in Fig. 1.3. The top-ranking central male, Dano, performed 18 of these copulations and his two coalition partners, Kovu and Pua, 11 and 8. The other males reached mating frequencies of 8 (Kula), 8 (Mdomo), and 0 (Mark). So, the three central males together mated more than twice as often as the three rival males together. Aggressive sexual competition was quite pronounced in this troop, and according to the investigators, the resulting mating frequencies illustrate the advantage of the central coalition: 'Kovu, an old male whose teeth were worn level with his gums and who was individually the least dominant adult male in the group, was nevertheless second only to the most dominant male in copulations completed at the time of maximal swelling in the female. Since ovulation occurs during the period of maximal swelling, it is likely that he was therefore one of the most effective breeders in the group (p. 76).'

AREAS OF INTEREST

In the years following the above pioneering work, only a handful of systematic studies on primate coalitions were published (e.g. Bramblett 1970; Masserman *et al.* 1968; Struhsaker 1967). This changed by the end of the 1970s, when a stream of quite thorough analyses appeared in the scientific journals. It was during this period that the methodology of ethological research on coalitions was developed and that the main issues of interest were defined.

In this section, we review primatological studies up to the mid 1980s. Studies are divided into those dealing with evolutionary issues such as reciprocity and the role of kinship; description of characteristic communication and interaction patterns; strategy and cognition, and experimental approaches. This division is obviously somewhat arbitrary, as many studies cover more than one aspect.

Evolutionary explanations

If interventions in favour of another individual are costly to the performer, evolutionary theory predicts a significant kin bias in the direction of this behaviour as well as a reciprocal distribution among non-kin. Yet, as discussed above, not all acts of support conform to this condition. Many acts seem free of risk while benefiting both performers and recipients. Hence, a signifiant proportion of the coalitions observed in primates may be explained on the basis of simultaneous cooperation, or mutualism.

Kinship effects were confirmed in macaques and baboons. Not only do these monkeys support relatives more often than non-relatives, they also support close relatives more often than distant ones. The kin bias is particularly pronounced in the more risky types of interventions, such as defence against dominant attackers (Cheney 1977; Kaplan 1978; Kurland 1977; Massey 1977; de Waal 1977; Watanabe 1979).

While kin selection may explain the evolution of selective support for matrilineal relatives, it tells us little about its proximate causation because monkeys probably have no concept of genetic relatedness. Kurland (1977, p. 50), observing preferential association among close relatives in Japanese macaques, proposed the following mechanism: 'If undispersal and relatedness correlate positively, then an individual who directs more altruism and less selfishness towards animals who are more often around it is in effect biasing its behaviour in favor of closely related individuals'.

From the beginning, investigators have tried to link kin-based alliances to the phenomenon of dependent rank and the matrilineal hierarchy in cercopithecine monkeys (Cheney 1977; Datta 1983; Horrocks and Hunte 1983; de Waal 1977; Walters 1980). The most common hypothesis is that offspring achieve positions closely below their mothers because of the support received from both their mother and other relatives. If each kinship unit acts as a large alliance, alliances will be ordered according to their relative power. As a result, the members of each alliance will share similar rank positions. This is not to say that matrilineal alliances necessarily manifest themselves with high frequency. In captive long-tailed macaques (*Macaca fascicularis*) mothers intervened in only 2.2 per cent of aggressive incidents involving their juvenile offspring (de Waal 1977). It might be that this is sufficient for mothers to teach other group members about their offspring's matrilineal status, yet Horrocks and Hunte (1983) drew a different conclusion from a similarly low intervention tendency in free-ranging vervet monkeys (*Cercopithecus aethiops*), namely that interventions may be more important for the maintenance than for the acquisition of dominance. Resolution of this issue requires an experimental approach.

Reciprocity of supportive relations was first investigated for baboons by Packer (1977) and Cheney (1977), for rhesus monkeys by Kaplan (1978), and for chimpanzees by de Waal (1978). In the same period, Seyfarth (1977) formulated his influential model concerning the possible exchange of grooming by subordinate monkeys for agonistic support by dominant monkeys. Although none of these studies can be said to have produced unequivocal statistical evidence for reciprocal altruism, they did suggest its existence and as such stimulated the development of a promising area of research.

Finally, intragroup alliances can be expected only in species with rather exclusive groups. Although there undoubtedly exist many factors that determine the size, compactness, and general structure of primate groups, intergroup competition has from the beginning played a central role in evolutionary models of social organization (e.g. Wrangham 1980). With groups probably functioning as macro-alliances in resource competition with other groups, we may assume a close connection between inter- and intragroup alliances. A broad ecological perspective will therefore add to our understanding of the evolution of primate coalition behaviour.

Experimental approaches

Varley and Symmes (1966) demonstrated that a rhesus monkey's priority of access to grapes or peanuts depended on the presence of allies. Individual A might dominate B in a small group that includes C, but the A–B relation would reverse upon C's removal. The system of dependent ranks was investigated by paired tests and systematic manipulations of group composition. The investigators concluded that the most dominant individual sets the overall pattern of the rank-order first by establishing an alliance with a subordinate and thus raising its status, second by forcibly suppressing the power of one or more other monkeys.

Similarly, Tokuda and Jensen (1968) removed the most dominant male from a group of pigtail macaques (*Macaca nemestrina*) to measure the effect on the general aggression level in the group, thereby attributing the observed effect to the absence of control activities. Stynes *et al.* (1968) demonstrated that the dominance relations between the females of two different macaque species depended on the presence of a conspecific male. Marsden (1968), finally, experimented with the composition of rhesus groups to induce rank changes in both mother and offspring. All studies confirmed that removals or introductions of particular individuals affected the relations among others in the group, which strongly suggests alliances as the mediating mechanism.

Thus, Stynes *et al.* (1968) describe how the entrance of the male bonnet macaque (*Macaca radiata*) in a mixed-species group of females resulted in a sudden and short-lived outbursts of aggression in which bonnet females threatened the females of the other, larger species with appeal behaviour towards the male. The male typically responded by supporting the females of his own species. As a consequence, the bonnet females moved about the pen more freely than in the absence of a conspecific male, and were able to claim preferred sitting places.

Behaviour specific to the coalition context

A variety of vocal and facial communication appears to be specific to triadic and polyadic agonistic interactions. The most widespread category of signals serves the recruitment of support. For example, Wolfheim and Rowell (1972,

p. 244) described how talapoin monkeys (*Miopithecus talapoin*) 'gang up' on others: 'While threatening their victim with open-mouth grins and lunges they repeatedly look at and pant chirp at the other group members, who respond by joining the attack'. Other qualitative descriptions of recruitment behaviour in primates were provided by Angst (1974), Hall and DeVore (1965), Klein and Klein (1971), Kummer (1957), Oppenheimer and Oppenheimer (1973), and Rowell (1966). This type of behaviour is not limited to the primate order, however; not even to mammals. When a magpie (*Pica pica*) pair encounters another bird the female may hop to and fro between her mate and the intruder, uttering begging calls to her mate while threatening the intruder with harsh notes (Baeyens 1979).

Systematic descriptions of the structure of multi-party interactions among primates and characteristic communication patterns were provided by de Waal *et al.* (1976) for long-tailed macaques, and by de Waal and von Hooff (1981) for chimpanzees (*Pan troglodytes*). Multi-animal interactions were videotaped and analysed as to the order of events, the structure of the encounter and the occurrence of so-called *side-directed* communication. This refers to communication directed by participants in an agonistic encounter to non-opponents (the main direction of communication being between the opponents themselves). The term 'recruitment' covers only a fraction of the possible functions of this behavioural category. De Waal and van Hooff (1981) demonstrated seven distinct forms of side-directed behaviour in the chimpanzee:

- Taking refuge, i.e. seeking protection near a dominant third party.
- Seeking reasurrance and calming contact from a third individual (Figure 1.4)
- Enlistment, i.e. an attempt to activate and recruit a third party. In its most extreme form this is done by almost dragging the other to the scene while raising the voice against the opponent.
- Instigation, i.e. stimulating the third party to 'act on one's behalf' by sitting behind or next to this individual while gesticulating and barking at the opponent. A typical situation is a female 'egging on' a male against a female rival.
- Requesting support, i.e. stretching out a hand toward a third individual while intermittently vocalizing to both the opponent and the potential supporter.
- Taking courage, i.e. a dramatic change in the attack/flight balance, such as a lunge by a previously fearful individual, resulting from contact with an ally.
- Third-party appeasement. In an apparent attempt to forestall intervention, kissing and embracing of uninvolved bystanders is biased towards individuals normally supportive of the opponent.

Fig. 1.4. Before side-stepping his threatening adversary (right) an adult male seeks reassurance from an uninvolved female (left) by screaming at her and putting a finger in her mouth. This is one of the chimpanzee's many forms of side-directed communication. From de Waal (1982).

In macaques, facial and vocal threat displays vary with the agonistic context in which they are used. Fully dominant individuals usually act on their own, showing silent threats with staring eyes, open mouth, and ears moved forward. Individuals in a coalition (or attempting to mobilize one), on the other hand, usually threaten by lifting their eyebrows, moving back their ears, and drawing attention by noisy grunting and looking-around movements with exaggerated jerky head-turns (de Waal *et al.* 1976). Angst (1974) called the second threat pattern *Hetzen*, a German term originally used by Lorenz (1941) for the incitement behaviour of ducks and geese.

Another illustration of the sophisticated communication involved in coalition formation among primates was provided by a study of free-ranging rhesus monkeys (*Macaca mulatta*). An analysis of the acoustical features of screams uttered by attacked or threatened monkeys, and a comparison with behavioural and contextual information, demonstrated five distinct scream signals each with its own meaning. Each type of scream was associated with a particular level of aggressive intensity and a particular class of opponent (e.g. kin or non-kin). In theory, therefore, rhesus monkeys can derive quite specific information about agonistic incidents from scream vocalizations alone. In support of this hypothesis, the investigators found marked differences in response during playback experiments of scream vocalizations (Gouzoules *et al.* 1984).

Side-directed behaviour serves to make aggressive encounters 'public', i.e.

open to all kinds of social influences. Because of the rich variety of communication, polyadic confrontations among primates can be quite confusing, with each party exchanging signals with a number of others and looking around to draw in even more parties. Side-directed behaviour demonstrably facilitates intervention by third individuals (de Waal *et al.* 1976; de Waal and van Hooff 1981), and these interventions can be shown to affect the course and outcome of the original confrontation by making the supported individual 'bolder' and the opponent more likely to withdraw or submit (Cheney 1977; Kaplan 1977; Struhsaker 1967; de Waal 1977; de Waal and van Hooff 1981).

Table 1.1 illustrates that purely dyadic aggressive encounters, without side-directed behaviour towards third parties or other potential influences, is a relatively rare interaction type in captive groups of macaques and chimpanzees. Even though the chimpanzees lived on a large island whereas the macaques were held in a pen, the proportion of triadic and polyadic encounters was higher among the chimpanzees. In both species, there existed a snowball effect in the growth of agonistic interactions: the probability of new participants joining in increased with the number of individuals already involved (de Waal *et al.* 1976; de Waal and van Hooff 1981).

The fact that many primate species possess communication patterns of

Table 1.1. Agonistic interactions in captive long-tailed macaques and chimpanzees classified according to the number of participants, i.e. individuals performing or receiving agonistic behaviour. Dyadic interactions may be 'impure' because of the performance of side-directed behavior towards an uninvolved third individual, or a non-agonistic interference by such an individual. The total number of interactions analysed was 1036 in the macaques and 387 in the chimpanzees. Data from de Waal *et al.* (1976) and de Waal and van Hooff (1981). All figures are percentages.

Agonistic interaction	Macaques		Chimpanzees	
Dyad		75.7		60.2
'Pure'	20.9		34.4	
'Impure'	54.8		25.8	
Triad		14.2		16.5
Polyad		10.1		23.3
4 participants	4.8		10.3	
5 participants	1.9		3.4	
6 participants	1.2		4.1	
7 participants	1.0		1.3	
8 participants	0.3		0.8	
9 participants	0.2		2.0	
10 or more participants	0.7		2.3	

which the function appears specific to polyadic agonistic interaction indicates the great importance that this type of interaction must have had during evolutionary history.

Cognitive and political strategies

Coalition research raises important questions about social cognition. In view of the above observational and experimental data on coalitions and alliances, it was no exaggeration when Kummer (1971, p. 148) concluded that 'Success requires that a monkey know and integrate the status of all group members present, and their alliances and antagonisms toward him and among each other'.

Rather than with innate action patterns, we are dealing with complex social skills acquired through repeated interactions among more than two individuals at a time. A comparison of socially deprived with socially reared rhesus monkeys demonstrated serious deficiencies in agonistic interaction patterns in the first group. Complete triadic interactions, defined by the exchange of agonistic or side-directed signals between all three parties, were observed in experienced monkeys only. Only these monkeys appeared to operate with an understanding of the dominance relations among the other participants (Anderson and Mason 1974).

Around the beginning of the 1980s, a number of investigators switched 'from a concern with the structure of hierarchies and with dyadic relationships, to looking at the process of attaining/maintaining a given rank position in the social nexus', as Walker Leonard (1979, p. 156) phrased it. They described apparently intelligent tactics of primates, such as seeking affiliative relationships with strategically important individuals; interventions in fights to fortify their own positions; testing the reactions of potential supporters; undermining hostile alliances through disruptive behaviour, and so forth (Chance *et al.* 1977; Nishida 1983; Riss and Goodall 1977; Seyfarth 1977; de Waal 1978, 1982; Walker Leonard 1979).

If we follow Lasswell's (1936) classical definition of politics as 'who gets what, when, and how' there is no reason why dominance strategies and coalitions among non-human primates should not be labelled as such: the behaviour clearly affects the distribution of resources (e.g. male coalitions determine sexual access to females; Hall and DeVore 1965; Nishida 1983; Packer 1977; de Waal 1982). The behaviour may even fit more restrictive definitions, such as: 'Political behavior, properly so called, would seem to be those actions in which the rivalry for and perpetuation of social dominance impinges on the legal or customary rules governing a group' (Masters 1975, p. 34). The existence of aggressively enforced 'social norms' in primate groups was already observed by Hall (1964).

Some primate studies explicitly stressed the connection with politics, even Machiavellianism, by judging each individual's objectives and decision-making from a history of social manoeuvres and by adopting an intentionalistic vocabulary to describe these processes (e.g. de Waal 1982). Although partly of an anecdotal nature, and often speculative, these were serious attempts to come to grips with the function of what appeared to be remarkably intelligent behaviour of primates in the social domain. Opening of this area of inquiry was stimulated by the general thesis of Chance and Mead (1953), Humphrey (1976), and Jolly (1966) that social problem-solving is the original function of primate (including human) higher mental capacities from which, evolutionarily speaking, all other applications of intelligence derive.

RECENT DEVELOPMENTS

Most of the issues and approaches treated above are recognizable in the contemporary coalition research reviewed and presented in the remainder of this volume. These ethological studies are taking us an important step further by their greater methodological sophistication, new theoretical developments, and the inclusion of nonprimates. The only area which has received relatively little attention in recent years, perhaps because of its largely descriptive character, is the analysis of interaction structure and communication patterns. This area is therefore hardly represented in the present volume.

The question of the degree to which non-human primates are political animals is still under debate (e.g. Falger 1990; Schubert 1986; Schubert and Masters 1991), as is the general issue of reciprocal altruism in the animal kingdom (Taylor and McGuire 1988). The rapidly developing field of cognitive ethology deals with issues of direct relevance to coalition formation, such as the knowledge of primates about one another's relationships (reviewed by Byrne and Whiten 1988; Cheney *et al.* 1986; Cheney and Seyfarth 1990).

In short, ethological research on coalitions and alliances has stimulated the development of, and brought together, some of the most exciting trends in contemporary ethology, from evolutionary to cognitive perspectives. It has been part of primatology since its relatively recent inception, and if it is true that the formation of coalitions elevates societies to a higher degree of complexity than otherwise possible, research in this area will no doubt continue to be an essential component of ethology and evolutionary biology.

SUMMARY

1. In ethology, a coalition is defined as cooperation in a competitive or aggressive context. The usual interaction pattern is that of one individual

interfering in an ongoing confrontation by supporting one of the two parties. It is this well-coordinated 'us' against 'them' character that sets coalition formation apart from other types of cooperation. Coalitions in animals share fundamental characteristics with human coalitions as defined by social scientists.

2. Pioneering studies on coalitions in animals, particularly non-human primates, were conducted in the 1950s and 1960s. These studies demonstrated that: (1) the dominance relation between two individuals may depend on the presence of a third party, (2) individuals support one another in fights, and (3) they may actively recruit support from outsiders. The studies also indicated that coalitions determine access to resources.
3. Many species possess communication signals of which the function appears specific to coalition formation and multi-party agonistic interaction. This means that this type of interaction must have played an important role in evolution.
4. Ethological research on coalitions has intensified over the last two decades. Two foci of interest have been the evolution of cooperation, particularly acts that benefit the recipient while involving a cost to the performer, and the role of social cognition and calculation in social strategies.

ACKNOWLEDGMENTS

Writing was supported by grant RR-00167 of the National Institute of Health to the Wisconsin Regional Primate Research Center. This is publication No. 30-012 of the WRPRC.

REFERENCES

Anderson, C. O. and Mason, W. A. (1974). Early experience and complexity of social organization in groups of young rhesus monkeys (*Macaca mulatta*). *Journal of Comparative and Physiological Psychology*, **87**, 681–90.

Angst, W. (1974). Das Ausdrucksverhalten des Javaneraffen (*Macaca fascicularis* Raffles). *Fortschritte der Verhaltensforschung*, **15**, Parey, Berlin.

Baeyens, G. (1979). Description of the social behavior of the magpie (*Pica pica*). *Ardea*, **67**, 28–41.

Boissevain, J. (1971). Second thoughts on quasi-groups, categories and coalitions. *Man*, **6**, 468–72.

Bramblett, C. A. (1970). Coalitions among gelada baboons. *Primates*, **11**, 327–33.

Byrne, R. W. and Whiten, A. (1988). *Machiavellian intelligence*. Clarendon Press, Oxford.

Chance, M. R. A. and Mead, A. P. (1953). Social behavior and primate evolution. *Symposia of the Society for Experimental Biology, Evolution*, **7**, 395–439.

Chance, M., Emory, G., and Payne, R. (1977). Status referents in long-tailed Macaques (*Macaca fascicularis*): precursors and effects of a female rebellion. *Primates*, **18**, 611–32.

Cheney, D. (1977). The acquisition of rank and the development of reciprocal alliances among free-ranging immature baboons. *Behavioral Ecology and Sociobiology*, **2**, 303–18.

Cheney, D. L. and Seyfarth, R. M. (1990). *How monkeys see the world*. University of Chicago Press, Chicago.

Cheney, D. L., Seyfarth, R. M. and Smuts, B. B. (1986). Social relationships and social cognition in nonhuman primates. *Science*, **234**, 1361–6.

Darwin, C. (1859). *On the origin of species*. (Revised edition: Harvard University Press, Cambridge, MA, 1964.)

Datta, S. (1983). Patterns of agonistic interference. In *Primate social relationships* (ed. R. Hinde), pp. 289–97. Sinauer, Sunderland, MA.

Falger, V. S. E. (1990). The Arnhem Zoo chimpanzee project: a political scientist's evaluation. *Social Science Information*, **29**, 33–54.

Gamson, W. (1961). A theory of coalition formation. *American Sociological Review*, **26**, 373–82.

Gouzoules, S., Gouzoules, H., and Marler, P. (1984). Rhesus monkey (*Macaca mulatta*) screams: representational signalling in the recruitment of agonistic aid. *Animal Behaviour*, **32**, 182–93.

Hall, K. R. L. (1964). Aggression in monkey and ape societies. In *The natural history of aggression* (ed. J. Carthy and F. Ebling), pp. 51–64. Academic Press, London.

Hall, K. R. L. and DeVore, I. (1965). Baboon social behaviour. In *Primate behaviour* (ed. I. DeVore), pp. 53–110. Holt, Rinehart and Winston, New York.

Hamilton, W. (1964). The genetic evolution of social behaviour, I, II. *Journal of Theoretical Biology*, **7**, 1–52.

Horrocks, J. and Hunte, W. (1983). Maternal rank and offspring rank in vervet monkeys: an appraisal of the mechanism of rank acquisition. *Animal Behaviour*, **31**, 772–82.

Humphrey, N. K. (1976). The social function of intellect. In *Growing points in ethology* (ed. P. Bateson and R. Hinde), pp. 303–21. Cambridge University Press, Cambridge.

Imanishi, K. and Altmann, S. A. (1965). *Japanese monkeys: a collection of translations*. Emory University, Atlanta.

Jolly, A. (1966). *Lemur behavior*. University of Chicago Press, Chicago.

Kaplan, J. (1977) Patterns of fight interference in free-ranging rhesus monkeys. *American Journal of Physical Anthropology*, **47**, 279–88.

Kaplan, J. (1978). Fight interference and altruism in rhesus monkeys. *American Journal of Physical Anthropology*, **49**, 241–9.

Kawai, M. (1958). On the system of social ranks in a natural troop of Japanese monkey. I: Basic rank and dependent rank. II: Ranking order as observed among the monkeys on and near the test box. *Primates*, **1**, 111–48.

Kawamura, S. (1958). Matriarchal social ranks in the Minoo-B troop: A study of the rank system of Japanese monkeys. *Primates*, **1**, 148–56.

Klein, L. and Klein, D. (1971). Aspects of social behaviour in a colony of spider monkeys (*Ateles geoffroyi*) at San Francisco Zoo. *International Zoo Yearbook*, **11**, 175–81.

Kropotkin, P. (1902). *Mutual aid: a factor of evolution* (Revised edition: New York University Press, New York 1972).

Kummer, H. (1957). *Soziales Verhalten einer Mantelpavian-Gruppe*. Verlag Hans Huber, Bern.

Kummer, H. (1967). Tripartite relations in hamadryas baboons. In *Social communication among primates* (ed. S. Altmann), pp. 63–72. Chicago University Press, Chicago.

Kummer, H. (1971). *Primate societies*. Aldine Press, Chicago.

Kurland, J. (1977). Kin selection in the Japanese monkey. *Contributions to Primatology*, **9**. Karger, Basel.

Lasswell, H. (1936). *Who gets what, when and how*. McGraw-Hill, New York.

Lorenz, K. (1941). Vergleichende Bewegungsstudien an Anatiden. *Supplement to Journal of Ornithology*, **89**, 194–294.

Marsden, H. (1968). Agonistic behavior of young rhesus monkeys after changes induced in social rank of their mothers. *Animal Behaviour*, **16**, 38–44.

Masserman, J., Wetchkin, S., and Woolf, M. (1968) Alliances and aggressions among rhesus monkeys. *Science and Psychoanalysis*, **12**, 95–100.

Massey, A. (1977). Agonistic aids and kinship in a group of pigtail macaques. *Behavioural Ecology and Sociobiology*, **2**, 31–40.

Masters, R. D. (1975). Politics as a biological phenomenon. *Social Science Information*, **15**, 7–63.

Nishida, T. (1983). Alpha status and agonistic alliance in wild chimpanzees. *Primates*, **24**, 318–36.

Norris, K. S. and Schilt, C. R. (1988). Cooperative societies in three-dimensional space: On the origins of aggregations, flocks, and schools, with special reference to dolphins and fish. *Ethology and Sociobiology*, **9**, 149–80.

Oppenheimer, J. R. and Oppenheimer, E. C. (1973). Preliminary observations of *Cebus nigrivittatus* on the Venezuelan Llanos. *Folia primatologica*, **19**, 409–36.

Packer, C. (1977). Reciprocal altruism in *Papio anubis*. *Nature*, **265**, 441–3.

Riss, D. and Goodall, J. (1977). The recent rise to the alpha-rank in a population of free-living chimpanzees. *Folio primatologica*, **27**134–51.

Rothstein, S. I. and Pierotti, R. (1988). Distinctions among reciprocal altruism, kin selection, and cooperation and a model for the initial evolution of beneficent behavior. *Ethology and Sociobiology*, **9**, 189–210.

Rowell, T. E. (1966). Hierarchy in the organization of a captive baboon group. *Animal Behaviour*, **14**, 430–43.

Sade, D. S. (1967). Determinants of dominance in a group of free-ranging rhesus monkeys. In *Social communication among primates* (ed. S. Altmann), pp. 99–114. Chicago University Press, Chicago.

Schubert, G. (1986). Primate politics. *Social Science Information*, **25**, 647–80.

Schubert, G. and Masters, R. D. (1991). *Primate Politics*. Southern Illinois University Press, Carbondale.

Seyfarth, R. (1977). A model of social grooming among adult female monkeys. *Journal of Theoretical Biology*, **65**, 671–98.

Shaw, M. E. (1971). *Group dynamics*. McGraw-Hill, New York.

Struhsaker, T. (1967). Social structure among vervet monkeys (*Cercopithecus aethiops*). *Behaviour*, **29**, 83–121.

Stynes, A. J., Rosenblum, L. A., and Kaufman, I. C. (1968). The dominant male and behavior within heterospecific monkey groups. *Folio primatologica*, **9**, 123–34.

Taylor, C. E. and McGuire, M. T. (1988). Reciprocal altruism: 15 years later. Special issue of *Ethology and Sociobiology*, **9**, 67–72.

Tokuda, K. and Jensen, G. D. (1968). The leader's role in controlling aggressive behavior in a monkey group. *Primates*, **9**, 317–20.

Trivers, R. (1971). The evolution of reciprocal altruism. *Quarterly Review of Biology*, **46**, 35–59.

Trivers, R. (1985). *Social evolution*. Benjamin/Cummings, Menlo Park, CA.

Varley, M. and Symmes, D. (1966). The hierarchy of dominance in a group of macaques. *Behaviour*, **27**, 54–75.

de Waal, F. B. M. (1977). The organization of agonistic relations within two captive groups of Java-monkeys (*Macaca fascicularis*). *Zeitschrift für Tierpsychologie*, **44**, 225–82.

de Waal, F. B. M. (1978). Exploitative and familiarity-dependent support strategies in a colony of semi-free living chimpanzees. *Behaviour*, **66**, 268–312.

de Waal, F. B. M. (1982). *Chimpanzee politics*. Jonathan Cape, London.

de Waal, F. B. M. and van Hooff, J. A. R. A. M. (1981). Side-directed communication and agonistic interactions in chimpanzees. *Behaviour*, **77**, 164–98.

de Waal, F. B. M., van Hooff, J. A. R. A. M. and Netto, W. J. (1976). An ethological analysis of types of agonistic interaction in a captive group of Java-monkeys (*Macaca fascicularis*). *Primates*, **17**, 257–90.

Walker Leonard, J. (1979). A strategy approach to the study of primate dominance behaviour. *Behavioral Processes*, **4**, 155–72.

Walters, J. (1980). Interventions and the development of dominance relations in female baboons. *Folia primatologica*, **34**, 61–89.

Watanabe, K. (1979). Alliance formation in a free-ranging troop of Japanese macaques. *Primates*, **20**, 459–74.

Wilson, E. O. (1975). *Sociobiology: the new synthesis*. Belknap, Cambridge, MA.

Wolfheim, J. H. and Rowell, T. E. (1972). Communication among captive talapoin monkeys (*Miopithecus talapoin*). *Folia primatologica*, **18**, 224–55.

Wrangham, R. W. (1980). An ecological model of female-bonded primate groups. *Behaviour*, **75**, 262–300.

PART I

Coalitions, alliances, and the structure of society

Coalitions, alliances, and the structure of society: Introduction

At one and the same time, a society is its individuals, their interactions, and their relationships, and it affects these different levels (Hinde 1976, 1979). Societies sometimes appear to act as integrated organisms. However, unlike organisms, the integration results not from any superordinate organizing mechanisms but from the often conflicting actions and goals of the individuals in the society. Humans are to some extent an exception. We can have as our goal the creation of a certain type of society, and rules imposed by society can govern an individual human's behaviour. Nevertheless, much of our daily behaviour probably operates independently of such societal goals and rules.

This section aims to elucidate the two-way interaction between individuals and society when individuals cooperate in conflict. How do individuals, their interactions—in the form of coalitions—and their relationships—in the form of alliances—influence the structure of the society? How does the society's structure influence the individuals, their coalitions, and their alliances?

Almost any analysis of the structure of a society probably incorporates two fundamental explanatory principles, competition and consanguinity: competitive ability and cooperation between kin determines an individual's own and its relatives' access to resources (Darwin 1859; Hamilton 1964). One of the more important findings in the study of animal social structure was that of the system of 'inheritance of dominance rank' among female cercopithecine monkeys. In social groups of many vertebrate and invertebrate taxa, individuals can be arranged in a hierarchy of competitive ability. In a number of cercopithecines, closely related adult females have adjacent ranks in a group's dominance hierarchy, with the result that whole families, matrilines, not just individuals, are ranked with respect to one another.

In Chapter 2 **Chapais** reviews the fundamental influence that coalitions and alliances have on this type of dominance ranking system. Support in contests, especially support from kin, determines the ability of individuals to acquire and maintain a dominance rank that is quite largely independent of their own unaided competitive ability. Chapais demonstrates this both with observational data and, more importantly, with data from the most detailed experimental manipulation of social groups yet done.

Chapais argues that in a society in which rank depends heavily on support from others, individuals have to work hard to maintain their own supportive alliances and prevent others from forming rival alliances. Competitive coalitions are best prevented by alliances with dominant group members, including non-kin. We thus have the seemingly anomalous situation in which simply to maintain their own status, animals are forced to ally with their closest competititors, so reinforcing those competitors' rank, rigidifying the hierarchy, and diminishing their own ability to rise in rank.

Following the work of Seyfarth (1977), many primatologists have interpreted attempts to ally with dominant animals as a form of reciprocal cooperation: subordinates perform services now for needed support later from a powerful group member. However, Chapais' analysis, along with that of Noë (this volume, pp. 285–321), suggests that the potential benefits of the cooperation are often immediate and, therefore, hint that reciprocation might be less of an issue than sometimes supposed. This possibility is relevant to application of Boyd's models to primate social behaviour (this volume, pp. 473–89), and to interpretations of primate behaviour in both Harcourt's (pp. 445–71) and de Waal's (pp. 233–57) chapters.

Coalitions and alliances influence individuals' competitive ability and hence the nature of the dominance hierarchy. The number of potential allies in a group is influenced by births, deaths, and dispersal, in other words by demographic parameters, which are themselves determined by the environment. Societies in poor environments can thus have a different social structure to those in good habitat, not for any adaptive functional reasons, but simply because demographic processes influence group composition and therefore social interactions (Altmann and Altmann 1979; Dunbar 1988). **Datta** (Chapter 3) shows with computer simulation models that this is precisely the case in the specific instance of the availability of allies and their influence on the dominance rank system of females in primate groups. Two cercopithecine primate groups that differ only in the age of sexual maturity of females, the interbirth interval and mortality rate, and consequently in the availability of suitable allies, have very different ranking systems as a result. The group in the rich environment has a matrilineal inheritance of rank system with mothers dominating daughters, and younger daughters outranking older; in the poor environment group, elder daughters occasionally outrank their mother, and almost always outrank younger daughters. Lee and Johnson (this volume, pp. 391–414) makes a slightly different, but related, point in their suggestion that the effects of maternal rank on sons' rank is most evident in seasonal breeders, where differences in age among peers at least, and therefore extra support in a contest can have a significant effect on the outcome.

Datta points out that the differences that she sees between computer simulations of groups as a result simply of realistically changing demographic parameters, and hence number of allies, match what had hitherto been interpreted as interspecific differences in dominance systems. Of course, animals, and perhaps especially primates, are highly opportunistic; so, while the ranking system itself might not be an adaptive response, the tactics and strategies individuals use when allying probably change, Datta suggests, with the changing availability of allies.

Chapais and Datta concentrate on social structure as reflected in the interactions and relationships of females. This is appropriate since females are the resident core of cercopithecine groups (Pusey and Packer 1987), while males are almost transient hangers-on (cf. Clutton-Brock 1989; Emlen and Oring 1977; van Hooff and van Schaik (this volume, pp. 357–89). This being the case, males and females can be expected to use coalitions and alliances in different ways. In Chapter 4 **Ehardt and Bernstein** review the nature and distribution of coalitions and alliances of adult male cercopithecines, and compare them to those of females. In a detailed analysis of their data on captive rhesus macaques, they suggest that adult males intervene in contests largely to control the behaviour of their main future potential rivals, adolescent males. This form of aggression, they argue, leads to the emigration of the young males, not through aggressive expulsion, but through breaking of the males' ties with their female kin. While emphasizing the several functions of interventions, they suggest that the process does not simply oust rivals, but also removes potentially disruptive elements of society, and thus contributes to maintenance of a functioning society.

Lee and Johnson (pp. 391–414) also discuss sex differences in cooperative and competitive interactions and relationships, but of immatures and from a different viewpoint. Whereas Ehardt and Bernstein concentrate on the effect of intervention on social structure, Lee and Johnson relate differences between the sexes to their different competitive regimes and life history trajectories.

Zabel, Glickman, Frank, Woodmansee, and Keppel (Chapter 5) consider what might be thought of as the most basic influence on the structure of society, namely how characteristics of the group members themselves and differences between them in competitive ability influence their interactions and hence the structure of the social group. After a brief review of the use of coalitions and alliances by carnivores, they turn to their own investigation of the factors that correlate with participation in coalitions among juvenile hyaenas. In a group of similarly aged peers, relative competitive ability is the main correlate: immature hyaenas tend to join dominant group members in attacks on subordinates. The effect, they argue, is to reinforce the existing hierarchy, as Chapais argued for female cercopithecines, although occasionally coalitions and alliances of subordinates can overthrow a dominant. In some contrast to the papers on primates in Part II and those by Connor and Harcourt in Part III, they reject the notion that complex tactical decisions are involved in such a choice of coalition partners, and suggest instead a generalized tendency to attack animals that are losing fights. Behaviour that is advantageous, because it suppresses the competitive ability of others, and that might appear to be goal-oriented, thus results from a very simple rule that requires no great cognitive complexity (cf. Kummer *et al.* 1990).

So far, the immediate social group has been considered as 'the society', but any society is also composed of interacting and often competing social groups. Indeed, were it not for competition between groups, societies and the social groups of which they are composed might not have the complexity of social structure that they do (Alexander 1974; Brown 1987, Chapter 8; Vehrencamp 1983; Wrangham 1980; van Hoof and van Schaik this volume, pp. 357–89). The theme of both of the next two chapters is the relation between intergroup competition and intragroup cohesiveness and cooperation.

Boehm (Chapter 6) suggests that intergroup competition is associated with increased intragroup cohesion and cooperation, at the time of the conflict, in similar and homologous ways in the two species. Mobilization of supporters and secrecy in movement, for example, is identifiable during raids into neighbouring territories and attacks on smaller neighbouring groups in both species. Boehm does not mention any species other than humans and chimpanzees, but it might be useful to bear in mind that some comparable behaviours are also shown by non-primates (see editorial to part III, pp. 351–55 and Harcourt this volume, pp. 445–71). Boehm goes further in his suggestion that the cognitive assessment of relative power is similar in humans and chimpanzees. The main difference between the two species, of course, is that in humans, culture and language bring to bear a far greater degree of coordination than is possible in chimpanzee society, with the result that the conflict can escalate into warfare in humans, but not, Boehm argues, in chimpanzee society.

Since the early 1900s, and especially since the rise of Nazism and antisemitism in the 1930s psychologists have attempted to explain, as **Rabbie** puts it in Chapter 7, 'One of the best documented findings in social psychology ... that people tend to display more favourable attitudes and stereotypes about their own group and its members than about an out-group and its members and are inclined to allocate more economic and symbolic rewards to members of their own group than to members of another group.' The attempts cover a number of different levels of explanation. The one Rabbie and his coworkers have concentrated on is 'realistic group conflict theory'. It is probably the body of theory with which biologists would feel most familiar, for it concerns the effects of actual or perceived conflict between groups on intragroup cooperation, along with effects on attitudes to in-group and out-group members. Until Rabbie's work, it was thought that intergroup competition alone was sufficient to induce intragroup cohesion. He shows that the crucial factor is perceived interdependence, the perception by individuals that cooperation is necessary for a favourable outcome to be achieved.

The use of reciprocal altruism as a competitive strategy requires that defection be detected and punished, at the least by cessation of cooperation (Trivers 1971; Axelrod and Hamilton 1981; Boyd this volume pp. 473–89). The possibility of reciprocal cooperation is an important thread running through many of the chapters in this book, and is the basis of de Waal's (pp. 233–57) and Boyd's (pp. 473–89) chapters. Rabbie closes with a review of some experiments which effectively test individuals' and groups' reactions to cheating, as opposed to merely competing. The results show that cheating (reneging on a promise to cooperate) is indeed reacted to more unfavourably than is competition alone, and that groups react more severely than do individuals.

REFERENCES

Alexander, R. D. (1974). The evolution of social behavior. *Annual Review of Ecology and Systematics*, **5**, 325–83.

Altmann, S. A. and Altmann, J. (1979). Demographic constraints on behavior and social organization. In *Primate Ecology and Human Origins*. (eds I. S. Bernstein and E. O. Smith), pp. 48–63. Garland STPM Press, New York.

Axelrod, R. and Hamilton, W. D. (1981). The evolution of cooperation. *Science*, **211**, 1390–6.

Brown, J. L. (1987). *Helping and communal breeding in birds*. Princeton University Press, Princeton.

Clutton-Brock, T. H. (1989). Mammalian mating systems. *Proceedings of the Royal Society of London*, B, **236**, 339–72.

Darwin, C. (1859). *On the origin of species*. John Murray, London.

Dunbar, R. I. M. (1988). *Primate social systems*. Croom Helm, London.

Emlen, S. T. and Oring, L. W. (1977). Ecology, sexual selection, and the evolution of mating systems. *Science*, **197**, 215–23.

Hamilton, W. D. (1964). The genetical evolution of social behaviour. *Journal of Theoretical Biology*, **7**, 1–52.

Hinde, R. A. (1976). Interactions, relationships, and social structure. *Man* **11**, 1–17.

Hinde, R. A. (1979). *Toward understanding relationships*. Academic Press, London.

Kummer, H., Dasser, V. and Hoyningen-Huene, P. (1990). Exploring primate social cognition: some critical remarks. *Behaviour*, **112**, 84–98.

Pusey, A. E. and Packer, C. (1987). Dispersal and philopatry. In *Primate societies* (eds B. B. Smuts, D. L. Cheney, R. M. Seyfarth, R. W. Wrangham, and T. T. Struhsaker), pp. 250–66. University of Chicago Press, Chicago.

Seyfarth, R. M. (1977). A model of social grooming among adult female monkeys. *Journal of Theoretical Biology*, **65**, 671–98.

Trivers, R. L. (1971). The evolution of reciprocal altruism. *Quarterly Review of Biology*, **46**, 35–57.

Vehrencamp, S. L. (1983). A model for the evolution of despotic versus egalitarian societies. *Animal Behaviour*, **31**, 667–82.

Wrangham, R. W. (1980). An ecological model of female-bonded primate groups. *Behaviour*, **75**, 262–300.

2

The role of alliances in social inheritance of rank among female primates

BERNARD CHAPAIS

INTRODUCTION

Agonistic alliances in non-human primates may have various and far-reaching consequences for all the participants, such as granting allies temporary access to resources otherwise controlled by dominant individuals (e.g. Bercovitch 1988; Packer 1977), or providing them with effective protection against aggressors. But the most important consequence of alliances probably relates to their role in the formation of dominance relationships. An individual's dominance status is indeed a long-lasting attribute which may have profound and diverse influences on many aspects of that individual's social life (for reviews and general discussions of the concept of dominance, see Bernstein 1981; Dunbar 1988; Fedigan 1983; Hinde 1978; Silk 1987; de Waal 1986). This paper is about the influence of alliances on the formation and maintenance of dominance relations among females in non-human primates. It focuses on a specific system of rank relations known as matrilineal dominance or matrilineal rank inheritance. This particular ranking system is observed in many species of cercopithecines, one of the two sub-families of Old World monkeys.

In all species that form matrilineal hierarchies, males leave their natal group around puberty and attempt to join other groups, while females are philopatric and genetically related. Dominance status is transmitted across generations from a mother to her daughters, with the result that female kin rank adjacently in the dominance hierarchy, rather than distributing themselves according to individual attributes such as their age and size. Japanese researchers were the first, some 30 years ago, to describe this type of dominance system in free-ranging groups of Japanese macaques (*Macaca fuscata*, Kawai 1958; Kawamura 1958). Since then, 11 other species or sub-species of cercopithecines have been found to exhibit a similar ranking structure (references below).

The most direct evidence for the role of alliances in rank relations comes from observations of rank changes that are temporally connected to agonistic interventions and coalitions. Such events happen spontaneously in natural groups, but infrequently, for reasons that are examined at length below. Rank changes can also be induced experimentally by manipulating the composition of a group and its network of alliances. Such experiments make it possible to conduct a detailed analysis of the role of alliances in the dynamics of rank relations.

A second, less direct, but equally significant category of evidence for the role of alliances in dominance relations is provided by observational data on agonistic interventions. The role of alliances in the determination of rank can be inferred by comparing the observed patterning of interventions with the pattern expected under the hypothesis that alliances do determine rank. The predictions of such a hypothesis must be clearly stated, however, and the data must be classified so as to make the test meaningful, two requirements that are often difficult to meet.

In this paper both categories of evidence (direct and indirect) and both sources of data (observational and experimental) for the role of alliances in the acquisition and maintenance of rank among females are reviewed, and their complementary nature emphasized. The discussion bears upon the proximate causes and evolutionary aspects of alliance formation in this type of hierarchy.

STRUCTURE OF MATRILINEAL DOMINANCE RELATIONS

Matrilineal rank relations have been reported for wild chacma baboons, *Papio cynocephalus ursinus* (Cheney 1977), yellow baboons, *P. c. cynocephalus* (Hausfater *et al.* 1982; Lee and Oliver 1979; Pereira 1988; Samuels *et al.* 1987; Walters 1980), olive baboons, *P. c. anubis* (Johnson 1987), wild geladas, *Theropithecus gelada* (Dunbar 1980), wild and captive vervet monkeys, *Cercopithecus aethiops* (Bramblett *et al.* 1982; Horrocks and Hunte 1983; Fairbanks and McGuire 1986; Lee 1983), and for seven species of free-ranging or captive macaques: Japanese macaques (Kawai 1958; Kawamura 1958; Koyama 1967), rhesus macaques, *Macaca mulatta* (Berman 1980; Bernstein and Williams 1983; Chapais and Schulman 1980; Datta 1983*a*; Missakian 1972; Sade 1967, 1972; de Waal and Luttrell 1985), Barbary macaques, *M. sylvanus* (Kuester and Paul 1988; Paul and Kuester 1987), long-tailed macaques, *M. fascicularis* (Angst 1975; Netto and van Hooff 1986; de Waal 1977), stumptail macaques, *M. arctoides* (Estrada *et al.* 1977; Gouzoules 1975); bonnet macaques, *M. radiata* (Silk *et al.* 1981), and pigtail macaques, *M. nemestrina* (Bernstein 1969). Thus, matrilineal hierarchies are widespread in macaques, baboons, and vervets, and have been observed in various settings: in the wild, in free-ranging, provisioned groups, in corral populations, and in the laboratory. Such consistency of the rules governing matrilineal rank orders across situations entails that observational and experimental studies can potentially complement each other.

In matrilineal hierarchies, as a rule, any adult female is dominant to all the females that are subordinate to her mother, and she is subordinate to all the females that are dominant to her mother. These rank relations are well defined and strongly asymmetric, with submissive behaviours flowing unidirectionally upwards in the hierarchy. Aggressive behaviours generally flow downwards in the hierarchy, with some exceptions, e.g. when a subordinate female threatens a dominant one in defence of a relative (Datta 1983*b*; Lee 1983; de Waal and Luttrell 1985).

Females, however, do not assume their mother's rank from the time they are born. Of the females subordinate to her mother, an immature female is dominant only to the subset of females that are of her age or younger. This subset thus increases as the female becomes older until she reaches maturity, at which time she is dominant over all the females subordinate to her mother, regardless of their age. This interaction between maternal rank and relative age in the structure of female rank relations is well documented for Japanese macaques (Koyama 1967), rhesus macaques (Bernstein and Williams 1983; Datta 1983*a*; Sade 1967; de Waal and Luttrell 1985), long-tailed macaques (de Waal 1977), Barbary macaques (Paul and Kuester 1987), vervet monkeys (Lee 1983), and savanna baboons (Johnson 1987; Lee and Oliver 1979; Pereira 1988; Walters 1980). The age at which females become dominant to most females that are subordinate to their mother varies across species, from 3 years in vervets, to 4 years in rhesus macaques, 5 years in savanna baboons, and 6 years in Barbary macaques (references above). This variation may relate to interspecific differences in the time it takes to achieve maturity, and the relative size of individuals of differing ages.

The period of time during which an immature female born into a high-ranking matriline is still subordinate to a larger/older female born into a lower-ranking matriline is referred to as her period of *rank acquisition*. That period ends with the *rank reversal* between the two females, and is followed by the period of *rank maintenance*. In this paper, *rank inheritance* refers to the whole process of rank acquisition and rank maintenance. During rank acquisition, the female from the higher-ranking matriline is said to have a higher *maternal* rank, although she may be subordinate in dyadic interactions with the larger female from the lower-ranking matriline. Depending on their relative maternal ranks, females can be said to be higher-born or lower-born in relation to each other (Datta 1983*a*), of higher or lower matriline (de Waal and Luttrell 1985), or of higher or lower family (Netto and van Hooff 1986). In the present chapter, the terms highborn and lowborn are used.

The period of rank acquisition is a stepwise process. For example, the female first goes through a phase during which she is not involved in any agonistic interactions. This periods lasts for about 5 months in vervets (Horrocks and Hunte 1983) and 6 months in rhesus (Berman 1980). It is followed by a phase of instability during which the highborn female directs aggression to the older lowborn female, which often resists by returning aggression. This sub-period has been referred to as the period of *challenge* (of the lowborn female by the highborn female) (Datta 1983*a*), and the lowborn females have been said to be *targeted* by the highborn ones (Walters 1980). Note that the onset of challenge is more a matter of the relative versus absolute age of the highborn and lowborn females. In rhesus monkeys, for example, challenges are more likely between females close in age (Datta 1983*a*). See Datta (1983*a*), Horrocks and Hunte (1983), de Waal (1977), and Walters (1980) for more details and some degree of interspecific variation in the developmental stages of rank relations.

This paper is concerned with rank relations among females only. The rank relations of males (before they leave their natal group) may obey different principles, and the extent of this gender difference varies across species (Lee and Johnson this volume, pp. 407–10). In rhesus and Japanese macaques, and probably vervets, the rank of immature males in relation to each other and to immature females correlates with their maternal rank, as is the case for females (Koyama 1967; Sade 1967). For this reason, data on the rank relations of immature males and females are lumped in some studies. In baboons and Barbary macaques, in contrast, immature males rank among each other and in relation to females according to their relative age/size, irrespective of their maternal rank (Johnson 1987; Kuester and Paul 1988; Lee and Oliver 1979; Pereira 1988). Furthermore, male baboons and Barbary macaques outrank adult females irrespective of their maternal rank at a younger age than male rhesus and Japanese macaques do (Kuester and Paul 1988; Pereira 1988). Thus, immature males and females do not form a homogeneous category, as far as rank relations are concerned, and this is probably true for rhesus and Japanese macaques as well, though to a lesser extent.

OBSERVATIONAL EVIDENCE FOR THE ROLE OF ALLIANCES IN RANK INHERITANCE

We now move on to the first category of evidence for the role of alliances in the acquisition and maintenance of matrilineal rank. *Agonistic interventions* (hereafter interventions) refer to situations where an individual enters a conflict by siding with one of the opponents. Three roles can be defined, that of *supporter*, *recipient* of support, and *target*. The supporter and the recipient are called *allies*. An individual is said to *solicit* the support of a potential ally when it directs to the latter typical behaviours which result in the solicited individual joining the soliciting one (scream threat, Shirek-Ellefson 1972; barkscreaming, appeal-aggression, pointing, de Waal *et al.* 1976; head flagging, Hausfater 1975; various scream vocalizations, Gouzoules *et al.* 1984; etc.). *Coalitions* refer to instances where two individuals simultaneously threaten or attack a third party, e.g. when an individual supports another which is already directing aggression to the target. Finally, *alliances* refer to longer-term interindividual relationships in which one partner supports the other unilaterally, or the two support each other mutually.

Consider a dyad composed of a highborn female subordinate to a larger/older lowborn female, and the hypothesis that interventions by third parties are responsible for the highborn female ultimately outranking the lowborn female and maintaining her rank above her. The effect of such interventions could be mediated through a number of mechanisms, and could have a number of consequences, which we will refer to as the behavioural correlates of rank inheritance. Seven such correlates are defined below, based on a review of the literature and on reasoning. Note that these need not all be at work in the same species.

1. If interventions in favour of the highborn female are a necessary condition for rank acquisition, they should occur before the highborn female reverses rank with the lowborn one. If interventions do play a role in prompting the period of aggressive challenge, they should take place even sooner, i.e. before the highborn female begins directing aggression to the lowborn female.
2. During the period of rank acquisition the highborn female may be expected to receive less aggression from the lowborn female, and to direct more aggression to her, when she is physically near, rather than far away from her potential allies. She should also submit less often to the lowborn female, and succeed in inducing her to submit more often, when she is near her allies.

The other correlates apply to both the period of rank acquisition and of rank maintenance.

3. The highborn female's allies may be more active (i.e. more prompt to interfere) than the lowborn female's allies (regardless of the number of allies).
4. The highborn female may have more allies than the lowborn female (whether kin or non-kin, and highborn or lowborn in relation to the target).
5. The highborn female may have more powerful allies (i.e. allies of a higher matrilineal rank, or allies otherwise dominant to those of the lowborn female).
6. If correlate 5 is verified, the consequences of interventions are expected to differ for highborn and lowborn targets. More powerful allies can be

expected to induce the targets to submit at a higher rate and/or intensity level.
7. The highborn female may be more active than the lowborn female in the utilization of her allies. For example, she could solicit them more often, or be more aggressive towards the target when she receives support, than a lowborn recipient of support.

In order to test these 'predictions', data on agonistic interventions must be analysed according to a number of principles. First, ideally only the interventions in conflicts between females should be considered, since the rank relations of males and the interventions they receive may obey different principles (see above). Second, correlate 1 requires that interventions be classified according to their timing in relation to the onset of the period of challenge and the rank reversal. Third, in order to test correlate 2, the aggressive interactions between highborn and lowborn females should be classified according to the proximity of each female to her potential allies. Fourth, correlate 3 requires that the propensity to give support be measured in relation to the number of opportunities for doing so (i.e. in relation to the number of conflicts). Fifth, correlates 4 and 5 require that the degrees of genetic relatedness, the matrilineal rank, and the relative age/size of all participants be considered.

Needless to say these conditions are rarely met in any single study, for at least three reasons: (1) information on genetic relatedness is often lacking, or at best incomplete; (2) frequencies of interventions may be so low as to require the lumping of different temporal or structural categories; (3) the research questions of the authors were often not defined as above (even if the data set would have so allowed in a good number of cases). The following review is thus based on a limited number of studies concerned specifically with the process of rank inheritance, or in which interventions were classified in a way that made it possible to assess their effect on rank relations. These studies bear on rhesus macaques (Berman 1980; Chapais 1983*a*; Datta 1983*a*, *b*, *c*), Japanese macaques (Chapais *et al.* 1991), long-tailed macaques (Netto and van Hooff 1986), vervet monkeys (Cheney 1983; Horrocks and Hunte 1983; Hunte and Horrocks 1986), and savanna baboons (Cheney 1977; Pereira 1989; Walters 1980).

Correlate 1

According to this correlate, interventions are expected to take place before the highborn recipient reverses rank, and even before it begins challenging the lowborn. In most studies, interventions were not classified in relation to the timing of the rank reversal, or the onset of challenge. The available information, however, supports correlate 1. Berman (1980) reported that when threatened, infant rhesus monkeys could be protected (supported or recuperated) at a time when they had not yet begun to challenge other individuals (i.e. over their first 6 months of life). Datta (1983*a*) observed the period preceding the onset of challenge in 45 highborn–lowborn dyads of rhesus monkeys (females and males included). Agonistic interventions taking place before the onset of challenges were observed in 30 of the 45 dyads. The absence of data in the 15 remaining dyads may be accounted for partly by the fact that interventions are rare events. Cheney (1977), Walters (1980), and Pereira (1989) studied agonistic interactions between juvenile female baboons and the lower-born adult females they targeted, i.e. during the period of challenge. Interventions in favour of the highborn females did occur over that period.

Correlate 2

Correlate 2 states that the proximity of allies should affect the amount of aggression and submission exchanged by highborn and lowborn females. The available data accord with that correlate. Kawai (1958) was among the first to note that a highborn female Japanese macaque could submit to a lowborn female in the absence of certain individuals (e.g. her mother), but be dominant over the same lowborn female in the presence of these individuals. His concept of 'basic' and 'dependent' rank relate respectively to these two situations. Walters (1980) noted that during the period of challenge, highborn juvenile baboons directed aggression to lowborn females when their mother was nearby. Datta (1983*a*) reported that before the period of challenge, highborn rhesus monkeys received less aggression from lowborn individuals when they were near their protectors. Horrocks and Hunte (1983) found that maternal rank predicted 94.1 per cent of the outcome of aggressive interactions between juvenile vervets when the mother was present, but only 69.4 per cent when she was absent.

Correlate 3

The available evidence supports correlate 3. The allies of highborn females are more active than those of lowborn females. In baboons, Cheney (1977) observed that the rate of maternal interventions was higher for highborn than lowborn offspring. Horrocks and Hunte (1983) also observed that high-ranking vervet mothers intervened in a greater proportion of their offspring's conflicts than lower-ranking mothers. Netto and van Hooff (1986) similarly reported that in long-tailed macaques highborn mothers interfered more often (10 per cent of their offspring's conflicts with lowborn adult females) than did lowborn mothers (0.5 per cent of their offspring's conflicts with highborn adult females). Hunte and Horrocks (1986) stated that in vervet monkeys interventions in favour of kin were much more frequent when the supporter and recipient ranked above the target (84.7 per cent) rather than below it (5.3 per cent), i.e. the highborn was supported by its kin more often than the lowborn. In Japanese macaques, Chapais *et al.* (1991) observed that the relatives of the highborn opponent interfered both more frequently and more consistently for their kin compared to the relatives of the lowborn opponent. In rhesus monkeys, Datta (1983*a*) reported that in interventions occurring before the period of challenge and in those taking place before the rank reversal, the highborn individual received greater support. Whether this difference was attributable to the highborn individual's allies being more active or being more numerous was not stated, however. In interventions occurring during the period of rank maintenance the highborn and lowborn were supported equally frequently (Datta 1983*b*).

Correlate 4

According to this correlate, highborn females might have more individuals supporting them than do lowborn females. Again, this possibility is confirmed by the available data. Considering kin allies first, Berman (1980) reported that infant rhesus monkeys from the top-ranking lineage were protected by their mother and other close female relatives during their first 6 months, whereas no infants in the two other lineages were protected by close female relatives besides the mother. Datta (1983*c*) observed that lowborn rhesus monkeys were supported only by their close relatives (mother and siblings) against highborn individuals; less close relatives were more likely to support their kin when it was highborn in relation to its opponent. Pereira (1989) stated that in baboons, mothers invariably assisted their daughters against lowborn individuals but not against highborn ones. Overall, these observations indicate that highborn females have more kin intervening for them, not necessarily because highborn matrilines are larger than lowborn ones, but because highborn females have a larger proportion of their kin prepared to support them.

Regarding distantly related or non-kin allies, an important constant is that females are supported by non-kin (juvenile and adult females) against lowborn females, whereas the latter are infrequently or never supported by non-kin against higher-born females (Chapais 1983*a*; Chapais *et al.* 1991; Cheney 1983; Hunte and Horrocks 1986; Netto and van Hooff 1986; Pereira 1989; Silk 1982; Walters 1980). In summary, these data indicate that any highborn female has more allies (kin and non-kin) than any lowborn female.

Correlate 5

Correlate 5 states that highborn females may have more powerful allies. The available evidence also support this hypothesis. In a matrilineal hierarchy, kin occupy adjacent ranks. The fact that females support their kin entails in itself that highborn females have more powerful (dominant) kin allies.

Second, highborn females have more powerful non-kin allies as well. Theoretically, any non-kin could rank either below, between, or above the two opponents. It is found, however, that non-kin allies are almost always dominant to the lowborn target. Non-kin females that are subordinate to both the highborn and the lowborn opponents rarely or never interfere in their conflicts. Non-kin females ranking between the highborn and lowborn opponents almost always support the highborn against the lowborn, and non-kin females ranking above both the highborn and lowborn opponents support the highborn against the lowborn significantly more often than the reverse (Chapais 1983*a*; Chapais *et al.* 1991; Hunte and Horrocks 1986; Netto and van Hooff 1986). Other studies similarly found that females were supported against lower-born individuals by non-kin females dominant to the targets, and that they were not supported by non-kin against higher-born females (Cheney 1983; Datta 1983*a*; Pereira 1989; Silk 1982; Walters 1980). In these studies, however, whether the supporter ranked above or below the recipient was not specified. Overall, these data indicate clearly that non-kin preferentially support higher-born females.

Correlate 6

Since correlate 5 is verified (highborns have more powerful allies), lowborn targets are expected to be more prone to submit, or to submit more intensely, in the course of an intervention, compared to highborn targets. Only data on the frequency of submission were available, and they confirm this correlate. Cheney (1977) observed that when an immature baboon was the object of aggression, the probability that an intervention by its mother would reverse the direction of the threat was greater for higher-ranking females. Such support was said to be 'successful'. Berman (1980) reported that the protectors of highborn infant rhesus monkeys never demonstrated fearful and submissive behaviours, whereas the protectors of lowborn infants did. She stated that the interventions in favour of highborn individuals were successful as a result. Datta (1983*a*) also reported that in rhesus monkeys support was 'effective' for the highborn and not for the lowborn. In interventions occurring before the period of challenge, the allies of the highborn induced submissiveness in the lowborn target in 91 of 98 cases, whereas the allies of the lowborn did so in only 2 out of 11 cases. The same pattern was observed for interventions taking place during the period of challenge. Similarly, in the interventions occurring during the period of rank maintenance, support for the highborn was effective in 87.2 per cent of 102 cases while support for the lowborn was effective in only 9.2 per cent of 120 cases (Datta 1983*b*).

Correlate 7

Whether highborns are more active than lowborns in the utilization of their allies (solicitations or aggressiveness) could not be tested adequately. Although many authors mention that highborn females do solicit their allies (Datta 1983*a*; Pereira 1989; de Waal 1977; Walters 1980), there are no quantitative comparisons of the rates at which highborn and lowborn females do so. It is noteworthy, however, that no author has reported that females solicited their allies against highborn females. This is to be expected since females do not receive effective support against the latter (previous correlate). Nor are there data that would permit us to compare the reaction of highborns and lowborns when they receive support.

Table 2.1 summarizes the results pertaining to the above correlates. Considering the species together, all correlates are confirmed, except for the last one, which could not be tested. Thus, under the hypothesis that interventions are responsible for the matrilineal transmission of rank, a female would acquire and maintain her mother's rank because (1) she is supported at a higher rate, (2) by a greater number of kin and non-kin allies, (3) which are more powerful (i.e. rank higher than those of lowborn females), and (4) give her, as a result, more successful or effective support. Some of these correlates remain to be verified in many species.

Some further observational data provide a different type of (negative) evidence for the role of alliances in rank acquisition. In the absence of interventions, females should fail to acquire their mother's rank. There are two categories of evidence here. The first concerns the fate of female orphans. Johnson (1987) reported that three female baboons aged less than 3 years had fallen in rank following the death of their mother, whereas another three females, which were older, had been able to retain their rank. Four other juvenile females whose mother had fallen in rank also fell in rank. Hasegawa and Hiraiwa (1980) also reported that in a group of Japanese macaques many female orphans had been outranked by lower-born females.

The second category of evidence concerns the rank relations of natal males. Pereira (1989) has shown that adult female baboons intervened differently in the conflicts of juvenile males and those of juvenile females. While they strongly biased their support of juvenile females in the sense of the matrilineal rank order, they did not do so for juvenile males (interfering infrequently, and regardless of the maternal rank of opponents), thus leaving the males unconstrained to assume ranks based on individual attributes. This would explain why juvenile male baboons do not rank according to their maternal rank, but in relation to their relative age/size.

Both categories of evidence offer additional support for the role of interventions in rank acquisition and maintenance.

EXPERIMENTAL EVIDENCE FOR THE ROLE OF ALLIANCES IN RANK ACQUISITION

Rank acquisition between females differing in age is an ontogenetic phenomenon extending over months or years. It is therefore not usually possible to observe adequately the whole process whereby an individual acquires the rank of its allies above lower-born targets—but see Chapais (1983b) for a description of how a natal alpha male was able to recover his dominance status following an illness, with the support of his matriline, in free-ranging rhesus monkeys. Fortunately, the experimental approach makes it possible to analyse some aspects of the process of rank inheritance in 'time lapse' form.

Early experiments by Maslow (1936) and Varley and Symmes (1967), carried out on small bisexual groups of unrelated rhesus monkeys ($N = 6$), had shown that the direction of dominance between any two monkeys could change according to whether they were tested in groups of two, three, four, etc. It was concluded that the rank relations of monkeys were affected by their relationships with the other members of the group. The role of alliances in some of the rank changes was also described. Until recently, very little experimental work had been carried out on matrilineal hierarchies as such. The first study was that of Marsden (1968) who described rank changes among 10 rhesus macaques, subsequent to the replacement of the alpha male or the removal of an adult female: when a mother gained or lost rank in the hierarchy, her immature daughters could likewise gain or lose rank. Few details on the processes involved were given, however.

More recently, Chapais (1988*b*) carried out experiments on a group of Japanese macaques composed of one adult male and three unrelated matrilines with similar age-sex compositions. In each matriline, the mother (named A, B, or C) had three immature daughters, born in 1981 (A1, B1, C1), 1982 (A2 and B2; there was no C2), and 1983 (A3, B3, C3), and one son born in 1984 (A4, B4, C4).

For example, in one series of experiments, a peer subgroup composed of A3, B3, and C3 (aged 2 years), and their mothers (A, B, and C) was formed and maintained for 2 days. The six individuals maintained their original matrilineal rank order. A and B were then removed, so that the peer subgroup was left with mother C. In this situation, C outranked A3 and B3 in dyadic interactions, and C3 soon assumed her mother's rank above the two previously dominant peers. At the end of a 4-day period, C was removed and replaced by mother B. In this situation, B3 acquired her mother's rank above A3 and C3. Similarly, 4 days later, B was replaced by mother A for a 4-day period, and A3 recovered her matrilineal rank above her peers.

Five such experiments were carried out (Chapais 1988*b*; Chapais and Larose 1988), in which each of the subgroups had a tripartite composition: a

Table 2.1. Role of agonistic interventions in rank acquisition and rank maintenance: summary of the data pertaining to the first six correlates

Species and authors	Correlates						
	1 Support precedes rank reversal	2 Highborn is more aggressive near its allies	3 Allies of highborn are more active	4 Highborn has more allies		5 Highborn has more powerful allies	6 Highborn receives more effective support
				Kin	Non-kin		
M. mulatta							
Berman (1980)	Yes			Yes			Yes
Datta (1983a)	Yes	Yes	Yes			Yes	Yes
Datta (1983b)			No				Yes
Datta (1983c)				Yes			
Chapais (1983a)					Yes	Yes	Yes[3]

M. fuscata							
Chapais *et al*. (1991)			Yes		Yes	Yes	
M. fascicularis							
Netto and van Hoof (1986)			Yes[1]		Yes	Yes	Yes[3]
C. aethiops							
Horrocks and Hunte (1983)		Yes	Yes[1]				Yes
Hunte and Horrocks (1986)			Yes[2]		Yes	Yes	Yes[3]
Cheney (1983)					Yes	Yes	Yes[3]
Papio sp.							
Cheney (1977)	Yes		Yes[1]	?	?	Yes	Yes
Walters (1980)	Yes	Yes			Yes	Yes	Yes[3]
Pereira (1989)	Yes			Yes[1]	Yes	Yes	Yes[3]

[1] maternal interventions only
[2] kin interventions
[3] could be deducted from the results relating to correlate 5

subordinate immature (called *dependent subject*), its adult kin (*ally*), and one or more individuals (*targets*). The targets were subordinate in dyadic interactions to the ally but dominant to the dependent subject. Each of these situations, therefore, created an opportunity for the dependent subject to acquire its ally's rank above the targets. The ally was the mother in four experiments, an older sister in one experiment; the dependent and target individuals were immature females in four experiments, immature males in one.

The five experiments resulted in 53 instances of experimental rank acquisition (a rank reversal between a dependent subject and a target) out of a possible maximum of 54 rank reversals. These results demonstrated clearly that rank can be matrilineally transmitted through the formation of alliances, and that any female can assume a dominant or subordinate position in the hierarchy depending on her relative alliance power. The 53 rank changes were analysed separately for the two following categories: rank changes among non-experienced individuals ($N = 31$), and rank changes among experienced individuals ($N = 22$). Non-experienced individuals had not found themselves in previous experimental situations performing the same roles in relation to each other (ally, dependent subject, or target). Experienced individuals, in contrast, had had the opportunity to learn from previous rank changes, in the course of which they had occupied the same roles. These separate analyses made it possible to control for the effect of learning on experimental rank acquisition.

Among non-experienced individuals, 68 per cent of rank acquisitions occurred through an aggressive intervention of the ally in favour of her dependent kin. The remaining cases took place through the spontaneous submission of the target shortly after the formation of the experimental subgroup. Thus, in the majority of cases where dependent subjects were experiencing an advantage in relative alliance power for the first time, they obtained their ally's rank following an agonistic intervention by the latter. In these situations they did not solicit their ally against the target, or challenge the latter. This result complements the observations relating to correlate 1 (see above), suggesting that offspring of high-ranking females may not be able to solicit their allies, or challenge lower-born individuals, before they have been supported at least once.

Only one aggressive intervention, or one short series of consecutive interventions (mostly contact-aggression), was needed before the targets began submitting to the protected subjects. Thus, interventions were highly effective, suggesting that the rarity of interventions in stable groups is not incompatible with their preponderant role in rank inheritance.

Among experienced individuals, in contrast, rank acquisition occurred mainly through the spontaneous submission of the target to the dependent subject (71 per cent). Thus, targets that had previously experienced the negative effects of having less alliance power rapidly applied this experience in similar situations. By doing so they probably minimized the risk of injuries, readily submitting to dependent subjects that had more alliance power. This adaptive response demonstrates the capability of monkeys to recognize kin alliances (see also Cheney and Seyfarth 1986).

In one experiment, older sisters (aged 4 years) proved to be effective allies (i.e. allies capable of inducing rank acquisition). Although more experiments are needed before firm conclusions can be drawn, this suggests that the presence of the mother may not always be a necessary condition for the matrilineal transmission of rank. The fact that orphans are sometimes able to acquire their mother's rank (e.g. Sade 1972; Walters 1980) may be due to the presence of other kin, and possibly non-kin allies, as discussed later. The age of the orphans is also an important variable (Hausfater *et al.* 1982; Johnson 1987: see above).

EXPERIMENTAL EVIDENCE FOR THE ROLE OF ALLIANCES IN RANK MAINTENANCE

The experimental approach was also utilized to uncover the role of alliances in the maintenance of rank, and to assess the conditions affecting the stability of matrilineal hierarchies. We shall examine separately the evidence relating to the idea that rank maintenance requires allies, and the results concerning the composition of such alliances.

Rank maintenance requires allies

In order to examine whether females seek to rise in rank when given adequate opportunities, and whether they depend on allies for maintaining their rank in the face of such challenges, Chapais (1988*a*) carried out a series of experiments with Japanese macaques. These experiments tested the capability of single highborn females (i.e. females deprived of their kin) to maintain their rank above lowborn females. Three types of experimental subgroups were formed. The single highborn females were placed: (1) with a single, but older/larger lowborn female; (2) with two or three lowborn sisters; or (3) with a complete subordinate matriline (mother and four offspring). Fifty-eight such experimental subgroups were formed, totalling 148 highborn–lowborn dyads. Of the 148 dyads, 56 per cent reversed rank. The capability of a highborn female that had been separated from her matriline to maintain her rank above lowborn females was affected by two factors: her absolute age and the number of lowborns. Highborn females aged 3 years or less were outranked in the majority of cases (80 per cent of 65 dyads). In contrast, highborn females aged 4 years or more were outranked much less frequently (37 per cent of 83 dyads). These older highborn females were never outranked by a single larger lowborn female, rarely outranked by a group of lowborn sisters (11 per cent of 26 dyads), but frequently outranked by a complete subordinate matriline (58 per cent of 48 dyads).

These results indicated, first, that competition for rank is a fundamental feature of matrilineal hierarchies, with lowborn females constituting a potential and constant threat for any highborn female, immature or adult. Second, that kin can form revolutionary coalitions. Third, that the maintenance of rank is conditional upon the high-ranking females having enough alliance power to counter such potential coalitions. And fourth, that although competition for rank does occur, it also appears somewhat constrained. In all of the 58 experimental subgroups, actual relative power was in favour of the subordinates, i.e. the latter were either larger or more numerous than the single highborn female. Despite this, subordinate females outranked the single dominant female in only 56 per cent of the 148 dyads tested. It is noteworthy that they outranked them only in situations where the power asymmetry (relative size or relative alliance power) was pronounced. This was referred to as a 'minimal risk strategy' of competition for dominance (Chapais 1988*a*).

Further evidence for the minimal risk constraint on rank competition came from the experiments on rank acquisition described in the previous section. In three of the five experiments, the ally was removed 1 or 2 days after its dependent kin had outranked a group of peers. Thus, peers with experimentally reversed ranks were left together (e.g. B3 dominant to A3 dominant to C3). In the course of the three experiments, five such subgroups of peers with reversed ranks were formed and maintained for 2 to 4 days. In all cases the newly dominant individuals were able to maintain their rank above their peers even though their ally was absent (Chapais and Larose 1988). Contrary to expectations, these rank relations went unchallenged (unidirectional submission and aggression) even though the experimentally dominant individuals were facing peers with a long ontogenetic experience of dominance over them. Theoretically, the latter might have used this experience to recover their former rank, but they abided by the rank order set by the ally of the dominant peer. These results illustrate the minimal risk constraint on competition: for matrilineal rank reversals to occur, the subordinate challenger must either be significantly larger than the dominant, or have significantly more alliance power.

Composition of alliances affecting the maintenance of rank

The foregoing evidence indicates that any female, even as an adult, is dependent on allies for maintaining her rank above lower born females. On theoretical grounds, such allies can be kin or non-kin. The primary role of kin in the matrilineal acquisition of rank was reviewed above. But do non-kin also play a role in the maintenance of rank?

Chapais *et al.* (1991) conducted experiments on alliances among non-kin in a group of Japanese macaques composed of three unrelated matrilines (A, B, and C; see above). The experiments tested the existence of mutual support between the A and B matrilines against the C matriline. The rationale was the following. Knowing that a member of the B matriline was outranked when left alone with the C matriline, it could be tested whether this same individual could maintain its rank above the C matriline in the presence of non-kin only, in this case the A matriline. Similarly, knowing that a single member of the A matriline was unable to maintain its rank when left alone with the C matriline, it could be checked whether the presence of the B matriline could prevent this outranking.

Twelve subjects were tested, six from the B matriline and six from the A matriline (the study group described above had grown from its original 15 members to include 22 members). When each of the 12 subjects was placed with the C matriline only, 11 were outranked soon after the formation of the subgroup. It was clear, therefore, that these individuals were dependent on allies (kin or non-kin) to maintain their rank in the intact study group. On the other hand, when the six subjects from the B matriline were placed singly with the A and C matrilines, each was able to maintain its rank above the C matriline for as long as the experimental subgroup was maintained (between 1 and 6 days, depending on the experiment). Similarly, the six subjects from the A matriline, which were placed singly with the B and C matriline, did maintain their rank above the C matriline. Thus, the presence of dominant non-kin did prevent the formation of revolutionary coalitions among subordinates.

In 11 of the 12 cases, the subject at once adopted a dominant attitude towards the members of the C matriline, and the latter responded submissively most of the time. One exception was a B subject, which first submitted to the C individuals, but later outranked them after receiving intense support from the A matriline. Of the 12 subjects, 11 received support from the dominant matriline against the C matriline. These interventions were infrequent, however, and they most often occurred at times when the rank relations between the subject and the C matriline already appeared well established. Thus, both the confident attitude of the subjects upon their introduction, and the low frequencies of support they received subsequently, point to the prior existence of mutual support between the A and B matrilines in the intact study group. This hypothesis was confirmed by the analysis of interventions ($N = 258$) by non-kin in conflicts between matrilines, collected over a 52-month study period. This analysis revealed a clear pattern of preferential support among the members of the A and B matrilines against the C matriline (Chapais *et al.* 1991). It was concluded that alliances among non-kin contributed to the maintenance of matrilineal rank relations in this group.

DISCUSSION

Alliances and the proximate causes of rank acquisition

The patterning of agonistic interventions (summarized in Table 2.1) and the findings on the role of allies in experimental rank acquisition provide strong and complementary evidence for the preponderant role of alliances in the social inheritance of rank among female cercopithecines. The most general principle emerging from the present review is perhaps that any female seeks to outrank any other female against whom she is given adequate alliance power. In a matrilineal dominance hierarchy, it is generally the case that such asymmetry in alliance power favours the highborn females (for reasons that are discussed later), and this explains why maternal rank is transmitted across generations. According to that principle, then, a female could challenge any target (lower- or higher-born) against whom she obtains sufficient support (from any ally). We shall now examine evidence showing that this is indeed the case.

The exceptions to the pattern of matrilineal rank transmission nicely confirm the idea that females can use any advantage in alliance power to outrank any target. In all cases where it was reported that a female challenged or outranked a higher-born female it was also found that she had received strong and unusual support against the latter. Such support was provided either by dominant males or by females ranking higher than the target female. An alliance whereby a female rises in the hierarchy with help coming from above the target is referred to as a *bridging alliance*. Such alliances occur in various situations.

Carpenter (1942) observed that female rhesus monkeys in sexual consort with males could become temporarily dominant to higher-born females. Datta (1983b) also mentioned that female rhesus in sexual consort could threaten highborn females and be actively supported by their consort. Homosexual consortships between females can also lead to the formation of a bridging alliance, and induce rank reversals. For example, a low-ranking female in consort with a high-ranking one temporarily but clearly outranked unrelated females ranking between them (Chapais and Mignault 1991).

Bridging alliances formed with an alpha male may also play a major role in matrilineal rank reversals (long-tailed macaques, Chance *et al.* 1977; Japanese macaques, Gouzoules 1980; reviewed in Chapais 1985). A particularly clear example of the latter was observed in a captive group of Japanese macaques composed of one adult male and three matrilines (personal observation). The matrilineal hierarchy had been stable for several years (Chapais *et al.* 1991). When the alpha male began consistently to support some young lowborn individuals against certain members of the top-ranking matriline, lowborn females began to challenge highborn females, and many rank reversals soon occurred. In order to confirm the role of the alpha male in these rank changes, the latter was removed temporarily. The hierarchy returned to its traditional state within the hour following the removal of the male: the highborn females recovered their rank above all lowborn individuals by forming coalitions, and all challenges to the matrilineal order stopped at once. When the alpha male was later reintroduced, the previous pattern of instability returned and continued until the male reduced its rate of support.

High-ranking females can also form bridging alliances. For example, immature females have been observed to challenge and outrank their mothers with the support of unrelated females ranking higher than their mother, in both Japanese and rhesus macaques (Chapais 1985; Marsden 1968). Samuels *et al.* (1987) mentioned that a female baboon probably contributed to the rise in rank of her mother and older sister above females ranking below herself, by attacking these females before or after they were challenged by her mother or sister. Similarly, Pereira (1989) reported the case of a female baboon from a low-ranking matriline, which dominated or challenged as many as 10 higher-born females. This female was the only one to receive support from adult females against females from matrilines ranking higher than her own. She also received much support from her 'primary adult male associate'.

The foregoing evidence supports the idea that the matrilineal inheritance of rank is only one manifestation of a more general principle governing the dynamics of alliance formation in matrilineal hierarchies. Females compete for rank as conditional opportunists. They are opportunist in that they are prompt to use any type of advantage in alliance power against any target, not only lower-born ones, and their opportunism is conditional upon their receiving adequate support against the target, either from individuals ranking above it (bridging alliances) or from coalition partners ranking below the target but clearly outnumbering it (revolutionary alliances). In so doing, they exhibit a minimal or low-risk strategy, in contrast, for example, to male chimpanzees which seem willing to incur higher risks when competing for rank through coalitions (e.g. Nishida 1983; Riss and Goodall 1977; de Waal 1982).

Critique of the role of alliances in rank acquisition

The role of alliances in rank acquisition has been questioned on the basis of the arguments that rank acquisition can take place in the absence of kin allies (Walters 1980), that mechanisms other than interventions may better explain it, and that interventions are rare events (Johnson 1987; Lee and Oliver 1979; Walters 1980). These arguments will now be briefly examined.

In his study of rank acquisition in baboons, Walters (1980) found that by considering the rank of juvenile females at the time they were born (their birth rank), it was possible to predict correctly 97 per cent of their actual rank relations with highborn and lowborn adult females. This figure dropped to 87 per cent when the rank of the females' kin at the time of the study was considered instead. On this basis Walters rejected the 'ally rank model' in favour of the 'birth rank model', stating that interventions were of secondary importance, and proposing instead (following Kawai 1958), that observational learning better accounted for the matrilineal inheritance of rank. However, it is noteworthy that Walters excluded non-kin allies from his ally rank model, while emphasizing at the same time that non-kin support to dominants was more frequent than kin support. Had he included kin and non-kin allies, the explanatory power of the ally rank model might have improved significantly. The present review indicates that any model of rank acquisition should include all allies, kin and non-kin, which were active during the whole period between the female's birth and the time at which she assumed her rank in the hierarchy.

Moreover, the observational learning hypothesis, conceived as a sufficient (rather than complementary) explanation for rank acquisition does not withstand closer examination. According to this hypothesis, highborn and lowborn females would learn their respective ranks by observing the aggressive and non-aggressive interactions of their mother *before* they themselves become involved in agonistic interactions. Lowborn females would thus be expected to submit passively to highborn females, whereas they actually resist being outranked, and submit only after receiving intense aggression (Datta 1983*a*; Horrocks and Hunte 1983; Pereira 1988; Walters 1980). Moreover, the fact that interventions appear to precede, rather than to follow, the first agonistic interactions of infants is difficult to reconcile with the idea that rank is learned independently of interventions (Chapais 1988*b*; Datta 1983*a*).

Another mechanism has been proposed to explain rank acquisition. Berman (1980) observed that highborn infants received fewer threats from adult females and males compared to lowborn infants, and suggested that 'infants may learn their rank through differences in amounts and sources of agonistic interactions directed towards them' (Berman 1980, p. 165). Horrocks and Hunte (1983) went further by proposing that such aggression by highborn adult female to lowborn females and juveniles represent a third mechanism of rank acquisition and 'the behaviour primarily responsible for controlling the acquisition of rank by offspring' (Horrocks and Hunte 1983, p. 773). They stated that such aggression, beginning before the yearlings exchanged aggression and submitted to highborn females, led to submission *among* the yearlings. While it is clear that aggression directed by highborn adult females to lowborn infants would lead to the latter submitting to the former, it is not clear why such aggression would lead to the lowborn submitting to a highborn peer or to a younger/smaller highborn female. To make sense of this, Horrocks and Hunte had to invoke 'the constant threat of maternal interventions by the high-ranking mother'. Their proposed mechanism therefore did include interventions as a necessary component of rank acquisition, and so it does not appear to be a separate third mechanism.

Finally, interventions are rare events. For example, Johnson (1987) in her study of olive baboons saw mothers supporting their daughters only three times in 17 months. However, one should consider the support given by all categories of allies, kin and non-kin. For example, Chapais *et al.* (1991) found that in Japanese macaques intermatriline conflicts gave rise to more support by non-relatives of the opponents than by their relatives. Moreover, infrequent interventions may nevertheless be highly effective (cf. experimental data above). Interventions may also have long-lasting effects, especially if the allies maintain proximity (Chapais 1988*b*) and exchange grooming, possibly as a reminder of their alliance to the other group members (Chapais 1983*a*; Johnson 1987).

In conclusion, the evidence for the preponderant role of agonistic alliances in rank inheritance appears overwhelming. This is not to say, however, that other factors (observational learning, differential agonistic experience) do not play a significant and complementary role in rank inheritance (Chapais 1988*b*; Harcourt and Stewart 1987).

Alliances and the proximate causes of rank maintenance

Matrilineal hierarchies are remarkably stable over periods of several years (Bramblett *et al.* 1982; Hausfater *et al.* 1982; Sade 1972) but may sometimes go through periods of instability (Ehardt and Bernstein 1986; Samuels *et al.* 1987). Since the acquisition and maintenance of rank are intimately linked to the dynamics of alliances, it is likely that the factors regulating the stability of rank relations have much to do with those regulating the formation of stabilizing and destabilizing alliances. Why is it that lowborn females, acting singly or collectively, do not challenge highborn females more frequently? The above review suggests that this is because highborns have more alliance power, based on kin or non-kin support. If the key to the stability of matrilineal hierarchies is kin support, highborn females would be expected to belong to larger kin groups on average, i.e. to have more potential kin allies. If this is not the case, the stability of matrilineal hierarchies is likely to rest on the direction of non-kin support.

If the first possibility was indeed the case there would exist a positive correlation between matriline rank and matriline size. Although some authors have reported positive correlations between the rank of females and their reproductive output (e.g. Dittus 1979; Dunbar 1980; Sade *et al.* 1976; Silk *et al.* 1981; Wilson *et al.* 1978), other authors have found no such correlation (e.g. Cheney *et al.* 1981; Fedigan *et al.* 1986; Gouzoules *et al.* 1982; Wolfe 1984). This inconsistency suggests that the relation between matriline rank and matriline size is context-dependent (Gouzoules *et al.* 1982). If this is so, it remains to be shown that the hierarchy's stability is also context-dependent, i.e. rank is less stable in the absence of a correlation between matriline rank and size. This does not seem to be the case, however, since hierarchies including a high-ranking matriline smaller than lower-ranking ones may nevertheless be stable (e.g. Chapais 1983*a*).

On the other hand, the role of non-kin alliances in the maintenance of rank has been established above. Data on the patterning of agonistic interventions (see correlates 4 and 5 and Table 2.1), as well as experimental evidence, show that unrelated females ally against lower-born females, thus reinforcing the existing rank relations between matrilines.

In summary, matrilineal hierarchies would be stable for as long as highborn females have more alliance power than lowborn females. This asymmetry would rest not so much on a difference in the extent of kin support, but on a difference in the extent of non-kin support. Evolutionary and functional aspects of kin and non-kin alliances will now be examined.

Evolutionary aspects of alliance formation in matrilineal hierarchies

Evolution of kin support

Why do females expend energy to support their kin and transmit their rank to their offspring? Are they altruistic in doing so, or do they derive immediate or future benefits? Consider a hypothetical hierarchy of females in which rank relations would be established strictly in the context of dyadic aggressive encounters (i.e., kin and non-kin support are non-existent). Such a hierarchy would likely be age/size-graded. Suppose that a female begins to support her daughters consistently against the females that she dominates. By extending her maternal investment into the area of dominance relationships she would be protecting her offspring at a time when they are most vulnerable. Moreover, by ensuring that they rank adjacent to her, she would spare them a period of time at the bottom of an otherwise age-graded hierarchy (Chapais and Schulman 1980).

In addition to these benefits for her offspring, the female would probably gain personal advantage as well. In the long run she would be building a coalition of adjacently-ranked partners (her daughters and their offspring) that would support each other mutually, although not necessarily symmetrically. Support among kin is frequent in cercopithecines (e.g. Bernstein and Ehardt 1985; Chapais 1983*a*; Datta 1983*c*; Massey 1977; Watanabe 1979). Such a kin group would represent a cohesive force, both in the face of challenges from subordinate non-kin and as a potential revolutionary coalition which could overpower any single dominant female not building a similar familial coalition. Dunbar (1980) has similarly argued that in gelada baboons a female forms coalitions preferentially with her daughters in order to increase her average rank, because this is 'the only bond of sufficient strength to provide the basis for an investment which is asymmetric in the short-term and reciprocal only over the length of a life-time' (Dunbar 1980, p. 253; see also Pereira, 1992).

Returning to our hypothetical age-graded hierarchy, once a female begins to support her daughters, all females should do the same, or risk being outranked. When such a pattern is established we obtain a hierarchy of matrilines, i.e. a nepotistic, matrilineal hierarchy, but one lacking any non-kin support. The stability of such a rank order would therefore be conditional upon the matrilines (kin coalitions) not being too dissimilar in size (see previous section). However, this is without considering the effect of non-kin support.

Evolution of non-kin support

Why do non-kin females interfere at all and why do they consistently support the most dominant opponent? The answer is likely to be that by so doing females prevent the formation of the two types of alliances through which they could be outranked: revolutionary alliances, formed among lower-ranking females, and bridging alliances, formed between a female ranking below the target and one ranking above her. Consider the type of hierarchy just mentioned (nepotistic and lacking non-kin support), and suppose that it consists of four matrilines (A, B, C, and D). Such a hierarchy would probably be evolutionarily unstable (*sensu* Maynard Smith 1982) because the formation of destabilizing coalitions among non-kin would always be possible. For example, females from matriline C might be able to outrank females from the B matriline by forming alliances with either females from the A matriline (bridging alliances) or with those of the D matriline (revolutionary alliances). The core of the present argument is that the targets of such destabilizing alliances (in this case the B females) could prevent the formation of these alliances and maintain their rank by exhibiting the observed patterns of support in favour of dominant non-kin (Chapais and Schulman 1980).

Consider, first, the possibility of the formation of a bridging alliance between females C1 and A1 against female B1. How could such an alliance begin? The easiest way is perhaps for C1 to begin to 'offer her help' to A1, i.e. to consistently join A1 when the latter directs aggression to B1. Female A1 would benefit from that support, even if she is already dominant to B1, because it would then be considerably more difficult for B1 to eventually challenge A1. For this reason, A1 can be expected to support her new ally C1, given, for example, that the latter is attacked by B1. At this stage, therefore, A1 and C1 are supporting each other mutually.

If bridging alliances can indeed form that way, both B1 and C1 would have an interest in allying with A1. In other words, B1 and C1 would be in competition for A1's support. What individual, then, should succeed? It is suggested that in a nepotistic hierarchy it is much easier for the higher-born female (B1) to succeed. The reason is that females, being conditional opportunists (see above), are expected to challenge a target only if they have *already* been protected successfully against that target. In our hypothetical nepotistic hierarchy, any female would be used to receive effective support from her kin against lowborn targets, and consequently, would be able to challenge the latter with little risk of retaliation. For example, should B1 challenge C1 by joining A1, she would incur no risk of retaliation by C1. In contrast, should C1 challenge B1, either on her own or by joining A1, she would be punished by B1's relatives. Thus, although B1 and C1 both have an interest in allying with A1, B1 would be able to do so at little cost, while C1 would incur a significant risk.

This reasoning shows that females could counter the threat of bridging alliances by supporting the females ranking above themselves against those ranking below; this is indeed the pattern observed in matrilineal hierarchies (cf. correlate 5).

Consider now the possibility of the formation of revolutionary alliances, e.g. of C1 allying with D1 against B1. In order to prevent the occurrence of such an alliance, B1 could attempt to form an alliance with a higher-ranking female (A1; case described above), but she could also form one with the most dominant of the two low-ranking challengers (C1). By consistently supporting C1 in her conflicts with D1, B1 would induce a state of dependency in C1, and with it, the inhibition of revolutionary alliances (Chapais 1988*a*). This pattern of intervention (supporting the most dominant of two lower-ranking females) is not independent of the one described above (supporting higher-ranking females against lower-ranking ones). If C1 already supported B1 against D1 in order to counter the threat of a bridging alliance, she could hardly, at the same time, challenge B1 with the help of D1: these patterns of support would be incompatible. In other words, alliances formed to counter the threat of bridging alliances eliminate the possibility of revolutionary alliances.

To summarize, the pattern of non-kin support characterizing matrilineal hierarchies would be a response to the threat of the two types of destabilizing alliances. Once this pattern is established, it would be evolutionarily stable and rank relations would be 'frozen'. An important consequence of the formation of non-kin alliances is that a female losing kin allies could nevertheless maintain her rank by conserving non-kin allies. The present hypotheses may also account for the attractiveness of high-ranking females, observed in cercipithecine societies, with females grooming preferentially unrelated, higher-ranking females and attempting to form alliances with them (e.g. Cheney 1977, 1983; Seyfarth 1977, 1983).

Alliances among non-kin: mutualism versus reciprocal altruism

Another issue is that of the evolutionary basis of non-kin support. Hunte and Horrocks (1986) have interpreted non-kin interventions among vervets in terms of reciprocal altruism (Trivers 1971) on the grounds that intervening 'appears to involve considerable risk . . .' and 'in most cases there were no immediate proximate causes' (Hunte and Horrocks 1986, p. 262). Reciprocal altruism implies indeed that supporting non-kin partners is costly to the supporter, who gains no immediate benefits but is repayed later by the recipient. However, when siding with non-kin against lower-born individuals, the interferer appears to incur little risk (retaliation rarely occurs) and to gain an easy victory. Because non-kin partners appear to be mutually dependent for maintaining their rank (see above), such victories would include important benefits in terms of rank maintenance.

Mutualism (or cooperation) provides a more satisfactory explanation of non-kin interventions, compared to reciprocal altruism. In a mutualistic interaction both partners gain immediate benefits (see Axelrod 1984 and Wrangham 1982 for general discussions of the evolution of cooperation). This interpretation would probably come to mind immediately if non-kin allies initiated their attacks against the target simultaneously. It is the alteration of support that gives the impression that the recipient is gaining, while the supporter is not. Group hunting provides an analogy. Every partner serves its own interest (given that meat is shared) even though one participant may be joining others already engaged in the pursuit. A similar reinterpretation has been proposed for coalitions formed by male baboons for access to estrous females. Initially, these coalitions had been explained in terms of reciprocal altruism (Packer 1977). Recent data on similar coalitions have been reinterpreted in terms of mutualistic cooperation (Bercovitch 1988).

ACKNOWLEDGEMENTS

I thank Sandy Harcourt, Daniel Pérusse, and Frans de Waal for their helpful comments on the manuscript, and I am grateful to Frans de Waal for suggesting the expression 'bridging alliance'. The prepration of this paper was facilitated by grants from the Natural Sciences and Engineering Research Council of Canada, and the Fonds FCAR of the province of Québec.

SUMMARY

1. In many species of cercopithecine monkeys, females achieve ranks below their mother in the dominance hierarchy. The evidence for the role of alliances in the formation of such hierarchies of matrilines appears overwhelming. Observational data indicate that in conflicts between females from different matrilines, the higher-born female is more likely to receive support, from a larger number of kin and non-kin allies, which rank higher than the allies of lower born females, and give her, as a result, more

successful or effective support. Moreover, the core of the process of rank acquisition can be reproduced experimentally. Any female can assume a dominant or subordinate status depending on the rank of her allies, and agonistic interventions can induce the matrilineal transmission of rank.

2. Matrilineal hierarchies probably originated from the extension of maternal investment into the realm of dominance. By transmitting her rank to her offspring, a female is protecting them at a time when they are most vulnerable, and sparing them a period of time at the bottom of an otherwise age-graded hierarchy. In the long run, she is also building a coalition of adjacently-ranked partners (her daughters and their offspring), which could overpower any single dominant female not forming such kin alliances.
3. The remarkable stability of matrilineal hierarchies reflects a dynamic process. The occurrence of spontaneous rank reversals, as well as experimental evidence, both indicate that females seek to outrank higher-ranking females when they are given the opportunity to do so. They can form either 'bridging alliances' (with individuals ranking above the target), or 'revolutionary alliances' (with individuals ranking below the target). Consequently, the stability of matrilineal hierarchies must be conditional upon the higher-born females having more alliance power than lower-born females.
4. Since highborn females do not necessarily belong to larger and more powerful kin groups, the key to the stability of matrilineal hierarchies appears to rest on non-kin support. Unrelated females do indeed support each other against lower-born females. By supporting unrelated females ranking higher than herself against those ranking below her, a female may prevent the formation of bridging alliances; and by supporting the most dominant of lower-ranking opponents, she may prevent the latter from forming revolutionary alliances. Thus, the need to counter the formation of destabilizing alliances would 'freeze' the matrilineal rank order.
5. Non-kill alliances have been interpreted in terms of reciprocal altruism. This hypothesis implies that support to non-kin partners is at first costly to the supporter, who gains later through returned support. Mutualism, however, offers a more parsimonious explanation. Since non-kin allies appear to be mutually dependent for maintaining their rank, they may gain immediate benefits when intervening for each other.

REFERENCES

Angst, W. (1975). Basic data and concepts in the social organization of *Macaca fascicularis*. In *Primate behavior*, vol. 4 (ed. L. A. Rosenblum). Academic Press, New York.

Axelrod, R. (1984). *The evolution of cooperation*. Basic Books, New York.

Bercovitch, F. B. (1988). Coalitions, cooperation and reproductive tactics among adult male baboons. *Animal Behaviour*, **36**, 1198–209.

Berman, C. M. (1980). Early agonistic experience and rank acquisition among free-ranging infant rhesus monkeys. *International Journal of Primatology*, **1**, 153–70.

Bernstein, I. S. (1969). Stability of the status hierarchy in a pigtail monkey group (*Macaca nemestrina*). *Animal Behaviour*, **17**, 452–8.

Bernstein, I. S. (1981). Dominance: the baby and the bathwater. *The Behavioral and Brain Sciences*, **4**, 419–57.

Bernstein, I. S. and Ehardt, C. L. (1985). Agonistic aiding: kinship, rank, age and sex influences. *American Journal of Primatology*, **8**, 37–52.

Bernstein, I. S. and Williams, L. E. (1983). Ontogenetic changes and the stability of rhesus monkey dominance relationships. *Behavior Processes*, **8**, 379–92.

Bramblett, C. A., Bramblett, S. S., Bishop, D. and Coelho, A. M. Jr. (1982). Longitudinal stability in adult hierarchies among vervet monkeys (*Cercopithecus aethiops*). *American Journal of Primatology*, **2**, 10–19.

Carpenter, C. R. (1942) Sexual behavior of free-ranging rhesus monkeys (*Macaca mulatta*). I. Specimens, procedures and behavioral characteristics of estrus. *Journal of Comparative Psychology*, **33**, 113–42.

Chance, M. R. A., Emory, G. R. and Payne, R. G. (1977). Status referents in long-tailed macaques (*Macaca fascicularis*): precursors and effects of a female rebellion. *Primates*, **18**, 611–32.

Chapais, B. (1983*a*). Dominance, relatedness and the structure of female relationships in rhesus monkeys. In *Primate social relationships: an integrated approach* (ed. R. A. Hinde), pp. 208–19. Blackwell, Oxford.

Chapais, B. (1983*b*). Matriline membership and male rhesus reaching high rank in their natal troop. In *Primate social relationships, an integrated approach* (ed. R. A. Hinde), pp. 171–5. Blackwell, Oxford.

Chapais, B. (1985). An experimental analysis of a mother–daughter rank reversal in Japanese macaques (*Macaca fuscata*). *Primates*, 407–23.

Chapais, B. (1988*a*). Rank maintenance in female Japanese macaques: experimental evidence for social dependency. *Behaviour*, **104**, 41–59.

Chapais, B. (1988*b*). Experimental matrilineal inheritance of rank in female Japanese macaques. *Animal Behaviour*, **36**, 1025–37.

Chapais, B. and Larose, F. (1988). Experimental rank reversals among peers in *Macaca fuscata*: rank is maintained after the removal of kin support. *American Journal of Primatology*, **16**, 31–42.

Chapais, B. and Mignault, C. (1991). Homosexual incest avoidance among females in captive Japanese macaques. *American Journal of Primatology*, **23**, 171–83.

Chapais, B. and Schulman, S. R. (1980). An evolutionary model of female dominance relations in primates. *Journal of Theoretical Biology*, **82**, 47–89.

Chapais, B., Girard, M. and Primi, G. (1991). Non-kin alliances, and the stability of matrilineal dominance relations in Japanese macaques. *Animal Behaviour*, **41**, 481–91.

Cheney, D. L. (1977). The acquisition of rank and the development of reciprocal alliances among free-ranging immature baboons. *Behavioral Ecology and Sociobiology*, **2**, 303–18.

Cheney, D. L. (1983). Extra-familial alliances among vervet monkeys. In *Primate social relations: an integrated approach* (ed. R. A. Hinde), pp. 278–86. Blackwell, Oxford.

Cheney, D. L. and Seyfarth, R. M. (1986). The recognition of social alliances by vervet monkeys. *Animal Behaviour*, **34**, 1722–31.

Cheney, D. L., Lee, P. C. and Seyfarth, R. M. (1981). Behavioral correlates of non-random mortality among free-ranging female vervet monkeys. *Behaviural Ecology and Sociobiology*, **5**, 263–78.

Datta, S. (1983*a*). Relative power and the acquisition of rank. In *Primate social relationships, an integrated approach* (ed. R. A. Hinde), pp. 93–103. Blackwell, Oxford.

Datta, S. (1983*b*). Relative power and the maintenance of rank. In *Primate social relationships, an integrated approach* (ed. R. A. Hinde), pp. 103–12. Blackwell, Oxford.

Datta, S. (1983*c*). Patterns of agonistic interference. In *Primate social relationships, an integrated Approach* (ed. R. A. Hinde), pp. 289–97. Blackwell, Oxford.

Dittus, W. (1979). The evolution of behaviours regulating density and age-specific sex ratios in a primate population. *Behaviour*, **69**, 265–301.

Dunbar, R. I. M. (1980). Determinants and evolutionary consequences of dominance among female gelada baboons. *Behavioral Ecology and Sociobiology*, **7**, 253–65.

Dunbar, R. I. M. (1988). *Primate social systems*. Cornell University Press, Ithaca.

Ehardt, C. L. and Bernstein, I. S. (1986). Matrilineal overthrows in rhesus monkey groups. *International Journal of Primatology*, **7**, 157–81.

Estrada, A., Estrada, R. and Ervin, F. (1977). Establishment of a free-ranging colony of stumptail macaques (*Macaca arctoides*): Social relations I. *Primates*, **18**, 647–76.

Fairbanks, L. A. and McGuire, M. T. (1986). Age, reproductive value, and dominance-related behaviour in vervet monkey females: cross-generational influences on social relationships and reproduction. *Animal Behaviour*, **24**, 1710–21.

Fedigan, L. (1983). Dominance and reproductive success in primates. *Yearbook of Physical Anthropology*, **26**, 91–129.

Fedigan, L. M., Fedigan, L., Gouzoules, S., Gouzoules, H. and Koyama, N. (1986). Lifetime reproductive success in female Japanese macaques. *Folia Primatologica*, **47**, 143–57.

Gouzoules, H. (1975). Maternal rank and early social interactions of infant stumptail macaques, *Macaca arctoides*. *Primates*, **16**, 405–18.

Gouzoules, H. (1980). A description of genealogical rank changes in a troop of Japanese monkeys (*Macaca fuscata*). *Primates*, **21**, 262–7.

Gouzoules, H., Gouzoules, S. and Fedigan, L. M. (1982). Behavioural dominance and reproductive success in female Japanese monkeys (*Macaca fuscata*). *Animal Behaviour*, **30**, 1138–50.

Gouzoules, S., Gouzoules, H. and Marler, P. (1984). Rhesus monkey (*Macaca mulatta*) screams: representational signaling in the recruitment of agonistic aid. *Animal Behaviour*, **32**, 182–93.

Harcourt, A. H. and Stewart, K. (1987). The influence of help in contests on dominance rank in primates: hints from gorillas. *Animal Behaviour*, **35**, 182–90.

Hasegawa, T. and Hiraiwa, M. (1980). Social interactions of orphans observed in a free-ranging troop of Japanese monkeys. *Folia Primatologica*, **330**, 129–58.

Hausfater, G. (1975). Dominance and reproduction in baboons (*Papio cynocephalus*). In *Contributions to primatology*, vol. 7. Karger, Basel.

Hausfater, G., Altmann, J. and Altmann, S. A. (1982). Long-term consistency of dominance relations among female baboons (*Papio cynocephalus*). *Science*, **217**, 752–5.

Hinde, R. A. (1978). Dominance and role: two concepts with dual meaning. *Journal of Biological and Social Structures*, **1**, 27–38.

Horrocks, J. and Hunte, W. (1983). Maternal rank and offspring rank in vervet monkeys: an appraisal of the mechanisms of rank acquisition. *Animal Behaviour*, **31**, 772–82.

Hunte, W. and Horrocks, J. (1986). Kin and non-kin interventions in the aggressive disputes of vervet monkeys. *Behavioral Ecology and Sociobiology*, **20**, 257–63.

Johnson, J. A. (1987). Dominance rank in juvenile olive baboons *Papio anubis*: the influence of gender, size, maternal rank and orphaning. *Animal Behaviour*, **35**, 1694–708.

Kawai, M. (1958). On the system of social ranks in a natural troop of Japanese monkeys (1): Basic rank and dependent rank. *Primates* **1**, 111–48. Also in *Japanese monkeys, a collection of translations* (eds. K. Imanishi and S. A. Altmann), pp. 66–86. Emory University, Atlanta, 1965.

Kawamura, S. (1958). Matriarchal social ranks in the Minoo-B troop: a study of the rank system of Japanese monkeys. *Primates* **1**, 148–56. Also in *Japanese monkeys, a collection of translations* (eds. K. Imanishi and S. A. Altmann), pp. 105–12. Emory University, Atlanta, 1965.

Koyama, N. (1967). On dominance rank and kinship of a wild Japanese monkey troop in Arashiyama. *Primates*, **8**, 189–216.

Kuester, J. and Paul, A. (1988). Rank relations of juvenile and subadult natal males of Barbary macaques (*Macaca sylvanus*) at Affenberg Salem. *Folia primatologica*, **51**, 33–44.

Lee, P. C. (1983). Context-specific unpredictability in dominance interactions. In *Primate social relationships, an integrated approach* (ed. R. A. Hinde), pp. 35–44. Blackwell, Oxford.

Lee, P. C. and Oliver, J. I. (1979). Competition, dominance and the acquisition of rank in juvenile yellow baboons (*Papio cynocephalus*). *Animal Behaviour*, **27**, 576–85.

Marsden, H. M. (1968). Agonistic behavior of young rhesus monkeys after changes induced in the social rank of their mothers. *Animal Behaviour*, **16**, 38–44.

Maslow, A. H. (1936). The role of dominance in the social and sexual behavior of infra-human primates: IV. The determination of hierarchy in pairs and in a group. *Journal of Genetic Psychology*, **49**, 161–98.

Massey, A. (1977). Agonistic aids and kinship in a group of pigtail macaques *Behavioral Ecology and Sociobiology*, **2**, 31–40.

Maynard Smith, J. (1982). *Evolution and the theory of games*. Cambridge University Press, Cambridge.

Missakian, E. A. (1972). Genealogical and cross-genealogical dominance relations in a group of free-ranging rhesus monkeys (*Macaca mulatta*) on Cayo Santiago *Primates*, **13**, 169–80.

Netto, W. J. and van Hooff, J. A. R. A. M. (1986). Conflict interference and the develoment of dominance relationships in immature *Macaca fascicularis*. In

Primate ontogeny, cognition and social behaviour (ed. J. G. Else and P. C. Lee), pp. 291–300. Cambridge University Press, Cambridge.

Nishida, T. (1983). Alpha status and agonistic alliance in wild chimpanzees (*Pan troglodytes schweinfurthii*). *Primates*, **24**, 318–36.

Packer, C. (1977). Reciprocal altruism in *Papio anubis*. *Nature*, **265**, 441–3.

Paul, A. and Kuester, J. (1987). Dominance, kinship and reproductive value in female Barbary macaques (*Macaca sylvanus*) at Affenberg, Salem. *Behavioral Ecology and Sociobiology*, **21**, 323–31.

Pereira, M. E. (1988). Agonistic interactions of juvenile savanna baboons I. Fundamental features. *Ethology*, **79**, 195–217.

Pereira, M. E. (1989). Agonistic interactions of juvenile savanna baboons II. Agonistic support and rank acquisition. *Ethology*, **80**, 152–71.

Pereira, M. E. (1992). The development of dominance relations prior to puberty in cercopithecine societies. In *Aggression and non-aggression in primates* (ed. P. Gray and J. Silverberg). Oxford University Press, Oxford.

Riss, D. C. and Goodall, J. (1977). The recent rise to the alpha rank in a population of free-living chimpanzees. *Folia primatologica*, **27**, 134–51.

Sade, D. S. (1967). Determinants of dominance in a group of free-ranging rhesus monkeys. In *Social communication among primates* (ed. S. A. Altmann), pp. 99–111. University of Chicago Press, Chicago.

Sade, D. S. (1972). A longitudinal study of social behavior of rhesus monkeys. In *The functional and evolutionary biology of primates* (ed. R. Tuttle), pp. 378–98. Aldine Press, Chicago.

Sade, D. S., Cushing, K., Cushing, P., Dunaif, J., Figueroa, A., Kaplan, J. R., Lauer, C., Rhodes, D. and Schneider, J. (1976). Population dynamics in relation to social structure on Cayo Santiago. *Yearbook of Physical Anthropology*, **20**, 253–62.

Samuels, A., Silk, J. B. and Altmann, J. (1987). Continuity and change in dominance relations among female baboons. *Animal Behaviour*, **35**, 785–93.

Seyfarth, R. M. (1977). A model of social grooming among adult female monkeys. *Journal of Theoretical Biology*, **65**, 671–98.

Seyfarth, R. M. (1983). Grooming and social competition in primates. In *Primate social relationships: an integrated approach* (ed. R. A. Hinde), pp. 182–90. Blackwell, Oxford.

Shirek-Ellefson, J. (1972). Social communication in some old world monkeys and gibbons. In *Primate patterns* (ed. P. Dolhinow), pp. 297–311. Holt, Rinehart and Winston, New York.

Silk, J. B. (1982). Altruism among female *Macaca radiata*: explanations and analysis of patterns of grooming and coalition formation, *Behaviour*, **79**, 162–88.

Silk, J. B. (1987). Social behavior in evolutionary perspective. In *Primate societies* (eds B. B. Smuts, D. L. Cheney, R. M. Seyfarth, R. W. Wrangham, and T. T. Struhsaker), pp. 318–29. University of Chicago Press, Chicago.

Silk, J. B., Clark-Wheatley, C. B., Rodman, P. S. and Samuels, A. (1981). Differential reproductive success and facultative adjustment of sex ratios among captive female bonnet macaques (*Macaca radiata*). *Animal Behaviour*, **29**, 1106–20.

Silk, J. B., Samuels, A. and Rodman, P. (1981). Hierarchical organization of female *Macaca radiata* in captivity. *Primates*, **22**, 84–95.

Trivers, R. L. (1971). The evolution of reciprocal altruism. *Quarterly Review of Biology*, **46**, 35–57.

Varley, M. and Symmes, D. (1967). The hierarchy of dominance in a group of macaques. *Behaviour*, **27**, 54–75.

de Waal, F. B. M. (1977). The organization of agonistic relations within two captive groups of Java monkeys (*Macaca fascicularis*). *Zeitschrift für Tierpsychologie*, **44**, 225–82.

de Waal, F. B. M. (1982). *Chimpanzee politics*. Harper and Row, New York.

de Waal, F. B. M. (1986). The integration of dominance and social bonding in primates. *Quarterly Review of Biology*, **61**, 459–79.

de Waal, F. B. M. and Luttrell, L. M. (1985). The formal hierarchy of rhesus macaques: an investigation of the bared-teeth display. *American Journal of Primatology*, **9**, 73–85.

de Waal, F. B. M., van Hooff, J. and Netto, W. (1976). An ethological analysis of types of agonistic interaction in a captive group of Java monkeys (*Macaca fascicularis*). *Primates*, **17**, 257–90.

Walters, J. (1980). Interventions and the development of dominance relationships in female baboons. *Folia primatologica*, **34**, 61–89.

Watanabe, K. (1979). Alliance formation in a free-ranging troop of Japanese macaques. *Primates*, **20**, 459–74.

Wilson, M. E., Gordon, T. P. and Bernstein, I. S. (1978). Timing of births and reproductive success in rhesus monkey social groups. *Journal of Medical Primatology*, **7**, 202–12.

Wolfe, L. (1984). Female rank and reproductive success among Arashiyama B Japanese macaques (*Macaca fuscata*). *International Journal of Primatology*, **5**, 133–43.

Wrangham, R. W. (1982). Mutualism, kinship and social evolution. In *Current problems in sociobiology* (ed. King's College Sociobiology Group), pp. 269–89. Cambridge University Press, Cambridge.

3

Effects of availability of allies on female dominance structure

S. B. DATTA

INTRODUCTION

Many primates live in social groups which consist of one or more groupings of female kin, together with one or more associated, usually unrelated, adult males. Group females act as a coalition against females in other groups, increasing the chances of gaining access to contested resources (Wrangham 1980). Group living may also have other advantages, such as cooperative defence against predators (van Schaik 1983), or against infanticidal males (Hrdy 1977). But while there may be benefits to living in a group there are usually also costs, due to intragroup competition for food, water, safe resting places, and so on. Such intragroup competition is often manifested as agonistic dominance relationships, in which some females regularly displace or defeat others in contests. It is possible that high rank enhances the reproductive success of females, since although a positive, significant, correlation between high female rank and reproductive success has not been found in all cases, significant correlations tend to be positive rather than negative (Harcourt 1987; Silk 1987). Certainly, in many primate species individuals behave as if high rank is important, since it appears to be the object of considerable competition (de Waal 1987, review by Chapais this volume, pp.29–61).

It is well established that in many species of primates success in the competition for high rank depends both on the characteristics of individuals relative to one another (for example their relative ages or sizes), and on alliances between individuals. Changes in one or both of these factors would be expected to have considerable impact on individual prospects of high rank, as well as on the overall pattern of dominance relationships. The possibility that demographic factors, by altering the age structure of a population, can alter rank relationships was explored by Hausfater *et al.* (1982) using a simulation approach. However, they stopped short of considering the impact of demographic factors on the availability of allies. The latter was taken into account in a recent paper by Datta and Beauchamp

(1991), who were concerned with the effects of demographic variation specifically on dominance relationships between mothers and daughters, and between sisters. Following on from these attempts, the present paper attempts an overview of the kinds of effects demographic variation could have on alliance availability, dominance relationships and alliance strategies among female-bonded primates.

UNDERLYING MECHANISMS AND PROPENSITIES

Underlying the simulation of Datta and Beauchamp (1991) were some realistic assumptions about the behavioural propensities of females, and about the mechanisms underlying the acquisition and maintenance of rank. These assumptions were based on available information about the behavior of female cercopithecines, in particular rhesus macaques (*Macaca mulatta*), Japanese macaques (*M. fuscata*), and savannah baboons (*Papio cynocephalus*) (e.g. Varley and Symmes 1966; Koyama 1967, 1970; Marsden 1968; Cheney 1977; Chance *et al.* 1977; Kaplan 1977; Gouzoules 1980; Walters 1980; Welker *et al.* 1980; Datta 1983*a*, *b*, 1988; Chapais 1985, 1988, this volume pp. 29–59; Ehardt and Bernstein 1986; Samuels *et al.* 1987; de Waal 1987).

Briefly, they made the following assumptions.

1. Females prefer to achieve and maintain as high a rank as possible.
2. Dominance relationships are based on relative power. Thus older, larger females tend to dominate younger, smaller females, but younger females can dominate older, larger females, provided that they have powerful allies (see 4 below) against the latter, and that the size difference between opponents is not too great.
3. Dominants tend to be challenged and outranked in circumstances in which their power is unambiguously reduced relative to the challenger, for example when an individual has an ally feared by (dominant to) the dominant, or the dominant has lost a powerful ally, or the dominant has suffered a clear decline in fighting ability, for example due to old age or illness.
4. Allies dominant to (feared by) a female's opponent are most likely to be powerful or effective in helping her to improve or maintain rank (e.g. Cheney 1977; de Waal 1977; Berman 1980; Walters 1980; Datta 1983*a*, *b*; Netto and van Hooff 1986; Chapais 1988). This reflects the fact that, in most circumstances, relative dominance status will be a reliable indicator of asymmetries in relative power between individuals or subgroups. In these circumstances, an offensive coalition of subordinate females against dominants is very likely to be ineffective and costly, and hence is unlikely to form. This is the situation we consider below. This is not to suggest that

an offensive coalition of subordinate females against dominants cannot arise in some circumstances (Ehardt and Bernstein 1986; Chapais this volume, pp. 29–59.

5. On the whole, females show a marked solidarity with close kin (who are normally close associates) against more distant and non-kin (Kaplan 1977; Massey 1977; Watanabe 1979; Cheney and Seyfarth 1987), although, as befits opportunists, females sometimes betray even their closest relatives for an ally, such as a high-ranking female or male, with whose help they can improve or maintain their own rank (see below). There is also a tendency to defend vulnerable individuals, in particular closely related infants and young juveniles, in circumstances in which the latter might risk injury. This may partly explain why younger sisters are supported against their older sisters from birth (Datta 1988).

INTRA-FAMILY RANK RELATIONS

Datta and Beauchamp (1991) considered the effect on mother–daughter and sister–sister dominance relations of differences in demographic parameters, leading to differences in population age structure and the availability of allies. Using realistic values where possible, they simulated two different populations. Females in population A reached sexual maturity earlier, had shorter interbirth intervals, a lower mortality rate, and hence a higher fecundity than females in population B. This meant that population A, in which the average female produced 2.8 daughters in her lifetime, was an expanding one, while population B, in whih the average female produced only 0.61 daughters, was in decline. Population A resembles rapidly expanding populations, such as that of the protected, artificially provisioned rhesus macaques on Cayo Santiago (Sade *et al.* 1976; Rawlins and Kessler 1986), while population B is similar to declining wild populations of some primates, such as rhesus macaques in the foothills of the Himalaya (Melnick and Pearl 1987), or yellow baboons in Amboseli (Altmann 1980; Hausfater *et al.* 1982).

Details of the simulation procedure and analysis are given in Datta and Beauchamp (1991). Briefly, for each population, each run of the simulation was based on the lifespan and reproductive career of 100 females (matriarchs) who had reached sexual maturity, using the parameters relevant to each population. The sex of infants born to matriarchs was randomly assigned, assuming a birth sex ratio of 50:50. Male infants were not considered further in the simulation. Daughters were assigned a lifespan using the relevant population parameters. The above procedures resulted in a number of matrilines in both populations two generations deep, in which the lifespans of all females in a matriline, and age-differences between them, were known. Dominance relationships between the females (i.e. between mothers and daughters, and between sisters) over time were then determined as described below.

Sister–sister relations

Based where possible on available empirical data (e.g. Kawamura 1965; Koyama 1970; Missakian 1972; Sade 1972; Datta 1988), the following rules for sister–sister dominance contests were applied to the two populations.

1. An older sister, O, dominates a sexually immature younger sister, Y. (Females become sexually mature at 4 years in population A and at 6 years in population B).
2. The mother and any other sisters, as long as they are dominant to O, help the subordinate younger sister Y against O. They never support O against Y.
3. On reaching sexual maturity, Y outranks O, provided that at least one family member (mother or sister) dominates O at the time.
4. Once Y has reversed dominance with O she remains dominant, even if other family members dominant to O die.
5. If by old age O is still dominant to one or more Y, the highest-ranked Y can outrank O dyadically, without the help of a dominant mother or sister, provided that Y is fully-grown, and that O has no fully-grown daughter to help her against Y.

(For this analysis, females were considered old at 15 years in population A and at 13 years in population B, and were considered fully grown at 6 years in population A and 8 years in population B.) The effect of mother–daughter reversals on sister–sister dominance relations was also taken into account (Datta and Beauchamp 1991), since a mother–daughter reversal prevents the mother from raising younger daughters above the daughter who has outranked her.

Sister–sister dominance was monitored at three different times: (1) t1, just after the mother's death; (2) t2, just after the younger sister reached sexual maturity; (3) t3, just before the death of the highest-ranked sister from t2 to live to old age. Any reversals by t1 were due to the mother's help; those between t1 and t2 were due to help from other sisters, and those between t2 and t3 were due to successful dyadic challenges of aging older dominant sisters by full-grown younger subordinate sisters. Figure 3.1 shows the proportion of sister dyads in which the younger sister dominated the older sister, in the two populations, at the three points in time. Just after the mother's death, at t1, sisters ranked in inverse order of age in a significantly higher proportion of dyads in population A than in population B. This difference largely reflected the fact that in population B younger sisters were significantly less likely to have reached sexual maturity by the time the mother died than in population A. By t2, when the youngest sister was

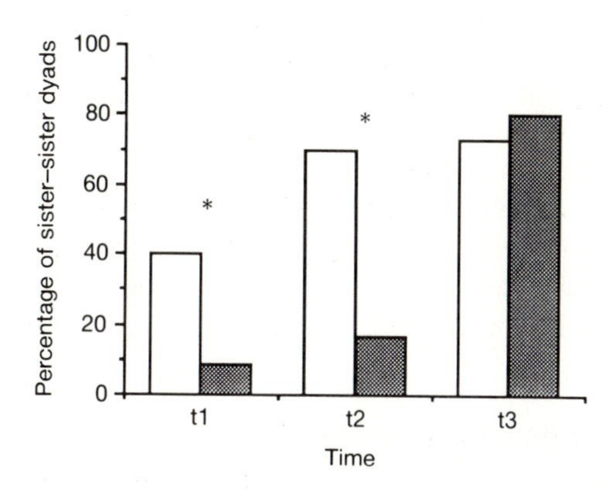

Fig. 3.1. The median proportion of sister–sister dyads, in Population A (open bars) and in Population B (solid bars), in which the younger sister dominates the older by sexual maturity at three points in time: t1, just after the mother's death; t2, just after the youngest sister reaches sexual maturity (4 years in Population A, 6 years in Population B); t3, just before the death of the highest-ranked sister from t2 to live to old age. Population A is expanding and Population B is in decline. *:Mann–Whitney U test, $z < -3.2$, $p < 0.001$, two-tailed.

sexually mature, the proportion of dyads in which sisters ranked in inverse order of age had increased in both populations, but in population A sisters were still significantly more likely to rank in inverse order of age than in population B. The increase from t1 to t2, which occurred after the mother's death, was due wholly to the effect of other sisters on sister–sister dominance. It occurred when a Y reaching sexual maturity had at least one sister who dominated O, and who therefore helped Y outrank O. Note that, although sibships in population B were smaller, the effect of other sisters on YO reversals was relatively greater than in population A. This was because a greater proportion of sister dyads were unreversed at t1, and hence available for other sisters to reverse by t2, in population B. At t3 there was no longer a significant difference between the two populations in the proportion of dyads in which sisters ranked in inverse order of age (Fig. 3.1). In population B, unlike in population A, there was a significant increase from t2 to t3 in the proportion of reversed dyads. This was partly because few unreversed (OY) dyads were present in population A at t2, so the scope for change from t2 to t3 was smaller in this population. However, it was also due to differences in the likelihood that O would be outranked in those OY dyads that did exist. In these dyads O was significantly more likely to be outranked by Y in population B than in population A. This was because the old Os were much less likely to have a full-grown daughter to help them retain rank against a challenging Y in population B than in population A.

To summarize, the cause and time course of sister–sister reversals was significantly dependent on population demography. In the expanding population A, the great majority of sister dyads ranked in inverse order of age by the time (t2) the youngest sister was sexually mature, in marked contrast to the declining population B. These reversals were due to help given to younger sisters by mothers and other sisters, so that in population A allies played the more prominent part in sister–sister reversals. In population B, on the other hand, it was the ageing and physical decline of an older dominant sister that was most likely to give a younger subordinate sister the chance to outrank her.

Mother–daughter dominance relations

Empirical information about alliances and other factors affecting mother–daughter dominance relations is comparatively rare. Nevertheless, available information (e.g. Marsden 1968; Missakian 1972; Moore 1978; Chikazawa *et al.* 1979; Datta 1981; Hausfater *et al.* 1982; Dunbar 1980, 1988; Chapais 1985; A. Simpson personal communication; I. S. Bernstein personal communication) suggested the following rules for mother–daughter contests:

1. A daughter has no allies dominant to the mother, and must therefore challenge and defeat the mother on her own.
2. The highest-ranking daughter challenges the mother, but only if the mother is old.
3. The mother cannot be defeated if she has one or more allies (here limited to sisters dominant to her), to help her against a rebellious daughter. It is assumed that the mother needs only one ally to defeat the daughter, because the daughter herself has no allies. Note that any of the mother's allies dominant to the mother will also be dominant to the daughter, on the assumption that the daughter attains her mother's rank relative to others in the group by sexual maturity.
4. A daughter who outranks her mother remains her ally against other daughters, so inhibiting their rise. In effect this means that only one daughter can rank above the mother at a given time.
5. Once outranked, a mother remains outranked, till her own death or that of the outranking daughter, whichever occurs first.

Matrilines in which the mother was susceptible to outranking by a daughter were therefore those in which (1) the mother reached old age; (2) during the mother's old age a daughter reached rebellious age; and (3) during the mother's old age the rebellious aged daughter was the highest-ranked daughter. A challenge was assumed to occur in every year in which these conditions held. To determine whether a susceptible old mother, M, was actually outranked it was determined whether, in every year of her old age, M had a dominant sister to help her against a rebellious daughter. A reversal occurred if she did not (Datta and Beauchamp 1991). Since empirical data on the age at which a female can be considered old or her daughter able to outrank her dyadically are generally lacking, a variety of ages were used in the simulation. For each matriline in which M reached old age, it was determined (1) whether M was susceptible to outranking; (2) whether or not M was outranked; (3) in which year M was outranked; and (4) which of M's daughters outranked her.

Irrespective of the age at which M became old, the proportion of old females susceptible to outranking, that is, who had a highest-ranked daughter of rebellious age, was always significantly higher in population A than in population B (Fig. 3.2(a)). This was because old females in population A were always significantly more likely to have daughters of rebellious age (kept constant for each population for this analysis—see legend, Fig. 3.2). Susceptible old females were, however, always significantly *less* likely to be outranked by a daughter in population A than in population B (Fig. 3.2(b)). Whether a susceptible old female was outranked depended only on whether or not she had a dominant sister to help her against a rebellious daughter, and Ms in population B were much less likely to have such allies.

In both populations, the proportion of old females susceptible to outranking decreased significantly as daughter's age at rebellion was increased (Fig. 3.3(a)). (For this analysis, age at old age was held constant in each population—see legend, Fig. 3.3). However, the reasons for this decline differed in the two populations. In population B, it occurred because females were less likely to have daughters of rebellious age as that age was increased, since fewer daughters survived to older ages. In population A, however, the decline occurred because the older the rebellious daughter, the more likely she was to have been outranked by a younger sister, and the less likely she therefore was to be highest-ranked. Old females in population A therefore escaped susceptibility to a significant extent by ranking their daughters in inverse order of age. Nevertheless, a significantly higher proportion were susceptible to outranking than in population B (Fig. 3.3(a)), simply because they were much more likely to have daughters, and these daughters were more likely to survive to rebellious age. As before (Fig. 3.2(b)), however, those old females who were susceptible were significantly *less* likely to be outranked by a daughter in population A than in population B (Fig. 3.3(b)), because in population A they were more likely to have a surviving dominant sister to help them against a rebellious daughter.

To summarize, a significantly higher proportion of old females were susceptible to outranking in population A than in population B (Figs 3.2(a) and 3.3(a)), mainly because old females in population A were more likely to have surviving daughters of rebellious age. However, of those old females who *were* susceptible, a significantly smaller proportion were outranked in population A than in population B (Figs. 3.2(b) and 3.3(b)), due wholly to the fact that old mothers in population A were more likely to have a surviving dominant sister to help them against a rebellious daughter.

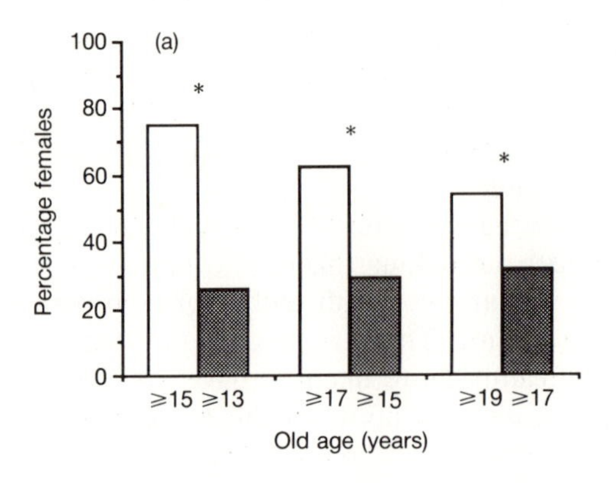

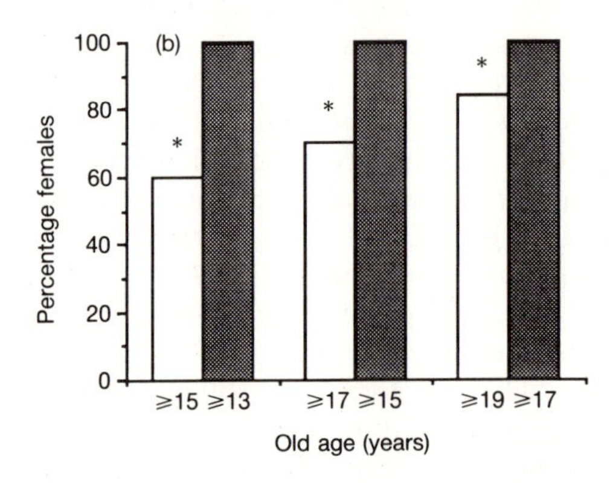

Fig. 3.2. The effect of changing the age at which a female reaches old age, in Population A (open bars) and in Population B (solid bars), on (a) the median proportion of old females susceptible to outranking, that is, whose highest-ranked daughter was at least the age at which a female can defeat an old female on her own (⩾6 years in Population A, ⩾8 years in Population B); and (b) the median proportion of susceptible old females outranked, that is, who did not have a dominant sister to defend them against a rebellious daughter. Interpopulation comparisons; *Mann–Whitney *U* test, $z < -3.2$, $p < 0.001$, two-tailed. In intrapopulation comparisons, the trends were significant, at the above level, for Population A in Figs. 3.2(a) and (b).

INTERMATRILINEAL DOMINANCE RELATIONS

A formal consideration of the effects of demographic factors on intermatriline dominance will not be attempted here, since there is still considerable uncertainty about the determinants of intermatriline rank (Ehardt and

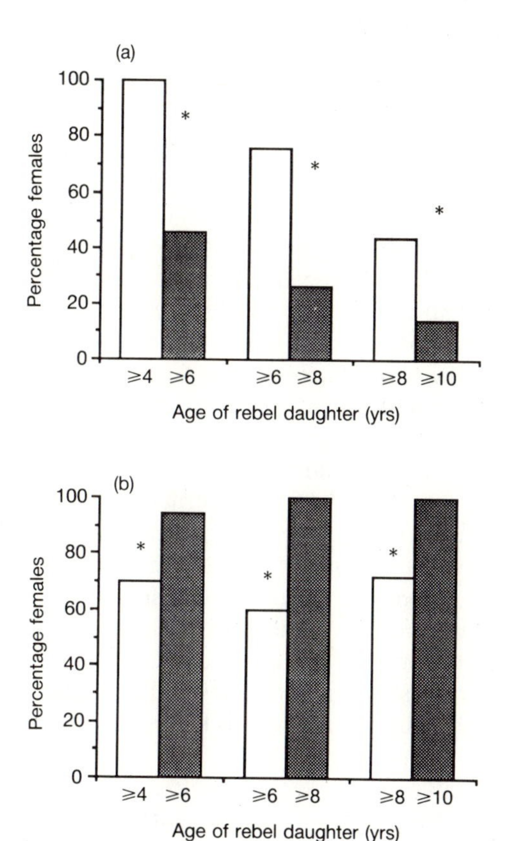

Fig. 3.3. The effect of changing the age at which a daughter can defeat an old mother unaided, in Population A (open bars) and in Population B (solid bars), on (a) the median proportion of old females susceptible to outranking, that is, whose highest-ranked daughter was at least the age at which an unaided female can defeat an old female (≥15 years in Population A, ≥13 years in Population B); and (b) the proportion of susceptible old females outranked, that is, who did not have a dominant sister to defend them against a rebellious daughter. **: Interpopulation comparisons; *Mann–Whitney U test, $z < -3.2$, $p < 0.001$, two-tailed. In intrapopulation comparisons, the trends were significant, at the above level, in both Population A and Population B, in Fig. 3.3(a).

Bernstein 1986; Samuels *et al.* 1987). However, using the same arguments as for individuals, it seems plausible that demographic factors could affect both the stability of intermatriline ranks, and the inheritance of family or maternal rank by offspring.

The stability of intermatriline ranks

The dominance relationship between two matrilines will probably vary according to their relative power. Note that the power of a matriline depends not only on the numbers and kinds of individuals in it, but also on alliances with other matrilines (Datta 1983*c*; Chapais 1983, 1988). If the power of a dominant matriline or coalition of matrilines relative to that of a subordinate matriline or coalition of matrilines clearly declines, we would expect the dominant females to be challenged and overthrown. Indeed, intermatriline reversals often appear to be accompanied by changes in relative power, due to the death or incapacitation of females in a higher-ranked matriline or coalition of matrilines, the acquisition of powerful allies by a subordinate matriline, or an increase in the relative size of a subordinate matriline (Koyama 1970; Gouzoules 1980; Samuels and Henrickson 1983; Ehardt and Bernstein 1986; Cheney and Seyfarth 1987, Datta in prep.).

If relative power is important in intermatriline dominance relations, instability in the hierarchy of matrilines might be expected to increase as matriline size decreases. This is because, as matriline size decreases, the proportion of the power of the matriline represented by a single individual increases. Hence random fluctuations of birth sex ratio and mortality are likely to have a greater impact on the relative power of small matrilines, resulting in more reversals in relative power. The probability of intermatriline reversals would thus be expected to be greater in population B than in population A.

A number of factors could act to stabilize an intermatriline hierarchy. First, coalitions between higher-ranked matrilines against lower-ranked matrilines (for example A + B against C) could reduce the impact of stochastic factors on relative power. Such coalitions are known to exist (e.g. Chapais 1983; Cheney 1983). Their stabilizing effect is expected to operate in population B as well as in population A, since there is no evidence that the number (as opposed to the size) of matrilines in a group is smaller in population B than in population A (Melnick and Kidd 1983). However, because coalitions will be smaller in population B, the relative stability of the hierarchy may still be lower in population B. Here it should be pointed out that the stabilizing effect of intermatriline coalitions may not extend to the relationship between alpha and beta matrilines, since the alpha matriline has no higher-ranked matrilineal ally. Indeed alpha matrilines appear to be susceptible to overthrow by beta matrilines (Datta in prep.), and given the relatively greater impact of stochastic factors, in population B they would be expected to be more susceptible than in population A.

Second, high-ranking matrilines could consolidate matriline power by having female-biased sex ratios, which have been observed in some groups (Altmann *et al.* 1988; Gomendio *et al.* 1990; Simpson and Simpson 1982). Producing daughters could benefit any female in the long term by increasing her matriline size and power, but if daughters are more expensive to produce and raise than sons (Simpson and Simpson 1982; Silk 1983) then only high-ranking females may be able to afford a lot of them. At present, however, it is not clear whether observed sex-ratio variation reflects chance effects or whether it is a genuine option available to females (e.g. Altmann *et al.* 1988; though see Gomendio *et al.* 1990). Third, high-ranking females might be able to manipulate matriline size through harassment of subordinate females and their offspring, which could decrease fertility and increase offspring mortality among subordinate females (e.g. Dunbar and Dunbar 1977; Silk *et al.* 1981; Wasser 1983). If there is high intergroup competition, however, such a strategy could be counterproductive, since it might decrease effective group size (Cheney and Seyfarth 1987; Vehrencamp 1983). Finally, the integrity of a subordinate matriline could be effectively undermined if dominant females formed alliances with some members of the subordinate matriline against others, which would consolidate their own power and that of the betraying individual, while reducing that of the lower-ranked matriline. There is some evidence that this occurs (Chikazawa *et al.* 1979; Datta 1981; Ehardt and Bernstein 1986; Chapais 1985, 1988). All the above strategies might be particularly valuable where there is a paucity of allies, as in population B. But whether they are more likely to occur in those conditions is not clear (but see below). If they do, then hierarchical stability in population B might be greater than expected on stochastic grounds.

The transmission of family rank

Immatures need allies to help them outrank members of lower-ranked families, particularly those older and larger than themselves (e.g. Datta 1983*a*). Allies are generally close relatives such as the mother and other siblings, but because of reciprocal alliances between families, less close relatives as well as non-relatives are also likely to intervene in interfamily fights (Berman 1980; Chapais 1983; Datta 1983*c*). Following the same logic as for intermatriline relations (see above), we would expect stochastic factors to have a greater impact on the alliance base of an immature where matrilines are small. Thus violations of the maternal inheritance rule might be more common in population B than in population A. In population A, for example, an orphan is more likely to have other surviving close relatives or family friends to help it inherit family or maternal rank. Thus in group J on Cayo Santiago, a pre-adolescent male rhesus macaque (566), orphaned in infancy and without siblings, was actively helped by his higher-ranked uncle and aunts to achieve his mother's rank relative to members of other matrilines (Datta 1981). In population B, orphans are less likely to have such surviving allies, and may therefore find it much more difficult to acquire or maintain maternal rank (e.g. Johnson 1987; Cheney and Seyfarth 1987).

Two other factors could add to this effect. First, where matrilines are small, dominant females may have insufficient backing to enforce submission in subordinate females, which could hamper the transmission of rank to offspring. In a possible example of this among captive rhesus monkeys, the daughters of a weakly dominant alpha female outranked females in the lower-ranked matriline younger than, or the same age as, themselves, but, unusually, failed to outrank the low-ranking adult female by sexual maturity. The alpha female seemed unable to intimidate the low-ranking female when the latter attacked the alpha female's daughters; she also behaved as if wary of escalation, and eventually stopped supporting her older daughters against the low-ranking female (personal observation). Second, as age and size differences between individuals increase (as in population B), high-born immatures may be relatively more dependent on alliances to overcome older lowborn opponents. Yet it is precisely in these conditions that there is likely to be a dearth of allies, and this could lead to an increase in violations of the maternal inheritance rule. This effect is expected to be particularly pronounced in male–male and younger female–older male dyads, because, for a given age difference, there tends to be a greater asymmetry in body size (intrinsic power) in these dyads than in female–female or younger male–older female dyads (Lee and Johnson this volume, pp. 391–415. Datta and Janus in prep.). A larger intrinsic asymmetry needs a correspondingly large asymmetry in alliances to counteract it. However, this is less likely to materialize in conditions where allies are at a premium. Thus, in population B, the direction of dominance in male–male and younger highborn female–older lowborn male dyads among immatures may be determined by intrinsic asymmetries, rather than by family rank (see also Lee and Johnson this volume, pp. 391–415).

AVAILABILITY OF ALLIES AND ALLIANCE STRATEGIES

A possibility that needs more consideration is that the kinds of alliances which play a part in rank acquisition and maintenance may differ as a consequence of constraints on the availability of allies (Dunbar 1988). As suggested by rule 5 (see 'Underlying mechanisms and propensities', above), females can be highly opportunistic in the competition for high rank. In some circumstances, this could lead to alliances rare in other circumstances, as the following examples suggest.

The first concerns the outranking of mothers by daughters, which affects both intramatriline ranks, and, since it violates the 'rule' of the inheritance of maternal rank, intermatriline ranks. Here we are not concerned with the outranking of old mothers by adult daughters, which can be explained by the fact that a mother's intrinsic fighting ability declines with age so that an adult daughter can defeat her one to one, but with outrankings of a mother in the prime of life by a pre-adult daughter. Since these daughters cannot match their mothers in intrinsic fighting ability, they appear to rely on an ally dominant to the mother, such as a high-ranked female (e.g. rhesus macaques, Marsden 1968; Chikazawa *et al.* 1979; see below; Japanese macaques, Chapais 1985). That subordinate females are more likely to be outranked by their daughters than high-ranking females is strikingly shown by data from captive rhesus macaques at Madingley, Cambridge. In the old Madingley colony, all seven low-ranking females were outranked by at least one daughter, but only one out of nine high-ranking females was. Daughters who reversed rank with their mothers received support from higher-ranked females both before and during challenge to mothers.

These data raise the question of why the phenomenon is absent in other situations, for instance among free-ranging rhesus macaques on Cayo Santiago, where mother–daughter reversals are exceedingly rare (Sade 1967; Missakian 1972). Available data from Cayo show that high-ranking females do not ally with daughters of low-ranking females against the latter. Indeed, mother–offspring disputes as a whole are characterized by a striking lack of agonistic intervention by others (Datta 1981; I. S. Bernstein personal communication). A major difference between the two situations which may be relevant to the difference in dominance outcomes is that matrilines on Cayo Santiago (which resembles population A) are extended, while those in the old Madingley colony were not. Several high-ranking females at Madingley lacked daughters or other close female relatives. Apparently as a result, they formed closed affiliative relationships with the young daughters of some low-ranking females. As these daughters matured, they used the support of their high-ranking friends to challenge and outrank their mothers. These alliances obviously benefited the daughters, but they may also have benefited the high-ranking females, who in effect increased the size and power of their own alliance base, while diminishing those of a low-ranked matriline, which could have helped them maintain their own rank in the longer term (cf. 'Intermatriline dominance relationships' above).

There is evidence of a similar phenomenon in other groups, when circumstances are appropriate. Thus Datta (1981) found that the only daughter to attempt rebellion against her mother in group J on Cayo Santiago had had a close relationship since infancy with a high-ranked female who had several sons but no daughter, and de Waal (personal communication) noted that an alpha female who lacked offspring formed a close affiliative relationship with a young lower-ranked female, enabling the latter to outrank her own and several other matrilines. It is conceivable that the appropriate circumstances occur more often under some demographic conditions (e.g. population B) than in others, which may lead to differences in the kinds of alliances seen, as well as differences in the outcome of mother–daughter dominance contests. At present, however, data on mother–daughter reversals in different populations are insufficient to judge whether the above makes a significant contribution to mother–daughter reversals in population

B. (See also Dunbar's (1989) analysis of differences between social relationships in captivity and the wild and their relation to availability of partners.)

Another possible outcome of ally availability may be on the degree to which close relatives contribute to the acquisition and maintenance of rank by immatures. For instance, where matrilines are extended, immatures appear to depend rather passively on close relatives such as mothers and siblings to help them attain and maintain rank. These relatives are very active in supporting immatures and usually take the initiative in doing so. Perhaps as a result, immatures very rarely solicit help from members of higher-ranked matrilines (Datta 1981). Where immatures have few relatives (close or otherwise) in a social group, however, they may need more help from high-ranking, often unrelated, individuals, from whom they actively solicit alliances (e.g. Cheney 1977; Walters 1980).

DISCUSSION

Patterns of dominance relationships among females in female-bonded species can vary within a species (e.g. Chikazawa *et al.* 1979) and between species (e.g. Hrdy and Hrdy 1976; Moore 1978; Hausfater *et al.*). Intraspecies differences are usually explained by invoking differences in circumstances, for example the presence or absence of extended matrilines, while differences between species are often ascribed to underlying differences in the agonistic propensities of females.

While the potential role of species differences, or indeed of interindividual differences, should not be underestimated, there is a need to explore more fully the effects on dominance patterns of variation in the circumstances relevant to rank determination, especially in view of the highly conditional nature of rank contests (cf. Chapais this volume pp. 29–59). This paper gives an overview of possible effects of demographic circumstances on the outcome of dominance contests, and on the alliance strategies pursued by individuals to acquire and maintain high rank. The results are potentially relevant to differences between populations of the same species which differ in demographic status. They are also likely to be relevant to differences between populations of different species, since many interspecies comparisons are based on data from populations which differ in demographic status.

The kinds of effects expected can be explored with the help of simulations using realistic assumptions about the rules and propensities underlying dominance contests and alliance formation. Such a simulation was used to compare the outcome of dominance contests between sisters, and between mothers and daughters, in two populations which differed in demographic status (Datta and Beauchamp 1991). One was an expanding population (population A), in which females became sexually mature early, interbirth intervals were short, and the mortality rate was relatively low, while the other (population B) was a declining one in which females reached sexual maturity later, interbirth intervals were longer on average, and the mortality rate was relatively high. In both populations females were assumed to be competing actively for high rank. The same rules for interindividual contests and alliance formation were applied to mother–daughter and sister–sister dominance contests in both populations. Essentially, these were as follows.

1. When no allies exist, older females dominate younger females, except that females who have reached old age can be defeated by fully grown younger females.
2. An old female (mother or older sister) can retain her rank over a full grown female (daughter or younger sister) if she has an ally dominant to the latter. In the case of the mother this was a sister of the mother, and in the case of the older sister it was an adult daughter of the older sister.
3. A younger sister who is not fully grown (though sexually mature) can outrank an older sister as long as she has an ally (mother or other sister) dominant to the older sister.

An important consequence of differences in demographic factors between the two populations was that females in population A had a relative abundance of allies, while many females in population B were stripped of their designated allies. In contests for high rank, therefore, population B females were thrown back much more on individual fighting ability than population A females, and the pattern of ranks among females reflected this. In population B, rank relationships among females resembled an 'age-graded' pattern (cf. Hrdy and Hrdy 1976; Dunbar 1980, 1988), with full-grown females in their prime likely to dominate old females (mothers and older sisters) as well as younger ones (younger sisters). In other words, full-grown daughters were often able to outrank old mothers, and older sisters in their prime were often able to dominate sexually mature younger sisters. In population A, on the other hand, old mothers were often able to maintain their rank against full-grown daughters, and older sisters were usually outranked by sexually mature younger sisters.

Similarly, the paucity of allies in population B relative to population A could lead to several differences in intermatriline dominance outcomes between the two populations. Since the alliance base for intermatriline dominance contests is larger in population A, it may be less susceptible to stochastic variation than in population B. This would allow more stable dominance relationships between matrilines, since the death or maturation of a single female in a large matriline would be expected to have a proportionately smaller effect on the power of that matriline than a similar event in a small matriline. It would also promote the inheritance of maternal or family rank by immatures, since, for example, an orphan is more likely to have other surviving relatives to help it achieve family rank in population A than in population B.

The above differences between populations A and B are potentially significant for several reasons. First, they suggest that the consequences for individuals of competing for high rank may differ depending on population status. In population B, where female rank is much more closely tied to age, females will, on average, have more similar lifetime rank trajectories than females in population A. If dominance rank affects reproductive success, then variance in lifetime rank-related reproductive success may be smaller among females in population B than among females in population A.

Second, hypotheses about selective forces underlying the relationship between female age and patterns of dominance (e.g. Hrdy and Hrdy 1976, Schulman and Chapais 1980) need to be re-evaluated in light of the possibility that demographic factors can have a marked effect on the coincidence between age and rank. For instance, Hrdy and Hrdy (1976) contrasted the age-graded pattern of dominance (cf. population B) found among Hanuman langur females (*Presbytis entellus*) with the genealogical pattern found, for example, among macaques (cf. population A). The age-graded pattern could lead to a more egalitarian outcome (see above), which led them to suggest a functional explanation for the difference. They argued that, in the single-male breeding system of langurs, average relatedness among females would be higher than in multi-male groups of macaques, and the inclusive fitness of langur females might therefore be better served by 'altruistically' deferring to female relatives of higher reproductive value than by 'selfishly' maintaining rank. While it is possible that the underlying propensities of langur females differ from those of macaques (McKenna 1979), it is also possible that langur females, which live in a population resembling population B, are constrained in the competition for high rank by a dearth of allies, leading to an age-graded hierarchy (see above). More information on the mechanisms and propensities underlying dominance relationships among langur females are needed to clarify this issue.

Third, and related to the above, it suggests that some observed differences in dominance patterns between populations of different species could be due to differences in population status rather than to species-specific factors. For example, savannah baboons appear to differ from rhesus and Japanese macaques in showing a higher incidence of daughters outranking old mothers, and a lower incidence of younger sisters outranking older sisters (Moore 1978; Hausfater *et al.* 1982). However, the traditional macaque model is based on data from expanding populations (population A), while the traditional baboon model is based on populations which are declining (population B). In the latter, rhesus-like propensities and strategies are expected to make it more likely that daughters will be able to outrank old mothers, and to make it less likely that younger sisters will be able to outrank older sisters by sexual maturity, exactly what is seen in savannah baboons. This interpretation is strengthened by evidence that the mechanisms and propensities underlying dominance relations in savannah baboon females are very similar to those among rhesus females (e.g. Walters 1980; Samuels *et al.* 1987).

The above assumes that the same 'rules' for alliance formation apply in populations irrespective of their demographic status. Indeed, one of the points of this paper is to consider the possibility that surface dominance structure can vary solely as a result of demographic variation. However, could the 'rules' themselves alter to some extent with demographic conditions? For instance, it is assumed above that mothers and sisters in both populations ally with younger sisters against older sisters. Consequently, differences in the incidence of sister–sister reversals in the two populations are due mainly to differences in the availability of allies to the younger sister. However, what if mothers (for example) had a different alliance strategy in different populations? Thus Dunbar (1988) suggests that a mother interested in maintaining as high a rank as possible for herself in the group would do best to shift her strategy from allying with her younger daughter in population A, to allying with her older daughter in population B. The result of this shift in strategy would give results similar to those obtained above, namely that younger sisters would be more likely to outrank older sisters in population A than in population B—but for completely different reasons. Similarly, mothers may be more likely to be outranked by daughters in population B than in population A because they lack allies against rebellious daughters, as assumed in the simulation, or because daughters are more likely to find high-ranked female allies willing to help them against mothers in population B (see 'Availability of allies and alliance strategies' above). These examples raise the possibility that demographic variation could have even more complex and varied repercussions for female strategies. In the absence of rigorous data on differences in female strategies in different populations, however, they must remain speculative at present.

SUMMARY

1. Dominance relationships between female monkeys can be affected by both individual characteristics (such as age and size) and also by the characteristics and number of their allies. The availability of allies depends on the age, size and kinship structure of the social group. These factors in turn are influenced by demographic parameters, such as age at first birth, intervals between births, and mortality rate (which are affected by the environment). Thus demographic parameters are expected to have a marked effect on female dominance patterns and strategies in matrilineal groups, those groups in which daughters achieve a dominance rank close to that of their mother, and families thus rank with respect to one another.
2. An expanding and a declining population (populations A and B, respectively) were simulated, and the same realistic rules for the acquisition and

maintenance of rank within families applied to both. The rules include, for example, tendencies on the part of close kin to support younger sisters against older ones, and to support mothers against daughters who threaten them.

3. The results suggest that in expanding populations, in which females achieve sexual maturity early, interbirth intervals are short, mortality is low, and hence suitable allies are relatively plentiful, younger sisters will usually be able to outrank older sisters by sexual maturity, and older mothers will often be able to remain dominant to full-grown daughters. However, in declining populations (population B), in which females are older at sexual maturity, interbirth intervals are relatively long, mortality rate is high, and there is a paucity of allies, younger sisters will be less likely to outrank older sisters by sexual maturity, and old mothers will be less able to remain dominant to full-grown daughters.
4. Demographic factors are also expected to affect dominance relationships between matrilines, since stochastic events may have a greater impact on the availability of allies in a declining population than in an expanding one. In general, the inter-matriline hierarchy may be relatively stable, and immatures highly likely to achieve and maintain family or maternal rank in population A. In population B, on the other hand, matrilines may switch rank more often, and immatures, particularly orphans, may often find it difficult to achieve family rank, because of lack of suitable allies.
5. Since females can be highly opportunistic in their behaviour, the possibility exists that demographic differences could lead to differences in alliance strategies, which could in turn affect dominance patterns.
6. There are at least two implications of the influence of demography on availability of allies and hence dominance systems and alliance strategies. First, how successful a female is in pursuing fitness biasing strategies such as the acquisition and maintenance of high rank may be significantly affected by the state of the population. Thus females in the declining population B, in which age is a relatively more important determinant of rank, are expected to have more similar lifetime trajectories than are females in population A, where female age is a less important determinant rank. Second, since the state of a population could affect dominance pattern, observed differences in dominance systems among females are not necessarily due to adaptive differences in underlying female propensities and strategies. At least some apparent differences between primate species in dominance pattern among females could be due simply to demographic differences between the species.

REFERENCES

Altmann, J. (1980). *Baboon mothers and infants*. Harvard University Press, Cambridge.

Altmann, J., Hausfater, G. and Altmann, S. A. (1988). Determinants of reproductive success in savannah baboons, *Papio cynocephalus*. In *Reproductive success: studies of individual variation in contrasting breeding systems* (ed. T. H. Clutton-Brock), pp. 404–18. Chicago University Press, Chicago.

Berman, C. M. (1980). Early agonistic experience and rank acquisition among free-ranging infant rhesus monkeys. *International Journal of Primatology*, **1**, 153–70.

Chance, M., Emory, G., and Payne, R. (1977). Status referents in long-tailed macaques (*Macaca fascicularis*): Precursors and effects of a female rebellion. *Primates*, **18**, 611–32.

Chapais, B. (1983). Dominance, relatedness, and the structure of female relationships in rhesus monkeys. In *Primate social relationships: an integrated approach* (ed. R. A. Hinde), pp. 209–19. Blackwell, Oxford.

Chapais, B. (1985). An experimental analysis of a mother-daughter rank reversal in Japanese macaques (*Macaca fuscata*). *Primates*, **26**, 407–23.

Chapais, B. (1988). Rank maintenance in female Japanese macaques: experimental evidence for social dependency. *Behaviour*, **104**, 41–59.

Cheney, D. L. (1977). The acquisition of rank and the development of reciprocal alliances among free-ranging immature baboons. *Behavioral Ecology and Sociobiology*, **2**, 303–18.

Cheney, D. L. (1983). Extra-familial alliances among vervet monkeys. In *Primate social relationships: an integrated approach* (ed. R. A. Hinde), pp. 278–86. Blackwell, Oxford.

Cheney, D. L. and Seyfarth, R. M. (1987). The influence of intergroup competition on the survival and reproduction of female vervet monkeys. *Behavioral Ecology and Sociobiology*, **21**, 375–86.

Chikazawa, D., Gordon, T., Bean, C., and Bernstein, I. S. (1979). Mother-daughter dominance reversals in rhesus monkeys (*Macaca mulatta*). *Primates*, **20**, 301–5.

Datta, S. B. (1981). Dynamics of Dominance among Free-ranging Rhesus Females. Ph.D. dissertation, Cambridge University.

Datta, S. B. (1983*a*). Relative power and the acquisition of rank. In *Primate social relationships: an integrated approach* (ed. R. A. Hinde), pp. 93–102. Blackwell, Oxford.

Datta, S. B. (1983*b*). Relative power and the maintenance of rank. In *Primate social relationships: an integrated approach* (ed. R. A. Hinde), pp. 103–11. Blackwell, Oxford.

Datta, S. B. (1983*c*). Patterns of agonistic interference. In *Primate social relationships: an integrated approach* (ed. R. A. Hinde), pp. 289–98. Blackwell, Oxford.

Datta, S. B. (1988). The acquisition of rank among free-ranging rhesus monkey siblings. *Animal Behaviour*, **36**, 754–72.

Datta, S. B. (in preparation). Matriline size and the stability of inter-matriline ranks.

Datta, S. B. and Beauchamp, G. (1991). Effects of group demography on dominance relationships among female primates 1: Mother–daughter and sister–sister relations. *American Naturalist*, **138**, 201–26.

Datta, S. B. and Janus, H. (in preparation). Gender and the acquisition of rank among rhesus monkey siblings.

Dunbar, R. I. M. (1980). Determinants and evolutionary consequences of dominance among female gelada baboons. *Behaviural Ecology and Sociobiology*, **7**, 253–65.

Dunbar, R. I. M. (1988). *Primate social systems*. Croom Helm, London.

Dunbar, R. I. M. (1989). Social systems as optimal strategy sets: the costs and benefits of sociality. In *Comparative socioecology* (eds. V. Standen and R. A. Foley), pp. 131–49. Blackwell, Oxford.

Dunbar, R. I. M. and Dunbar, E. P. (1977). Dominance and reproductive success among gelada baboons. *Nature*, **266**, 351–2.

Ehardt, C. L. and Bernstein, I. S. (1986). Matrilineal overthrows in rhesus monkey groups. *International Journal of Primatology*, **7**, 157–81.

Gomendio, M., Clutton-Brock, R. H., Albon, S. D., Guinness, F. E., and Simpson, M. J. (1990). Mammalian sex ratios and variation in costs of rearing sons and daughters. *Nature*, **343**, 261–3.

Gouzoules, H. (1980). A description of genealogical rank changes in a troop of Japanese monkeys (*Macaca fuscata*). *Primates*, **21**, 262–7.

Harcourt, A. H. (1987). Dominance and fertility in female primates. *Journal of Zoology*, **213**, 471–87.

Harcourt, A. H. and Stewart, K. J. (1987). The influence of help in contests on dominance rank in primates: hints from gorillas. *Animal Behaviour*, **35**, 182–90.

Hausfater, G., Altmann, J. and Altmann, S. A. (1982). Long-term consistency of dominance relations among female baboons (*Papio cynocephalus*). *Science*, **217**, 752–5.

Hausfater, G. J., Cairns, S. J., and Lewin, R. N. (1987). Variability and stability in the rank relations of non-human primate females: Analysis by computer simulation. *American Journal of Primatology*, **12**, 55–70.

Horrocks, J. A. and Hunte, W. (1983). Rank relations in vervet sisters: a critique of the role of reproductive value. *American Naturalist*, **122**, 417–21.

Hrdy, S. B. (1977). *The langurs of Abu*. Harvard University Press, Cambridge.

Hrdy, S. B. and Hrdy, D. B. (1976). Hierarchical relations among female hanuman langurs (Primates: Colobinae, *Presbytis entellus*). *Science*, **193**, 913–15.

Johnson, J. A. (1987). Dominance rank in juvenile olive baboons, *Papio anubis*: the influence of gender, size, maternal rank and orphaning. *Animal Behaviour*, **35**, 1694–708.

Kaplan, J. R. (1977). Patterns of light interference in free-ranging rhesus monkeys. *American Journal of Physical Anthropology*, **49**, 241–50.

Kawamura, S. J. (1965). Matriarchal social ranks in the Minoo-B troop: A study of the rank system of Japanese monkeys. In *Japanese monkeys* (eds K. Imanishi and S. A. Altmann), pp. 105–12. Emory University, Atlanta.

Koyama, N. (1967). On dominance rank and kinship of a wild Japanese monkey troop in Arashiyama. *Primates*, **8**, 189–216.

Koyama, N. (1970). Changes in dominance rank and division of a wild Japanese monkey troop in Arashiyama. *Primates*, **11**, 335–90.

Marsden, H. M. (1968). Agonistic behaviour of young rhesus monkeys after changes induced in the social rank of their mothers. *Animal Behaviour*, **16**, 38–44.

Massey, A. (1977). Agonistic aids and kinship in a group of pigtail macaques. *Behavioral Ecology and Sociobiology*, **2**, 31–40.

McKenna, J. J. (1979). The evolution of allomothering behaviour among colobine monkeys: function and opportunism in evolution. *American Anthropologist*, **81**, 818–40.

Melnick, D. J. and Kidd, K. K. (1983). The genetic consequences of group fission in a wild population of rhesus monkeys (*Macaca mulatta*). *Behavioral Ecology and Sociobiology*, **12**, 229–36.

Melnick, D. J. and Pearl, M. C. (1987). Cercopithecines in multimale groups: Genetic diversity and population structure. In *Primate societies* (eds B. B. Smuts, D. L. Cheney, R. M. Seyfarth, R. W. Wrangham and T. T. Struhsaker), pp. 121–34. University of Chicago Press, Chicago.

Missakian, E. A. (1972). Genealogical and cross-genealogical dominance relations in a group of free-ranging rhesus monkeys (*Macaca mulatta*) on Cayo Santiago. *Behaviour*, **45**, 224–40.

Moore, J. (1978). Dominance relations among free-ranging female baboons in Gombe National Park, Tanzania. In *Recent advances in primatology* (eds D. J. Chivers and J. Herbert), pp. 67–70. Academic Press, London.

Netto, W. J. and van Hooff, J. A. R. A. M. (1986). Conflict interference and the development of dominance relations in immature *Macaca fascicularis*. In *Primate ontogeny, cognition and social behaviour* (eds J. G. Else and P. C. Lee), pp. 291–300. Cambridge University Press, Cambridge.

Rawlins, R. G. and Kessler, M. J. (1986). *The Cayo Santiago macaques: History, behaviour and biology*. State University of New York Press, Albany.

Sade, D. S. (1967). Determinants of dominance in a group of free-ranging rhesus monkeys. In *Social communication among primates* (ed. S. A. Altmann), pp. 99–114. University of Chicago Press, Chicago.

Sade, D. S. (1972). A longitudinal study of the social behaviour of rhesus monkeys. In *The functional and evolutionary biology of primates* (ed. R. H. Tuttle), pp. 378–98. Aldine Press, Chicago.

Sade, D. S., Cushing, K., Cushing, P., Dunaif, J., Figueroa, A., Kaplan, J. R., Lauer, C., Rhodes, D., and Schneider, J. (1976). Population dynamics in relation to social structure on Cayo Santiago. *Yearbook of Physical Anthropology*, **20**, 253–62.

Samuels, A. and Henrickson, R. V. (1983). Outbreak of severe aggression in captive *Macaca mulatta*. *American Journal of Primatology*, **5**, 277–81.

Samuels, A., Silk, J. B., and Altmann, J. (1987). Continuity and change in dominance relations among female baboons. *Animal Behaviour*, **35**, 785–93.

van Schaik, C. P. (1983). Why are diurnal primates living in groups? *Behaviour*, **87**, 120–44.

Schulman, S. R. and Chapais, B. (1980). Reproductive value and rank relationships among macaque sisters. *American Naturalist*, **115**, 580–93.

Silk, J. B. (1983). Local resource competition and facultative adjustment of sex ratios in relation to competitive abilities. *American Naturalist*, **121**, 56–66.

Silk, J. B. (1987). Social behaviour in evolutionary perspective. In *Primate societies* (eds B. B. Smuts, D. L. Cheney, R. M. Seyfarth, R. W. Wrangham, and T. T. Struhsaker), pp. 318–29. University of Chicago Press, Chicago.

Silk, J. B., Samuels, A., and Rodman, P. S. (1981). The influence of kinship, rank, and sex on affiliation and aggression between adult female and immature bonnet macaques (*Macaca radiata*). *Behaviour*, **78**, 111–77.

Simpson, M. J. A. and Simpson, A. E. (1982). Birth sex ratios and social rank in rhesus monkey mothers. *Nature*, **300**, 440–1.

Varley, M., and Symmes, D. (1966). The hierarchy of dominance in a group of macaques. *Behaviour*, **27**, 54–75.

Vehrencamp, S. L. (1983). A model for the evolution of despotic versus egalitarian societies. *Animal Behaviour*, **31**, 667–82.

de Waal, F. B. M. (1977). The organisation of agonistic relations within two captive groups of Java-monkeys (*Macaca fascicularis*). *Zeitschrift für Tierspsychologie*, **44**, 225–82.

de Waal, F. B. M. (1987). Dynamics of social relationships. In *Primate societies* (eds B. B. Smuts, D. L. Cheney, R. M. Seyfarth, R. W. Wrangham, and T. T. Struhsaker), pp. 421–30. University of Chicago Press, Chicago.

Walters, J. R. (1980). Interventions and the development of dominance relationships in female baboons. *Folia Primatologica*, **34**, 61–89.

Wasser, S. K. (1983). Reproductive competition and cooperation among female yellow baboons. In *Social behaviour of female vertebrates* (ed. S. K. Wasser), pp. 35090. Academic Press, London.

Watanabe, K. (1979). Alliance formation in a free-ranging troop of Japanese macaques. *Primates*, **20**, 459–74.

Welker, C., Luhrmann, B., and Meinel, W. (1980). Behavioral sequences and strategies of female crab-eating monkeys, *Macaca fascicularis* Raffles, 1821, during group formation studies. *Behaviour*, **73**, 219–37.

Wrangham, R. W. (1980). An ecological model of female-bonded primate groups. *Behaviour*, **75**, 262–300.

4

Conflict intervention behaviour by adult male macaques: structural and functional aspects

CAROLYN L. EHARDT AND IRWIN S. BERNSTEIN

INTRODUCTION

Animals which typically live in social groups, especially groups as complex as those of many primates, are certain to experience conflicts and competition over resources. These group-living animals, however, have been observed to use behavioural mechanisms which modify the expression of aggression and promote continuance of social relationships. In non-human primates, such mechanisms have included formation of dominance relationships (e.g. Bernstein and Gordon 1980; Rowell 1974), restraint in the expression of aggression (Bernstein and Ehardt 1985*a*; Bernstein *et al.* 1983; Thierry 1985), and reconciliatory behaviour between former participants in agonistic encounters (Cords 1988; de Waal 1986; de Waal and van Roosmalen 1979; de Waal and Yoshihara 1983; York and Rowells 1988).

These mechanisms are essentially dyadic in that they may be expressed by and regulate the conflict between any two animals. As the current volume attests, however, agonistic interactions in group-living animals are often polyadic, involving more than two adversaries. Most characteristically, one or more animals may intervene in an originally dyadic agonistic interaction, supporting one or the other of the participants. The importance of such interventions in primates is suggested by observations that they are often accompanied by stereotyped forms of solicitation behaviour (Packer 1977; de Waal 1977, 1982; de Waal *et al.* 1976; Walters 1980; see also de Waal and Harcourt this volume, pp. 1–19), and that screaming or squealing victims often elicit aid (Bernstein and Ehardt 1985*c*; Cheney 1977; de Waal 1977). These polyadic interactions have been described and analysed in a number of primate species, including various macaques, baboons, and chimpanzees. In such species, both females and males intervene in conflicts within their social group, although the contexts which precipitate interference by one or the other appear to differ.

In the following discussion of conflict interference by males, predominantly adult male macaques, the contextual variables which appear to elicit agonistic intervention are summarized. Comparisons are made with adult females, and consideration is given to proposed functional aspects of this behaviour. Our observations of rhesus macaques (*Macaca mulatta*) at the Field Station of the Yerkes Regional Primate Research Center of Emory University have also contributed, we believe, to further understanding of the structure and function of male intervention in this species, and these studies are subsequently discussed.

MALE INTERFERENCE IN INTRAGROUP AGONISTIC EPISODES

As early as 1942, Carpenter suggested that adult male primates exhibit defensive behaviour in their social groups. His suggestion received further support from subsequent observations in the field and on captive populations, observations which suggested that particular high-ranking males were responding to external and intragroup disturbances in a predictable fashion. The high probability of particular behavioural responses by these animals, usually but not always the highest-ranking adult males, emitted under particular circumstances, led Bernstein (1964, 1966*a*, *b*) to introduce the term 'control animal' with reference to the role behaviour exhibited by these animals. The behaviour was characterized as a social role in that it was observed to consist of a related pattern of responses emitted under particular interactional conditions (i.e. a specified social context) by certain members of a social group, as originally discussed by Sarbin (1954). It was later further applied to primate societies by Bernstein and Sharpe (1966) and expanded upon by Hinde (1975, 1978). Such a role is perceived as a pattern of behavioural responses which may be shown by several animals, but not by all members of an age–sex category, and one which persists beyond the lifetime of any individual. Specifically in the case of the control animal role, certain individuals were consistently observed to respond aggressively to sources of disturbance external to the group (intergroup encounters, predators, human caretakers), and to control sources of disturbance within the group, particularly agonistic conflicts between group members. The latter expression of this behavioural role, conflict interference, has since been demonstrated to be present in some form or another in a wide range of primates.

Since the original formulation of the control role, a number of modifications have occurred in its conceptualization. As suggested by Eaton (1976), there may be a learned component that strongly influences performance of the role, and this learning aspect may allow continuation of appropriate role behaviour in the absence of physiological input, as demonstrated by the control role behaviour of an agonadal, alpha male rhesus (Bernstein *et al.* 1979). Equally important are the observations that females, as well as males, may fulfil the role by generalized intervention in intragroup agonistic episodes. Alpha adult female rhesus have been observed to control aggression in their groups (Reinhardt *et al.* 1986; Varley and Symmes 1966), and Fairbanks *et al.* (1978) defined both a control male role and a control female role in vervets (*Cercopithecus aethiops*). Wilson (1975) also further refined conceptualization of such role behaviour by correctly pointing out that the animal or animals most responsible for terminating episodes of intragroup conflict may or may not also show activity directed towards external sources of disturbance. Control role behaviour, as originally formulated, is therefore more accurately viewed as the expression of two distinct roles, which might be seen in the same or different animals.

Although Anderson *et al.* (1977) confirmed the effectiveness of adult male pigtail monkeys (*Macaca nemestrina*) in controlling intragroup aggression among adult females under specific circumstances, other authors have noted that the vigorous activities of even the highest-ranking males intervening in intragroup agonistic episodes may fail to control aggression effectively. This was most graphically demonstrated in our rhesus group at the Yerkes Primate Center Field Station during a series of violent overthrows of the highest-ranking matrilines in the group (Ehardt and Bernstein 1986). These overthrows were the result of the female and adolescent members of multiple matrilines jointly attacking the adult female and adolescent members of the ranking matriline. During these prolonged episodes, the high-ranking adult males vigorously defended the victims, pulling female aggressors off their targets and sometimes nipping attacking females with their incisors. This male interference, even by the apha, was functionally ineffective, as the attackers either continued their attacks, ignoring the males, or immediately and with impunity returned to pulling, biting, and dragging their victims. Although such mob aggression may be relatively rare, these events clearly demonstrated that even joint interference behaviour by many high-ranking males will not be effective in all situations.

Kaplan (1977) has also contributed to this view, suggesting that adult rhesus males are not only less active in agonistic interventions than adult females, but that adult females are actually more important and effective controllers of aggression due to their stronger tendency to aid victims rather than aggressors, and their willingness to interfere even when subordinate to their targets (see also Bernstein and Ehardt 1985*c*). In fact, observations of captive groups of Java monkeys (*Macaca fascicularis*) by de Waal (1977) suggested that even though adult males intervened to protect recipients of aggression most often, their efforts were not more effective than those of other protectors, such as genealogical relatives of the victims (also see Alexander and Bowers 1969; Kaplan *et al.* 1987). When these Java monkey males intervened in favour of aggressors, their behaviour did not shorten the conflict and sometimes actually led to intensification (see also de Waal *et al.* 1976).

While such observations do not nullify the existence of control role behaviour, or its functional significance in primate groups, it is clear that it may be a less common form of conflict intervention than interferences resulting from other factors. In particular, within long-term groups with extensive multi-generational matrilines, female intervention in intragroup conflicts may be far more frequent and effective than the interference of any male.

The structure of adult female intervention

There are clear differences in the expression of intervention behaviour by adult females and males, specifically in terms of the contexts in which each chooses to intervene and in terms of which participant is targeted. Delineation of these contexts involves consideration of the numerous variables present in a triadic agonistic interaction, including the age, sex, and genealogical relationships of the three participants, as well as their existing dominance relationships with one another. These factors interact to determine whether potential supporters intervene on behalf of aggressors or their victims, and they contribute to the consequences and functional significance of the interference. In the following discussion of the structure of agonistic interference, i.e. its description and contextualization, as defined by Tinbergen (1951), it should be kept in mind that a limited number of primate species, and even cercopithecine species, have been studied in this regard. Consequently, the information provided may be the result of detailed observations of only a few species; generalization to species not yet studied cannot be presumed.

Kinship, defined in multi-male, multi-female groups as matrilineal descent from an original adult female with paternity unknown, appears to be a strong predictor of which animals will interfere in an agonistic episode and accounts for the majority of interventions by females (rhesus macaques, Bernstein and Ehardt 1985*c*; Datta 1983; Kaplan 1977, 1978; Kaplan *et al.* 1987; Japanese macaques, *Macaca fuscata*, Alexander and Bowers 1969; Kurland 1977; Watanabe 1979; pigtail macaques, Massey 1977; long-tailed (Java) macaques, de Waal 1977, 1978*a*; de Waal *et al.* 1976; bonnet macaques, *M. radiata*, Silk 1982; yellow baboons, *Papio cynocephalus*, Walters 1980; chacma baboons, *Papio cynocephalus ursinus*, Cheney 1977; Rowell 1966; vervets, Hunte and Horrocks 1987; Struhsaker 1967; chimpanzees, *Pan troglodytes*, de Waal 1987*b*; de Waal 1984; gorillas, *Gorilla gorilla*, Harcourt and Stewart 1987). These studies have demonstrated that mothers are especially vigorous in support of their offspring, more so than the reverse, and that adult females support younger kin against older animals and will generally intervene on behalf of closer relatives against more distant kin. Moreover, the identity of the victim appears to be most salient. Although kinship with the aggressor is important (Datta 1983), kinship with the victim is a more powerful influence for females when viewed with regard to risk-taking (Bernstein and Ehardt 1985*c*; Kaplan *et al.* 1987). Adult female macaques may come to the aid of their immature kin against *any* third party, even when subordinate to the target of their intervention, whereas defence of non-kin is less often against dominant animals (Alexander and Bowers 1969; Bernstein and Ehardt 1985*c*; Chapais 1983; Kaplan 1977, 1978; Kaufmann 1967).

Although females in a number of primate species intervene on behalf of relatives most often, they also support unrelated females (macaques, Bernstein and Ehardt 1985*c*; Gouzoules 1980; Kaplan 1978, 1987; Massey 1977; Silk 1982; Takahata 1982; de Waal 1977; Watanabe 1979; baboons, Cheney 1977; Packer 1977; Walters 1980; *Theropithecus gelada*, Bramblett 1970; vervets, Fairbanks 1980; Hunte and Horrocks 1987; Kaplan 1987; *Erythrocebus patas*, Kaplan 1987; chimpanzees, de Waal 1987; gorillas, Harcourt and Stewart 1987). When vervet and macaque females support non-kin females against other non-kin females, there is a tendency to join the aggressor in targeting the lower-ranking animal (Cheney 1983; Kaplan and Manuck 1985; Watanabe 1979; but see Kaplan 1978; Kaplan *et al.* 1987), such that they are less likely to aid unrelated victims against attackers dominant to themselves (Bernstein and Ehardt 1985*c*; Chapais 1983, this volume, pp. 29–59). In addition, particular non-kin adult females have been observed to form coalitions in targeting other particular adult females, cases which have been described as 'scapegoating' (Bernstein and Ehardt 1986*c*; de Waal *et al.* 1976). What is perhaps most characteristic, however, is that females are especially active in forming coalitions to defend other non-kin adult females and immatures against adult males (macaques, Bernstein 1968; Bernstein and Ehardt 1986*c*; Bernstein and Sharpe 1966; Chance *et al.* 1977; Chapais 1981; Eaton 1976; Fedigan 1982; Kaplan 1977; Kurland 1977; Lindburg 1971; Watanabe 1979; baboons, Hall 1962; Smuts 1985; vervets, Lancaster 1972).

The structure of adult male intervention

A number of aspects of the intervention behaviour shown by the adult male cercopithecines which have been studied to date appear to contrast with the behaviour of adult females briefly reviewed above. Overall, adult female macaques and baboons are more active interveners than are adult males (Bernstein and Ehardt 1985*c*; Cheney 1977; Eaton 1976; Kaplan 1977; Kaplan *et al.* 1987; Kurland 1977; Massey 1977; but see de Waal 1977). Adult female intervention is also much more influenced by kinship than is the case for those adult male macaques which remain in their natal group (Bernstein and Ehardt 1985*c*; Eaton 1976; Kaplan 1977; Kaplan *et al.* 1987). When natal rhesus males intervene at risk to themselves (against a dominant animal), however, it is most characteristically when aiding a relative (Bernstein and Ehardt 1985*c*; Kaplan 1977). Tilford (1982) has also observed coalitions by high-ranking, young adult rhesus brothers against non-natal adult males, and Smuts (1985) reported that young adult male baboons defend their kin against adult males.

Given that adolescent or young adult male macaques and baboons typically leave their natal group around the time of puberty, opportunities to aid kin are not available to most adult males (see also Lee and Johnson this volume, pp. 391–414 for differences in male and female life histories). For rhesus macaques, at least, these adult males are more influenced by dominance relationships, rarely interfering against a dominant animal, unlike adult females who frequently intervene as subordinates (Bernstein and Ehardt 1985*c*; Kaplan 1977, 1978; Kaplan *et al.* 1987). These adult males also receive less aid than do females or any other age–sex class, and when they are supported, it is typically by other adult or adolescent males and against an adult or adolescent male opponent (Bernstein and Ehardt 1986*c*).

Several studies of captive adult male macaques suggest that they generally support older animals, predominantly females, more frequently than immatures (Bernstein and Ehardt 1985*c*; Kaplan 1977, 1978; Kaplan *et al.* 1987), and their interventions are less likely to favour victims than aggressors (Bernstein and Ehardt 1985*c*, 1986*c*; de Waal 1977). The latter observation is not entirely consistent across studies, however. Kaplan (1977) suggested that overall, rhesus males supported victims more than aggressors, but less consistently so than did females. Preference for victims was more apparent for natal males, however, and preferences for victims or aggressors did not appear to be significantly different for non-natal males. Watanabe (1979) also was able to make a distinction within this variable, in his case between the alpha male of his Japanese macaque group and the other adult males. Intervention by the alpha favoured victims, whereas other adult males typically supported aggressors. Such discrepancies in reported results may be a result of the fact that precise distinctions concerning the context of interference are not always made. For example, it may be that adult males support female and juvenile victims being attacked by adolescent or other adult males subordinate to the intervening adult male, but that these same adult males support aggressors against adolescent or adult male targets. Thus the age–sex classification of participants in fights may influence whether males more often support victims or aggressors, as will be discussed below. Additionally, however, we should not rule out species differences, even within a genus, or differences between groups of the same species. In the latter case, demographic factors may vary between groups and environmental conditions may differ in ways which influence opportunities for intervention, or affect the outcome of interventions within particular contexts (Harcourt 1989; Datta this volume, pp. 61–82).

Functional consequences of adult male intervention

Given the structure of conflict intervention behaviour by adult male macaques, what might be the functional consequences of their behaviour? If the observed consequences of their behaviour are beneficial to the males, can this suggest adaptive consequences which affect survival and reproduction, and therefore select for this behaviour in males? Some answers to these questions may be proposed, but as pointed out by Tinbergen (1951) and subsequently discussed by Kummer (1971) and Bernstein (1987), present function should not be confused with past evolution. Direct measures of costs and benefits of interference behaviour cannot be assessed easily, especially with regard to reproductive success of adult males, and it has proved difficult to sort out individual benefits from those accruing to others in the social group. Conflict intervention is a polyadic phenomenon which involves joint action and numerous relationships embedded within an even larger social matrix. Functional consequences in such a context may be difficult enough to deduce; determination of adaptive significance and evolution of the behaviour are clearly hypothetical at this point.

Control of social disruption

Where males target aggressors, or otherwise interfere in a manner that ameliorates or ends an agonistic episode, they exhibit behaviour that fits the control animal role as it was originally formulated. Their intervention reduces the threat of injuries to one or both participants and they contribute to controlling the stress and social disruption that potentially result from frequent or extreme aggression (Bernstein and Sharpe 1966; Boehm 1981; Imanishi 1957; Watanabe 1979). In investigations of this consequence, Erwin (1977), Oswald and Erwin (1976), and Tokuda and Jensen (1968) all utilized group manipulation, removal of adult males, to demonstrate that female aggression increased in frequency or intensity in the males' absence. These experiments can be and have been criticized by others (e.g. de Waal 1977) because removal of animals from groups may produce broader disruptions in social relationships than is presumed in the authors' assessment that absence of males *per se* leads to increased female aggression. In addition, as pointed out earlier, adult male macaques are not the most frequent interveners in normally constituted groups, nor has male interference been shown to be a more effective control of aggression than the intervention of adult females and immature kin. It should be remembered, however, that in sexually dimorphic species, at least, adult males are particularly well equipped, in terms of size and dentition, to inflict serious injury. In the absence of effective checks on their aggression, these males are potentially very formidable interveners, and when they do interfere they may be at less risk against any but their own age–sex class.

Kin selection and maintenance of social relationships

Females more characteristically support victims and immature animals, although these recipients are most frequently close relatives, especially young offspring. Female intervention, then, should contribute at least as much, if not more, to reduction of injuries and control of social disruption. Because their recipients are close kin, it is also more plausible to hypothesize selection of their behaviour on the basis of kin selection and inclusive fitness (Hamilton 1964), as have Datta (1983), Kurland (1977), Massey (1977), and others. But since adult females also support non-kin, as discussed earlier, Kaplan (1978) and others would have us temper this view somewhat. Kaplan (1978) has also specifically pointed out that this support of non-kin is typically not reciprocal, while reciprocation among kin occurs more frequently, suggesting that Trivers' (1971) reciprocal altruism hypothesis also cannot apply satisfactorily (but see Packer 1977 and de Waal and Luttrell 1988). He therefore argues that an inclusive fitness explanation cannot adequately account for the interference behaviour of rhesus females, and turns instead to a functional hypothesis less directly related to individual selection, emphasizing the potential role of conflict intervention in forming and maintaining social relationships and in reaffirming membership in a group.

In this view, females which typically spend their entire life in a single group are embedded in a series of nested social subunits: their close kin, their more distant kin, their female peers, other non-kin, etc., all of which are parts of a unit, the social group, which is distinct from groups of other conspecifics. Because it is the social group within which feeding, reproduction, and other essential activities occur for these females, bonds with females and other non-kin animals are adaptive in maintaining group cohesion, stability, and coordination within a particular ecological context. As Kummer (1971) pointed out, the primary ecological specialization of the primates is their sociality and their behaviour as groups. Nagel (1979) expanded this notion, recognizing that much of social behaviour is directed toward the formation and maintenance of social relationships, which then permit joint action on the environment (including the social environment) by two or more individuals acting in a coordinated manner. In this sense, agonistic interference is one example of a more general aspect of the behavior of social animals, and one that in Kaplan's (1977, 1978) view, and that of others (e.g. Bernstein 1976; Bernstein and Gordon 1974; Boehm 1981; de Waal 1977; Wade 1976; Watanabe 1979), promotes the very relationships that permit other forms of joint action.

Reinforcing dominance relationships, tension reduction, and ameliorating aggression

The intervention behaviour of adult male macaques, which is often in support of male aggressors against other adolescent or adult males and seldom targets a more dominant animal, also has been interpreted as functioning to establish or reinforce alliances with the aggressor (Bernstein and Ehardt 1985*c*; Kaplan 1977, 1978; de Waal 1977, 1978*a*; de Waal *et al.* 1976). The identity of the victim is not critical, as long as it is not of higher rank and therefore potentially able to reverse the direction of aggression (Bernstein and Ehardt 1985*c*). Support of males attacking other males could promote male alliances as well as maintain and consolidate dominance relationships (Packer 1977; de Waal 1977, 1978*a*), since adult males typically do not support aggressors unless the target is of low rank. These social skills in forming alliances could function more effectively in achieving and maintaining high rank than simple aggressivity (Bernstein 1976; Bernstein and Gordon 1974), and additionally could promote group cohesion and stability by minimizing conflict between higher-ranking males (Bernstein 1976).

The latter possibility may also relate to a functional consequence proposed by de Waal (1977). He interprets coalitions among high-ranking adult male Java monkeys as a tension-reduction mechanism. By redirecting their aggression toward lower-ranking 'scapegoats', these males are regulating the tensions that are likely to arise in males living in close proximity, while concurrently reaffirming their relationships through cooperative interaction (also see Kummer 1971; de Waal *et al.* 1976; Wade 1976). Kaplan *et al.* (1987) have also expressed this view, stating that female rhesus intervention functions predominantly to control aggression, while males (which seldom interfere at risk and often join aggressors) are more likely reducing intermale tensions with minimal physical risk. These functional consequences, tension reduction and reaffirming relationships, may be particularly critical under captive conditions, and the increased visibility and confinement under these conditions may also encourage minimization of risk by not intervening unless the target is subordinate and already losing (Kaplan *et al.* 1987).

Although the function of interventions in support of an aggressor against a subordinate target may relate to the relationship between the aggressor and the intervener, this does not necessarily preclude a protective function with regard to the target. An aggressive aider may substitute itself as the primary attacker of a victim when this intervener is dominant to both the original attacker and the victim. If the attack by the intervening animal is less vigorous than the attack by the original aggressor, then the target 'benefits'. For example, it is not unusual for a high-ranking female in our rhesus groups at the Yerkes Primate Center to displace a subordinate female that has cornered an animal and is biting her victim. The dominant intervener may then sit on the target or perhaps rough-groom it, thereby reducing the severity of aggression and effectively protecting the target from further bites, despite the aggressive nature of the intervention. By asserting themselves, such higher-ranking animals are pre-empting the attack of others by physically occupying space and by exhibiting less damaging forms of aggression.

At times, of course, several animals, including adult males, may aggressively support the original attacker and/or the dominant intervener. When this occurs, the target may receive severe aggression. The question is whether the aggressive support by particular higher-ranking animals more often ameliorates the original aggression, and not whether this sometimes escalates the attack. Function is best viewed in terms of the usual outcome, rather than as the explanation for each case.

Formation and maintenance of adult male–female relationships

Kaplan (1978) views the agonistic support behaviour of adult male macaques as more critically important in 'cementing' relationships with adult females, since these females are often recipients of their support (also see Bernstein and Ehardt 1986*c*; Watanabe 1979; Yamada 1966). Non-natal adult males must be accepted into groups of females and must become effectively integrated into these groups in order to reproduce successfully (Kaplan 1978; see also Bernstein 1976). The significant role of female choice in mating success and the necessity for males to ensure survival of their offspring suggest that male behaviour which facilitates and maintains relationships with females and promotes their tenure in a group may be adaptive. Agonistic support of females, therefore, could evolve through individual selection operating on males, rather than through other mechanisms, such as kin selection (Kaplan 1978).

Similarly, observations of baboons (Altmann 1980; Ransom and Ransom 1971; Seyfarth 1978*b*; Smuts 1985) and of Japanese macaques (Takahata 1982) have suggested that some adult males and females form particularly cohesive relationships, relationships also characterized by agonistic support of females by their male associates. Suggested benefits for the male members of these 'friendships' (Smuts 1985) include facilitation of acceptance into the social group when males immigrate, and increased opportunity for formation of consortships with estrous females (Smuts 1985). In several studies of baboons, however, no clear relationship has been demonstrated between previous affiliation and consort behaviour (see Collins 1981; Manzolillo 1982; Seyfarth 1978*a*, *b*), and observations by Smuts (1985) remain inconclusive without additional data. For Japanese macaques, the data suggest that male associates are actually *less* likely to be copulatory partners (see Baxter and Fedigan 1979; Enomoto 1975; Fedigan and Gouzoules 1978; Takahata 1982).

In terms of conflict interference, it has been reported that when adult male baboons agonistically support adult females or immatures, these males are often the particular 'friends' of the females and their offspring (Smuts 1985). Interpretations of this selective defence, however, must be made with care. Proximity of 'friends' may facilitate detection of aggression toward a female, or permit more rapid response. As such, a more general tendency for adult males specifically to support females (as a class) or immatures when attacked by other males cannot be ruled out. As discussed by Smuts (1985), not only might proximity play a significant role, but in addition, males did not always aid their 'friends' when opportunities arose (perhaps when the attacking male outranked them?); 19 per cent of defences by male 'friends' were by 'putative sons' of the female (thus presenting a confusion with kinship); and without data on number of supports given relative to number of opportunities, it is difficult to demonstrate preferential support. These qualifications aside, observations of adult male baboons and macaques indicate that they behave in ways which appear to promote relationships with adult females and facilitate their integration into the troop. Selective intervention in support of females and their offspring certainly contributes to these ends.

A multi-functional interpretation and the contribution of ontogenetic description

The functions of agonistic interference therefore may include defence of kin and other significant associates, establishment and maintenance of relationships through coalitions, promotion of group cohesion and stability, and tension reduction among high-ranking males. These functions have been suggested by the observed structure of intervention behaviour in various primates and by the social context in which they occur. Agonistic interference therefore may best be viewed as multi-functional, and one or more of the functions discussed above may apply to particular instances in particular groups living under varying environmental circumstances.

Although a synthesis of the existing literature supports this multi-functional nature of intervention behaviour, some of the studies which have tried to delineate structure and function of this behaviour have not clearly partialled out factors which could contribute to better or additional description and functional interpretation. As Boehm (1981) has commented, contextualization of interference behaviour has not always been adequate, particularly with regard to the form behaviour takes (impartial interference vs. scapegoating vs. support of aggressor or victim, etc.). It is also critical to know not just what class (age–sex, kin, rank) of antagonist receives support from what class of interferer, but also what the particular and relative characteristics of the target are in relation to each of the coalition partners. For example, if an observation is made that support given by adult males to aggressors may actually intensify and expand the aggression, it is necessary to know whether this occurs only in conflicts between males, or whether the aggressor is an adult female and the target an adolescent male, or whether both are females. This level of distinction may be very informative in the delineation of the various causes and functions which may or may not be associated with agonistic interference; complete description of the structure of support behavior must precede interpretations of cause and function.

One very powerful tool for deciphering cause and function is ontogenetic description. By examining changes in the expression of behaviour during the lifetime of individuals, and looking for concomitant biological and/or social factors which contribute to such changes, additional information relevant to causal or functional explanations is available which would be lost in a static or more restricted examination. It was precisely this approach which led us to propose an additional interpretation of the function of agonistic intervention in the rhesus macaques we have been studying at the Yerkes Regional Primate Research Center Field Station.

ONTOGENY, SOCIALIZATION, AND AGONISTIC INTERVENTION IN RHESUS MACAQUES

Details of the work to be summarized here may be found in a series of recent publications (Bernstein and Ehardt 1985*a*, *b*, *c*, 1986*a*, *b*, *c*; Ehardt and Bernstein 1986, 1987). To summarize our procedures briefly, the data were collected over a 2-year period from January 1982 to January 1984. An all-instances group scan procedure (Altmann 1974) was used to record 46 517 agonistic responses and 3774 episodes of agonistic support involving other subjects entering a dyadic encounter (2434 cases of defensive support of a victim and 1340 cases supporting an aggressor). All individuals in the rhesus group were individually recognized, and matrilineal genealogical relationships of all group members have been fully documented. Figure 4.1 illustrates the age–sex distribution and genealogical relationships of the group at the time of the study.

As described below, the data for agonistic responses and agonistic support are not randomly distributed. There are significant biases related to age, sex, and kinship, as well as to interactions among these variables. Within each of the various age–sex classes which were compared in our analyses, individual variations also occurred in the contribution of different members of these classes to the data. This is not unexpected, and examination of individual variation demonstrated that the results were not a consequence of inordinate contribution by only a few individuals within a class. We therefore described the overall pattern of agonistic responding and conflict interference within our rhesus group. Insofar as our group is not atypical of others of this species, we suggest the patterns observed may generalize to other socially-living rhesus groups, but with the recognition that at present $N = 1$.

Patterns of aggressive behaviour

As suggested by Bernstein and Gordon (1974), aggression need not be viewed as only or even essentially disruptive within social groups since it can serve very positive social functions. These beneficial consequences may include establishment and maintenance of social relationships and socialization of the young and new members of a social group. Hall's (1964)

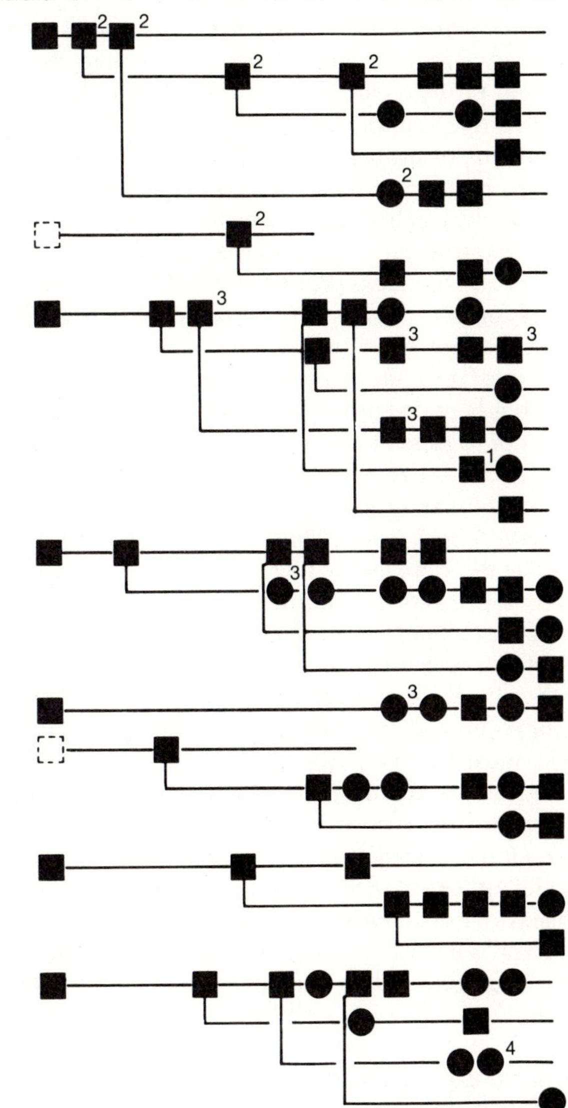

Fig. 4.1. Age–sex and genealogical relationships. Only animals alive and in the group during the study are shown; circles are males, squares are females. Broken line symbols are used for original matriarchs, now deceased, to indicate relationships. The numerals indicate that the animal died or was permanently removed: 1, spring 1982; 2, early fall 1982; 3, early winter 1983. Superscript 4 indicates that these two males are psychological twins; one was kidnapped from its natural mother within the first week of life. Only five removals were due to natural deaths; the others were removed due to fight wounds and were not returned to the group. (Adapted from Bernstein and Ehardt 1985*c*.)

designation of perceived violations of the social code as the single most common cause of aggression in a primate group suggests that aggression may serve as a powerful socialization mechanism, shaping the behaviour of others through negative reinforcement and avoidance training. The disproportionate amount of aggression directed toward kin in our rhesus groups, especially by mothers to their young offspring, supports this view of aggression as one of the mechanisms of socialization and suggests that rhesus matrilines may be particularly active in shaping appropriate social behaviour in immature rhesus monkeys (Bernstein and Ehardt 1986*a*).

If aggression is to serve socialization and other positive social functions, however, it cannot be expressed in a manner which would inflict serious injuries or escalate to the point of jeopardizing social relationships and group cohesion. It is here that other social factors operate to check and channel the expression of aggression. Dominance relationships may prevent escalation that could occur as a result of retaliation, or they may serve to avoid outright conflict. Inhibitions (Kummer 1978) or appeasements (McKenna 1978; Thierry 1985) may check the expression of aggression, and evidence suggests that there are marked restraints in the intensity of rhesus aggression when it is expressed (e.g. Bernstein and Ehardt 1985*a*; Bernstein *et al.* 1983). Agonistic interactions in our rhesus groups rarely include aggressive forms with the greatest potential for serious injury, such as biting or other forms of contact aggression. More than half of the expressed agonistic behaviour is submission, preventing escalation of conflict, and avoidance is the single most common response. Although the most extreme forms of contact aggression are most often directed toward kin, particularly younger animals, aggression received from kin declines in frequency and severity with maturity of the recipient. These observations support the socialization function hypothesis in that the expression of aggression becomes less severe as the behaviour of maturing animals is shaped into patterns which minimize conflicts and increase cooperative interaction. The immatures are also learning to respond to less intense agonistic signals which precede contact aggression.

While the positive social functioning of aggression is dependent upon controls in the expression of that aggression, it may be more important for the aggression of adult males in sexually dimorphic species such as macaques to be effectively restrained or redirected to targets outside the social unit. Their large size and elongated canine teeth give males the potential to inflict serious wounds with a single slash of the canines or a hard bite which produces deep punctures. Although there is considerable difference between studies as to whether adult males or adult females are the most aggressive sex (compare Anderson *et al.* 1977; Bernstein and Ehardt 1985*b*; Bernstein *et al.* 1983; Clark 1978; Drickamer 1975; Eaton *et al.* 1981; Mason *et al.* 1960; Nagel and Kummer 1974; Teas *et al.* 1980, 1982), at least some of the disagreement may have resulted from differences in the conditions under which aggression has been measured. When intragroup aggression is considered, our data indicate that distinct age–sex differences exist in the expression of agonistic behaviour, with adult females showing higher rates of intragroup agonistic behavior, while it is adult male participation which is most different from that of other age-sex classes (Bernstein and Erhardt 1985*b*, 1986*b*).

Whereas juvenile rhesus show the most submission, aggression increases steadily with age and most sharply in females. As females mature from adolescence to adulthood, their basic pattern of agonistic expression remains constant, while in contrast, the pattern for males alters markedly. Adolescent males receive significantly more aggression than any other age–sex class, whereas adult males receive significantly less. As aggressors, adolescent males are frequently vocal, significantly more so than adult males, and the targets of their vocal aggression are adolescent and adult males rather than females. At the same time, adult and adolescent males are the only animals to show a significant tendency to flee squealing animals that are either stationary or chasing them. Also, whereas adolescent males are frequent participants in chasing and threatening fleeing animals and are the targets of prolonged agonistic sequences, the participation of adult males in agonistic episodes is characteristically silent, brief, and rarely involves biting. The primary targets of adult male aggression are adolescent males and other adult males. Adult females, by contrast, are more frequent participants in agonistic interactions than adult males; these interactions are more prolonged and are more likely to include biting by the females, their targets are other adult females and adolescent males, and they are often very vocal. Finally, adolescent males stand out as the only age–sex class significantly more likely to be the target of joint aggression, and they are subjected to contact forms of aggression by third parties after being defeated by their initial antagonist.

This pattern of agonistic interaction, particularly the dramatic change in the expression of aggression by maturing males, suggested that selective aggression on the part of adult males toward adolescent males and other adult males may contribute to the changes seen in maturing males. More specifically, we have hypothesized that this selective aggressive targeting by adult males of adolescent and other adult males involved in intragroup conflicts decreases the participation of maturing males in intragroup aggressive encounters, changes the expression of their aggressive responses, channels their aggression into less serious forms, and redirects their aggression away from females and immatures. Evidence supporting this hypothesis can be found in the patterning of agonistic intervention in our rhesus group.

Rhesus males and patterns of agonistic interference

As seen in Table 4.1, the adult and adolescent rhesus male classes interact in agonistic encounters significantly more often among themselves and with

Table 4.1. Age–sex classes with which older males interacted significantly more (M) or less (L) often than expected. An M or an L entry indicates that this category occurred significantly more or less often, respectively, at $p < .0001$, than expected based on the proportion of the group in each age-sex class, the total frequency of the response by the actor class, and the number of interactions with the age-sex class. Ages are given in years. (Adapted from Bernstein and Ehardt 1986*b*.)

Response category	Adult male actor									Adolescent male actor								
	Male recipients				Female recipients					Male recipients				Female recipients				
	Adult	3.5–5.5	1.5–3.5	0.5–1.5	Adult	4.5–5.5	3.5–4.5	1.5–3.5	0.5–1.5	Adult	3.5–5.5	1.5–3.5	0.5–1.5	Adult	4.5–5.5	3.5–4.5	1.5–3.5	0.5–1.5
Total agonistic responses	M	M		L	L					M	M		L	L			L	L
Total aggression	M	M		L	L					M	M		L	L			L	L
Total contact aggression	M	M								M	M		L					L
Bites	M	M								M	M							
Manual contact aggression	M	M		L						M	M							
Chases	M		L	L					L	M	M							
Threats	M	M			L					M								
Vocal aggression	M				L					M	M			L				
Total submission	M		L		L	L		L		M				L			L	

each other. This effect is principally due to both adult and adolescent males showing significantly more aggression to adolescent males, and the high rate of agonistic interaction among these males is in contrast to the low rates of aggression by adult males towards adult females and immatures. However, these significant levels of agonistic interaction among adolescent and adult males cannot account alone for the observed ontogenetic changes in expression of aggression by maturing males. Indeed, this pattern could be hypothesized to be a function of frequent association of males with one another, presenting opportunities for the competitive conflicts likely to arise among animals most like one another (see Drickamer 1975; Southwick 1967). If, however, we hypothesize that not only do adult males show more aggression than expected toward adolescent and adult males, but also that they *selectively target* adult and adolescent males involved *specifically* in agonistic interactions with females and immatures, we may have a mechanism which channels male aggression away from the females and immatures and explains declining frequencies of male aggression toward these particular animals. This mechanism could also explain the shift in maturing males away from the noisy, prolonged agonistic episodes which continue to characterize female aggression, toward the silent, brief pattern of adult males (see Bernstein and Ehardt 1985*b*). Females and immatures are very vocal in their agonistic episodes, and squealing or screaming elicits support from third parties. If these females or immature animals vocalize when receiving aggression from adolescent males, and if adult males consistently respond by selectively targeting the aggressive male, this would encourage the quieter and briefer expression of aggression seen in adult males, especially when directed toward females. If contact aggression by males is most likely to elicit squealing, selective targeting by adult males could lead to limited expression of the more severe forms of contact aggression by older males, as is in fact the case.

Table 4.2 summarizes our data on selective interference in agonistic episodes in the Yerkes rhesus group. If we look at overall support patterns, examining only the recipient of the support and irrespective of the target, distinct male–female differences can be seen. As discussed above, adult females do most of the interference in agonistic interactions, and most of this is accounted for by their defence of immature close kin. As males mature, their support of kin consistently decreases, in strong contrast to the behaviour of females, and adult and adolescent males support non-kin significantly more often than do other age-sex classes. By adolescence, males are supporting non-kin adult or adolescent male aggressors most often.

When the final critical variable—particular identity of the target of the intervention—is also examined in conjunction with the identity of the recipient of support, the pattern of interference becomes much more specific, especially for males (Tables 4.3 and 4.4). Adult males are significantly likely to target adolescent males, and both adult and adolescent males

Table 4.2. Percentage of aiding done by each age–sex class in each of the categories of aiding. The first line gives the number of aids per member of the age–sex class. This corrects for different numbers in each age–sex class and can be used with the percent figure to allow the reader to compare differential frequencies among age–sex classes. The percentage figures shown otherwise indicate how each class divided those aids it did do according to the possible categories of aiding. (Adapted from Bernstein and Ehardt 1985*c*.)

	Males			Females				Group mean
	Adult	Adolescent	Juvenile	Adult	Subadult	Pubertal	Juvenile	
Mean number of aids performed per individual	31.6	29.0	19.4	119.1	59.1	36.6	23.4	43.9
Defence	48	42	46	74	68	55	50	64
Defence is of kin	11	55	79	83	76	67	78	76
Defence of kin is at risk	39	53	56	52	60	75	48	54
Aids at risk that are defence	60	74	67	98	98	92	73	91
Aids of aggressors are for kin	30	38	64	70	77	73	83	64
Aids of aggressor that are at risk	7	10	22	2	2	6	16	8
Aids aggressor at risk is for kin	25	17	64	77	100	80	87	66
Mean percentage of kin in group	3.6	11.9	12.6	12.7	13.2	12.4	11.9	11.3

Table 4.3. Number of times various age–sex classes aided either of two age–sex classes of antagonists when fighting with each other. Adults = 6.5+ years; adolescents = 3.5–6.5 years; juveniles = 3.5 years. (Adapted from Bernstein and Ehardt 1986c.)

Age–sex class of antagonists	Age–sex class of aider			
	Adult male	Adolescent male	Adult female	Adolescent female
Adult male vs. adolescent male	11:13	3:13	5:36*	3:2
Adult male vs. adult or adolescent female	5:60	9:37	24:207*	2:20*
Adult male vs. juvenile male or female	6:4	0:6	12:124*	3:19*
Adolescent male vs. adult or adolescent female	8:30*	7:11	50:78	12:12
Adolescent male vs. juvenile male or female	2:9	4:9	14:100*	5:23*
Adult female vs. juvenile male or female	34:25	6:17	86:378*	27:58
Adolescent female vs. juvenile male or female	7:11	6:4	65:133*	13:22

* Statistically significant bias at $p < .001$, on the basis of binomial probability.

Table 4.4. Comparison of interference against specific age–sex classes of antagonists with and without exclusion of kin. (Adapted from Bernstein and Ehardt 1986*c*.)

Antagonists	Age–sex class of aider		
	Adult male	Adolescent male	Adult female
Adult males vs. females or juvenile males	11:64*	9:43*	36:331*
Non-kin only	8:50*	8:14	23:56*
Adolescent males vs. females or juvenile males	10:39*	11:20	61:48
Non-kin only	9:29	7:72	9:24
Adult females vs. younger females or juvenile males	57:28	17:24	134:447*
Non-kin only	37:19	10:15	49:37

* Ratio significantly different from chance at $p < .001$.

support females by aggressing against adult males. Most critically, when agonistic episodes occur which involve an adult male and an adult female, every age–sex class, with the exception of juvenile males, supports the adult female when they interfere. Adult males are particularly active in these episodes, supporting the adult female without regard to kinship, whereas other animals are more likely to interfere if the female is a relative. Likewise, when adolescent males fight with females or immatures, adult males specifically target the adolescent male, even if unrelated to the participants (non-natal males or males of different matrilines). Adolescent males, by contrast, interfere significantly more often against adult males only on behalf of their kin.

These data demonstrate that the interference behaviour of adult females is principally in defence of kin, while that of adult males is in support of females and immatures against adult and adolescent males, regardless of kinship. Adult females will intervene in support of non-kin adult females or immatures against adult and adolescent male opponents, and natal adult males do show a greater proportion of their limited interventions against a dominant animal in support of kin, as discussed previously. However, adult female support of non-kin is secondary to their defence of kin, and adult male interference is most characteristically in support of any adult female or immature animal, whether victim or aggressor, through selective targeting of their adult or adolescent male opponents. It is also the case that adult males will attack lower-ranking adult or adolescent males merely in the vicinity of noisy conflicts within the group.

Functions of adult male rhesus selective interference

Our data suggest that the selective targeting of adult and adolescent male rhesus by adult male interveners accounts for the ontogenetic changes in the expression of male aggression. Unlike what might be predicted on the basis of rising androgen levels in maturing males, adolescent males do not become more aggressive within their social group. It is the adult female class which has the greatest rate of participation in aggressive expression in the group. This is not to say that females are always more aggressive than are males, or that rising androgen levels do not influence male tendencies to respond aggressively. What we do argue is that it is the selective response of adult males to the aggression of adult and adolescent males in encounters with other group members that channels their aggression away from females and vulnerable young animals, and relegates that expression to less serious, abbreviated forms. In conflicts with other males or when responding to extragroup disturbances, male aggression may reach its fullest expression.

In terms of function, the shaping and channelling of male aggressive behaviour through this selective interference may be viewed from the perspective of beneficial consequences for the adult male interveners. At the individual level, if an adult male can curb and channel the aggression of other adult males, it is of potential benefit to his reproductive succes. By inhibiting aggression of other large males against adult females, the male's behaviour functions to protect potential breeding partners. Similarly, if an adult male can prevent excessive aggression by other males against immature animals, his behaviour functions to protect the unknown subset of these individuals that are his offspring. Also, if in selectively targeting aggressive males the adult male injures the targeted male, it is a breeding rival which he injures.

Selective interference and its effect on male aggressive behaviour may also be regarded as an example of socialization processes within a rhesus monkey group, one that is independent of kinship. When adult males aggressively target adult or adolescent males engaged in conflicts with females or immature group members, they are responding to what Hall (1964) described as a violation of the social code; they are, in essence, punishing these males. If this selective interference affects the future behaviour of a maturing male, then it is a socialization process in the sense of Poirier (1972), who defined socialization as the influence of past social experience on future social behaviour.

This selective interference also may contribute to other observed changes in the behaviour of maturing males. Adolescent males become separated from female matrilineal kin and from other female and younger group members, while maintaining affiliative associations with other adolescent males, both kin and non-kin (Ehardt and Bernstein 1987). This general dissociation from kin and non-kin females and immatures, as well as specific avoidance of fights involving only females and immatures, may in part be due to the potential for aggressive targeting by adult males, since prolonged association increases opportunities for conflict. This is especially true for their association with kin, since kinship does not preclude such conflict (Bernstein and Ehardt 1986*a*). As adolescent males learn to aggress and leave, to aggress selectively against subordinate older males, and to be silent and avoid noisy aggression, they also may become less involved in coalitions with their matrilineal kin. These weakening ties to their matriline also decrease their ties to their natal group and may facilitate their emigration. If selective targeting by adult males thereby contributes to the eventual migration of other maturing males due to disruption of social relationships, these resident adult males may prolong their tenure in a social group and incidentally discourage breeding rivals. When they do transfer, these young adult males should exhibit the 'normal' pattern of adult male agonistic expression, and they should then become the socializing agents for the next generation of maturing males in their new group.

In a very positive sense, then, adult male rhesus significantly influence social organization through control of male aggression, and they may also thereby incidentally influence the means of genetic exchange between social groups. The functions of group cohesion and the establishment and maintenance of social relationships, proposed by Kaplan (1978) and others, may also be expected to follow from selective interference by adult males. If males immigrating into new groups are less inclined to exhibit aggressive behaviour toward females and immatures, their transfer and integration into a new troop may be facilitated. Within the group, male behaviour which is protective of females and young animals should contribute to formation of positive relationships with the female core of the group, which should also prolong the male's tenure in the group. If overall reproductive success is dependent on the number of mating seasons a male spends in a group, rather than on relative success in any one season, then males with the social skills which prolong their tenure by encouraging alliances and other positive relationships, especially with females and young, again may profit at an individual level of selection.

In general, the coalitions evident in agonistic interference are excellent examples of joint action on the social environment. It is through effective social behaviour that primates deal with environmental challenges in both the social and non-social environments (Kummer 1971), and in this sense agonistic interference is but one specific case of a more general mechanism (Bernstein 1984). If selective interference by adult males represents an effective social control on the level and direction of adult male aggression, then it should promote group cohesion and cooperative behavior at a very general level. In this regard, if we accept Kummer's (1971) view that the primary ecological specialization of the primates is found in the way they act as groups, then behaviour which contributes to efficient group organization should be selected for in the individuals dependent upon that group for their survival and reproduction.

SUMMARY

1. In a variety of primate species, particular adult males have been observed to exhibit role behaviour involving agonistic interference in intragroup conflicts. Even the highest-ranking adult males in a group, however, are not always effective in controlling intragroup aggression.
2. Overall, and especially in long-term cercopithecine groups with extensive multigenerational matrilines, female intervention in intragroup conflicts may be more frequent and more effective than the interference of males. Kinship accounts for the majority of interventions by females, and mothers will support their immature offspring even when subordinate to the target of their intervention. Females characteristically support victims and are especially likely to support other adult females and immatures against adult males.
3. Adult male macaques are less active interveners than are adult females and rarely intervene against a dominant animal (with the exception of

natal males supporting a relative). These males are more likely to support older animals, predominantly females, and have been observed to aid aggressors more than victims in particular contexts.

4. Intervention in intragroup conflicts may serve multiple functions. These include control of social disruption, reduction of injuries, formation and maintenance of social relationships and alliances, promotion of group cohesion and coordination, and tension reduction. For adult females, defence of kin may be particularly important; for adult males, the establishment and maintenance of relationships, and tension reduction, may be more salient.
5. Ontogenetic changes in the expression of aggression differ in male and female rhesus macaques. Aggression increases steadily with age, and most sharply in females, such that adult female intragroup aggression is more frequent, more prolonged, more vocal, and more likely to include contact forms such as biting. The intragroup aggressive expression of adult male rhesus is typically quieter, briefer, less serious in form, and directed to other males more often than to females and immatures.
6. Our data indicate that adult males selectively target adult and adolescent males specifically involved in agonistic interactions with females and immatures. This selective targeting may represent a socialization mechanism which leads to the ontogenetic changes in the expression of aggression by males and redirects their intragroup aggression away from females and immatures. Several adaptive consequences may be associated with this behaviour, both in terms of advantages to individual males and in terms of the maintenance of social groups in an order of mammals where the primary ecological specialization is joint action on the environment.

ACKNOWLEDGMENTS

The research reviewed in this chapter was supported in part by NIH grant RR-00165 from the National Center for Research Resources to the Yerkes Regional Primate Research Center. The Yerkes Center is fully accredited by the American Association for Accreditation of Laboratory Animal Care (AALAC).

REFERENCES

Alexander, B. K. and Bowers, J. M. (1969). Social organization of a troop of Japanese monkeys in a two-acre enclosure. *Folia Primatologica*, **10**, 230–42.

Altmann, J. (1974). Observational study of behavior: sampling methods. *Behaviour*, **49**, 227–65.

Altmann, J. (1980). *Baboon mothers and infants*. Harvard University Press, Cambridge, MA.

Anderson, B., Erwin, N., Flynn, D., Lewis, L., and Erwin, J. (1977). Effects of short-term crowding on aggression in captive groups of pigtail monkeys (*Macaca nemestrina*). *Aggressive Behavior*, **3**, 33–46.

Baxter, M. J. and Fedigan, L. M. (1979). Grooming and consort partner selection in a troop of Japanese monkeys (*Macaca fuscata*). *Archives of Sexual Behavior*, **8**, 445–58.

Bernstein, I. S. (1964). Role of the dominant male rhesus in response to external challenges to the group. *Journal of Comparative and Physiological Psychology*, **57**, 404–6.

Bernstein, I. S. (1966*a*). Analysis of a key role in a capuchin (*Cebus albifrons*) group. *Tulane Studies in Zoology*, **13**, 49–54.

Bernstein, I. S. (1966*b*). A field study of the male role in *Presbytis cristatus* (lutong). *Primates*, **7**, 402–3.

Bernstein, I. S. (1968). The lutong of Kuala Selangor. *Behaviour*, **32**, 1–16.

Bernstein, I. S. (1976). Dominance, aggression, and reproduction in primate societies. *Journal of Theoretical Biology*, **60**, 459–72.

Bernstein, I. S. (1984). The adaptive value of maladaptive behavior, or you've got to be stupid in order to be smart. *Ethology and Sociobiology*, **5**, 297–303.

Bernstein, I. S. (1987). The evolution of nonhuman primate social behavior. *Genetica*, **73**, 99–116.

Bernstein, I. S. and Ehardt, C. L. (1985*a*). Intra-group agonistic behavior in rhesus monkeys (*Macaca mulatta*). *International Journal of Primatology*, **6**, 209–26.

Bernstein, I. S. and Ehardt, C. L. (1985*b*). Age-sex differences in the expression of agonistic behavior in rhesus monkey (*Macaca mulatta*) groups. *Journal of Comparative Psychology*, **99**, 115–32.

Bernstein, I. S. and Ehardt, C. L. (1985*c*). Agonistic aiding: kinship, rank, age, and sex influences. *American Journal of Primatology*, **8**, 37–52.

Bernstein, I. S. and Ehardt, C. L. (1986*a*). The influence of kinship and socialization on aggressive behaviour in rhesus monkeys (*Macaca mulatta*). *Animal Behaviour*, **34**, 739–47.

Bernstein, I. S. and Ehardt, C. L. (1986*b*). Modification of aggression through socialization and the special case of adult and adolescent male rhesus monkeys (*Macaca mulatta*). *American Journal of Primatology*, **10**, 213–27.

Bernstein, I. S. and Ehardt, C. L. (1986*c*). Selective interference in rhesus monkey (*Macaca mulatta*) intragroup agonistic episodes by age-sex class. *Journal of Comparative Psychology*, **100**, 380–4.

Bernstein, I. S. and Gordon, T. P. (1974). The function of aggression in primate societies. *American Scientist*, **62**, 304–11.

Bernstein, I. S. and Gordon, T. P. (1980). The social component of dominance relationships in rhesus monkeys (*Macaca mulatta*). *Animal Behaviour*, **28**, 1033–9.

Bernstein, I. S. and Sharpe, L. (1966). Social roles in a rhesus monkey group. *Behaviour*, **26**, 91–103.

Bernstein, I. S., Gordon, T. P. and Peterson, M. (1979). Role behavior of an agonadal alpha-male rhesus monkey in a heterosexual group. *Folia Primatologica*, **32**, 263–7.

Bernstein, I., Williams, L., and Ramsey, M. (1983). The expression of aggression in Old World monkeys. *International Journal of Primatology*, **4**, 113–25.

Boehm, C. (1981). Parasitic selection and group selection: a study of conflict interference in rhesus and Japanese macaque monkeys. In *Primate behavior and sociobiology* (eds A. B. Chiarelli and R. S. Corruccini), pp. 161–82. Springer-Verlag, New York.

Bramblett, C. A. (1970). Coalitions among gelada baboons. *Primates*, **11**, 327–33.

Carpenter, C. R. (1942). Societies of monkeys and apes. *Biological Symposia*, **8**, 177–204.

Chance, M. R. A., Emory, G. R., and Payne, R. G. (1977). Status referents in long-tailed macaques (*Macaca fascicularis*): precursors and effects of a female rebellion. *Primtes*, **18**, 611–32.

Chapais, B. (1981). The Adaptiveness of Social Relationships among Adult Rhesus Monkeys. Ph.D. Dissertation, University of Cambridge.

Chapais, B. (1983). Dominance, relatedness, and the structure of female relationships in rhesus monkeys. In *Primate social relationships: an integrated approach* (ed. R. A. Hinde), pp. 209–19. Sinauer, Sunderland, MA.

Cheney, D. L. (1977). The acquisition of rank and the development of reciprocal alliances among free-ranging immature baboons. *Behavioural Ecology and Sociobiology*, **2**, 308–18.

Cheney, D. L. (1983). Extrafamilial alliances among vervet monkeys. In *Primate social relationships: an integrated approach* (ed. R. A. Hinde), pp. 278–86. Sinauer, Sunderland, MA.

Clark, T. W. (1978). Agonistic behavior in a transplanted troop of Japanese macaques: Arashiyama West. *Primates*, **19**, 141–51.

Collins, D. A. (1981). Social Behaviour and Patterns of Mating among Adult Yellow Baboons (*Papio c. cynocephalus* L. 1766). Ph.D. dissertation, University of Edinburgh.

Cords, M. (1988). Resolution of aggressive conflicts by immature long-tailed macaques, *Macaca fascicularis*. *Animal Behaviour*, **36**, 1124–35.

Datta, S. B. (1983). Patterns of agonistic interference. In *Primate social relationships: an integrated approach* (ed. R. A. Hinde), pp. 289–97. Sinauer, Sunderland, MA.

Drickamer, L. C. (1975). Quantitative observation of behavior in free ranging *Macaca mulatta*: methodology and aggression. *Behaviour*, **55**, 209–36.

Eaton, G. G. (1976). The social order of Japanese macaques. *Scientific American*, **235**(4), 97–106.

Eaton, G. G., Modahl, K. B., and Johnson, D. F. (1981). Aggressive behavior in a confined troop of Japanese macaques: effects of density, season, and gender. *Aggressive Behavior*, **7**, 145–64.

Ehardt, C. L. and Bernstein, I. S. (1986). Matrilineal overthrows in a rhesus monkey group. *International Journal of Primatology*, **7**, 157–81.

Ehardt, C. L. and Bernstein, I. S. (1987). Patterns of affiliation among immature rhesus monkeys (*Macaca mulatta*). *American Journal of Primatology*, **13**, 255–69.

Enomoto, T. (1975). The sexual behavior of wild Japanese monkeys: the sexual interaction pattern and the social preference. In *Contemporary primatology* (eds S. Kondo, M. Kawai, and A. Ehara), pp. 275–9. Karger, Basel.

Erwin, J. (1977). Factors influencing aggressive behavior and risk of trauma in the pigtail macaque (*Macaca nemestrina*). *Laboratory Animal Science*, **27**, 541–7.

Fairbanks, L. A. (1980). Relationships among adult females in captive vervet monkeys: testing a model of rank-related attractiveness. *Animal Behaviour*, **28**, 853–9.

Fairbanks, L. A. McGuire, M. T., and Page, N. (1978). Social roles in captive vervet monkeys (*Cercopithecus aethiops sabaeus*). *Behavioral Processes*, **3**, 335–52.

Fedigan, L. M. (1982). *Primate paradigms: sex roles and social bonds*. Eden Press, Montreal.

Fedigan, L. M. and Gouzoules, H. (1978). The consort relationship in a troop of Japanese monkeys. In *Recent advances in primatology*, vol. 1 (eds D. J. Chivers and J. Herbert), pp. 493–5. Academic Press, New York.

Gouzoules, H. (1980). A description of genealogical rank changes in a troop of Japanese monkeys (*Macaca fuscata*). *Primates*, **21**, 262–7.

Hall, K. R. L. (1962). The sexual, agonistic, and derived social behaviour patterns of the wild Chacma baboon, *Papio ursinus*. *Proceedings of the Zoological Society of London*, **13**, 283–327.

Hall, K. R. L. (1964). Aggression in monkey and ape societies. In *The natural history of aggression* (eds J. D. Carthy and F. J. Ebling), pp. 51–64. Academic Press, London.

Hamilton, W. D. (1964). The genetical evolution of social behaviour. I and II. *Journal of Theoretical Biology*, **7**, 1–52.

Harcourt, A. H. (1989). Social influences on competitive ability: alliances and their consequences. In *Comparative socioecology: the behavioural ecology of humans and other mammals* (eds V. Standen and R. A. Foley), pp. 223–42. Blackwell, Oxford.

Harcourt, A. H. and Stewart, K. J. (1987). The influence of help in contests on dominance rank in primates: hints from gorillas. *Animal Behaviour*, **35**, 182–90.

Hinde, R. A. (1975). Interactions, relationships and social structure in non-human primates. In *Proceedings from the symposia of the fifth congress of the International Primatological Society* (eds S. Kondo, M. Kawai, A. Ehara, and S. Kawamura), pp. 13–24. Japan Science Press, Tokyo.

Hinde, R. A. (1978). Dominance and role: two concepts with dual meanings. *Journal of Social Biological Structure*, **1**, 27–38.

Hunte, W. and Horrocks, J. A. (1987). Kin and non-kin interventions in the aggressive disputes of vervet monkeys. *Behavioral Ecology and Sociobiology*, **20**, 257–63.

Imanishi, K. (1957). Social behavior in Japanese monkeys, *Macaca fuscata*. *Psychologia*, **1**, 47–54.

Kaplan, J. R. (1977). Patterns of fight interference in free-ranging rhesus monkeys. *American Journal of Physical Anthropology*, **47**, 279–88.

Kaplan, J. R. (1978). Fight interference and altruism in rhesus monkeys. *American Journal of Physical Anthropology*, **49**, 241–9.

Kaplan, J. R. (1987). Dominance and affiliation in the Cercopithecini and Papionini: a comparative examination. In *Comparative behavior of African monkeys* (ed. E. L. Zucker), pp. 127–50). Alan R. Liss, New York.

Kaplan, J. R. and Manuck, S. B. (1985). Patterns of fight interference in one-male harem groups of cynomolgus macaques. *American Journal of Primatology*, **8**, 346.

Kaplan, J. R., Chikazawa, D. K., and Manuck, S. B. (1987). Aspects of fight interference in free-ranging and compound-dwelling rhesus macaques (*Macaca mulatta*). *American Journal of Primatology*, **12**, 287–98.

Kaufmann, J. H. (1967). Social relations of adult males in a free-ranging band of rhesus monkeys. In *Social communication among primates* (ed. S. A. Altmann), pp. 73–98. University of Chicago Press, Chicago.
Kummer, H. (1971). *Primate societies. Group techniques of ecological adaptation.* Aldine-Atherton, Chicago.
Kummer, H. (1978). On the value of social relationships to non-human primates: a heuristic scheme. *Social Science Information*, **17**, 687–705.
Kurland, J. A. (1977). Kin selection in the Japanese monkey. *Contributions to Primatology*, **12**, 1–145.
Lancaster, J. B. (1972). Play mothering: the relations between juvenile females and young infants among free-ranging vervets. In *Primate socialization* (ed. F. E. Poirier), pp. 83–104. Random House, New York.
Lindburg, D. G. (1971). The rhesus monkey in northern India: an ecological and behavioral study. In *Primate behavior: developments in field and laboratory research*, vol. II (ed. L. A. Rosenblum), pp. 1–106. Academic Press, New York.
Manzolillo, D. L. (1982). Intertroop Transfer by Adult Male *Papio anubis*. Ph.D. dissertation, University of California, Los Angeles.
Mason, W. A., Green, P. C. and Posepanko, C. J. (1960). Sex differences in affective social responses in rhesus monkeys. *Behaviour*, **16**, 74–83.
Massey, A. (1977). Agonistic aids and kinship in a group of pigtail macaques. *Behavioral Ecology and Sociobiology*, **2**, 31–40.
McKenna, J. (1978). Biosocial functions of grooming behavior among the common Indian langur monkey (*Presbytis entellus*). *American Journal of Physical Anthropology*, **48**, 503–10.
Nagel, U. (1979). On describing primate groups as systems: the concept of ecosocial behaviour. In *Primate ecology and human origins* (eds I. S. Bernstein and E. O. Smith), pp. 313–39. Garland, New York.
Nagel, U. and Kummer, H. (1974). Variation in cercopithecoid aggressive behavior. In *Primate aggression, territoriality, and xenophobia* (ed. R. L. Holloway), pp. 159–84. Academic Press, New York.
Oswald, M. and Erwin, J. (1976). Control of intragroup aggression by male pigtail monkeys (*Macaca nemestrina*). *Nature*, **262**, 686–7.
Packer, C. (1977). Reciprocal altruism in *Papio anubis*. *Nature*, **265**, 441–3.
Poirier, F. E. (1972). Introduction. In *Primate socialization* (ed. F. E. Poirier), pp. 3–28. Random House, New York.
Ransom, T. W. and Ransom, B. S. (1971). Adult male–infant relations among baboons (*Papio anubis*). *Folia Primatologica*, **16**, 179–95.
Reinhardt, V., Dodsworth, R., and Scanlan, J. (1986). Altruistic interference shown by the alpha-female of a captive troop of rhesus monkeys. *Folia Primatologica*, **46**, 44–50.
Rowell, T. E. (1966). Hierarchy in the organization of a captive baboon group. *Animal Behaviour*, **14**, 430–43.
Rowell, T. E. (1974). The concept of social dominance. *Behavioral Biology*, **11**, 131–54.
Sarbin, T. R. (1954). Role theory. In *Handbook of social psychology* (ed. I. G. Lindzey), pp. 488–567. Addison-Wesley, Reading, MA.
Seyfarth, R. M. (1978*a*). Social relationships among adult male and female baboons. I. Behaviour during sexual consort. *Behaviour*, **64**, 204–26.

Seyfarth, R. M. (1978*b*). Social relationships among adult male and female baboons. II. Behaviour throughout the female reproductive cycle. *Behaviour*, **64**, 227–47.

Silk, J. B. (1982). Altruism among female *Macaca radiata*: explanations and analysis of patterns of grooming and coalition formation. *Behaviour*, **79**, 162–88.

Smuts, B. B. (1985). *Sex and friendship in baboons*. Aldine Press, Hawthorne, NY.

Southwick, C. H. (1967). An experimental study of intragroup agonistic behaviour in rhesus monkeys (*Macaca mulatta*). *Behaviour*, **28**, 182–209.

Struhsaker, T. T. (1967). Social structure among vervet monkeys. *Behaviour*, **29**, 83–121.

Takahata, Y. (1982). Social relations between adult males and females of Japanese monkeys in the Arashiyama B troop. *Primates*, 23, 1–23.

Teas, J., Feldman, H. A., Richie, T. L., Taylor, H. G. and Southwick, C. H. (1982). Aggressive behavior in the free-ranging rhesus monkeys of Kathmandu, Nepal. *Aggressive Behaviour*, **8**, 63–77.

Teas, J., Richie, T., Taylor, H. and Southwick, C. (1980). Population patterns and behavioral ecology of rhesus monkeys (*Macaca mulatta*) in Nepal. In *The macaques: studies in ecology, behavior and evolution* (ed. D. G. Lindburg), pp. 247–62. Van Nostrand Reinhold, New York.

Thierry, B. (1985). Patterns of agonistic interactions in three species of macaque (*Macaca mulatta*, *M. fascicularis*, *M. tonkeana*). *Aggressive Behavior*, **11**, 223–33.

Tilford, B. (1982). Seasonal rank changes for adolescent and subadult natal males in a free-ranging group of rhesus monkeys. *International Journal of Primatology*, **3**, 483–90.

Tinbergen, N. (1951). *The study of instinct*. Clarendon Press, Oxford.

Tokuda, K. and Jensen, G. (1968). The leaders role in controlling aggressive behavior in a monkey group. *Primates*, **9**, 319–22.

Trivers, R. L. (1971). The evolution of reciprocal altruism. *Quarterly Review of Biology*, **46**, 35–57.

Varley, M. and Symmes, D. (1966). The hierarchy of dominance in a group of macaques. *Behaviour*, **27**, 54–75.

de Waal, F. B. M. (1977). The organization of agonistic relations within two captive groups of Java-monkeys (*Macaca fascicularis*). *Zeitschrift fur Tierpsychologie*, **44**, 225–82.

de Waal, F. B. M. (1978*a*). Join-aggression and protective aggression among captive *Macaca fascicularis*. In *Recent advances in primatology*, vol. 1 (eds D. J. Chivers and J. Herbert), pp. 577–9. Academic Press, London.

de Waal, F. B. M. (1978*b*). Exploitative and familiarity-dependent support strategies in a colony of semi-free-living chimpanzees. *Behaviour*, 66, 268–312.

de Waal, F. B. M. (1982). *Chimpanzee politics*. Jonathan Cape, London.

de Waal F. B. M. (1984). Sex differences in the formation of coalitions among chimpanzees. *Ethology and Sociobiology*, **5**, 239–55.

de Waal, F. B. M. (1986). Conflict resolution in monkeys and apes. In *Primates: the road to self-sustaining populations* (ed. K. Benirschke), pp. 341–50. Springer-Verlag, New York.

de Waal, F. B. M. (1987). Dynamics of social relationships. In *Primate societies* (eds B. B. Smuts, D. L. Cheney, R. M. Seyfarth, R. W. Wrangham, and T. T. Struhsaker), pp. 306–17. University of Chicago Press, Chicago.

de Waal, F. B. M. and Luttrell, L. M. (1988). Mechanisms of social reciprocity in three primate species: symmetrical relationship characteristics or cognition? *Ethology and Sociobiology*, **9**, 101–18.

de Waal, F. B. M. and van Roosmalen, A. (1979). Reconciliation and consolation among chimpanzees. *Behavioral Ecology and Sociobiology*, **5**, 55–66.

de Waal, F. B. M. and Yoshihara, D. (1983). Reconciliation and redirected affection in rhesus monkeys. *Behaviour*, **85**, 224–41.

de Waal, F. B. M., van Hooff, J. A. R. A. M., and Netto, W. J. (1976). An ethological analysis of types of agonistic interaction in a captive group of Java monkeys (*Macaca fascicularis*). *Primates*, **17**, 257–90.

Wade, T. D. (1976). The effects of strangers on rhesus monkey groups. *Behaviour*, **56**, 194–214.

Walters, J. (1980). Interventions and the development of dominance relationships in female baboons. *Folia Primatologica*, **34**, 61–89.

Watanabe, K. (1979). Alliance formation in a free-ranging troop of Japanese macaques. *Primates*, **20**, 459–74.

Wilson, E. O. (1975). *Sociobiology: the new synthesis*. Belknap, Cambridge, MA.

Yamada, M. (1966). Five natural troops of Japanese monkeys in Shodoshima Island. I. Distribution and social organization. *Primates*, **7**, 315–62.

York, A. and Rowell, T. E. (1988). Reconciliation following aggression in patas monkeys, *Erythrocebus patas*. *Animal Behaviour*, **36**, 502–9.

Juvenile spotted hyenas of Cohort One in the Berkeley Colony. Three females threaten a male as the rest of the group watches.

5

Coalition formation in a colony of prepubertal spotted hyenas

CYNTHIA J. ZABEL, STEPHEN E. GLICKMAN, LAURENCE G. FRANK, KATYA B. WOODMANSEE, GEOFFREY KEPPEL

INTRODUCTION

Alliances have been reported among free-ranging social carnivores, but not transitory coalitions. Alliances and their function have been described for lions, *Panthera leo* (Bertram 1975; Bygott *et al.* 1979; Packer and Pusey 1982), cheetahs, *Acinonyx jubatus* (Caro and Collins 1987), and coatis, *Nasura narica* (Russell 1983). Coalitions have been observed among captive wolves, *Canis lupus* (Fentress *et al.* 1986; Jenks 1988; Zimen 1976), and they are an important element of social behaviour in captive spotted hyenas (*Crocuta crocuta*). Although there do not appear to be any published field studies of coalitions among social carnivores, Mech (1970, p. 78) has observed group attacks among wild wolves that apparently involved coalition formation. In this chapter, we report on coalitions that formed in a colony of prepubertal spotted hyenas. Although these results were obtained under artificial circumstances, the existence of coalitions among spotted hyenas is supported by numerous field observations (Frank, personal observation). Kruuk (1972) described the 'parallel walk', when two hyenas threaten a third by approaching in attack posture, walking shoulder to shoulder almost touching one another. A target hyena can be a member of the same clan, or an intruding animal. This description fits our definition of a coalition as an aggressive interaction during which two or more hyenas join to threaten or attack a third animal.

Advantages of forming alliances among carnivores usually relate directly to reproductive benefits, as when groups of males are able to outcompete single males for females or their territories. Benefits gained from forming coalitions may relate directly to reproductive success (for example, in wolf packs where only a single female breeds). Alternatively, coalition formation may directly affect dominance status and indirectly modulate reproductive success by allowing dominant animals to control access to resources (in spotted hyenas, for example).

ALLIANCES AND COALITIONS AMONG SOCIAL CARNIVORES

Felids: lions and cheetahs

Male lions form alliances of 2–6 members in the Serengeti. They may be composed of siblings or related similar-aged cohort males which disperse together from their natal pride, or of unrelated males which join together subsequent to dispersal (Packer and Pusey 1982). Males which form alliances have a competitive advantage over single males in obtaining and maintaining possession of prides of females (Bertram 1975; Bygott *et al.* 1979; Packer and Pusey 1982). Male alliance partners are not ensured equal access to estrous females in a pride, but each male usually obtains some matings.

Lions are the only felid in which females are social. Most females remain in their natal pride for life, retaining close associations with 1–17 female relatives. Females occasionally leave their pride to avoid having their cubs killed by new males during pride takeovers. Females will support one another in aggressive interactions against males in efforts to prevent infanticide by immigrating males, and they may be wounded or even killed in these fights. Cooperatively defending cubs may be an important aspect of communal rearing among female lions (Packer and Pusey 1983).

Male cheetahs which form alliances (usually littermates) are more likely than single males to become territorial, and thereby gain access to solitary females within these territories (Caro and Collins 1987). Length of male tenure on territories is correlated with alliance size, and single territorial males are soon replaced by alliances of 2–4 males.

There is a higher incidence of single male cheetahs than single male lions in the Serengeti (41 per cent vs. 13 per cent). Caro and Collins (1987) suggest that this might reflect costs of sharing matings: some single cheetahs are able to defend a territory and mate, while single lions are highly unlikely to maintain possession of a pride, making mate sharing unavoidable among lions.

Wolves

Coalitions have been observed among captive wolves (Fentress *et al.* 1986; Jenks 1988), when individuals join ongoing dyadic aggressive encounters. In one colony of 10–12 wolves, including multiple males and females whose membership fluctuated from year to year, an average of 5.64 wolves participated in attacks (Fentress *et al.* 1986). The size of attacking groups ranged between 2–11 animals. In another wolf colony consisting of 8–10 animals, the number of aggressors ranged between 2–5 (Jenks 1988).

Other small carnivores

Coatis are a highly social species within the Procyonidae. Females and young live in bands, while adult males are solitary. Males are excluded from bands except during the mating season, apparently because they will opportunistically prey upon juveniles (Russell 1981). Pairs of females within bands preferentially groom each other, and also support one another during infrequent aggression that occurs (Russell 1983). Bands split when groups exceed 3–5 females, with fission occurring along previously established lines of affiliation. Russell (1983) concluded that '. . . individuals choose to invest in new relationships to a degree dependent upon their apparent potential for developing advantageous equitable relationships . . .', with kinship playing a secondary role to reciprocal support.

There are a number of social species among viverrids (Rood 1986), including the well-studied dwarf mongoose (*Helogale parvula*). Mongooses apparently associate in groups as a predator defence mechanism, allowing them to forage diurnally. The social system of the dwarf mongoose is similar to that of wolves: there are separate male and female dominance hierarchies, and only the alpha animals of each sex usually breed successfully. Other group members, including unrelated animals, may participate in the care of young in both dwarf (Rasa 1983; Rood 1978) and banded mongooses (*Mungos mungo*, Rood 1974). Dwarf mongooses may have preferred partners for grooming and resting (Rasa 1977). Male dwarf mongooses sometimes emigrate with other males and become breeders by taking over another pack (Rood 1990). There have been no observations that breeding females may be replaced by groups of intruding females, as occurs in males. Males in one solitary species, the slender mongoose (*Herpestes sanguineus*), can share home ranges and occasionally associate together (Rood and Waser 1978). Adult male slender mongooses have been seen denning, playing, travelling, and feeding amicably together (Rood 1989).

Among mustelids, the European badger (*Meles meles*) forms matrilineal groups which den in communal burrow systems ('setts') and share a common range (Kruuk 1989). However, mutually supportive relationships have not been described.

SPOTTED HYENAS

Female masculinization and social organization

Spotted hyenas cooperatively hunt ungulates, and feed competitively in large groups after a kill is made (Kruuk 1972). Juveniles are dependent on maternal support for access to prey (Tilson and Hamilton 1984), and this access is correlated with maternal rank (Frank 1986). Adult females and their dependent offspring are dominant to males in nearly all social situations. Females are heavier and more aggressive than males (Frank *et al.* 1989), and they display a unique syndrome of anatomical masculinization: they possess no 'normal' external female genitalia, but rather have a greatly hypertrophied clitoris that is fully erectile and nearly indistinguishable from the male penis (Matthews 1939). The vaginal labia are fused into a pseudoscrotum. The urogenital canal through which both mating and birth occur, traverses the length of the clitoris. This syndrome of 'masculinization' is associated with elevated levels of androgens in female spotted hyenas (Frank *et al.* 1985; Glickman *et al.* 1987; Lindeque and Skinner 1982; Racey and Skinner 1979).

The social system of spotted hyenas resembles that of many Old World primates (Frank 1986; Henschel and Skinner 1987; Mills 1985): females remain in their natal group ('clan') for life while males disperse to join other clans as adults. Maturing females acquire their mother's rank, resulting in stable dominance relations among matrilines over generations. Associations among related females are closer in higher-ranking matrilines (Frank 1986), presumably reflecting greater benefits that young females may accrue from the support of a high-ranking mother. Sons of high-ranking females are dominant to females which rank lower than their mother, and they remain in a clan longer before dispersing than do lower-ranking sons. Again, this may reflect advantages of maternal support in the highly competitive feeding situation of hyenas. After dispersing to new clans, high-ranking males have better prospects of mating than do subordinates.

The Berkeley study

Observations of coalitions among captive spotted hyenas occurred during the course of a long-term study of behavioural and hormonal development, focused upon the unique sexual differentiation of this species. We have been particularly interested in the development of sex differences in aggressive behaviour and dominance. However, because of the extremely tight association between maternal rank and the status/behaviour of offspring, any differences between males and females during early life could easily be obscured. This, in turn, would have interfered with our attempts to understand the proximal mechanisms of sexual differentiation. We therefore chose to study these animals in peer groups, without maternal influence.

The hormonal focus of our research required that a subset of hyenas have their ovaries or testes removed. Although one would not normally expect juvenile gonadectomy to influence sexually dimorphic behaviour during the juvenile or subadult period (see for example, LeBoeuf 1970; Mirsky 1955), the endocrine situation in hyenas is rather unusual. Female spotted hyenas have high circulating levels of the androgen androstenedione throughout the juvenile/prepubertal periods (Glickman *et al.* 1987; Lindeque *et al.* 1986).

Removal of the ovaries results in a marked decline in this hormone (Glickman *et al.* 1987). Since androstenedione has been found to facilitate aggressive behavior under certain circumstances in other species (Erpino and Chappelle 1971; Tsutsui and Ishii 1981), removing ovarian secretions through gonadectomy might reduce aggressiveness of female hyenas, including their participation in coalitions.

Coalitions, and possibly matrilineal alliances, may be an important factor in the near absolute dominance of female spotted hyenas over males in nature (Frank *et al.* 1989). In this chapter we report frequency of occurrence and duration of coalition attacks, and size of the groups involved in attacks. We analyse effects of age, sex, and gonadal secretion on coalitions; i.e., which hyenas initiate coalition attacks, join to support the initiator, and are targets of attacks. The relationship between dominance status and coalitions will be examined to determine if attacks are more likely to be initiated by dominant animals, and if subordinate animals are more likely to be targets.

Animals joining coalitions can either challenge or reinforce a dominance hierarchy. By supporting coalition attacks against lower-ranking animals, the existing hierarchy is reinforced. Alternatively, by joining coalitions against higher-ranking opponents, the dominance rank of a targeted animal may be challenged. Among primates, individuals are less likely to initiate or join in supporting aggression when they rank below a target (Bernstein and Ehardt 1985; Cheney 1977; Datta 1983; Ehardt and Bernstein this volume, pp. 86–8; Kaplan 1978; Walters 1980), presumably due to risks of retaliation (but see de Waal this volume, pp. 238–9). We present data for hyenas on frequency of initiating and joining coalitions against dominant animals, to determine whether their participation in coalition attacks reinforced or challenged the dominance hierarchy.

Finally, we suggest that coalitions in hyenas and other social mammals may reflect a more general tendency to exhibit socially facilitated behaviour, i.e. 'to do what other group members are doing'. Social facilitation of behaviour is found among many animal groups (Zajonc 1965), and may be particularly important among social carnivores. Social facilitation could serve to synchronize activities of individuals within a group (see for example, Lockwood 1976), which may make cooperative group hunters more effective.

Subjects

Two cohorts of 10 infants each were collected under permit in Narok District, Kenya in December 1984/January 1985 and November/December 1985. Infants were between 1 week and 2 months of age at the time of collection. When cohort I was collected, we were unable to sex the infants by visual inspection, and it consisted of seven females and three males. In addition to the biased sex ratio, there was a weight bias; two of the males showed slow growth and have remained unusually small as adults (Frank *et al.* 1989). The second cohort consisted of five females and five males.

Although there was daily contact with human caretakers, the hyenas lived in peer groups from the time of collection. Each cohort was housed in separate indoor–outdoor enclosures measuring 12 × 30 m at Berkeley, California.

Two females in cohort I and two females and two males in cohort II were bilaterally gonadectomized at 4–6 months of age. Mid-ranking animals, as determined by regular feeding competition tests (Frank *et al.* 1989), were selected for gonadectomy.

Behavioural sampling

Spontaneous behaviour of the older cohort was observed from September 1985 through August 1986 for this study. After September 1986 this cohort was divided into two separate groups, because of increasingly severe consequences of aggression. The younger cohort was sampled from March 1986 through August 1987, at which time it was subdivided for the same reason. Because of the nocturnal behaviour of the hyenas, observations were recorded between 1800 and 2200 hours using artificial light.

To assess the effects of age, data were subdivided into an early subadult age period (approximately 11.5–18 months of age; based on 62 h of observation for cohort I and 66 h for cohort II) and a late subadult age period (approximately 18–21.5 months of age; based on 19 h of observation for cohort I and 48 h for cohort II). The older age period coincided with the first indications of pubertly according to radioimmunoassay of various gonadal hormones (unpublished data). Male spotted hyenas are considered to be reproductively mature at approximately 2 years of age and females at 3 years (Kruuk 1972).

Critical incident sampling (Altmann 1974) of all occurrences of significant social behaviours (Martin and Bateson 1986) within each cohort was used. All instances of aggression, submission, affiliation, and specified social activities involving two or more animals were recorded at 30-s intervals in an actor–behaviour–recipient format. By definition, coalitions involved two or more hyenas (as actors) threatening/attacking a target (the recipient). Aggressive behaviours included threatening postures (a forward 'approach' movement with the ears and mane erect), one animal displacing another, bites and chasing/lunging attacks. Submissive behaviours included avoiding a threatening or attacking animal and appeasement postures, such as crawling on the carpals, with the head down or under and/or shaking the head from side to side, and an open mouth. Affiliative/prosocial behaviour were allogrooming, the meeting ceremony, and play.

Dominance ranks referred to in this study were determined from dyadic interactions during sampling of spontaneous social activities, when one member of a pair exhibited submissive behaviour. For each pair of animals, the hyena that received submissive displays during more 30-s intervals was the dominant animal for that dyad. This measure of dominance is highly correlated with ranks based on dyadic aggressive interactions, and with ranks determined from competitive interactions during group feeding at a cluster of meaty neck bones. In each cohort, the 10 hyenas were ranked according to the total wins and losses in dyadic interactions. A rank of 1 was attributed to the alpha hyena and 10 to the lowest-ranked animal in each cohort.

COALITIONS AMONG CAPTIVE SPOTTED HYENAS

Coalitions: frequency and rates of recruitment

Coalitions occurred when two or more hyenas joined in threatening or attacking a third hyena. Typically, this began with a dyadic encounter. One of the combatants was then joined by one or more allies. Frequencies of coalition attacks within the two cohorts are presented in Table 5.1. On average, attacks occurred 2–3 times per hour. The duration of attacks averaged about 1.5 to 2.5 30-s intervals. Mean sizes of coalitions that were involved ranged between 2.5 and 3.7 hyenas between the age periods we sampled. However, on some occasions all nine animals joined in attacking a single recipient. Almost half of the attacks were performed by two hyenas, with the occurrence of larger coalitions decreasing with their size (Fig. 5.1).

Table 5.1. Characteristics of coalition attacks in two cohorts of spotted hyenas during two age periods. Frequency, duration, and size of attacking coalitions. All values are means ± standard deviation. *N* is the total number of attacks.

Age	Early subadult	Late subadult
Cohort I	$N = 180$	$N = 51$
Attacks/h	2.9	2.7
Duration of attack*	2.6 ± 3.0	2.3 ± 1.6
Coalition size	3.2 ± 1.5	3.7 ± 1.7
Time to maximum coalition size*	1.7 ± 1.6	1.5 ± 0.8
Cohort II	$N = 165$	$N = 104$
Attacks/h	2.5	2.1
Duration of attack*	1.6 ± 1.2	1.3 ± 0.8
Coalition size	3.0 ± 1.3	2.5 ± 0.8
Time to maximum coalition size*	1.3 ± 0.9	1.1 ± 0.3

* Values for time are number of 30-s intervals.

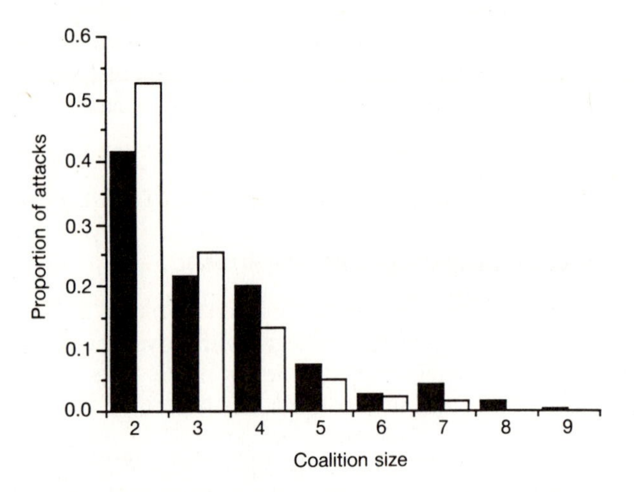

Fig. 5.1. Proportion of coalition attacks that occurred according to number of aggressors that were involved. Cohort I (solid bars), $N = 230$ attacks; cohort II (open bars), $N = 269$ attacks.

Attacking groups formed very quickly, and maximum coalition size tended to occur within one to two 30-s intervals.

Effects of age, sex, and gonadectomy on coalition attacks

Hyenas which participated in coalition attacks were characterized in one of three ways: (1) according to the animal which initiated an attack; (2) to the animals which joined or 'supported' the attack; and (3) to the animal which was selected as the recipient or 'target' of the attack. Rates that individual hyenas initiated, supported, or were targeted in attacks are presented separately by cohort in Figs. 5.2(a–c), according to the sex and ages of the animals. Individual three-way unbalanced factorial analyses of variance were computed for each dependent variable, with sex, age, and cohort as the three independent variables. The three analyses revealed only one significant effect: an interaction between cohorts and age in rates of supporting coalitions ($F = 5.01$; d.f. $= 1,16$; $p < 0.05$). This was presumably due to a decline in supporting coalition attacks with age in cohort II in contrast with a slight increase with age in cohort I. Thus, within the confines of this study, with its relatively small sample sizes, neither sex, age, or cohort influenced rates that hyenas initiated, supported, or received coalition attacks.

Effects of gonadectomy were tested by combining data from both cohorts. Individual three-way unbalanced factorial analyses of variance were computed for each dependent variable, with sex, age, and gonadectomy as the three independent variables. We found no significant effects of gonadectomy on rates of initiating, supporting, or receiving coalition attacks.

Absolute dominance rank and frequency of coalition attacks

Correlations between dominance rank and rates of initiating, supporting, or being targeted during coalition attacks are presented by cohort and age period in Figs. 5.2(a–c). There was a tendency for dominant animals to initiate attacks more frequently than subordinates, although statistical significance was achieved in only two of four correlations. In contrast, supporting coalition attacks was significantly correlated with rank in a more robust manner. Dominant animals supported attacks more frequently than subordinates in both cohorts during both age periods. Subordinate hyenas tended to be targeted during group aggression more frequently than dominants. There were significant correlations between low dominance rank and being targeted during three of four cohort-age intervals, with particularly consistent effects in cohort II.

We have found no consistent sex differences in dominance rank among the captive hyenas. Females do not dominate males, despite having higher aggression rates and weights (Frank *et al.* 1989). One exception to this lack of female dominance occurred in cohort I during the early subadult period. The three males in cohort I (which were the lightest in weight) ranked below all seven females at this age. Difference in mean rank was significant ($F = 12.99$; d.f. = 1,8; $p < 0.01$). Although there were no sex differences in dominance rank during the remaining age periods, we calculated semi-partial correlations, controlling for sex differences, between dominance rank and frequency of receiving attacks (Table 5.2). Even with sex differences removed statistically, negative correlations between dominance rank and frequency of receiving attacks were still significant during both age periods in both cohorts.

Table 5.2. Semi-partial correlations between dominance rank and rates of receiving coalition attacks, with sex of spotted hyenas controlled.

	Early subadult	Late subadult
Cohort I	−0.70*	−0.67*
Cohort II	−0.85**	−0.73**

* $p < 0.05$; ** $p < 0.01$.

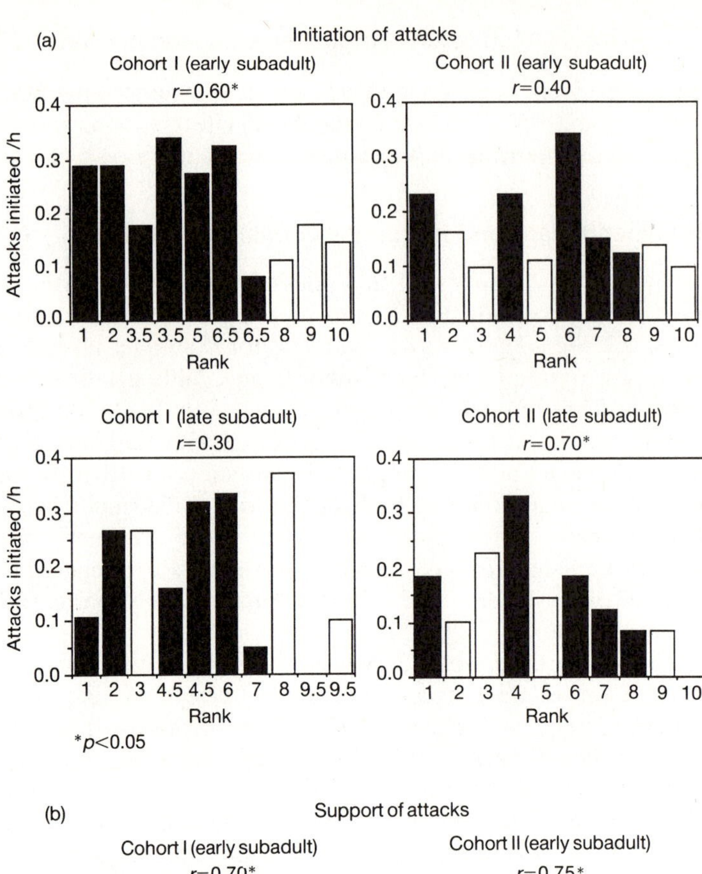
(a)
Initiation of attacks
Cohort I (early subadult)
r=0.60*
Cohort II (early subadult)
r=0.40
Cohort I (late subadult)
r=0.30
Cohort II (late subadult)
r=0.70*
Attacks initiated /h
Rank
*p<0.05

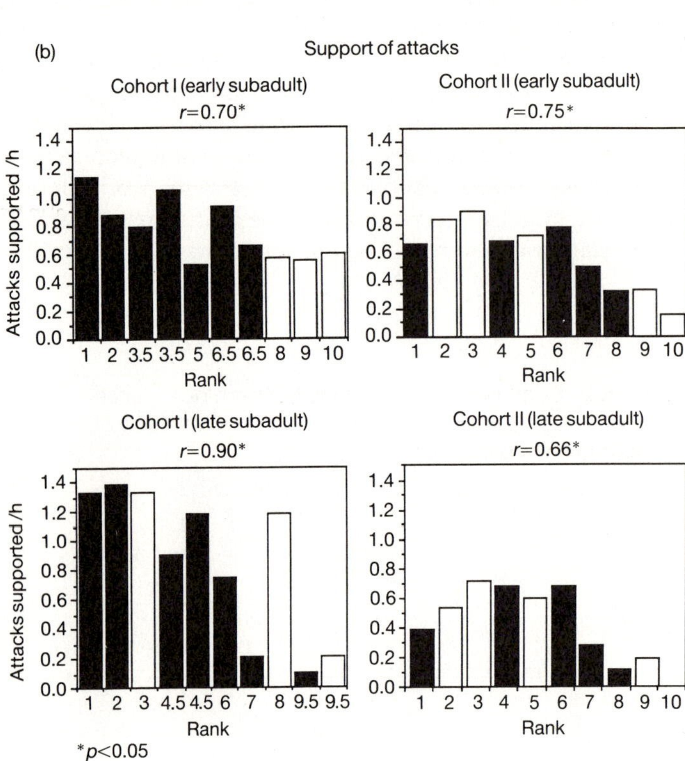
(b)
Support of attacks
Cohort I (early subadult)
r=0.70*
Cohort II (early subadult)
r=0.75*
Cohort I (late subadult)
r=0.90*
Cohort II (late subadult)
r=0.66*
Attacks supported /h
Rank
*p<0.05

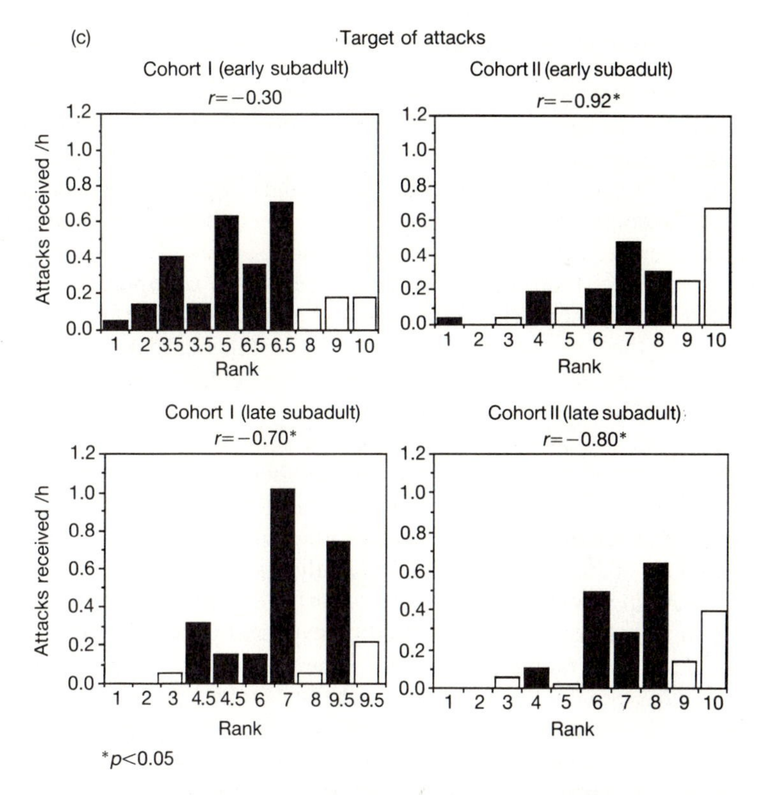

Fig. 5.2. Relationship between dominance rank and rates of initiating (a), supporting (b), and receiving (c) coalition attacks among prepubertal spotted hyenas. Rates were calculated from the number of 30-s intervals that each behavior occurred per hour of observation. Data are presented for each individual in both cohorts during two age periods. Spearman rank correlations (r) are provided between rank and rate of participation in attacks. Solid bars represent females and open bars males.

It should be emphasized that this report deals only with prepubertal animals. In the wild, all resident adult males which immigrate into a clan are subordinate to females and juveniles, suggesting that events related to male dispersal and immigration may be important in the development of female dominance (Frank *et al.* 1989). Furthermore, rank changes occurred among the captive females after they produced their first litters, and observations of maternal defensive aggression and coalition attacks by mothers on behalf of offspring suggest such behaviour may be important in the eventual dominance of females. Maternal aggression is apparently related to the process of maternal rank acquisition by juveniles in wild hyenas, and this process is currently under study.

Relative dominance rank between participants of coalition attacks

Many primates are less likely to initiate or join in supporting aggression when they rank below a target, due to risks of retaliation. To estimate frequencies of initiating or supporting attacks against dominant animals, data need to be corrected for the number of *potential* dominant opponents available. In cohort I, for example, the three males which occupied the bottom ranks of the dominance hierarchy in the early subadult period had fewer opportunities to attack subordinates than did the seven females which occupied the higher ranks. That is, the lowest-ranked male could attack no subordinate hyenas since he was at the bottom of the hierarchy; the next higher male could attack only one subordinate hyena, etc. Females in this cohort had increasingly more opportunities to attack subordinates.

In order to deal with this problem, we first eliminated data from hyenas occupying the top and bottom ranks since neither of these animals had an opportunity to choose between dominant and subordinate targets—the hyena in the first rank could initiate or join attacks only against subordinates, while the hyena in the last rank could do so only against dominant opponents. For the remaining eight hyenas in each cohort, we estimated the number of attacks against subordinates that would be expected purely on the basis of chance. The second ranking animal, for example, has one dominant target and eight subordinate targets. If only chance factors were operating in initiating or joining attacks against other hyenas, one would expect 1/9 of the total attacks participated in by this particular animal to be against dominant targets and 8/9 of the attacks to be against subordinate targets. Expected frequencies of attack were obtained by multiplying the appropriate chance ratio by the total number of attacks for each of the eight animals in both cohorts. Observed frequencies corrected for chance were calculated by subtracting expected frequencies from frequencies actually observed. Finally, we converted these corrected frequencies into rates (frequency/h) in order to compensate for different numbers of observation hours that were available for the two cohorts and different age periods.

When corrections for opportunities were calculated, no significant differences between male and female hyenas were found. Corrected data combined over sex are presented in Table 5.3, showing differences between observed and expected rates of initiating and joining attacks. Means in Table 5.3 are all positive, indicating a consistent tendency for hyenas to initiate and join attacks against subordinates. (A negative difference would indicate a tendency to attack dominants.) All of the means for cohort II are significantly different from zero, while only one mean for cohort I is significantly different from zero. Thus, hyenas in cohort I were equally as likely to initiate and support coalition attacks against animals that outranked themselves, as against lower-ranking animals during the early subadult age period.

Table 5.3. Differences between observed and expected rates of initiating and joining attacks (per hour) against subordinate hyenas. This analysis is corrected for the relative number of dominant and subordinate opponents. Each mean is based on eight subjects. All values are positive, indicating consistent tendencies to initiate and join attacks against subordinates.

		Initiating attacks		Joining attacks	
		Early subadult	Late subadult	Early subadult	Late subadult
Cohort I	Mean	0.012	0.073	0.041	0.233*
	S.D.	0.044	0.099	0.093	0.140
Cohort II	Mean	0.080*	0.067*	0.260*	0.168*
	S.D.	0.055	0.032	0.105	0.118

* $p < 0.01$; values are significantly different from zero.

Behaviours preceding attacks against dominant animals

Subordinate hyenas which initiated attacks against dominant animals were most frequently involved in a social activity with a third hyena just prior to the attack. In cohorts I and II, respectively, 71 per cent and 52 per cent of these interactions began after a subordinate attacker had been involved in non-aggressive, prosocial behaviour (such as play, or sniffing each other; N = 35 and 25 attacks in cohorts I and II, respectively). In all of these cases, the third hyena joined in the aggression and supported the lower-ranked initiator in the attack. In addition, on a few occasions a subordinate target successfully redirected a group attack against an initiator (in 17 per cent and 32 per cent of the attacks, in cohorts I and II, respectively). In these cases, hyenas which had joined the initial attack then supported that target in reversing the interaction and attacking the previous initiator.

Consequences of attacking dominant hyenas

Forming coalitions with other animals and attacking hyenas that were higher ranking than themselves may have been a mechanism for the three subordinate males in cohort I to rise in dominance rank. These three males all rose in rank: two by the conclusion of this study period, and one at a later date. During the late subadult age, one of these males rose from position 9 in the hierarchy up to position 3, while another rose from position 10 to 8.

DISCUSSION

Coalitions among social carnivores: spotted hyenas and wolves

Coalitions of 2–9 hyenas simultaneously attacking a target hyena were relatively common in both cohorts of prepubertal hyenas. Such coalitions have also been observed in nature (Frank unpublished data). Coalition threats/attacks observed in the present study are identical in appearance to those depicted by Kruuk (1972: Plates 38–40). Groups of resident hyenas gain advantages when they confront individual immigrants (Henschel and Skinner 1987).

Rates of coalition attacks appeared to be lower in wolf colonies compared to the hyena colony. During 435 h of observation, Fentress *et al.* (1986) observed 26 wolf attacks (0.06 attacks/h); rates of attack in the hyena colony were over 30 times higher. However, there were significant differences in group composition between the hyena and wolf colonies. The wolf colonies were more natural breeding groups, of mixed age and sex, while the hyena colony was composed entirely of immature animals during this study. This is particularly relevant since coalition attacks among wolves often involve mating competition. Nevertheless, there are similarities in coalitions between hyenas and wolves: in both species, transitory aggressive coalitions emerge during which both males and females threaten/attack a target animal. Animals participating in these coalitions shift from one attack to the next.

Effects of sex, age, and gonadectomy

Neither sex or age influenced rates that immature hyenas initiated, joined, or were targeted in coalition attacks. In contrast, female cercopithecine primates usually aid more frequently than males (Cheney 1977; Ehardt and Bernstein this volume pp. 85–8; Kaplan 1977; Kurland 1977; Massey 1977). It has been suggested that female aiding may help maintain the established social order of groups where they form the stable core (Bernstein and Ehardt 1985; Chapais this volume pp. 49–50). However, it seems likely that rearing immature hyenas in the absence of mothers may have obscured any sex differences that involve long-established matrilines. In nature, matrilines of females which live in the same area throughout their lives play a major role in defending territorial borders during disputes with other clans (Kruuk 1972). These same groups of females are also likely to confront individual intruders, including immigrating males. Until more coalition data are available from hyena clans in the field and from colony animals that include matrilines, it would be premature to draw any conclusions regarding differential behaviour of female and male hyenas compared to primates.

The function of coalition attacks may differ across species with different mating systems, and this may influence sex differences in coalition participation. For example, a single dominant female often monopolizes reproduction among wolf packs (Rabb *et al.* 1967; Zimen 1976). Associated with this pattern, females initiated 92 per cent of coalition attacks and were targeted in 96 per cent of attacks in one colony (Fentress *et al.* 1986). In contrast, all female hyenas within a clan can breed, so they may not engage in aggression to suppress reproductive behaviour of other females. However, daughters of high-ranking females may have higher reproductive success (Frank, Holekamp, and Smale unpublished data), and both male and female offspring of high-ranking hyena females have competitive advantages during feeding (Frank 1986). Thus, participation in group attacks by both male and female hyenas might optimize their competitive effectiveness within a group.

Gonadectomy of juvenile hyenas produced a marked reduction in plasma levels of androstenedione and testosterone (Glickman *et al.* 1987). However, there were no apparent effects of gonadectomy on initiating, supporting, or being the target of coalition attacks. This is consistent with data from dyadic aggressive interactions. Juvenile gonadectomy had no significant effects on individual levels of aggression or dominance rank during the subsequent subadult period (Frank *et al.* 1989). If there are activational effects of gonadal hormones during this time, they are not sufficiently robust to overcome the relatively small sample sizes in this study and enduring effects of social learning that may occur prior to gonadectomy. After puberty, there will be an even greater discrepancy between hormonal profiles of intact and gonadectomized subjects. It is possible that there will be effects of gonadectomy on aggressive interactions and coalitions during the postpubertal period.

Dominance rank and rules of coalition formation

In considering the relationship between dominance and participation in coalitions, it is necessary to distinguish between absolute dominance rank within a group and relative dominance rank between two individuals. Absolute dominance rank correlates with absolute frequencies of participation in coalition attacks. Dominant animals tended to initiate and support attacks more frequently than subordinates, and subordinate hyenas tended to be targeted during attacks more frequently than dominants. Similarly, among wolves, coalitions usually involved a group of dominant animals attacking a subordinate (Zimen 1976). Thus, coalition attacks among both spotted hyenas and wolves tend to reinforce the existing dominance hierarchy.

In order to understand the functional consequences of coalitions for individuals within a group, it is necessary to examine the relative dominance ranks of initiators and supporters in relation to the rank of targets. Individuals that join coalitions may benefit from closer association with dominants (Cheney 1977; Seyfarth 1977, 1980). If an animal that joins an interaction is subordinate to a targeted animal, it gains safety from the coalition in challenging its dominance, although with risks of retaliation. If an animal that joins an interaction is dominant to a targeted animal, it has safety from the coalition in reinforcing its own dyadic dominance over the target (Walters 1980; Datta 1983). Both tactics occurred among the captive hyenas.

All of the hyenas in cohort II, regardless of dominance rank, joined attacks when they were dominant to targeted animals during both age periods. Thus, coalitions in that cohort contributed to reinforcing the existing hierarchy, not challenging it. A similar situation existed during the late subadult period in cohort I. Hierarchy-reinforcing behaviours appear to be the rule among wolves as well. Dominant animals frequently participated in coalition attacks against subordinates in one wolf colony (Jenks 1988), and the dominant female in another colony participated in 82 per cent of attacks (Fentress *et al.* 1986).

Hyenas, wolves, and primates: aiding-at-risk

Cercopithecine primates (Bernstein and Ehardt 1985; Cheney 1977; Datta 1983; Kaplan 1978; Walters 1980), hyenas, and wolves are less likely to aid another individual when they rank below an opponent. Initiating or supporting attacks against dominant animals ('aiding-at-risk'; Bernstein and Ehardt 1985), occurred infrequently in the hyena colony. However, it did occur among hyenas in cohort I during the early subadult period: three low-ranking hyenas joined and initiated attacks against animals that outranked them. Even after data were appropriately corrected for the number of potential subordinate opponents available, hyenas in cohort I were equally likely to initiate and join attacks against dominants as against subordinates at this age. Hyenas in this cohort apparently gained safety from coalitions in challenging the dominance of targeted hyenas. Low-ranking hyenas in cohort I, which were all males during the early subadult period, seemed to use high female aggression and the readiness of females to attack other females to their own advantage for support in coalitions. Subordinates attacking or joining attacks against dominants may be infrequent but very important. Although these low-ranking males risked counterattack by targeting dominant hyenas, it appeared they were able to rise in the hierarchy by forming coalitions with other animals. These three males rose in rank during the late subadult period (Frank *et al.* 1989). After they rose in rank, they contributed to reinforcing the hierarchy by joining coalitions against subordinate animals more frequently than they challenged the dominance of higher-ranked animals. Thus, during the late subadult period, animals in cohorts I and II behaved similarly.

Wolves have also been observed challenging the rank of dominant animals by joining coalitions. A subordinate male in one colony initiated a successful coalition attack against his dominant father, and was supported by a number of animals which were aiding-at-risk (Jenks 1988). The dominant father subsequently fell in rank.

Primates will aid-at-risk while supporting offspring and other kin (Chapais 1983). Captive spotted hyena coalitions appeared to differ from those among primates in the relative frequency of aiding-at-risk: our hyenas usually supported dominant aggressors, rather than subordinate targets. However, this might have been due to the composition of the group, which consisted entirely of immatures without adults present, particularly mothers which might come to the defence of their offspring. Since these data were compiled, there have been a number of births in the hyena colony, and mothers and infants have been observed while they were being introduced into the main group. There was one dramatic occurrence of a female aiding-at-risk when a higher-ranking female attacked one of her offspring. The low-ranking mother subsequently attacked the dominant female which was biting and carrying her infant. The dominant female released her grip on the infant and turned her attack to the subordinate mother. Such maternal aiding-at-risk has also been observed in a colony of hyenas maintained at the Amsterdam Zoo (Kranendonk *et al.* 1983). As in most mammals, maternal defence of young takes precedence over hierarchical considerations.

Social facilitation and coalition formation

'Mob' behaviour and redirection in hyenas

An important aspect of hyena coalition behaviour is redirected aggression. When this occurs, an initial target of aggression attacks a lower-ranking third party, diverting the attention of the original attackers onto a 'scapegoat' (see deWaal this volume, pp. 240–1, and Ehardt and Bernstein this volume, pp. 90–2, for similar behaviour in primates). Because hyenas have a strong tendency to join attacks, this is often a successful tactic. At kills in the wild, it is common to see a chain of aggression: hyena A chases B, B chases C, C chases D, and D chases vultures. Among the captive hyenas, an initial victim often redirected aggression to a lower-ranking target. However, as noted above, there were a number of cases in which the attack was redirected against the initial aggressor. We have not yet analysed the phenomena of redirected aggression quantitatively. But it appeared as if supporting hyenas had strong tendencies to join an attack against whichever animal was losing, as long as they were higher ranking than the target. This is more suggestive of mob behaviour or social facilitation (Zajonc 1965) than of cognitively based assessments of relative rank or cost/benefit analyses of the political outcome of aggression. It appears to be similar to antipredator mobbing in this and other species. One must be cautious about inferring complex cognitive processes when simpler explanations will suffice.

Hyenas have a very strong tendency to 'do what other hyenas are doing'

(Woodmansee *et al.* 1991). The tendency of hyenas to join attacks against other hyenas could result from focusing on the stimuli provided by a subordinate hyena, rather than any active cognitive process of 'joining' a dominant attacking hyena(s). A general tendency to do what other group members are doing could then serve to maintain or improve the position of individuals within a social group.

Synchrony in social carnivores

Socially facilitated behaviour, or synchrony, was a prominent feature of activities in the hyena colony. One or more animals would often join an individual that was engaged in solitary activities, e.g. scent-marking, defecating, or intently surveying its surroundings. Social facilitation also occurred when individuals joined a group of hyenas that was engaged in social interactions; as when a third hyena approached and joined a pair that was playing, or engaged in the meeting ceremony (Kruuk 1972). (The meeting ceremony is an affiliative behaviour in this species, during which either sex may have an erection that other animal(s) will investigate.) Behaviours such as vigorous play, manipulating objects, shadowing a hyena (i.e. following at very close distances), group aggression, and even the meeting ceremony commonly recruited groups of three or more animals into the activity.

Synchronous activities have been reported in other social carnivores as well. For example, wolves engage in communal greeting, howling, and play prior to a hunt (Fox 1971; Mech 1970). Similarly, African wild dogs (*Lycaon pictus*) engage in group activities including scratching, yawning, communal greeting, and play prior to hunting (Estes and Goddard 1967; Kuhme 1965; Malcolm 1979). These group behaviours may reinforce mutual dependence and friendliness (Kuhme 1965), and may be characteristic of cohesive groups (Lockwood 1976). Synchronization of elimination has been noted in several carnivores, including spotted hyenas (Kruuk 1972; Woodmansee *et al.* 1991), African wild dogs (Kuhme 1965), and lions (Rudnai 1973). Among cooperative group hunters, it may be important that many group activities are coordinated. When social carnivores travel quickly, asynchronous animals might become separated from a group, and reduced cohesiveness could result in less effective hunting (Lockwood 1976).

Coalitions and alliances among social carnivores: future research

There are many gaps in our knowledge of coalitions and alliances among social carnivores. It is not clear whether an absence of data represents an absence of these behaviours among social carnivores, or rather a failure to report them. For example, there have been no studies of long-term alliances among spotted hyenas or wolves. However, anecdotal observations indicate that they do occur. Aggression among wild hyenas is more frequent between matrilines than within matrilines (Frank 1986), suggesting kin-related alliances may be occurring. In the Berkeley study, two captive female spotted hyenas which supported one another during aggressive interactions were the top-ranking animals in cohort I during the early subadult period. When this cohort was split into two subgroups of five animals at a later age, these two females were separated. Each of them subsequently dropped in rank to become nearly the lowest-ranked animals within their group. It appeared that when they were deprived of reciprocal support, they were unable to maintain their position of dominance. Similarly, two captive female wolves which supported one another in aggressive interactions more frequently than other pairs subsequently rose in rank (Jenks 1988).

Among free-ranging animals, there have been some significant studies of alliances, but no quantitative reports of transitory coalitions. It is our impression that interesting patterns of alliance and coalition formation would emerge if these behaviours were studied among wolves and hyenas in nature. This might extend to other social carnivores as well. Better comparisons between primates and carnivores await further studies on the latter group focused upon questions of reciprocity, relatedness, and the consequences of coalitions and alliances for reproductive success.

SUMMARY

Kin alliances and their function have been described in a few species of social carnivores, but there have been no quantitative field studies of coalitions. In this chapter we report on coalitions that formed in a colony of captive spotted hyenas (*Crocuta crocuta*) during their first 2 years of life. Spotted hyenas resemble cercopithecine primate groups in social organization, living in matrilineal 'clans' of up to 100 individuals. They hunt cooperatively, feed competitively, and den communally. Females are unique among mammals in being anatomically and behaviourally masculinized; they are more aggressive than males and dominant over them. In the wild, matrilineal associations of females are an important element of social organization, and probably a factor in the maintenance of individual females' rank and dominance over males.

In order to study sexual differentiation and anatomy, a captive colony was established. Research emphasis was on dominance relations and agonistic behavior among prepubertal hyenas in the absence of adults. Coalition attacks were a major feature of social interactions.

1. The mean number of hyenas involved in coalition attacks was about three, but as many as nine of ten animals were sometimes involved.

2. Neither sex, gonadectomy, age, or cohort influenced rates that hyenas initiated, supported, or received coalition attacks. Male and female hyenas were equally likely to engage in group attacks, unlike most primates. However, the lack of maternal influence on the juvenile hyenas, which is a powerful determinant of rank in the wild, may have significantly affected both the form and frequency of coalitions in this colony.

3. Captive females were not dominant over males during this age period.

4. Dominant animals were more likely than subordinates to both initiate and support attacks, and subordinates were more likely to be targets of group attacks.

5. Due to the tendency to join attacks against lower-ranking animals, coalition attacks among hyenas and wolves usually serve to reinforce existing dominance relations, as is sometimes the case among primates.

6. Attacking dominants ('aiding-at-risk') occurred infrequently. However, by forming coalitions with other animals and challenging the rank of dominants, hyenas may subsequently rise in rank.

7. In assessing the likelihood of attacking dominants or subordinates, it is necessary to adjust data for each individual according to the number of higher- and lower-ranking animals in a hierarchy.

8. The 'mob' nature of attacks, and the general tendency of hyenas to 'do what other hyenas are doing', suggest that group attacks may reflect a strong tendency to join any attack against lower ranking individuals. A general tendency to synchronize behaviour might be particularly significant to social predators, which must coordinate their actions in hunting, and in defence of both kills and territorial boundaries.

ACKNOWLEDGMENTS

We thank the Department of Wildlife Conservation and Management of the Government of the Republic of Kenya for permission to collect and export the hyenas which form the basis of our colony. This research is supported by grant MN39917 from the National Institute of Mental Health.

REFERENCES

Altmann, J. (1974). Observational study of behavior: sampling methods. *Behaviour*, **49**, 227–65.

Bernstein, I. S. and Ehardt, C. L. (1985). Agonistic aiding: kinship, rank, age, and sex influences. *American Journal of Primatology*, **8**, 37–52.

Bertram, B. C. R. (1975). Social factors influencing reproduction in lions. *Journal of Zoology*, London, **177**, 463–82.

Bygott, J. D., Bertram, B. C. R., and Hanby, J. P. (1979). Male lions in large coalitions gain reproductive advantages. *Nature*, **282**, 839–41.

Caro, T. M. and Collins, D. A. (1987). Male cheetah social organization and territoriality. *Ethology*, **74**, 52–64.

Chapais, B. (1983). Dominance, relatedness and the structure of female relationship in rhesus monkeys. In *Primate social relationships* (ed. R. A. Hinde), pp. 208–19. Blackwell, Oxford.

Cheney, D. L. (1977). The acquisition of rank and the development of reciprocal alliances among free-ranging immature baboons. *Behavioral Ecology and Sociobiology*, **2**, 303–18.

Datta, S. B. (1983). Relative power and the maintenance of dominance. In *Primate Social Relationships* (ed. R. Hinde), pp. 103–12. Sinauer, Sunderland, MA.

Erpino, M. J. and Chappelle, T. C. (1971). Interactions between androgens and progesterone in mediation of aggression in the mouse. *Hormones and Behavior*, **2**, 265–72.

Estes, R. D. and Goddard, J. (1967). Prey selection and hunting behavior of the African wild dog. *Journal of Wildlife Management*, **31**, 52–70.

Fentress, J. C., Ryon, J., McLeod, P. J., and Havkin, G. Z. (1986). A multidimensional approach to agonistic behavior in wolves. In *Man and wolf: advances, issues and problems in captive wolf research* (ed. H. Frank), pp. 253–73. Junk Publishers, Dordrecht.

Fox, M. W. (1971). *Behaviour of wolves, dogs and related canids*. Harper and Row, New York.

Frank, L. G. (1986). Social organization of the spotted hyaena (*Crocuta crocuta*). II. Dominance and reproduction. *Animal Behaviour*, **34**, 1510–27.

Frank, L. G., Davidson, J. M., and Smith, E. R. (1985). Androgen levels in the spotted hyaena: the influence of social factors. *Journal of Zoological Society*, London, **206**, 525–31.

Frank, L. G., Glickman, S. E., and Zabel, C. J. (1989). Ontogeny of female dominance in the spotted hyaena: perspectives from nature and captivity. In *The biology of large African mammals in their environment* (eds P. A. Jewell and G. M. O. Maloiy). *Symposium of the Zoological Society of London*, **61**, 127–46.

Glickman, S. E., Frank, L. G., Davidson, J. M., Smith, E. R., and Siiteri, P. K. (1987). Androstenedione may organize or activate sex reversed traits in female spotted hyenas. *Proceedings of the National Academy of Science*, **84**, 3444–7.

Henschel, J. R. and Skinner, J. D. (1987). Social relationships and dispersal patterns in a clan of spotted hyaenas, *Crocuta crocuta*, in the Kruger National Park. *South African Journal of Zoology*, **22**, 18–24.

Jenks, S. (1988). Behavioral Regulations of Social Organization and Mating in a Captive Wolf Pack. Ph. D. thesis, University of Connecticut, Storrs.

Kaplan, J. R. (1977). Patterns of fight interference in free-ranging rhesus monkeys. *American Journal of Physical Anthropology*, **47**, 279–88.

Kaplan, J. R. (1978). Fight interference and altruism in rhesus monkeys. *American Journal of Physical Anthropology*, **49**, 241–50.

Kranendonk, H. J., Kuipers, J., and Lensink, B. M. (1983). The management of spotted hyaenas, *Crocuta crocuta*, in Artis-Zoo, Amsterdam. The Netherlands. *Zool. Gart. N. F., Jena*, **53**, 339–53.

Kruuk, H. (1972). *The spotted hyena: a study of predation and social behavior*. University of Chicago Press, Chicago.

Kruuk, H. (1989). *The social badger*. Oxford University Press, Oxford.

Kuhme, W. (1965). Communal food distribution and division of labor in hunting dogs. *Nature*, **205**, 443–4.

Kurland, J. A. (1977). Kin selection in Japanese monkeys. In *Contributions to Primatology*, vol. 12, pp. 1–145.

LeBoeuf, B. J. (1970). Copulatory and aggressive behavior in the prepubertally castrated dog. *Hormones and Behavior*, **1**, 127–36.

Lindeque, M. and Skinner, J. D. (1982). Fetal androgens and sexual mimicry in spotted hyaenas (*Crocuta crocuta*). *Journal of Reproduction and Fertility*, **65**, 405–10.

Lindeque, M., Skinner, J. D., and Millar, R. P. (1986). Adrenal and gonadal contribution to circulating androgens in spotted hyaenas (*Crocuta crocuta*) as revealed by LHRH, hCG and ACTH stimulation. *Journal of Reproduction and Fertility*, **78**, 211–17.

Lockwood, R. (1976). An Ethological Analysis of Social Structure and Affiliation in Captive Wolves (*Canis lupus*). Ph.D. thesis, Washington University, St. Louis.

Malcolm, J. (1979). Social Organization and Communal Rearing in African Wild Dogs. Ph.D. thesis, Harvard University, Boston.

Martin, P. and Bateson, P. (1986). *Measuring behaviour: an introductory guide*. Cambridge University Press, Cambridge.

Massey, A. (1977). Agonistic aids and kinship in a group of pigtail macaques. *Behavioral Ecology and Sociobiology*, **2**, 31–40.

Matthews, L. H. 1939. Reproduction in the spotted hyaena, *Crocuta crocuta* (*Erxleben*). *Philosophical Transactions of the Royal Society, London* Series B, **230**, 1–78.

Mech, L. D. (1970). *The wolf*. Natural History Press, New York.

Mills, M. G. L. (1985). Related spotted hyaenas forage together but do not cooperate in rearing young. *Nature*, **316**, 61–2.

Mirsky, A. F. (1955). The influence of sex hormones on social behavior in monkeys. *Journal of Comparative and Physiological Psychology*, **48**, 327–35.

Packer, C. and Pusey, A. E. (1982). Cooperation and competition within coalitions of male lions: kin selection or game theory? *Nature*, **296**, 740–1.

Packer, C. and Pusey, A. E. (1983). Adaptations of female lions to infanticide by incoming males. *American Naturalist*, **121**, 716–28.

Rabb, G. B., Woolpy, J. H., and Ginsburg, B. E. (1967). Social relationships in a group of captive wolves. *American Zoologist*, **7**, 305–11.

Racey, P. A. and Skinner, J. D. (1979). Endocrine aspects of sexual mimicry in spotted hyaenas, *Crocuta crocuta*. *Journal of Zoology*, London, **187**, 315–26.

Rasa, O. A. E. (1977). The ethology and sociology of the dwarf mongoose (*Helogale undulata rufula*). *Zeitschrift für Tierpsychologie*, **43**, 337–406.

Rasa, O. A. E. (1983). A case of invalid care in wild dwarf mongooses. *Zeitschrift für Tierpsychologie*, **62**, 235–40.

Rood, J. P. (1974). Banded mongoose males guard young. *Nature*, **248**, 176.

Rood, J. P. (1978). Dwarf mongoose helpers at the den. *Zeitschrift für Tierpsychologie*, **48**, 277–87.

Rood, J. P. (1986). Ecology and social evolution in the mongooses. In *Ecological aspects of social evolution* (ed. D. I. Rubenstein and R. W. Wrangham), pp. 131–52. Princeton University Press, Princeton.

Rood, J. P. (1989). Male associations in a solitary mongoose. *Animal Behaviour*, **38**, 725–8.

Rood, J. P. (1990). Group size, survival, reproduction, and routes to breeding in dwarf mongooses. *Animal Behaviour*, **39**, 566–72.

Rood, J. P. and Waser, P. M. (1978). The slender mongoose, *Herpestes sanguineus*, in the Serengeti. *Carnivore*, **1**, 54–8.

Rudnai, J. (1973). *The social life of the lion*. Washington Square, Wallingford, PA.

Russell J. K. (1981). Exclusion of adult male coatis from social groups: protection from predation. *Journal of Mammalogy*, **62**, 206–8.

Russell, J. K. (1983). Altruism in coati bands: nepotism or reciprocity? In *Social behavior of female vertebrates* (ed. S. K. Waser), pp. 263–90. Academic Press, New York.

Seyfarth, R. M. (1977). A model of social grooming among adult female monkeys. *Journal of Theoretical Biology*, **65**, 671–98.

Seyfarth, R. M. (1980). The distribution of grooming and related behaviours among adult female vervet monkeys. *Animal Behaviour*, **28**, 798–813.

Tilson, R. L. and Hamilton, W. J. (1984). Social dominance and feeding patterns of spotted hyaenas. *Animal Behaviour*, **32**, 715–24.

Tsutsui, K. and Ishii, S. (1981). Effects of sex steroids on aggressive behavior of adult male Japanese quail. *General Comparative Endocrinology*, **44**, 480–6.

Walters, J. R. (1980). Interventions and the development of dominance relationships in female baboons. *Folia Primatologia*, **34**, 61–89.

Woodmansee, K. B., Zabel, C. J., Glickman, S. E., Frank, L. G., and Keppel, G. 1991. Scent marking ('pasting') in a colony of immature spotted hyenas (*Crocuta crocuta*): a developmental study. *Journal of Comparative Psychology*, **105**, 10–14.

Zajonc, R. B. (1965). Social facilitation. *Science*, **149**, 269–74.

Zimen, E. (1976). On the regulation of pack size in wolves. *Zeitschrift für Tierpsychologie*, **40**, 300–41.

6

Segmentary 'warfare' and the management of conflict: comparison of East African chimpanzees and patrilineal–patrilocal humans

CHRISTOPHER BOEHM

INTRODUCTION

Political coalitions among non-human primates are normally studied in the context of competition with communities of individuals who forage and socialize together. When such species collectively express hostility toward conspecific strangers, they may be said to operate through territorial '*macro-coalitions*'. That is, the entire territorial community, in spite of internal divisiveness, becomes one large coalition which cooperates to compete with other such macro-coalitions on a xenophobic basis.

Xenophobic behaviour is encountered rather frequently among the higher primates, including humans (Holloway 1974; LeVine and Campbell 1972). At a relatively simple level of social organization, gibbons live territorially in nuclear-family groups and exhibit behaviours hostile to outsiders. At the most complex level, modern humans living in socially complex nations or empires may direct their xenophobic enmities at people of different communities, nations, races, religions, classes, political ideologies, or, in some highly imaginative cases, even planets. I shall explore the relation of coalition formation to xenophobic behavior in highly male-bonded primate groups by comparing two species: (1) East African chimpanzees (*Pan troglodytes schweinfurth*; see for example Goodall 1979, 1986; Nishida 1979; Nishida *et al.* 1985); (2) humans (*Homo sapiens*) who live in egalitarian societies, this species' simplest form of political organization (Service 1975; see also Cashdan 1980; Woodburn 1979, 1982; Knauft 1991). For humans, I shall be concerned more specifically with egalitarian societies that are patrilineal and patrilocal.

A close comparison of macro-coalitions in East African chimpanzees and fraternally-oriented humans will be of interest not only in view of phylogenetic propinquity and similarity of social organization, but also in view of the significant differences in cultural level. Tools for analysis will include the extensive social dominance theory and coalition theory developed within ethology and primatology, but also anthropological theories of ethnocentrism formulated by LeVine and Campbell (1972), the fraternal-interest-group theory elaborated by Otterbein and Otterbein (1965), several ethnological theories of political segmentation, and an amplified theory of conflict management derived from political anthropology (see for example Gluckman 1955*b*; Hoebel 1967).

XENOPHOBIC FRATERNAL COALITIONS OF CHIMPANZEES

East African chimpanzees live in territorial communities with partial female transfer between communities but a very strong relative absence of male exchange (Nishida 1979; Goodall 1986). The result is a group of usually about half a dozen to two dozen adult males, very closely related genetically on the patrilateral side and closely related or distantly related through matrilineal ties, who must stay together for life, along with a mixed complement of natal and non-natal females. While the biological father–son relation appears to be devoid of any very individualized social consequences, and the brother–sister bond is not very strong, mothers bond strongly with their offspring and male siblings tend to develop associations (Goodall 1986).

In the wild, during intragroup competition brothers either join each other in political coalitions or at least abstain from joining a coalition hostile to their male sibling, while unrelated males also form coalitions and do so frequently (Goodall 1986). This last finding is corroborated by de Waal (1982) for captives. These potentially divisive coalitions routinely complicate the social dominance hierarchy, as males work singly or in partnership to advance their individual rank-positions.

However, in spite of this divisiveness within the community, wild chimpanzees, acting as well-unified territorial units, are among the most aggressively xenophobic non-human primates. At Gombe National Park in Tanzania, where Goodall has been working with the same study group for 30 years, the group's males form territorial macro-coalitions which tend to range in size from perhaps half of the adult males to all of them. They undertake collective forays ('patrols') during which strangers found individually or in small groups are stalked and savagely attacked; whereas, when two large patrols meet, they merely vocalize bellicosely and slowly withdraw (Goodall 1986).

In 16 months' experience of observing chimpanzees at Gombe, I have accompanied the males on several patrols, during which typically all individuals remained silent (see Goodall 1986). Within these special-purpose groups I have seen no evidence of agonistic behaviour that might call internal coalitions into action and disrupt the patrol or betray its presence. In terms of coalition theory, intracommunity competitiveness among individual males and coalitions of males appear to be suppressed and set aside, if only temporarily, in favour of a larger-scale competitive foray which requires temporary cooperation among intracommunity competitors.

This ability to form coalitions at two different levels provides political flexibility of a type that Durkheim (1933), referring to humans who live in 'tribes', has called *segmental*. Analyses of specific 'segmentary societies' such as the Nuer (Evans-Pritchard 1940) or the Tiv (Bohannan 1954), two African societies exhibiting structural features replicated elsewhere in the world, demonstrate that tribal societies having little political centralization can mobilize very flexibly even in times of internal political conflict, and do so at a number of different levels. For example, the only permanently mobilized political segments might be localized clans that feud (for example, Boehm 1986), but a cluster of competing clans might quickly mobilize as a 'tribe' which would gather to protect its common territory from another tribe, while a confederation of such tribes might be called together if any one of them was attacked by a different confederation. Characteristically, when the external pressure goes away the original conflict is resumed at a lower segmentary level.

Kummer (1971) has shown that the social organization of hamadryas baboons in Ethiopia (*Papio hamadryas*) is structurally similar to human segmentary-tribal political systems, in that a hierarchy of units operates so as to distribute hamadryas individuals into groups and subgroups that flexibly provide needed foraging effectiveness and predator protection. Basic is the harem, which constitutes mainly a response to open-country predator pressure combined with a need to spread out during foraging, while the largest segment, the troop, appears to be a response to predator pressure while sleeping.

With chimpanzees, according to Goodall (1986) the individual male or the individual mother with her sub-adolescent offspring are the essential permanent units (comparable to harems with the hamadryas). Then there are a multitude of *ad hoc* foraging groupings within the community that change very flexibly from day to day, and also temporary male–female consortships; adult political coalitions which may last for some time; and finally the males of the entire community as a recurrent territorial 'scouting' and defensive unit. With chimpanzees, at Gombe the segmentation into individual-adult based units would appear to be a maximally flexible response to foraging needs, while the largest units are directly related to competition between neighbouring communities. Because chimpanzee coalition formation also is rather similar to human political behaviour in this sphere, a more detailed comparison of the two will be made later.

XENOPHOBIC FRATERNAL COALITIONS OF HUMANS

Service (1962), Fried (1967), and Tiger (1969) have pointed out some interesting features of human society with respect to political division of labour by the sexes. With humans who live in small-scale societies that compare in size and complexity with chimpanzee territorial communities it is primarily the males who carry on the business of political competition within the community, be it at the 'band' or 'tribal' level, as one individual, household, lineage, or other coalition vies with another. Generally it is males who are group leaders and bear the primary physical responsibility for defence, raiding, feuding, and offensive warfare if this is present.

If one focuses on both social structure and coalition behaviour, the human groups most similar to East African chimpanzees are the warlike non-literate societies that feud, that is, those in which closely related males stay in their natal groups for life and develop very close bonding. Otterbein and Otterbein (1965) have shown that feuding, with its limited, reciprocating homicidal retaliation between groups, is most likely to develop among exogamous patrilineal groups with patrilocal postmarital residence. This arrangement ensures that closely-related males will remain coresident or live contiguously for life, while females are exchanged among various patrilineages or patri-clans. They have characterized these groups as 'fraternal interest groups', and have argued convincingly that a group of coresident males who are closely related (e.g. a clan within a tribe) are prone to act politically as a single unit. Societies based on fraternal interest groups readily develop a politically-competitive warrior mentality, and kin groups may compete as coalitions at many different levels.

Formally, the Otterbeins' theory is one of male bonding (see also Tiger 1969) among closely related kinsmen: a combination of contiguous residence, everyday cooperation, and a fraternally-conceived kinship ideology results in close bonding and also in a particularly 'aggressive' sense of social solidarity. This promotes mutual protection and assertive competition by intracommunity kin-coalitions and, when these set aside their differences and join together, against similar neighbouring communities.

The result is a male group, the members of which, while aggressively competitive among themselves as warriors, are bonded from childhood, are used to cooperating, are well-motivated to think of themselves as a single group in competition with similarly aggressive groups, and therefore are accustomed to setting aside their internal competitions in the interest of avoiding political divisiveness when faced with an external threat. Due to all of these factors, such groups can be extremely destructive in all-out warfare. They generally raid their neighbors and feud with them, and many of them also conduct either 'symbolic' (that is, ritualized) or intensive warfare. These different types of xenophobic behaviour will be discussed later in some detail.

A further interpretation of all feuding systems is that feuding controls intracommunity divisiveness and serves as social control in that certain antisocial behaviours likely to provoke a divisive, disruptive, feud become very costly (see Gluckman 1955*a*; Black-Michaud 1975). Another interpretation is that feuding constitutes a far less disruptive substitute for warfare at close quarters, and that the indigenous actors are aware of this functional benefit (Boehm 1986). Normally, hostile feuding groups exchange homicides quite systematically, with each group killing one or a few individuals at a time, by turns. By using this alternating system, which is based on custom, large-scale mutual military attacks can be avoided with their potentially very high costs to both sides. The assumption is that feuding systems arise because the humans involved realize that all-out warfare at very close quarters easily becomes disastrous for both sides: a minus-sum 'game', as it were. For this reason, tribesmen all over the world have developed very similar sets of rules to partially control their intergroup conflicts by feuding. Raiding, like feuding, tends to result in low-level conflict rather than warfare; but raiding is far less rule-bound.

Let us return now to the matter of political segmentation. In carrying on feuding and warfare, human communities may form segments at a variety of levels. Sometimes, as with chimpanzees, there is just the local community which sets aside internecine competition among political coalitions to confront its similar neighbour as a macro-coalition. But sometimes, as discussed earlier, such macro-coalitions themselves form segments within larger political segments, which may combine to form still higher-level, ephemeral special-purpose segments when external political pressures so dictate (see for example Evans-Pritchard 1940). With humans, the (usually patrilineally-based) principle of segmentation is rationalized genealogically as follows: you set aside a conflict with closer kin to close ranks against a group genealogically more distant (Bohannan 1954).

While human segmentation may balance power and help to keep a cluster of territorially competing tribes distributed efficiently over a piece of real estate in a fairly stable manner, the segmentary-lineage also can become an 'instrument of predatory expansion', particularly when it shares frontiers with societies that lack their own segmentary organization (Sahlins 1961). Later, I shall discuss expansionist aggression and its demographic effects, when these human behaviours are compared in more detail with certain behaviours of chimpanzees. But first, it will be useful to examine some other general similarities between the two species which are relevant to that comparison.

FUNDAMENTAL BEHAVIOURAL SIMILARITIES

I begin the two-species comparison by concentrating upon what I consider to be fundamental, homologous similarities of behaviour. These are similar genotypic dispositions, assumed to have been shared by a common African ancestor, which through adaptive modification (Kummer 1971) result in similar behaviours with similar functions. Some relevant examples are female exogamy, absence of female alliances, hostile relations between groups activated by males with stalking and attacking (see Wrangham 1987). Such homological reasoning is particularly appropriate when one is comparing chimpanzees with humans (Ghiglieri 1987). Let us begin with genotypic behavioural dispositions that are fundamental to the fraternal interest group as a well-bonded unit that comes into conflict with other such units.

Both species are capable of stalking and killing conspecifics (Wrangham 1987), but, perhaps more basically, males growing up in close proximity tend to bond strongly (see Tiger 1969). In both species, natural selection will favour individuals who enter into cooperative xenophobic activities (e.g. chimpanzee feeding forays into overlap zones or human raiding forays) because the participating individual directly increases his own individual fitness by acquiring access to extra food and, possibly, females. Ghiglieri (1987) has discussed these and other genetic ramifications for the two species. When chimpanzees exploit food resources used by other communities, it is to an individual's advantage to do so with a group too large to be attacked, particularly at times of food scarcity when the intensity of competition is likely to be high. If he can participate successfully in this pattern, not only will he help himself and his closer kin, a matter of inclusive fitness, but his entire group will benefit. If an individual male cannot fit into this pattern he will, especially in times of food shortage, be at a reproductive disadvantage compared to members of his own group who can join such feeding forays. He also could be at a direct breeding disadvantage, since currently available estrous females very often accompany patrols, and it is patrols that first copulate with reproductively attractive stranger females (Goodall 1986).

As at Gombe, at the long-term field site of Japanese scientists in Tanzania, in the Mahale Mountains, a smaller group has been all but exterminated by a larger group (Nishida *et al.* 1985). This apparently took place through recurrent attacks by larger groups on one or a few strangers, while male behaviours that suggest a similar territorial orientation were also observed with unprovisioned chimpanzees in Uganda's Kibale Forest (Ghiglieri 1984, 1988). Chagnon (1988) has documented similar occasional heavy attrition among humans, caused by conflict between groups.

There are some other general similarities. With respect to political segmentation as discussed above, in both species social dominance hierarchies are readily formed (see Nishida 1979; Goodall 1986; Omark *et al.* 1980) in which closely-related and closely-bonded adult males who are strongly oriented to political competition vie among themselves individually and as factional coalitions, but are labile enough to operate politically in macro-coalitions which unite an entire local community against similar communities.

Because both species are flexible in their behaviour, these large 'emergency coalitions', which in East African chimpanzees are xenophobic and essentially territorial and which in humans have significant territorial or xenophobic components, predictably take precedence over intracommunity conflicts, as discussed above. Thus, for chimpanzees, the two levels of political segmentation are coalition-based 'factions' within a territorial community and then that community as a whole, while among humans a frequent structure is to have the various localized patriclans within a tribal area vying as coalitions until an external threat unites the tribe.

In effect, continuously competing in small coalitions keeps the males 'in practice', as it were, as political competitors who learn to understand intergroup power plays quite well and who are also able to cooperate. Both factors make it possible to mobilize the entire community quickly and effectively so it can compete competently with other xenophobic communities that act as macro-coalitions. In the next section, I explore an important type of conflict management behaviour that helps to make such predictable mobilization into macro-coalitions possible.

INTEGRATIVE BEHAVIOURS THAT FACILITATE MACRO-COALITION FORMATION

In discussing human ethnocentrism, LeVine and Campbell (1971) emphasize that internal political divisiveness can often be set aside when an external threat is perceived. Undoubtedly an important factor in this capacity to neutralize lower-level internal dyadic conflicts and unite the entire community is the observed tendency of both chimpanzees and humans to resolve many of their internecine fights and other conflicts, sometimes intervening actively and purposefully with pacification in mind. An example would be the tribal Montenegrins in their Balkan refuge area (Boehm 1983), a human segmentary feuding society that was subject to massive external political pressure and loss of many women and children in a decisive defeat. Because feuding was endemic within the tribe and also between tribes, active conflict management was critical to forming defensive tribal alliances whenever there was the emergency need to act in concert, defensively. In segmentary tribal systems such as the Yanomamö (Chagnon 1983), who faced no powerful external military predator, active inter-tribal conflict management tends to be less prominent because fissioning that weakens the group, though costly in terms of self-defence, is usually less catastrophic.

Under the rubric of 'peacemaking', de Waal (1989) has discussed a set of primate conciliatory behaviours that have obvious and important implications for group social solidarity. For chimpanzees the focus was on reconciliation and reassurance behaviours, which are dyadic modes of 'patching things up' (see also Goodall 1986; de Waal and van Roosmalen 1979). Two individuals will touch hands, or else the dominant individual will touch the body of the anxious subordinate in reassurance. As field ethnographers know, such reassurance and reconciliation may be accomplished in human groups of comparable size by means of a combination of informal verbal and non-verbal communication, or by rituals. However, when ethnographers speak of 'conflict management', they tend to be thinking mainly about third-party mediation of internal conflicts. I extend the concept here to include not only triadic intervention in internal conflicts and reconciliation behaviour, but also purposeful manipulations of conflicts that involve two separate communities.

With chimpanzees, the other means of 'making peace' is for a third party to interfere in a dyadic conflict between adults. For many non-human primates, such interventions have been reported anecdotally; however, there are studies of macaques and chimpanzees in large captive groups (see for example, Kaplan 1977; de Waal 1978, 1982) which suggest that the effects of alpha male intervention can go in several directions, i.e. in tactics (support versus non-support) and in results (pacification versus agonistic exacerbation). Such interventions are discussed elsewhere in this volume by Ehardt and Bernstein, pp. 84–9, and de Waal, pp. 241–3).

With wild chimpanzees at Gombe, the majority of interventions in group-internal conflicts between adults take place when a higher-ranking individual, very often the alpha male, intervenes in a developing conflict or fight using a technique that suggests the intention is to stop the fight, and which has that effect. Such intervention takes place in about 10 per cent of the more serious agonistic bouts recorded. Because, as with humans, such interventions can so often be interpreted as carrying the intention to reduce the level of agonism, I shall refer to them as 'pacifying interventions'.

Among chimpanzees, a pacifying intervention characteristically proceeds as follows. Two adults have begun an agonistic bout, and the alpha male simply charges toward the protagonists (non-support tactics) and they scatter before he reaches them; or, if they are not engaged in physical contact, he may simply display between them (non-support tactics) with the same effect; alternatively, he may charge them and briefly attack the aggressor (support tactics) or briefly attack both of them (non-support tactics), and scatter them. His actions have two predictable effects. First, the protagonists become separated spatially. Second, their tendency to rejoin the conflict is inhibited by the presence of the intimidating interferer, who generally remains nearby but attacks neither party. Thus, the result is pacification.

In nearly half of the interventions that have pacifying effects, the overall strategy is what I have called 'impartial' (see Boehm 1981) and the tactics are those of non-support, i.e., the aggressive interfering behaviour is directed at both protagonists simultaneously, or sometimes briefly at each individual serially. In the remaining pacifying cases, the overall strategy is 'partisan': normally the victim is supported by a very brief display or attack directed at the aggressor, which is quickly cut off as soon as separation and inhibition are accomplished. This involves tactics of support, but in partisan pacifying interventions the strategy still appears to be to stop the conflict insofar as minimal force is exerted. Effective pacification takes place more frequently when the strategy is impartial and the tactics are those of non-support, but even partisan pacifying interventions have a high success rate, and in any event do not exacerbate levels of agonism.

These same high-ranking individuals occasionally enter into ongoing conflicts actively as fighters using tactics of support, and thereby exacerbate or decide the conflict, sometimes helping the aggressor. However, in wild chimpanzees such non-pacifying support interventions can be differentiated from interventions in which the higher-ranking individual mixes in a conflict with the apparent intention of reducing the level of agonism, be the tactics those of support or non-support. This is evident to an experienced observer watching the overall strategy, tactics, and especially the timing behaviour of the high-ranking interferer.

It can be generalized that such interventions effectively reduce levels of politically divisive agonism within the territorial community. Table 6.1 presents data collected by Goodall's Tanzanian field assistants at the Gombe Stream Research Centre over an 8-year period, during all-day follows of

Table 6.1. Pacifying non-support interventions at Gombe over 8 years. (Data courtesy of Gombe Stream Research Centre.)

Year	Alpha male	Other male	Females	Total
1976	—	3	2	5
1977	2	1	1	4
1978	6	2	1	9
1979	4	2	—	6
1980	1	1	—	2
1981	—	—	—	—
1982	5	3	2	10
1983	2	1	—	3
Totals	20	13	6	39

selected chimpanzee targets over all parts of their range. (During this period provisioning was moderate and individualized to avoid induced agonism taking place in camp; Goodall 1986.) In these instances, the Gombe chimpanzees intervened in what has been called a control role (Bernstein 1970; Ehardt and Bernstein this volume, pp. 84–9), but might better be defined as a 'pacifying control role', along the lines suggested for captive chimpanzees at the Arnhem Zoo (de Waal 1982; this volume, pp. 241–3). These impartial pacifying interventions were accomplished by acting dominantly in relation to an agonistic dyad that contained at least one adult; I emphasize that interferers did not single out just one member of the dyad to attack. Such intervention took place in about 10 per cent of the fights reported, and it predictably had the effect of interrupting the aggression and ameliorating the conflict.

Impartial pacifying interventions did not follow a kin-oriented strategy, nor did they favour dyads containing allies (see also de Waal, this volume, pp. 241–3). The behaviour appeared to be quite flexible, insofar as I have shown that it was by no means limited to a single set of tactics. It can also be seen that this tendency to intervene in a pacifying mode was not limited to alpha males, nor even to males, since females accounted for nearly one-sixth of such interventions and adult males ranking below the alpha accounted for another third; who intervenes seems to depend on the individual present at the agonistic encounter. According to my reading of these data, at Gombe interventions based on an impartial-pacifying strategy involved a flexible, purposeful behaviour that emerges in a variety of agonistic contexts, including incipient agonistic bouts without physical contact, but also ones with serious fighting.

These detailed written observations show that in this fission–fusion society, in which alpha males are not necessarily present in the community's sub-groups, other high-ranking individuals may flexibly assume a pacifying control role, as needed, when other adults become involved in agonistic bouts. There are additional data that are pertinent, collected on videotape by the author and four Tanzanian field assistants from 1987 to the present, with a research focus on developmental aspects of conflict intervention. Preliminary analysis suggested that they follow the same pattern: intervention by higher-ranking individuals can decisively interrupt developed fights (including ones likely to result in wounding), and it seems likely that pre-emptive interventions also are heading off some potentially serious conflicts that have not yet reached a stage of physical contact. However, some interventions appear to be directed at mere 'squabbles'.

In adult fights that involve physical contact, one strategy used by alpha male Goblin was to make brief serial attacks on two adult females that served to separate them; this was accomplished in pacifying a fight that began high in a tree. Goblin was also reported to be using this strategy several times in the period 1976 to 1983, before videotaping was possible; de Waal (1982,

p. 124) reports a similar intervention with a larger group fighting. In another intervention, observed by the author in 1985, just before videotaping began, the fight involved two adolescent males with protracted grappling, intense screaming by both parties, and what appeared to be a serious attack. In alpha male Goblin's absence the number two male Satan actually used the great power of his arms to slowly pry apart the two combatants, after his charging directly toward them had no effect. Such a prying tactic had never been observed previously by either Jane Goodall over 25 years or by the Gombe Stream Research Centre's Tanzanian field staff, some of whom had been conducting day-long 'follows' of individual chimpanzees for 10 years. However, de Waal (1982, p. 124) reports a similar intervention in a large captive group.

I have not detailed the full range of tactics observed. But it is clear that impartial pacifying intervention is a highly flexible conflict-management behaviour insofar as a range of tactics are employed. It is clear also that it appears to be purposive. After successfully damping the conflict, often an individual acting in the pacifying control role withdraws and sits down, or else returns to a prior activity such as feeding, but appears to be ready to intervene again should the conflict resume, while fights are regularly headed off, stopped, or ameliorated when these methods come into play.

Elsewhere (Boehm 1981) I have discussed possible motivations that could be operating in impartial pacifying interventions of non-human primates; while many strategies have been proposed, including protectiveness, helping allies who offer possibilities of reciprocation, and simply reinforcing the interferer's rank through intimidation, I also suggested that there could be some generalized hostility directed at the source of disturbance that was based on an appreciation of normal social harmony. In the case of high-ranking chimpanzees who intervene impartially, I think the more likely motives are protectiveness toward one or both protagonists and, quite possibly, some general aversion to agonistic disturbance.

With the Gombe chimpanzees, a protective motive seems likely since by preliminary analysis there appears to be a relatively higher incidence of interventions in conflicts of juveniles and infants, mostly resulting from play, than for adults. It is worth emphasizing that this protectiveness does not necessarily follow kin lines; mother chimpanzees sometimes protect other youngsters from their own larger or more aggressive offspring. It is also worth emphasizing that partisan pacifying interventions in adult conflicts, which protect victims but really protect both protagonists since mutual wounding is possible, did not favour socially-bonded kin or coalition-allies.

The second motivational possibility, more difficult to substantiate, is that interferers also may be reacting to agonistic encounters as an unwelcome perturbation of a perceived equilibrium in the social environment. Indeed, many of the videotaped chimpanzee interventions of juvenile/infant squabbles that I am currently analysing are stimulated by very noisy agonistic episodes judged not likely to lead to a severe fight, and the same seems to be true of certain adult conflicts.

With humans, protective motivations undoubtedly play a part in group-internal conflict management; but so apparently do 'quality of life' considerations (e.g. Fürer-Haimendorf 1967), as leaders of small groups try to pacify conflicts even before they become severe enough to threaten anyone physically. However, a different motivational hypothesis to explain these pre-emptive intervention behaviours in both chimpanzees and humans is that interferers intervene early as a practical matter, because fights are far easier to break up before they involve physical contact, and this eventual danger to the victim (and, potentially, to the aggressor) is readily anticipated. Obviously, in either species both pre-emptive protectiveness and annoyance could be present simultaneously.

With humans, internal conflicts in complex societies such as states or primitive kingdoms may be handled through a combination of non-support mediation and legal intervention based on coercive force (Service 1975). However, if the autonomous political unit remains small and egalitarian then intervention remains wholly non-coercive, even though strong group social pressure may be brought to bear upon the disputants and leaders may use their limited influence as peacekeepers, hoping to keep the group together. This is true even in nomadic bands that have very weakly developed leadership, and in which the solution for most serious conflicts is for one party simply to join a different band (for example, see Fürer-Haimendorf 1967). Reconciliation behaviour also plays its part in damping the effects of quarrels, but as de Waal (1989) points out, this aspect of human conflict management has been generally neglected by anthropologists. The literature on feuding and payment of blood-money is a special exception.

A major difference with humans is that when they live in small groups, people regularly arrive at a consensus about certain behaviours that are to be encouraged or discouraged through social control. Universally, it tends to be behaviours that cause social disruption and group divisiveness that are negatively sanctioned. Elsewhere (Boehm 1982), I have hypothesized that from the standpoint of cultural evolution, a chimpanzee-like ancestral triadic-intervention pattern that followed a pacifying intervention strategy gave rise to 'morality', when people moved from direct intervention in imminent or developed conflicts to a far more anticipatory approach to social control. Humans did this by labeling the behaviours that predictably caused conflicts, and by suppressing those behaviours generically by means of group-wide sanctioning.

By contrast, wild chimpanzees do not seem to control the behaviours of individuals through commonly agreed upon rules of conduct (Goodall 1982). In spite of this difference, there are some very important similarities between humans and chimpanzees in this area of protective intervention behaviours that tend to pacify conflicts, and which thereby contribute to social equilibrium and help to make possible the ready formation of politically effective macro-coalitions.

Such behaviours contribute to group size and fighting strength in humans by keeping group males from killing each other. They also do so by reducing emigration of individual males and, very important, they inhibit premature group fissioning, both of which further reduce group size and invite territorial vulnerability. Likewise with chimpanzees, in damping conflicts and thereby reducing overall levels of agonism, intervention behaviour seems to play an important part in keeping territorial communities numerically strong, but also in keeping chimpanzees within such communities prepared to quickly call off their internecine conflicts and make expedient allies out of recent enemies. It might also help to retard group fissioning as with humans. Such flexibly opportunistic political behaviour, common to chimpanzees and humans, is the essence of a segmentary-tribal way of life.

This flexibility usually enables macro-coalitions to coalesce in a unified military mode in confrontations with other groups, in spite of ongoing internal divisiveness. However, while with humans who live in segmentary tribes the conflict resolution is sometimes deliberately geared to patching up internal divisions for the explicit purpose of making a united stand against outsiders, given our present knowledge such a politically sophisticated motivation seems very unlikely in wild chimpanzees. They would appear to be pacifying conflicts because they wish to protect one or both protagonists, or because they are otherwise disturbed by the agonism.

DETAILED COMPARISON OF MACRO-COALITION BEHAVIOURS

I now compare macro-coalition xenophobic behaviours of the two species, measuring the behavior of chimpanzees with a rough typological yardstick developed by the many political anthropologists who study humans, that I have refined somewhat. Reference points (see Boehm 1986) will be warfare, 'symbolic' or ritualized warfare, raiding, feuding, and creation of macro-alliances which tend to balance power.

I shall begin with *Pan troglodytes*. However, before discussing specific behaviours, I would like to express some concerns about the two East African study sites on which I rely so heavily. Provisioning and its possible effects at Gombe (see Wrangham 1984) and Mahale pose potential problems for using these two sites as 'pristine' models for wild chimpanzee territorial behaviour. Resolution of this question will not be easy, given the fact that it takes so many years, even with provisioning, to understand a group well enough to specify its main territorial behaviours. In the absence of provisioning, obviously this time span is likely to lengthen due to slower habituation. Thus, the fact that Ghiglieri (1984, 1988) in his two-year unprovisioned study discerned all of the manifestations of territorial behaviour aside from actual defence of territory could be quite significant.

Ghiglieri (1988, pp. 258–9) has summed up the overall case for chimpanzees being territorial. He also has raised an issue that I shall try to resolve in making this comparison: a point on which I would significantly disagree has to do with this statement, referring not to Kibale but to Mahale and Gombe, that 'in Tanzania it was true warfare'. This seems to over-amplify Goodall's (1979, 1986) more modestly stated reaction to the savage attacks and community level annihilation at Gombe, which she refers to as being near the threshold of war.

Webster's Seventh New Collegiate Dictionary (1963, p. 1003) defines 'war' in the first instance as '1a: a state of usually open and declared armed hostile conflict between states or nations', and this wording indicates that both sides (by implication, the males) may be making lethal attacks at the same time, as does usage of the term in political science, history, and political anthropology. Ghiglieri's usage would appear to reflect a very generalized second sense of the word that is given '2a: a state of hostility, conflict or antagonism'. One thing that chimpanzees apparently lack, as inferred from behavioural descriptions so far by Goodall (1979, 1986) and Nishida (1979; see also Nishida *et al.* 1985), is *simultaneous* conflict between two entire neighbouring territorial groups. So even if we grant chimpanzees their formidable and potentially lethal jaws and teeth as 'arms', there remains the crucial question of whether there are simultaneous lethal attacks involving two entire groups.

Later in the chapter, because of the great philosophical importance of the issue of 'chimpanzee warfare', I shall further elaborate an anthropological definition of warfare on a more technical basis, since I am convinced that semantic fine points can be important when the aggressive acts of chimpanzee macro-coalitions are compared with those of humans. But I begin that discussion now by taking the position that a 'true warfare' pattern involves intensive, large-scale conflicts by fully mobilized male groups whose attacks are bilaterally aggressive, and involves lethal intentions and results. This can be defined as *intensive warfare* to differentiate it definitively from Webster's second meaning and to amplify the first. Such behaviour simply has not been observed so far in wild chimpanzees, since observed xenophobic lethal attacks have always involved lopsided unilateral aggressions.

There remains some ambiguity, also, in Ghiglieri's reference to Goodall's (1986) intepretations that chimpanzees have feelings and intentions sufficient to motivate killing when they attack isolated enemies caught by patrols. Chimpanzees seem to aim at mere 'serious disablement' both in hunting and in xenophobic attacks, and they appear to have a relatively crude appreciation of biological death when one considers the way that mothers behave with dead infants (see Goodall 1986). It may well be that in their savage and prolonged attacks on strangers (see Goodall 1986) they feel extreme hostility, but have not quite added up the issue of life and death cognitively.

At Gombe a second special consideration that has received far less attention than 'provisioning' is the fact that the study-group has been sandwiched between two other groups that for some time have had no other chimpanzee neighbours. Since the small rectangular park is circumscribed by a lake and now by cultivated lands, for the two end groups the central study group's area presents the only possibility for expansion. A likely effect would be to increase patrolling behaviour, particularly for the study group. A related factor is human diseases. Probably because tourists often do not avoid physical contact with curious juveniles, the study group has twice been subject to lethal epidemics (polio and pneumonia, respectively) that correlated with similar human epidemics in the area. Fatalities reduced population pressure on a finite area that allows no expansion unless humans are raided, but it is not known whether the diseases spread to the two unhabituated groups. Given these special problems, it is fortunate that the much larger and more isolated area in the Mahale Mountains has been studied so thoroughly. It is interesting that the similarities in territorial behaviour remain so striking; one might assume that some powerful force, such as a phylogenetic propensity, was driving that behaviour.

Territorial competition of this species involves an assortment of behaviours. To begin with, adult community members understand the territorial situation because they know who utilizes which natural resources. Secondly, they realize that individuals of neighbouring communities are dangerous. In spite of this, at Gombe certain peripheral females with offspring may live close to overlap areas, and males who take estrous females on consortship may use these areas, as may other individuals or small foraging groups (Goodall 1986). All these individuals experience some risk.

The territorial behaviour germane to the analysis of macro-coalition is, of course, the 'patrol'. Goodall (1986, p. 489) has made the important distinction between 'excursions' to peripheral areas, characterized by intention to feed, normal group composition, and the use of vocalizations to test for nearby enemies; and the 'patrols', which are characterized by silent, wary travel in a scouting mode on the part of all-male groups (sometimes accompanied by estrous females), and not necessarily an intention to feed.

The apparent immediate purposes of chimpanzees on patrol would be to monitor movements of a neighbour group, to search for nulliparous stranger females, to attack all other intruding single strangers in or near overlap zones, and sometimes, if the coast is clear, to safely utilize food resources at (or, sometimes, beyond) usual territorial 'boundaries'. Patrolling by chimpanzees operating in macro-coalitions also has the probable effect of defending habitually-exploited food resources from enemy excursions and patrols, and it seems possible that there may be some immediate awareness of these stakes.

Table 6.2, adapted from Goodall (1986, p. 491), is based on nest-to-nest continuous 'follows' that, unless targets are lost, last up to 12 h. These data establish that over an 8-year period, out of 758 long follows (6 h or more),

Table 6.2. Gombe chimpanzee forays to boundary areas 1976–83. (Adapted from Goodall 1986.)

Year	1977	1978	1979	1980	1981	1982	Totals
Follows	109	224	155	93	100	77	758
Boundary visits	19	32	35	21	11	16	134
Patrols	5	10	11	7	3	2	38
Excursions	7	14	20	12	6	6	65
Meet stranger female(s)	3	6	12	6	2	2	31
Meet stranger male(s)	n/a	9	7	7	5	9	37

134 (17 per cent) involved visits to boundaries, while 38 (5 per cent) also involved patrols. Over the same period, there were 129 encounters with stranger females, and 182 cases of visual or vocal contact with stranger males. It seems clear from these figures, that continued use of natural resources is constantly being 'negotiated' by group members on both sides who go on excursions and patrols.

As defined in Webster's Seventh New Collegiate Dictionary (1963, p. 618) the military term 'patrol'—as a verb meaning to 'traverse' or 'guard'—aptly describes the general behaviour of Gombe groups as they travel warily toward enemy territory. But, as will be seen below, exploitation of food sources, capturing or accepting females (and sometimes even their offspring; see Nishida and Hiraiwa-Hasegawa 1985), attacks on strangers, and confrontations with large groups may be better analogized to other types of human political behaviour.

Let us now turn to our own species. As a manifestation of conflict management emerging at more of a macro-level, humans who live in small-scale, locally autonomous 'tribal' societies often manage to avoid bilateral large-scale combat with close neighbours. They do this by restricting a conflict to feuding, by restricting attacks to raiding, by engaging sometimes in ritualized conflict that is mostly 'symbolic', and, very importantly, by forming alliances in a way that balances power. Keeping in mind that feuding, raiding, and warfare are nothing more than useful basic 'types', and that frequently there exist variants of actual behaviour that blend these types (see Boehm 1986), I shall compare each of these three patterns for humans with chimpanzee patrolling parties and their confrontations.

First, there is the case of *feuding*. Elswhere, I have emphasized that feuding, because it is so highly rule-bound, is consciously designed by humans to keep some control over local retaliatory conflicts so that full-scale warfare does not erupt at close quarters. Thus, in a feuding pattern humans express both their tendency to retaliate violently, and their ability to manage the resulting conflict by setting up a code of behaviour to follow. This can be considered a form of internal conflict management.

Chimpanzees, too, are capable of vindictiveness (de Waal and Luttrell 1988; de Waal this volume, pp. 236–9) and they have excellent memories (Goodall 1986). Furthermore, adjacent communities from time to time stalk and kill solitary neighbours or neighbours in very small groups, and they appear to do so with a generalized feeling of strong enmity toward strangers, if not a sense of specific retaliation directed at targeted individuals. On the face of it, one might say that this pattern of limited and recurrent 'panicide' by two political communities is roughly comparable to feuding of humans. However, this chimpanzee behaviour differs significantly from feuding, in which retaliatory hostilities are exchanged between two corporate social groups with each side alternately taking the role of aggressor or defender, and relying upon careful score-keeping combined with rituals designed to interrupt or end the conflict (Boehm 1986). Indeed, by definition of the physical situation members of a chimpanzee community are not likely to learn about it when isolated members of their group are picked off, so any very specific, rule-bound system of alternating retaliation would be logistically impossible even if chimpanzees could somehow duplicate it with their available cultural tools.

While chimpanzees do not appear to feud, their xenophobic macro-coalition behaviour is quite similar to many aspects of human raiding and warfare. Human *raiding* can involve motives of enmity or vengeance, and it shades into warfare. But essentially, a raid is conducted into enemy territory by a relatively weak force which relies upon coordination, stealth, and an element of surprise, either to quickly kill a few of the enemy or to quickly take away readily transportable commodities such as livestock or, not infrequently, captives (especially women). Other prizes include manufactured items, and items of ritual importance such as human heads (e.g. Harner 1973). The idea, in raiding, is always to minimize risk to the raiders. But sometimes raiding also involves some careful 'conflict management', as when Bedouin raid for camels and take great pains not to kill people in doing so (Sweet 1970) for fear of beginning a feud.

Human *warfare* mobilizes large populations for genocide, conquest, or some other form of political domination: taking of captives or booty, use of particular natural resources, reprisal for raiding, or other grudges. In political anthropology, the definition of 'primitive warfare' has been refined to include a state of hostility, seen as legitimate in terms of sovereign interest, of one autonomous community toward another (Meggitt 1970, p. 10). It also has been significantly amended to include a state of intergroup hostility that is viewed to be temporary, insofar as a return to peace is contemplated, but with some political advantage achieved (see Bohannan 1983). Among the non-literate people who are roughly comparable to chimpanzees in their social organization, whereas a raid (involving only part of the community) tends to be limited in objectives, unilateral and highly ephemeral, warfare tends to be bilateral and of longer duration: both sides may simultaneously mobilize all their warriors for offence, defence, or both, usually with the knowledge that there exist mechanisms to deliberately set aside the state of war and make peace. This, too, amounts to conflict management, in that it seems to parallel dyadic reconciliation of individuals but at the group level.

An important subtype of warfare that is essentially symbolic, referred to earlier, might be called 'bluffing warfare'. In many populations that live territorially, warfare appears to have been ritualized, in the sense that it is carried out in such a way that all-out combat between two large, evenly-matched groups does not take place and casualties are kept minimal. A well-known battlefield example is 'counting coups' as practised by various American Indians, in which after one enemy is slain heroic 'points' are acquired not by killing further enemies, but by running up to the slain enemy, touching his body, and safely getting away. Another is found in New Guinea, where variants of what Meggitt (1977, p. 17) describes as a 'great fight' are widespread. Two large groups fully mobilize for warfare, preparing themselves both ritually and with weapons, and meet for a confrontation in which, along with bragging and taunting, arrows are fired and spears are thrown, at a distance. When it is time for the confrontation to end, everyone simply packs up and goes home. Meggitt's detailed description makes it clear that this type of conflict is not irretrievably 'ritualized': a rough balancing of power may be a critical factor, insofar as when the two opposing armies pose and 'manoeuvre', if one should gain a very significant tactical advantage a real attack might take place.

Thus, non-literate humans sometimes spare themselves the losses of life and injuries that intensive warfare can bring when both groups are fairly large, rather evenly matched, and live in close proximity. As habitual but discreet risk-takers, they do this by setting up rules for feuding, by merely raiding, or by restricting communities mobilized for warfare to bluffing conflicts that are, by custom, largely symbolic. By the amplified definition I am using here, all three behaviours qualify as conflict management even though the conflicts are external.

I mentioned earlier an additional feature of human tribal political systems which has a similar effect, and which in a sense also 'manages' conflicts. Setting up of *alliances* between segments, be they clans, settlements, villages, tribes, or confederations of tribes, tends to follow a pattern such that weak political positions of certain groups become compensated through advantageous alliances. 'Balance of power' strategies are exhibited by single localized political units, for example one Yanomamö village allying itself with another (Chagnon 1983), but also by multi-village tribes joining together against others, for example the non-literate Montenegrins (Boehm 1983), who could also set aside tribal differences to consolidate all of their two dozen tribes into a single confederation, and then form external alliances with nineteenth-century European great powers against other great powers. Humans go to war because they hope they can win or because they want to defend themselves; they avoid being attacked by finding allies, or by trying to turn certain enemies into allies.

In comparing the xenophobic macro-coalition behaviors of chimpanzees with these human patterns, I have shown already that chimpanzees do not seem to have anything resembling the blood feud; nor do they engage in all-out warfare, in which the mobilized males of one group attack another group as a whole, or in which two such groups deliberately meet on the battlefield. However, they do clearly mobilize most (or all) of the males of a community on a purposeful basis, to go out on sorties which clearly involve scouting and intimidation of the enemy, and they utilize peripheral resources located in no-man's-land or sometimes even in a neighbouring group's core territory. Because of anatomical and cultural differences, they eat the food rather than carrying it away, but they also take away new females for sexual purposes and kill enemies found alone or in small numbers. In this matter of limited and immediate goals, they seem most similar to humans when we raid: however, with chimpanzees it often is the 'entire army' that goes out on the raid.

When patrols are composed of most or all of the adult males of a group, these territorially-oriented 'raiders' sometimes end up in large-scale confrontations. This seems particularly likely when food resources in an overlap zone are just ripening, which attracts the patrols to the same place at the same time. When two large patrols do meet, vocalize hostilely in bluffing confrontations, and then withdraw, then the similarity to a Mae Enga symbolic 'great fight' is striking.

As with human raiding and feuding, chimpanzees may repeatedly decimate an enemy population through small-scale attacks, and thereby eventually take over some or all of their territory. When humans living in small-scale societies merely feud and raid, there is some question as to how aware they become of the gradual territorial changes that may accrue (e.g. Turton 1977). By contrast, when they also make intensive attacks (warfare) under conditions of territorial crowding, it would appear that they are well aware of the possible economic consequences (e.g. Meggitt 1977; see also Vayda 1989).

It is very difficult to evaluate chimpanzee awareness in this area. Guarding peripheral food resources and acquisition of estrous females involves immediate purposes that may be readily interpreted with respect to motive: chimpanzees appear to know what they are doing, and their behaviour patterns suggest that their action is purposefully cooperative. However, the stalking and killing of strangers is more difficult to evaluate. Quite possibly, they are perceived as competitors for food. But stranger males might be taken to be rivals in mating. Males might even be perceived somehow as potential competitors for dominance status, but this is speculation. Xenophobic behaviours undoubtedly have deeper roots, as well.

This matter of chimpanzee 'intentions' merits further consideration, since it is germane to understanding fully the basis on which the ever-competitive males within a chimpanzee community cooperatively mobilize into a macro-coalition to go on patrol. It also is germane to making a fuller comparison with humans.

MACRO-COALITION DECISIONS AND LEADERSHIP

Elsewhere, using detailed field descriptions by Kummer (1968, 1971), I have compared the ecological decision processes of hamadryas baboons and non-literate humans (Boehm 1978). Kummer's (1971) conclusion, for large troops, was that while ultimately one of the more experienced males determined the common direction of travel, other individuals were contributing their own 'suggestions' by making tentative moves. In effect, the entire macro-segment was able to combine the knowledge of its more experienced individuals in making important decisions about where to forage on a given day.

Structurally similar nested hierarchies that function politically are found widely among social mammals, including geese and turkeys (see Wilson 1975, p. 287), and also dolphins (Connor *et al.* this volume, pp. 426–30). A human segmentary society is structured similarly, and this structure channels territorially relevant decision-making and planning, as well as action. This behaviour involves both strategic calculations, discussed above, and some mode of group coördination. In agonistic encounters, I think it is safe to say that both tribal humans and wild chimpanzees depend heavily upon their intelligence and collective experience to judge the strength of their own and opposing community-wide macro-coalitions, and to effectively evaluate the wisdom of avoiding their enemies, setting up for defence, or making a bluffing or direct attack on them.

The actual strategies are somewhat easier to identify with humans, since sometimes all the warriors gather to discuss their future (or past) political possibilities using a symbolic language that is at least partially understood by observers, and occasionally such meetings are recorded and interpreted in detail (e.g. Turton 1977; see also Boehm 1983). However, chimpanzees on patrol appear to be evaluating their neighbours very carefully. They stay silent and scan a great deal of the time as they approach a peripheral area, while on sighting neighbours they regularly retreat silently if they are few or vocalize hostilely and withdraw slowly if both sides are in force (Goodall 1986). Members of the other community behave the same way. It is emphasized that if both sides are in sufficient force to vocalize, then both communities will gradually retreat without fighting; but if one side can successfully stalk and attack a very small opposing group or an individual, it does so. These are strategic decisions.

At Gombe, a typical patrol involves a cautious deep foray that ends with feeding, possibly an attack on strangers, or a major confrontation involving displays, vocalizations, and retreating behaviour on both sides. In confronting territorially competitive macro-coalitions, I believe that alpha male leadership plays an important role in decision-making, in that the alpha male (with other senior males) provides cues that are followed by the entire group. Based on limited data so far, I believe I can make the preliminary case that something more than very immediate cueing and automatic one-way social facilitation is operative.

This can be illustrated anecdotally, by referring to one patrol that was exceptionally well-videoed under unusually favourable conditions of observation. To my knowledge, this was the first territorial confrontation to be documented in this way. The camera work was done in 1987 by Yahaya Almasi, a senior Tanzanian field assistant at the Gombe Stream Research Centre who was trained in ethological field techniques in the late 1970s by Jane Goodall, and was trained by myself in 1986 in a 2-week intensive program on field videotaping that I conducted at Gombe. (This footage was shown publicly at the international conference on chimpanzee behavior that was held at the Chicago Academy of Sciences in November, 1987; it is available from the Academy as part of the videotaped presentation.) A majority of the study group's adult males had gone out on an extended patrol, walking purposefully in single file and stopping periodically to scan. (Scanning tends to be done by one or more males, with individuals sometimes positioning themselves in nearby trees; see Goodall 1979, 1986.) Their behaviour while travelling was typical of other 'follows' that involved patrolling. The patrol finally stopped on a ridge top, just in sight of territory used by the southern Kalande community. The following description was derived from frame-by-frame analysis of the videotape.

As the patrol rests, several of the members are out of the camera's purview. Three males are feeding intermittently on a few sparse bushes on the ground near the edge of a semi-open ridge overlooking the valley to the south. They do not appear to be scanning constantly, but the two adults are oriented squarely to that valley, which runs from east to west. [The field assistant is filming back away from the edge of the ridge, probably out of sight of the distant group, which is unhabituated. The ridge is relatively open.]

0 seconds. For the first time an apparently large, distant group of stranger males to the south is heard to vocalize, combining a variety of calls that include roars and a 'squealing' call that sounds like an extended pant-hoot climax. Because the calls are of the intercommunity agonistic type, it would appear that the study group may have been spotted by an enemy it cannot see. The three individuals sit as before but do some scanning or look fixedly to the south.

14 seconds. A few seconds after the calls no longer are audible on tape, adult male Evered turns his head around sharply from facing directly south toward alpha Goblin, who is some distance behind him, and emits a soft vocalization, surely not audible to

the distant group, that sounds like the beginning of an intragroup pant-hoot when it is being suppressed or is broken off. The distant calls resume.

21 seconds. As Goblin runs south toward a vantage point on the ridge, he is intercepted by young adult male Mustard who approaches him whimpering with a fear-grin, tries to embrace Goblin as Goblin passes very quickly, then follows Goblin and sits close as they look to the south. Goblin's hair is partly erect, while Mustard's hair appears sleek. To judge from Goblin's head movement, he is scanning back and forth (in a broad arc but oriented directly toward the valley below or the ridge beyond) as if he cannot see the group that is vocalizing. The distant calls continue.

40 seconds. Distant calls cease.

43 seconds. Goblin stops scanning south and turns his head to look for 2 seconds in the direction of Evered, who is about 5 yards distant, and Satan, who is further away. (Satan is the number two male; Evered is the oldest.) Then Goblin orients his head to the south again and the camera pans to Evered and Satan, who continue to sit on the ground facing south.

49 seconds. Goblin grunts loudly and begins a display, vocalizing, to the east along the ridge top. A young male and eventually Evered join Goblin, Mustard and another male as they all vocalize loudly and display moving to the east. Satan stays in place and vocalizes directly toward the south. They produce a variety of vocalizations, including roars, squealing sounds reminiscent of the pant-hoot at its climax phase, and eventually some classical pant-hoots.

This begins an extended bout of similar vocalizations exchanged between the two groups over a number of minutes, followed by retreat of the study group vocalizing as it withdraws.

The decision process took nearly a minute, and it would appear that neither Goblin nor the group was committed to a response until the situation was carefully evaluated. It is important to note that the alpha male, Goblin, appeared to be sizing up the situation himself and also sizing up the reactions of two other high-ranking males, Evered and Satan. It is also noteworthy that initially Evered suppressed a vocalization as he turned to see what Goblin was going to do. It appears that the other males were waiting for Goblin to make a decision, while at the same time he seemed to be taking some of their reactions into account. Once he began his display and vocalization, however, they followed his cues immediately. I am at a loss to interpret the fact that Satan did not join the group in displaying, but continued to sit on the ground facing south even though he joined in with the vocalizations. But he and Goblin were rivals.

On the basis of this one episode, which seemed consistent with others observed or recorded in far less ideal conditions of observation, aggressive high-ranking male chimpanzees who possess superior political experience seem to be important in catalyzing the predictable, aggressive-yet-prudent decisions of their macro-coalitions, when they face territorially competing groups that are strong. But in spite of the alpha male's high position in the dominance hierarchy, and his ability to intimidate and control other males of his group in certain contexts (e.g. in feeding, or sometimes in monopolizing an estrous female), in territorial macro-coalitions he can lead only by example and by giving cues that are followed at the option of his followers. Like human leaders of bands or small tribes, he cannot coerce his followers to follow his lead once he decides to confront the enemy. But, when the adult and adolescent males act as a macro-coalition, he does play an important role in processing information on his own and in assessing any deliberately communicated or otherwise obvious dispositions of other individuals before he tries to catalyse a course of action for his community.

The reason that humans who live in small-scale fraternal interest groups allow their leaders no real authority (e.g. Chagnon 1983; Meggitt 1977; Boehm 1986) is that they are egalitarian; that is, they carefully suppress their own potential dominance hierarchy and create what amounts to a 'reverse dominance hierarchy' (Boehm 1984). Thus, like the chimpanzee alpha leader in this episode, in a decision of war or peace a human egalitarian leader normally can only consult, trying to facilitate a consensus, or try to lead by example. However, a possibly important difference between the two species is the absence of any authoritative control by a chimpanzee patrol leader in an actual situation of confrontation, since egalitarian humans sometimes allow their war leaders a temporary but strict control on the battlefield, once the decision to fight has been made consensually (Boehm 1991). Since chimpanzees avoid mass combat, it is difficult to compare the two species any further in this instance.

COMPARISON OF CAPACITIES FOR STRATEGIC ACTION

It may seem that I have played down the significant differences that exist between chimpanzees and humans, in the important matter of arriving at and implementing political strategies. However, chimpanzees depend heavily upon group sociocultural traditions that are learned observationally, through experience, and so do humans. In both species, also, group political behaviours often are triggered by immediate acts of leadership and are spread by social facilitation; indeed, in egalitarian bands and tribes the headmen are very often obliged to lead merely by setting an example (e.g. Chagnon 1983), since like chimpanzee alpha males they have no power to coerce the entire group. In fact, their position is even weaker: it is difficult for them even to dominate individual rivals, because of the collectively enforced egalitarian ethic.

It is easy to lose sight of these basic similarities in light of the marvels of symbolically-assisted culture. When an entire human group gathers to cope with commonly perceived problems of opportunities, symbolic communication permits not only detailed *in situ* discussion of decision alternatives, but careful advance planning. A group that has successfully talked its way through to a consensus is better prepared to act with unanimity toward a single purpose. Such an approach is universal among egalitarian tribesmen.

By contrast, chimpanzees are masters at reading (and sharing) an immediate context in order to solve a problem collectively, and if needed they can communicate effectively on an immediate basis, in order to take coordinated action. However, this 'immediacy' of communication does not mean that their shared understanding of the context of decision making is entirely 'of the moment'. Captive chimpanzees exhibit an excellent longitudinal understanding of long-term political competition between coalitions (de Waal 1982; see also de Waal and Luttrell 1988 for a limited discussion of captive chimpanzee political cognition), and the same would appear to be true in the wild (see Goodall 1986). Furthermore, chimpanzee communication in the wild sometimes seems to involve what Hockett (1963) calls displacement, in that certain vocalizations, gestures, or postures seem to refer to past behaviours or to things or events that are out of sight (see Boehm 1989).

The fact that all male members of chimpanzee patrols study the situation so carefuly as they go along supports the hypothesis that a general sense of situational context is shared and that the communication described above could easily have involved some displacement: assuming that leaders were 'referring' to a neighbouring group that was heard but not seen by members of the study group, this 'inferential' communication was facilitated by the fact that all group members had been carefully collecting information about their geographic position and trying to evaluate visual, olfactory, and aural cues about the presence or absence of strangers for some time. In considering the pooled strategic wisdom of these macro-coalitions, it must be kept in mind that our understanding of chimpanzee communication in the wild is very limited (see Boehm 1989) in comparison with what can be reliably inferred in experiments with laboratory subjects, or what can be determined for humans on the basis of informant statements and participant-observation techniques.

My point is that the difference in cultural level is significant, but far from absolute. With respect to many of the group-level agonistic behaviours involved, such as seeking either females or external food resources, or appreciating the dangers of large-scale combat, for both species problem-solving proceeds on a rather similar basis. There appear to be comparable goals, comparable fears of being lethally attacked, and more generally comparable cognitive evaluations of the strategic political possibilities and exigencies at hand. In making such assumptions, I am setting up interpretations which by stricter ethological standards may seem quite 'liberal'. However, I agree with the case that Griffin (1976) has advanced for making positive assumptions about psychological awareness and cognitive sophistication of non-human animals where data warrant doing so. I believe his strictures can be applied with greatest confidence to the great apes.

Thus, in comparing the political motives, perceptions, and behaviours of chimpanzees and humans, I am taking into consideration both a close phylogenetic connection and the many sophisticated problem-solving capabilities documented in other areas for wild and captive chimpanzees by Goodall (1986). It is worth mentioning, too, that my comparative evaluation has been tempered by more than 16 months of field work with one well-habituated chimpanzee study group, observing them in all phases of their wild life including territorial patrols. I have also resided for 2 years with a quasi-tribal fraternal-interest-group human society in which the clans of the once-territorial tribe I lived in retained much of their earlier rivalry but the original feuding, raiding, and warfare patterns were long suppressed (see Boehm 1986).

To sum up this discussion, which in some important aspects must remain exploratory, chimpanzees on patrol do appear to operate according to calculated strategies. They seek information intensively along the way, can estimate enemy strength visually or by inference from vocalizations they hear, and then make decisions with some mutual consultation as a politically cohesive group. These individuals obviously share a sense of mission, take cues from one another, and sometimes communicate directly on matters relevant to decision-making, such as fear, with a shared and well-informed sense of situational context. Even though humans often scout the enemy first and then plan their major forays as entire groups, using language with its advantages of highly specific communication with displacement, chimpanzees seem to manage quite well without such an elaborate cultural tool: it seems likely that patrolling enables chimpanzees of adjacent communities to monitor enemy male strength or numerical weakness on a longitudinal basis. Like humans, they also seem to know enough to take over enemy territory if a radical pattern of weakness is perceived; this has happened at both Mahale (Nishida *et al.* 1985) and Gombe (Goodall 1986) study sites, which are situated less than 200 miles apart in East Africa.

CULTURAL LEVEL AND ADAPTIVE MODIFICATION

I emphasized above that presently-studied chimpanzees and many human groups of comparable social scale exhibit striking structural and behavioural similarities, and that both species may be seen as operating with cultural traditions and as communicating to make culturally-routinized group decisions with the facilitation of leaders. I now turn to certain differences in macro-coalition behaviour that are important for this comparative analysis.

To begin, we know less about the overall variety in social organization of chimpanzees. I have established that fraternal interest groups resulting from male non-exogamy and also male territorial patrols have been observed in two not very distant provisioned groups (Nishida 1979; Nishida *et al.* 1985; Goodall 1979, 1986), and behaviours highly suggestive of a similar pattern have been observed in an unprovisioned group somewhat farther away (Ghiglieri 1988), all in East Africa. However, while West African chimpanzee social organization generally seems similar, Sugiyama (1988) reports a rather different organization among a population in Guinea that was exceptionally isolated by human encroachment, and many regions await adequate reporting, especially on intergroup behaviour. It cannot be ruled out that in natural settings marked differences in population density might actually lead to presence *or absence* of competing fraternal-territorial communities; there also are chimpanzees who live in arid, open country (see Moore 1986), where possible inability of solo females with offspring to survive predator pressure might enter into the social-structural picture. But this is speculation, and the basic chimpanzee group structure seems to be territorially oriented, and fraternal.

By comparison, human adaptive modifications are better studied. We know that not all band-level or tribal humans who hunt and gather live in fraternal interest groups exhibiting a strong tendency to behave territorially (Knauft 1991), and that even in such groups, patterns of political conflict vary: for example, obligatory retaliatory killing may be absent, or be merely individualized rather than group-wide (see Black-Michaud 1975), while intensive warfare may be absent or present (Otterbein and Otterbein 1965; Ember 1978; see also Knauft 1991). Furthermore, our species is adaptively modifiable in several other directions. There are matrilineal–matrilocal communities in which females stay put and males marry in. These sometimes can be territorial with feuding and warfare, but they tend to be less prone to homicidal intergroup conflict than fraternal interest societies. The same is true of the many 'bilateral' kinship systems, in which mothers' and fathers' kindreds are of equal importance to their children and postmarital residence is often neolocal; there, localized groups have no special unilineal kinship basis and their membership tends to shift, yet some of them exhibit some form of feuding (e.g. Balikci 1989).

This human 'social flexibility', which makes possible the far wider variety of environmental niches that humans inhabit, would appear to stem from several sources. Probably one is a rather different set of genetic behavioural preparations, less specific in their effects than with other animals. Another, treated above, is verbal-symbolic language; this makes possible cultural traditions based on detailed information exchange, and highly explicit social understandings shared and sanctioned by the entire group.

Universally, these social understandings involve a special kind of tyranny, that of the human group as a morally judgemental sanctioning entity that can become highly assertive in controlling the behaviours of its own individual members (see Service 1962, 1975; Fürer-Haimendorf 1967; Boehm 1982, 1984), and which ostracizes or executes its members for certain acts of non-conformity (see Gruter and Masters 1986; Otterbein 1986). A relevant effect of human norms and social control is that male humans perform what might be called 'territorial duties' (feuding, raiding, warfare) partly (and importantly) because they are manipulated by a combination of patriotic ideology and social pressure to behave courageously (see Campbell 1972, 1975). So far as we know, chimpanzees do not, although at a more fundamental level both species readily develop strong xenophobic feelings and behaviours. I shall return to this important difference later.

How large a difference does a human symbolically-advanced cultural level and a presumably greater degree of 'adaptive modifiability' make for the present analysis? As we have seen, apes do develop closely-bonded fraternal interest groups. But they do so without the benefit of ideologically rationalized patrilineal kin loyalty, rules of exogamy based on morally reinforced incest prohibitions, and systematic arrangements to exchange women as wives. Rather, they are obliged to stay together because of their basic territorial setup: the males cannot migrate without being attacked and killed.

Thus, the basis for group life itself does show significant differences. However, chimpanzees are not held together by fear alone, nor are humans held together by 'culture' alone. Both species naturally form bonds, and they learn to cooperate, by playing together as youngsters who sometimes are attached to the same mother, by cooperating as adults who are not necessarily very closely related in the location and acquisition of food, by sharing a common leader, and by forming political macro-coalitions. Both species do this in spite of individual competitiveness. These are important similarities.

There remains also the issue of group internal conflict management. Elsewhere (Boehm 1986), I have demonstrated for a human segmentary-tribal group in which cultural values strongly promoted warlike aggressiveness, and in which unavoidable internal divisiveness was viewed to be dangerous in the face of serious external threats, that a great deal of conflict resolution behaviour was exhibited within the lower-level political segments as a consciously developed antidote to serious internal division. While it seems very unlikely that chimpanzees intervene in conflicts with such sophisticated long-range objectives in mind, their attempts at pacification are salient and usually effective and do promote group unity: generally, virtually all of the males can set aside their differences to make a strong showing on patrol. At a functional level, this is an important area of similarity, and one that holds in spite of differences in cultural level and political insight.

WHAT ABOUT INTENSIVE WARFARE?

The planning and coordination that human communication facilitates may well be a prerequisite for successful all-out warfare, a complex and dangerous undertaking which involves the possibility of severe casualties on both sides. This difference may help to explain why chimpanzees do not develop anything like full-scale battles (see Goodall 1986). But perhaps equally important, I have suggested that humans can be motivated to assume enormous risks on the basis of norms favouring patriotic self-sacrifice and a quest for personal social position, in the form of valour in combat. By contrast, the individualized chimpanzee quest for status remains limited to *intracommunity* competition among individuals and micro-coalitions.

It is probably right on the battlefield that human military decision behaviour becomes closest to that of chimpanzees, that is, very immediate. For example, individualistic Montenegrin tribesmen, when they needed to act in concert, were obliged to use just a few signals in combat to 'play the situation by ear' (Boehm 1983) because there was no way in the heat of battle to assemble the tribe and talk things over. Therefore, I can see no definitive reason to hold that chimpanzees with their relatively immediate tools for communication and coordination could not cooperate well enough to do battle in large groups, if two opposing communities were so motivated. With respect to on-the-spot planning and coordination, it is worth considering the quite different behavioural context of hunting, which nevertheless can produce a situation something like a battle. It is interesting that, acting as macro-coalitions, chimpanzees on an impromptu basis sometimes coordinate their attacks fairly well (Goodall 1986). For example, I observed an episode in which a large group of colobus monkeys (*Colobus badius*) was cornered and several adult male monkeys stood their ground and made defensively very effective biting attacks, but nevertheless four monkeys (a high count) were taken because the chimpanzees seemed to be working together rather effectively.

In thinking about the absence of warfare, then, it seems possible that with chimpanzees an attack of one large patrol on another is not entirely out of the question, in terms of cognitive ability. Rather, we must look to psychological motivation as the critical proximate cause. It is individual fear, amplified by social facilitation, that wins out over any tendency to attack a stranger-group that is large and that contributes to the predictable bluffing response. In overcoming comparable individual fears, humans are helped by a well-discussed vicarious familiarity with a previous warfare pattern that has been successful in aggression or has been useful for defence, by confidence born of an ability to plan in advance, and by the aforementioned symbolically communicated cultural emphasis on patriotism and bravery which is so important motivationally.

A less obvious difference, which enables humans to sustain a warfare pattern without inflicting heavy casualties very frequently, is the ability of larger segments to make far-reaching strategic decisions of alliance with like groups, and the function of such behaviour in *balancing power* (see Falger this volume, p. 329, for such alliances at the international level). I pointed out earlier that egalitarian humans can be organized into three or more levels of nested hierarchy, while, so far as is known, chimpanzees never join together to fight in alliances between different communities. Thus, chimpanzees are prone to balance power within the community (see de Waal 1982, Goodall 1986), and also in effect seem to do this on a very immediate, bluffing basis, when the males of two communities meet, but they are unable to compensate for a weak territorial position by allying with one neighbour against another. When humans do this effectively, in effect they can protect their territory or other resources over a relatively long period of time without risking loss of life (e.g. Chagnon 1983). In being without such a political tool, chimpanzees might have to do battle very frequently if they were not otherwise inhibited from doing so.

With respect to the application of political power by macro-coalitions that raid and bluff, the case for general similarities between chimpanzees and fraternal interest group humans has now been established. The behaviour similarities appear to be quite pervasive, except for the absence of feuding and intensive warfare. With respect to the absence of intensive warfare, it will be useful at this point to summarize the differences between the two species that were discussed above, along with a few others that could be important to this comparison.

1. *Moral control.* Chimpanzees lack a morally induced positive and negative social incentive system that motivates individuals to overcome their fear when they sight a large enemy group. Human fears of all-out warfare are strongly countered by cultural norms that carry sanctions of moralistic group rejection of cowards.

2. *Language*. Chimpanzees obviously lack the human type of communication system that makes possible the detailed discussion of previous large-group conflict so that this becomes vicariously familiar and predictable, and that permits laying out of specific plans, e.g. battle strategies. Both serve to give human warriors confidence.

3. *Cognition and motivation*. Language also permits stimulation of xenophobic feelings in a patriotic direction, and the resulting positive ideological commitment also helps as a counter to individual fears.

4. *Technology*. Humans fight with fashioned lethal weapons; chimpanzees essentially use their teeth. Otterbein (1970) suggests that as weapons technology increases, so do casualty rates. But humans nevertheless continue to overcome their fears.

5. *Intention to kill*. Chimpanzees appear to lack a cognitive tradition that includes any very focused implicit conceptualization of death and, therefore, of killing as the immediate cause of death. It may be that the lack of any practical need to develop a definitive *coup de grace* in hunting helps to leave chimpanzees cognitively unprepared to be efficient killers conspecifically; i.e., prey animals, once disabled, are killed incidentally in the process of being consumed as meat, while adult conspecific strangers who are savagely attacked and then abandoned to die of their wounds (or possibly recover) are not eaten (Goodall 1986). It may be also that some genotypic disposition helps to inhibit killing of adult conspecifics, even during savage inter-community attacks when the aggression appears to be so unbridled.

6. *Culturally-based conventions*. Humans control their intercommunity conflicts by rather complicated cultural inventions such as setting up rules for feuding or balancing power through forming higher level alliances. Chimpanzees do neither.

7. *War and peace*. Bohannan (1983) makes the important point that when humans go to war they know that this includes also the idea of making peace: in effect, the two groups are able to manage their conflict without third-party intervention, as chimpanzees do at the level of individuals. But chimpanzees, who dyadically reconcile their conflicts rather effectively within the community, exhibit no such tool when it comes to conflict between communities.

The differences in this general area of 'warfare' capacity are considerable, and significant. However, my vote still must go with basic similarity. This is because where fraternal interest groups are present, hierarchically-nested alliancing behaviour in both species permits the formation of xenophobic macro-coalitions whose similar predatory and defensive behaviours have important territorial implications, and because in both species there exists a strong potential for behaviour that can result functionally in one group's eliminating the other. Furthermore, in both species there also exist behaviours, a few of which seem extremely similar (e.g. large-group bluffing), that effectively mitigate the potential for mutually detrimental mass conflict at close quarters. It would appear that with humans, these behaviours are merely effective much of the time, while with chimpanzees as studied so far, they appear to be effective all of the time even though near-genocide can still take place.

This discussion about warfare leads to a conclusion that has philosophical implications. What it comes down to, in my opinion, is that while both species are prone to bluff when they confront xenophobic macro-coalitions of conspecifics, humans have invented ways of stimulating individual males to overcome their fear and to go beyond bluffing even when they face numerous homicidally-inclined enemies bearing effective weapons that are lethal at a distance. If this is the case, we are presented with a paradox. It would appear that the same capacity for moral sanctioning that *within* the group helps to guide our human conduct strongly in the direction of avoiding or reducing homicide and resolving conflicts, also stimulates us to move from bluffing behaviour to mass homicidal behaviour when two different communities end up in a warlike confrontation. This is because one of the uses of morality (see LeVine and Campbell 1971) is to spur on or justify xenophobic slaughter of morally-denigrated enemies. This ethnocentric potential for genocide provides some very special food for thought, if we wish to reflect dispassionately on the 'advantages' of morality and symbolic culture.

CONCLUSIONS

For both chimpanzees and egalitarian humans who live in fraternal interest groups, I have demonstrated some striking similarities in the ways that the adult males purposefully conduct their political business as permanent residents of locally autonomous territorially-oriented groups that are xenophobic. I have documented these similarities not by comparing chimpanzees with modern humans, but by comparing them with non-literate tribesmen whose societies are much like chimpanzee communities in scale and kin structure. The similarities include individual male competition, existence of divisive fraternal and other coalitions within the local group, prominent conflict management resolution behaviours within the group which control this divisiveness, and the capacity to orm a community-wide coalition. The result is a highly workable unit in a territorial system that is based essentially on bluffing, but can also involve lethal physical conflict.

Because internal fights are not infrequent, the impressive capacity for making peace (by reconciliation or intervention) when internal conflicts threaten to get out of hand is functionally important. I have emphasized this even though the selection mechanisms that support such behaviours cannot easily be hypothesized. Such conflict management behaviour has surely coevolved with the competitive behaviours it manages; indeed, it is useful to see them as functioning in a complementary manner (Bohannan 1967). Together the two make possible a social dominance hierarchy that lends itself to the formation of macro-coalitions. To understand exactly how they have coevolved, I believe that some very complicated modelling of individual selection processes will be necessary (Boehm 1981), and possibly some interdemic modelling as well (Wilson 1980).

In comparing the activities that involve these segmentary macro-coalitions, I have demonstrated the greatest similarity to be between chimpanzee patrolling behaviour and human raiding and bluffing warfare, while close chimpanzee behavioural homologies to feuding or full-scale, intensive warfare are not apparent. However, chimpanzees can repeatedly *decimate* a neighbouring community over a period of a few years, killing one or two individuals at a time, so the political–territorial and selection consequences of their macro-coalition 'raiding-and-bluffing' can become functionally similar to those of human warfare when this is intensive or genocidal.

While the fraternal interest group is far from being universal among humans, similar patterns of feuding, raiding, bluffing warfare, and intensive warfare also can occur with matrilineal or bilateral societies. Thus, in terms of potentiality, the general similarities of territorial behaviour between the two species may be rather pervasive. The study of chimpanzee social–territorial organization is not yet complete, and there may be variants that are less comparable to xenophobic fraternal-interest humans, and are more comparable to other human types or to no human types. However, the parallels I have discussed would appear at the very least to be ones of homology between behavioural central tendencies of two unusually labile species, while the general similarities in male competitiveness, ability to pacify internal conflicts, ability to form macro-coalitions, and in the shared tendency to stalk and kill neighbours are striking.

It is only natural that we humans should be highly interested in *intensive warfare*, a controversial and dangerous behaviour that still looms prominently in our political life today, but which seems to have been optional, if widespread, in the non-literate world. The fact that it appears to have been optional raises fundamental questions about human nature itself, and in most of us engenders biases with respect to what we would like that nature to be (see Bohannan 1967). If we look to the macro-coalition behaviour of wild chimpanzees for 'the answer', we are likely to be disappointed no matter which predictable set of biases drives us.

The 'hawks' among us will be disappointed because, if one defines warfare carefully, chimpanzees do not exhibit intensive warfare as it has been defined here. They do live in what amount to sovereign communities, and the males pursue their common interests as whole communities that know how to fight, but they do not actually join in battle. On the other hand the 'doves' will be disappointed because when the raiding forays of chimpanzees result in systematic killing of individuals from neighbouring groups, it seems dubious that such behaviour could be taking place solely under the influence of unnatural provisioning. Nature, too, sets up edible prizes for competition, and can do so in dense patches.

What seems likely, although further research must confirm this hypothesis, is that wherever chimpanzees populate large, unrestricted areas whose carrying capacity is similar to study sites in East Africa such as Mahale, Gombe, and Kibale, and do so for a very long time with a chance for population growth, fraternal interest groups quite comparable to those of humans are likely to form, and that such groups will be xenophobic and will exhibit signs of territorial behaviour acting as macro-coalitions. My conclusion, however, is that they will be found to practice something more like a combination of raiding-patrols and bluffing warfare, rather than intensive warfare.

This basic question about intensive warfare is not likely to be resolved in the very near future. Even if chimpanzee patrols actually do conquer their fears once in a long while and join in full-scale battle, as humans do and as patrilineal–patrilocal tribal or band-level humans are especially prone to do, it may require many more decades of arduous field work to finally observe this. Until that time, it is more consistent with the facts to say that xenophobic macro-coalitions, acting as patrols, involve themselves in raiding attacks and in bluffing episodes that fall short of true (intensive) warfare.

In conclusion, I return to Bohannan's (1983) definition of human warfare that was cited earlier, which suggests that insofar as peace and war are part of the same overall process, human conflict management capacity, broadly defined, extends to conflicts between macro-coalitions at higher segmentary levels. I have emphasized that chimpanzees do have effective means for managing their agonistic bouts at both the individual and micro-coalition levels, in the form of intra-community dyadic reconciliation and triadic intervention; however, such mechanisms are lacking for conflicts at the macro-coalition level. By contrast, humans not only have the capacity to build stabilizing alliances at the macro-coalition level; they also understand that, under many circumstances, once war is begun one can sue for peace as well as continue to wage war, and that one can even turn an enemy into a friend. Chimpanzees, it would seem, are always 'at war' with all of their neighbours; yet they never, or very seldom, engage in battles.

SUMMARY

Coalition theory is extended to an analysis of macro-coalitions that unite entire territorial communities against like communities. Such segmentation takes place in spite of divisiveness fostered by internal coalition behaviour. In comparing East African chimpanzees and patrilineal–patrilocal humans, the following points are made.

1. Structural similarity is significant: both species develop fraternal interest groups that are subject to divisive internal quarrels.
2. In both species, effective management of internal conflicts helps to make possible the formation of community-wide macro-coalitions that have important territorial functions.
3. Humans develop feuding systems between groups, while chimpanzee revenge behaviour remains at the individual level.
4. Acting as macro-coalitions both species go raiding for sustenance and breeding partners, and sometimes kill their enemies.
5. Acting as macro-coalitions both species also engage in bluffing confrontations of entire communities, in which at the individual level of motivation aggressive tendencies and fear appear to stay in equilibrium. Chimpanzees do this very predictably; for patrilineal humans it is one frequently manifested option of several.
6. Patrilineal humans not infrequently engage in intensive warfare between two large groups. Apparently, East African chimpanzees do not.
7. Communities of humans often 'manage' such intensive external conflicts by making external alliances that balance power, and by ending their wars with peace treaties. Chimpanzees do neither.
8. In comparing the two species, it is found that while social and territorial structure are very similar, as are xenophobic tendencies and capability to

mobilize communities at the macro-coalition level, humans are unique in being able to motivate individual males to engage in mass attacks, in situations in which the lethally-armed males of two entire communities face one another. Human warriors are moved to engage in mass combat by a combination of patriotic ideology and negative sanctioning of cowards, two features of macro-coalition competition that chimpanzees lack.

ACKNOWLEDGEMENTS

Support for research on long-distance vocal communication of wild chimpanzees came from the L. S. B. Leakey Foundation. Support for 2 years' investigation of conflict intervention behaviour at Gombe National Park in Tanzania came from the H. F. Guggenheim Foundation, Northern Kentucky University, and the government of Tanzania, including Tanzanian National Parks and the Serengeti Wildlife Research Institute. All of this support is gratefully acknowledged. I also am deeply indebted to Jane Goodall and to Hilali Matama, Eslom Mpongo, Yahaya Almasi, and Hamisi Mkono, for collaboration in research and to Jane Goodall for generously making available the Gombe Stream Research Centre's facilities and the earlier research data on conflict interference. Support for ethnographic research on Upper Moracha Tribe in Montenegro, Yugoslavia, came from the Ford Foundation, The National Institutes of Mental Health, Northwestern University, and the Yugoslav government. I also wish to thank the editors of this book for substantial and useful suggestions.

REFERENCES

Alexander, R. D. (1974). The evolution of social behavior. *Annual Review of Ecology and Systematics*, **5**, 325–383.

Balikci, A. (1989). *The Netsilik Eskimo*. Waveland Press, Prospect Heights, IL.

Bernstein, I. S. (1970). Primate status hierarchies. In *Primate behavior: developments in field and laboratory research, vol. 1* (ed. L. A. Rosenblum), pp. 71–109. Academic Press, New York.

Black-Michaud, J. (1975). *Cohesive force: feud in the Middle East and Mediterranean*. St. Martin's, New York.

Boehm, C. (1978). Rational preselection from hamadryas to Homo sapiens: the place of decisions in adaptive process. *American Anthropologist*, **80**, 265–96.

Boehm, C. (1981). Parasitic selection and group selection: a study of conflict interference in rhesus and Japanese macaque monkeys. In *Primate behavior and sociobiology*. (eds A. B. Chiarelli and R. S. Corruccini), pp. 161–82. Springer-Verlag, Berlin.

Boehm, C. (1982). The evolutionary development of morality as an effect of dominance behavior and conflict interference. *Journal of Social and Biological Structures*, **5**, 413–22.

Boehm, C. (1983). *Montenegrin social organization and values*. AMS Press, New York.

Boehm, C. (1984). Can hierarchy and egalitarianism both be ascribed to the same causal forces? *Politics and the Life Sciences*, **1**, 34–7.

Boehm, C. (1986). *Blood revenge: the enactment and management of conflict in Montenegro and other tribal societies*. University of Pennsylvania Press, Philadelphia.

Boehm, C. (1989). A research strategy for studying chimpanzee vocal communication in isolation from postural and gestural modes of communication. In *Understanding chimpanzees* (eds P. Heltne and L. Marquardt), pp. –59. Harvard University Press, Cambridge, MA.

Boehm, C. (1991). Human intentions, human nature, and egalitarian society: an evolutionary anomaly explained. *American Anthropologist*, (in press).

Bohannan, P. J. (1954). The migration and expansion of the Tiv. *Africa*, **24**, 2–16.

Bohannan, P. J. (1967). Introduction. In *Law and warfare* (ed. P. J. Bohannan), pp. xi–xiv. Natural History Press, Garden City, NJ.

Bohannan, P. J. (1983). Some bases of aggression and their relationship to law. In *Law, biology and culture: the evolution of law* (ed. M. Gruter and P. J. Bohannan), pp. 147–58. Ross-Erikson, Santa Barbara.

Cashdan, E. A. (1980). Egalitarianism among hunters and gatherers. *American Anthropologist*, **82**, 116–20.

Campbell, D. T. (1972). On the genetics of altruism and the counter-hedonic component of human culture. *Journal of Social Issues*, **28**, 21–37.

Campbell, D. T. (1975). On the conflicts between biological and social evolution and between psychology and moral tradition. *American Psychologist*, **30**, 113–26.

Chagnon, N. (1983). *Yanomamö: the fierce people*. Holt, Rinehart and Winston, New York.

Chagnon, N. (1988). Life histories, blood revenge, and warfare in a tribal population. *Science*, **239**, 985–92.

Durkheim, E. (1933). *The division of labor in society*. Free Press, New York.

Ember, C. R. (1978). Myths about hunter-gatherers. *Ethnology*, **4**, 439–48.

Evans-Pritchard, E. E. (1940). *The Nuer: a description of the modes of livelihood and political institutions of a Nilotic people*. Clarendon Press, Oxford.

Fried, M. H. (1967). *The evolution of political society: an essay in political anthropology*. Random House, New York.

von Fürer-Haimendorf, C. (1967). *Morals and merit: a study of values and social controls in south Asian societies*. University of Chicago Press, Chicago.

Ghiglieri, M. (1984). *The chimpanzees of Kibale Forest*. Columbia University Press, New York.

Ghiglieri, M. (1987). Sociobiology of the great apes and the hominid ancestor. *Journal of Human Evolution*, **16**, 319–57.

Ghiglieri, M. (1988). *East of the Mountains of the Moon: chimpanzee society in the African rain forest*. Free Press, New York.

Gluckman, M. (1955*a*). Peace in the feud. In *Custom and conflict in Africa* (ed. M. Gluckman), pp. 1–26. Free Press, Glencoe, IL.

Gluckman, M. (1955*b*). The judicial process among the Barotse of Northern Rhodesia. University of Manchester Press, Manchester.

Goodall, J. (1979). Life and death at Gombe. *National Geographic*, **155**, 592–622.

Goodall, J. (1982). Order without law. *Journal of Social and Biological Structures*, **5**, 349–52.

Goodall, J. (1986). *The chimpanzees of Gombe*. Harvard University Press, Cambridge, MA.

Griffin, D. (1976). A possible window on the minds of animals. *American Scientist*, **64**, 530–5.

Gruter, M. and R. D. Masters (eds) (1986). *Ostracism: a social and biological phenomenon*. Elsevier, New York.

Harner, M. (1973). *The Jivaro: people of the sacred waterfalls*. Natural History Press, Garden City, NJ.

Hockett, C. F. (1963). The problems of universals in language. In *Universals of language* (ed. J. H. Greenberg). MIT Press, Cambridge, MA.

Hoebel, E. A. (1967). Song duels among the Eskimo. In *Law and warfare* (ed. P. J. Bohannan), pp. 255–65. Natural History Press, Garden City, NJ.

Holloway, R. L. (ed.), (1974). *Primate aggression, territoriality and xenophobia: a comparative perspective*. Academic Press, New York.

Kaplan, J. R. (1977). Patterns of fight interference in a free-ranging band of rhesus monkeys. *American Journal of Physical Anthropology*, **49**, 241–9.

Knauft, B. M. (1991). Violence and sociality in human evolution. *Current Anthropology*, 32, (in press).

Kummer, H. (1968). *Social organization of hamadryas baboons: a field study*. University of Chicago Press, Chicago.

Kummer, H. (1971). *Primate societies: group techniques of ecological adaptation*. Aldine Press, Chicago.

LeVine, R. A. and Campbell, D. T. (1972). *Ethnocentrism: theories of conflict, ethnic attitudes, and group behavior*. Wiley, New York.

Meggitt, M. (1977). *Blood is their argument*. Mayfield, Palo Alto.

Moore, J. (1986). Arid country chimpanzees. *Anthroquest*, **36**, 8–10.

Nishida, T. (1979). The social structure of chimpanzees of the Mahale Mountains. In *The great apes* (eds D. A. Hamburg and E. R. McCown), pp. 73–122. Benjamin/Cummings, Menlo Park.

Nishida, T., and Hiraiwa-Hasegawa, M. (1985). Responses to a stranger mother–son pair in the wild chimpanzee: a case report. *Primates*, **26**, 1–13.

Nishida, T., Hiraiwa-Hasegawa, M., Hasegawa, T., and Takahata, Y. (1985). Group extinction and female transfer in wild chimpanzees in the Mahale National Park, Tanzania. *Zeitschrift für Tierpsychologie*, **67**, 284–301.

Omark, D. R., Strayer, F. F., and Freedman, D. G. (eds) (1980). *Dominance relations: an ethological view of human conflict and social interaction*. Garland STPM Press, New York.

Otterbein, K. (1970). *The evolution of war: a cross-cultural study*. HRAF Press, New Haven.

Otterbein, K. (1986). *The ultimate coercive sanction: a cross-cultural study of capital punishment*. HRAF Press, New Haven.

Otterbein, K. and Otterbein, C. S. (1965). An eye for an eye, a tooth for a tooth: A cross-cultural study of feuding. *American Anthropologist*, **67**, 1470–82.

Sahlins, M. D. (1961). The segmentary lineage: Instrument of predatory expansion. *American Anthropologist*, **63**, 322–45.

Service, E. R. (1962). *Primitive social organization: an evolutionary perspective*. Random House, New York.

Service, E. R. (1975). *Origin of the state and civilization: the process of cultural evolution*. Norton, New York.

Sweet, L. E. (1970). Camel raiding of North Arabian Bedouin: A mechanism of ecological adaptation. In *Peoples and cultures of the Middle East, vol. 1: Cultural depth and diversity* (ed. L. Sweet), pp. 265–89. Natural History Press, Garden City, NJ.

Sugiyama, Y. (1988). Grooming interactions among adult chimpanzees at Bossou, Guinea, with special reference to social structure. *International Journal of Primatology*, **9**, 393–408.

Tiger, L. (1969). *Men in groups*. Random House, New York.

Tiger, L. and Fox, R. (1971). *The imperial animal*. Delta, New York.

Turton, D. (1977). War, peace and Mursi identity. In *Warfare among East African herders*. (eds K. Fukui and D. Turton), pp. 179–211. National Museum of Ethnology, Osaka.

Vayda, A. P. (1989). Explaining why Maring fought. *Journal of Anthropological Research*, **45**, 159–77.

de Waal, F. (1978). Join-aggression and protective-aggression among captive *Macaca fascicularis*. In *Recent Advances in Primatology* (ed. D. J. Chivers and J. Herbert), pp. 577–99. Academic Press, New York.

de Waal, F. (1982). *Chimpanzee politics*. Jonathan Cape, London.

de Waal, F. (1989). *Peacemaking among primates*. Harvard University Press, Cambridge, MA.

de Waal, F. and Luttrell, L. M. (1988). Mechanisms of social reciprocity in three primate species: symmetrical relationship characteristics or cognition? *Ethology and Sociobiology*, **9**, 101–18.

de Waal, F. and van Roosmalen, A. (1979). Reconciliation and consolation among chimpanzees. *Behavioral Ethology and Sociobiology*, **5**, 55–6.

Woodburn, J. (1979). Minimal politics: The political organization of the Hazda of North Tanzania. In *Politics in leadership: a comparative perspective* (eds W. A. Shack and P. S. Cohen), pp. 244–66. Clarendon Press, Oxford.

Woodburn, J. (1982). Egalitarian societies. *Man*, **17**, 431–51.

Wilson, D. S. (1980). *The natural selection of populations and communities*. Cummings, Menlo Park.

Wilson, E. O. (1975). *Sociobiology: The new synthesis*. Harvard University Press, Cambridge, Mass.

Wrangham, R. (1984). Artificial feeding of chimpanzees and baboons in their natural habitat. *Animal Behavior*, **22**, 83–93.

Wrangham, R. (1987). African apes: the significance of African apes for reconstructing social evolution. In *The evolution of human behavior: primate models*. (ed. W. G. Kinzey), pp. 51–71. SUNY Press, Albany.

7

The effects of intragroup cooperation and intergroup competition on in-group cohesion and out-group hostility

JACOB M. RABBIE

INTRODUCTION

One of the best-documented findings in social psychology has been that people tend to display more favourable attitudes and stereotypes about their own group and its members than about an out-group and its members, and are inclined to allocate more economic and symbolic rewards to members of their own group than to members of another group. This pervasive phenomenon has been labeled ethnocentrism (Sumner 1906), the 'we–they' partition (Sherif 1966), in-group bias (Rabbie and Horwitz 1969), in-group favouritism (Tajfel *et al.* 1971) or in-group–out-group differentiation (Rabbie 1982). There is only one important exception to this tendency: members of underprivileged minority groups who have a relatively low status in our society like Jews, blacks, immigrant workers, and females have often more negative opinions and stereotypes about their in-group than about the more powerful out-groups in our society, leading sometimes to low self-esteem or even to self-hatred (Lewin 1948). A stereotype is defined here as a set of beliefs about the personal attributes of a group or social category (Stroebe and Insko 1989).

Tajfel (1981) has made a useful distinction between numerical and psychological minorities. A psychological minority is a social category or segment of society that feels bound together by common traits that are held in low esteem by the more powerful majority. In some cases, psychological minorities, like the blacks in South Africa and females in our society, are actually numerical majorities. In this paper we use the term in a psychological rather than in a numerical sense.

In-group cohesion and out-group hostility have been explained at different 'levels of analysis' (Stroebe and Insko 1989). At a societal level of analysis, the origins of the positive attitudes about the in-group and the negative attitudes (and prejudices) toward underprivileged out-groups have been described in terms of socialization processes by means of the mass media, schools, peer groups, parents, and other socializing agencies. This societal approach stresses the importance of social learning processes in the development of in-group–out-group differentiation. At the intergroup level of analysis the in-group bias has been seen as a result of realistic (instrumental) or symbolic (relational) conflict between groups (e.g. ethnocentrism, realistic group conflict theory). At the individual level of analysis there are some motivational and cognitive approaches which attempt to account for ingroup cohesion and outgroup hostility in terms of intraindividual motives and personality traits (e.g. scapegoat theory, the authoritarian personality) or focus mainly on the limitations of the information-processing or cognitive capabilities of members in a group (e.g. social categorization theory and social identity theory).

In the last 15 years, research in social psychology about stereotyping and prejudice has almost exclusively been guided by social-cognitive approaches which view stereotypes as mental representations or cognitive schemas of groups and social categories and seek to understand how these cognitive structures influence information processing, social perception, (and consequently) interpersonal and intergroup behaviour (Hamilton and Sherman 1989). These dominant cognitive approaches have been reviewed elsewhere, e.g. Stephan (1985), Brewer and Kramer (1985), Stroebe et al. (1988), Bar-Tal *et al.* (1989) and Messick and Mackie (1989). In this chapter, the focus will be mainly on the motivational processes within and between individuals and groups which may account for the in-group–out-group bias. These motivational approaches emphasize the importance of perceived outcome interdependence and social conflict in intra- and intergroup relations in an attempt to explain the origins of the in-group–out-group differentiation.

By social conflict, we mean a perceived divergence of interests, goals, values, and behavioural tendencies within and between individuals, groups, organizations, nations, and larger social systems which cannot be realized simultaneously (Pruit and Rubin 1986). Social conflict can be resolved by competition or by cooperation or may result in a deadlock between the conflicting parties.

In competitive situations, the fates of the parties are negatively linked or correlated within one another: the greater the movement of one party to reach one's goals, the less the likelihood that the other parties will reach their goals (negative or 'contrient' goal interdependence, Deutsch 1982). In cooperation there is a positive correlation between the goal attainments between the parties: the parties have to coordinate their efforts since the success of one party contributes to the success of the other(s) (positive or 'promotive' goal interdependence, Deutsch 1982). People may not only perceive themselves interdependent on each other with respect to attaining their goals but may also rely on each other's actions as a means for achieving their goals (means interdependence, Raven and Rubin 1983).

In our behavioral interaction model, two types of competition and cooperation have been distinguished, dependent on the nature of the goals the actors or parties wish to achieve (Rabbie 1987, 1989*b*; in press Rabbie *et al.* 1989; Rabbie and Lodewijkx 1991). In instrumental competition the aim is to compete with others in order to achieve more economic or other tangible outcomes than another party. Instrumental cooperation occurs when the interdependence structure between the parties is such that it is better in the long run to cooperate with others to obtain tangible and material outcomes than to compete with them (Pruitt and Kimmel 1977). In social or relational cooperation the aim is to achieve a mutually satisfying relationship with the other as an aim in and of itself rather than as an instrument to reach an external tangible goal. The goal is to attain a relational understanding with the other in an attempt to explore the possibility whether the relation may grow in depth or will stay at a more superficial level (Rabbie and Schot 1989*b*). The goal of social or relational competition is aimed at differentiating oneself, one's own group, organization, or nation from comparable others in an effort to achieve more symbolic or intangible rewards like prestige (Sherif 1966), status (Lewin 1948), or a 'positive social identity' i.e. that aspect of a person's self-concept which is derived from her or his perceived membership in a social category (Tajfel and Turner 1986). Social or relational competition between groups may contribute to a positive or a negative group identity: the unique way in which members perceive their own group relative to other relevant comparison groups.

The in-group–out-group differentiation is based on in-group cohesiveness and out-group antagonism. It is our contention that the development and maintenance of in-group cohesion is a function of instrumental and relational intragroup cooperation among individual members of a group. The development of intergroup hostility and out-group antagonism is facilitated by instrumental and relational intergroup competition, particularly when the in-group members assume that the competing out-group has obtained an unfair advantage over them (Brown 1986).

Our position implies that we should search for intraindividual and interpersonal factors which facilitate the development of intragroup cooperation among individuals. At the intergroup level of analysis, conditions have to be identified which seem to favour intergroup competition or intergroup cooperation which respectively enhances or reduces the 'we–they' or 'in-group–out-group' demarcation. Finally, from a societal point of view, intergroup relations always occur in a specific sociocultural context. Therefore, the current values and ideologies should be taken into account which exalt or diminish the significance of interpersonal and intergroup competition or cooperation in particular cultures. (For an extensive discussion of this point of view, see Kohn 1986.) In our opinion, a multi-level approach is needed to provide a full explanation of the complex phenomenon of in-group cohesion and out-group hostility.

THEORETICAL PERSPECTIVES ON IN-GROUP–OUT-GROUP DIFFERENTIATION

The 'classic' approaches to the 'we–they' demarcation differ from each other in the level of analysis they use in their research. The ethnocentrism theory of Sumner (1906) emphasizes the importance of intergroup competition in the development of intergroup differentiation. Lewin (1948) argues for a multi-level approach to intergroup conflict (in particular anti-Semitism). He pays attention to the needs of individuals for status and security, the permeability of intergroup boundaries and sociocultural factors; for example the tendency of the dominating groups in the society to keep the underprivileged minority groups in their place. The psychodynamic approaches focus mainly on intraindividual motives and social learning processes to account for prejudice and racial discrimination.

Ethnocentrism

Around the turn of the century, Sumner (1906), an American sociologist, developed a functional theory of ethnocentrism. Strongly influenced by the social darwinism of his time, he believed that the differentiation of people into distinctive ethnic groups originated as a function 'of the struggle for existence' or intergroup competition. The psychological consequences of this differentiation both reflect and sustain the basic state of conflict between the in-group (or 'we-group') and outgroups (or 'others-groups'). He wrote

> The insiders in a we-group are in a relation of peace, order, law, government, and industry, to each other. Their relation to all outsiders, or others-groups, is one of war and plunder, except so far as agreements have modified it. ... the relation of comradeship and peace in the we-group and that of hostility and war toward others-groups are correlative to each other . . . The closer the neighbors, and the stronger they are, the more intense is the warfare, and then the more intense is the internal organization and discipline of each. Sentiments are produced to correspond. Loyalty to the group, sacrifice for it, hatred and contempt for outsiders, brotherhood within, warlikeness without—all grow together, common products of the same situation. . . . *Ethnocentrism* is the technical name for this view of things in which one's own group is the centre of everything, and all others are scaled and rated with reference to it . . . Each group nourishes his own pride and vanity, boast itself as superior, exalts its own divinities, and look with contempt on outsiders' (Sumner 1906, pp. 12–13).

Thus, in this view the own group is seen as the main frame of reference for judging and evaluating other groups, an idea that is elaborated by other researchers (e.g. Hyman 1942; Sherif and Sherif 1964).

In Sumner's view, the origin of in-group–out-group differentiation is the product of intergroup competition: positive 'sentiments' or attitudes about the in-group and the corresponding negative attitudes about the out-group serve the dual function of preserving in-group solidarity and cohesion and justifies the exploitation of out-groups (Brewer 1979).

Although Sumner gives many ethnographic illustrations of ethnocentric perceptions, it has been pointed out by Brewer (1986) that his reliance on anecdotal evidence allowed him considerable flexibility to focus selectively on different cases and does not constitute an adequate test of his working hypothesis.

In an effort to provide a more systematic empirical basis for Sumner's theory, Brewer (1986) conducted a large-scale questionnaire survey among ethnic groups in East Africa in 1965. She was particularly interested in the relationship between physical proximity and intergroup hostility. On the basis of Sumner's theory, it was assumed that neighboring ethnic groups, presumably because of overlapping territories, would engage in more intense intergroup conflict about scarce resources and would therefore show more intergroup differentiation than groups which lived farther apart from each other. The evidence for this hypothesis was conflicting. Support or rejection of the hypothesis appeared to depend on the different attitudinal dimensions of the intergroup relations which were assessed.

On a dimension of mutual attraction based on questionnaire items involving liking, familiarity, and perceived social distance, it was found, contrary to Sumner's hypothesis, that the average mutual attraction was highest between adjacent groups that shared a common boundary but dropped off with more remote groups. This measure of intergroup attraction was also highly positively correlated with actual and perceived similarity between groups, suggesting that the affective relationship between groups reflects the opportunity for contact and ease of interaction between in-group and out-group members. This finding is consistent with the massive evidence that a greater similarity among individuals, on a variety of dimensions, leads to an increase in mutual attraction (Byrne 1971).

A second measure of intergroup attitudes involved evaluative ratings about the trustworthiness and moral virtues of in-groups and out-groups. Although groups rated themselves uniformly higher than out-groups on this measure, no positive relationship was obtained between physical proximity and the evaluative ratings of out-groups. On the contrary, there was a tendency to ascribe more negative traits such as 'hot tempered' and 'quarrelsome' to neighbouring groups than to more remote groups. These negative characterizations of out-groups occurred more frequently among those proximate groups that shared a history of warfare and overt conflict with each other. Thus, although physical proximity may provide more opportunities for mutual social interaction, it will depend on the nature of the interaction—cooperation or competition—whether it will result in a mutual attraction or mutual antagonism.

A third dimension, involving trait attributions of perceived achievement or respect, in-groups rated themselves higher than out-groups but not nearly as strongly as on the second evaluative dimension of trustworthiness and moral virtue. In fact, 10 of the 30 groups in the survey gave somewhat higher positive ratings to at least one out-group than to the in-group. Of course, this finding goes against Sumner's hypothesis that in-groups have invariably more positive attitudes or 'sentiments' about the in-group than about the out-group. Again, it seems to depend on the nature and outcome of the intergroup interaction in the past (e.g. having won or lost of the other group) whether out-groups elicit attraction, antagonism, or respect. In any case, no evidence was found for the hypothesis that outgroups always evoke negative attitudes, as is often assumed in the literature (e.g. Turner 1981).

Sumner's theory is perhaps more of an (inaccurate) description than an explanation of the in-group–out-group phenomenon and where it is an explanation, perhaps more of a description of the maintenance and hardening of the dichotomy rather than an account of the origins of the differentiation. Additional assumptions about the effects of perceived positive and negative interdependence within and between groups, the conflict of interests between them, the ensuing perceptions of social injustice, and the experience of having won or lost to the other group are probably needed to explain the origins of the in-group cohesion and the out-group hostility.

Obviously, ethnocentrism theory cannot account for the *out*-group bias among members of underprivileged 'minority' groups whereby the out-group is favored over the ingroup. That is the main topic of Lewin's work.

Intergroup conflict and group belongingness

Kurt Lewin, a former professor of the University of Berlin, has been one of the major figures in the development of experimental and applied social psychology. As a Jewish refugee from Nazi Germany, he wrote a number of essays between 1935 and 1946, which were published in 1948, a year after his death, under the title *Resolving social conflicts*. (In our brief summary of his main ideas I use the papers published in the 1948 publication. The first date after his name refers to the year of publication of the original paper.)

In the section 'Intergroup conflicts and group belongingness', Lewin (1948) repeatedly referred to the persecution of Jews during the Nazi regime in Germany and in other parts of Europe and what this threat could mean for the maintenance and development of Jewish group identity. He emphasized the crucial importance group belongingness has for the individual. In his own words,

> The group to which an individual belongs is the ground on which he stands, which gives or denies him social status, gives or denies him security and help. The firmness or weakness of this ground might not be consciously perceived, just as the firmness of the physical ground is not always thought of. Dynamically, however, the firmness and clearness of this ground determine what the individual wishes to do, what he can do,

and how he will do it. This is true equally of the social ground as of the physical (Lewin 1940, 1948; p. 174).

Thus, in Lewin's view, membership in a group may lower or heighten the social status of an individual member and is not always a source of a 'positive social identity'. When the group cannot satisfy the basic security and status needs of the individual members, people will be inclined to leave the group when they have the opportunity to do so.

The first essay in this section was written in 1935, 2 years after Hitler had come to power in Germany. In this paper Lewin compared the life of the Jewish group during and after the Ghetto period. During this period the Jews were physically and socially separated by strong, almost impassible boundaries from the non-Jewish groups. At that time the individual Jew may have been exposed to especially high pressures when acting outside the group but on the other hand there was for him some region in which he felt 'at home', in which he could act freely as a member of his group. After the Ghetto period, with the intermingling of the Jewish and non-Jewish groups, the Jew was more often exposed to the pressure directed to him or her as an individual which, in the Ghetto period, was directed to the Jewish group as a whole. In addition, during the Ghetto period there were no desires on the part of the individual to become a member of the majority group. Even if some individual had some secret wish to cross the boundary of his group, the character of this boundary as a strong and practically impassable barrier destroyed all such hopes at once. After the Ghetto period, however, the intergroup boundaries weakened considerably, leading to an increase of the force away from one's own group and toward other non-Jewish groups.

However, as Lewin (1939, 1948) pointed out in a later paper,

It is well to realize that every underprivileged minority group is kept together not only by forces among its members but also by the boundaries the majority erects against the crossing of an individual from the minority to the majority group. It is in the interest of the majority to keep the minority in its underprivileged status. ... What then is the situation of a member of a minority group kept together merely by the repulsion of the majority? The basic factor in his life is his wish to cross this insuperable boundary. Therefore, he lives almost perpetually in a state of conflict and tension. He dislikes or even hates his group because it is nothing than a burden to him (p. 164).

An additional factor in the self-hatred and low self-esteem of a member of a minority group is that members of the lower social strata tend to accept the fashions, values, and ideals of the higher strata. In the case of an underprivileged minority group, it means that their opinions about themselves are greatly influenced by the low esteem the majority has for them. Lewin refers here to the important distinction other researchers have made between 'membership groups' and 'reference groups' which was first used by Hyman (1942) and elaborated by Sherif and Sherif (1964). Reference groups are those groups or sets of people to which a person belongs or aspires to belong which serves as a standard or reference point for judging one's self-accomplishment or self-esteem (Sherif 1966). In the case of a member of a majority group, the membership groups and reference group coincide; for a member of underprivileged groups these types of groups may be different.

Lewin (1939, 1948) was very insistent in his view that anti-Semitism is not an individual but an intergroup problem:

> If it has ever been a question whether the Jewish problem is an individual or a social one, a clear-cut answer was provided by the SA men in the streets of Vienna who beat with steel rods any Jew irrespective of his past conduct or status' (Lewin 1948, p. 161).

In another statement, he stressed the importance of contextual factors for the study of prejudice and stereotypes (Lewin 1948)

> Recent research findings have indicated that the ideologies and stereotypes which govern inter-group relations should not be viewed as individual character traits but that they are anchored in cultural standards, that their stability and change depend largely on happenings in groups as groups' (pp. 208–9).

In his definition of a group, Lewin (1939, 1948) made a sharp distinction between a 'social category', which is based on some kind of similarity among a collection of individuals, and a 'social group' which is characterized by perceived interdependence among their members. He argued

> that similarity between persons merely permits their classification, their subsumption under the same abstract concept, whereas belonging to the same social group means concrete, dynamic interrelation between persons . . . it is not similarity or dissimilarity that decides whether two individuals belong to the same or to different groups, but *social interaction or other types of interdependence*. A group is best defined *as a dynamic whole based on interdependence rather than on similarity*. (Italics in the original).

Addressing himself to the parents of Jewish adolescents in the United States who felt uncertain about their belongingness to the Jewish group, he wrote:

> . . . belonging or not belonging to the Jewish group is not a matter mainly of similarity or dissimilarity, nor even one of like or dislike. He will understand that regardless of whether the Jewish group is a racial, religious, national or cultural one, the fact that it is classified by the majority as a distinct group is what counts . . . He will see that the main criterion of belongingness is *interdependence of fate*. (Italics in the original; Lewin, 1940, 1948, p. 184)

It should be noted that Lewin wrote these words in 1940, the same year in which the victorious German armies had occupied great parts of Europe. Lewin was very concerned about the fate of the Jewish population in these areas and felt that the Jews in America, as uncertain they might be about their group belongingness should be 'sufficiently fact-minded to see clearly their interdependence of fate with the rest of the American Jews and indeed with the

Jews all over the world' (p. 184). Thus, in his view while individual Jews may be factually or 'objectively' interdependent, they only become a social group to the extent that they are perceived by others and *perceive* themselves to be mutually dependent on each other to some degree. To him, (perceived) interdependence of fate is the main defining characteristic of a social group, which is as applicable to a small face-to-face group as a 'compact unit'—in which members can directly interact with each other—as to a 'loose mass' which may be a large-scale grouping such as 'the Jews all over the world'. Thus, for Lewin, intergroup relations rather than intraindividual motives should be the starting point for studying prejudice and discrimination against minority groups. We shall see that many of Lewin's ideas were incorporated in later approaches, although that has not always been properly acknowledged.

Psychodynamic approaches

In contrast to the intergroup perspectives of Sumner (1906) and Lewin (1948), the psychodynamic approaches view the prejudice of negative attitudes towards the outgroup as a sign of some intraindividual conflict or maladjustment.

The *scapegoat theory* was first developed by Hovland and Sears (1940). It was later expanded by Allport (1954) in his classic study of prejudice and discrimination. The theory assumes that prejudice and aggression against minority groups occurs as result of a displacement of aggression from a powerful frustrator to a powerless minority group. The scapegoat theory was derived from the frustration aggression hypothesis of Dollard *et al.* (1939). In this theory an attempt was made to integrate the Freudian postulate about a universal aggressive drive with constructs drawn from learning theory in order to explain the particular circumstances in which a universal aggressive drive becomes operative (specifically in response to 'frustrating stimuli' (cf. Condor and Brown 1988). In this view, social conflict was seen to result from the simultaneous exposure of a number of individuals to the same frustrating events. Dolard *et al.* postulated that aggression, the motive or drive to injure others, is always a reaction to some frustration or interference with goal attainment. If people are blocked or frustrated in reaching their goals, they react with aggression, which is usually directed towards the source of the frustration. If this source is too powerful or cannot be identified, than the aggression will be displaced toward some less powerful, safer, clearly visible, and readily available target to which already some dislike exists (Berkowitz 1962). The scapegoat hypothesis adds the assumption to the frustration–aggression theory that members of minority groups are frequently used as targets of such displaced aggression, with the displacement being rationalized by blaming the minorities for the frustration or by attributing negative attributes to them.

Although the scapegoat theory has received some support (e.g. Hovland

and Sears 1940; Miller and Bugelski 1948), the original theory does not explain what kind of characteristics a group should have in order to be chosen as a scapegoat, like the Jews in Nazi Germany or the blacks in South Africa, except for the provision that they should be too powerless to retaliate effectively (Stroebe and Insko 1989).

The theory of the *authoritarian personality* represents another psychodynamic, individualistic explanation of in-group cohesion and out-group hostility. In dismay at the cruel treatment that Jews had received in prewar Nazi Germany, a vast research program was initiated in the 1940s in the United States by Adorno and his associates in an effort to explore the roots of anti-Semitism (Adorno *et al.* 1950). Adorno and Frenkel-Brunswick were Jewish refugees from Nazi Germany, while Levinson and Sanford were born in the United States. Their 'individual difference' approach, guided by psychoanalytic theory, was directed towards the identification of particular individuals who seemed receptive to antisemitic ideology: 'the potential fascistic individual' (Condor and Brown 1988). In the course of their research, they suggested that a particular personality type exists that not only dislikes Jews (and glorifies the own group) but also tends to dislike all minority groups. The dislikes and prejudices of negative attitudes of these individuals about minority groups were measured by an 'ethnocentrism' or E-scale. A high score on this E-scale correlated positively with a high score on the Fascist or F-scale, which was designed to measure the 'authoritarian personality'. This term was used since individuals scoring high on the F-scale tended to obey their leaders in a blind and uncritical way. Authoritarian persons are characterized by a rigid adherence to conventional norms and values which are often endorsed by religious and political leaders. They have a hierarchical view of social and political relations, glorify toughness and ruthlessness, are cynical about human nature, tend to project (or displace) their own aggressive impulses against their fathers on members of minority groups, and show a preoccupation with sexuality (Meloen 1983). Consistent with a social learning perspective, interviews indicated that children learn their basic attitudes towards authority and towards minority groups from their parents (Frenkel-Brunswick 1954). College students with high F-scores tend to have parents who also score high on the F-scale (Byrne 1965). Authoritarian parents are also more likely than non-authoritarian parents to emphasize discipline, conventionalism, and submission to authority in their child-rearing practices. This evidence suggests that authoritarianism is self-perpetuating: authoritarian parents raise authoritarian children who will tend to have prejudiced attitudes (Gergen and Gergen 1981). However, it has been shown that if one's later social group punishes authoritarian tendencies, these tendencies may be greatly reduced (Griffitt and Garcia 1979).

The relationship between authoritarian and in-group–out-group differentiation has often been studied by asking people about their attitudes about existing ethnic groups (Meloen 1983). In these studies it is difficult to ascertain whether the glorification of the in-group and the hostility about the out-group by highly authoritarian people is due to social learning processes, e.g. by the current hostile stereotypes about out-groups which are transmitted by their (authoritarian) parents, like-minded peers, or prejudiced mass media, or can be attributed to the particular personality structure of these individuals. Downing and Monaco (1979) conducted an experiment to obtain some information on this issue. Subjects attending ski-classes were given abbreviated forms of the F-test. Following a procedure developed by Rabbie and Horwitz (1969), they were randomly assigned to Blue or Green groups for the alleged purposes of ski instruction. The subjects had to rate the performance of individual skiers wearing blue or green ribbons as they skied down the hill at one time. As compared with subjects low in authoritarianism, the high authoritarians rated own-colour significantly better, rated other-colour skiers significantly worse, and differentiated more strongly between their ratings of own- and other-colour skiers. The low authoritarians showed little or no bias in favour of own-colour skiers over other-colour skiers. The results indicate that under conditions of weak in-group–out-group distinctiveness among *ad hoc*, temporary groups, high authoritarians showed a greater in-group–out-group differentiation than low authoritarians who appear to see others primarily as individuals rather than as group members.

Apart from the methodological criticism which has been levelled against this 'individual difference approach' (e.g. Brown 1965; Cherry and Byrne 1977; Christie and Jahoda 1954), the obvious argument against such an individualistic approach is that intergroup conflict by definition, involves not just particular individuals, but is often a pervasive characteristic of whole societies like Nazi Germany or South Africa (Pettigrew 1958). Nevertheless, as Downing and Monaco (1979) have shown, authoritarianism appears to contribute significantly to the in-group–out-group bias and cannot be missed in a multi-level approach to understanding and explaining this complex intergroup phenomenon. Authoritarianism may become particularly important when individuals have to act as representatives of their group or nation. It is likely that authoritarian leaders may have a strong influence on the belligerence or peacefulness of their relationships with other groups or nations as the behaviour of Adolf Hitler and Saddam Hussein would suggest (Bar-Tal 1989).

Realistic group conflict theory

Realistic conflict theory can be seen as a reaction to the individualistic psychodynamic approaches discussed earlier. As LeVine and Campbell (1972) have pointed out:

> Most who have rejected psychological explanations have espoused a point of view that is called here realistic-group-conflict theory. This theory assumes that group conflicts are rational in the sense that groups do have incompatible goals and are in competition

for scarce resources. Such 'realistic' sources of group conflict are contrasted with the psychological theories that consider intergroup conflicts as displacements or projective expressions of problems that are essentially intragroup or intra-individual in origin (p. 29).

Among those who have articulated this point of view, LeVine and Campbell (1972) mention Sumner (1906), Sherif and Sherif (1953), and Coser (1956).

Realistic group conflict theory views in-group cohesion and out-group hostility as results of competition about scarce resources. This point of view is reflected in the following propositions of LeVine and Campbell (1972, pp. 30–33):

Real conflict of interests, . . . and or presence of hostile, threatening, and competitive outgroup neighbors, which collectively may be called 'real threat', causes the perception of threat' which then results in hostility to the sources of threat.

LeVine and Campbell assume a correlation between the degree of threat and the degree of hostility. Referring to the work of Sherif (1966) they state: 'the more important the goal competed for, the greater the value that is being threatened, and the greater the perceived interference with goal attainment, the greater the hostility'. According to the theory 'real threat' causes 'in-group solidarity', 'increases the awareness of own group identity', (reflected in the exaggeration of in-group virtues and out-group vices), 'increases the tightness of group boundaries', 'reduces defection from the group', 'increases punishment and rejection of defectors', and 'creates punishment and rejection of deviants'.

Thus, in this view, realistic intergroup conflict causes in-group cohesion and out-group hostility. There is little or no attention to the possibility that intragroup cooperation, in and out of itself, may contribute to the in-group bias (Rabbie 1982).

It is important to note that it is not the actual or 'objective' threat but the *perceived* threat which causes ingroup solidarity and outgroup hostility, regardless of whether these perceptions are correct or not. Other writers (e.g. Tajfel and Turner 1986) have incorrectly emphasized the 'objective' aspects of realistic conflicts in their summary of the theory.

Coser (1956), in his reformulation of the conflict theory of Simmel (1908; translated 1955) discussed the function of realistic conflict in the formation of defensive coalitions among groups. He wrote:

Coalitions and temporary associations, rather than more permanent and cohesive groups, will result from conflicts where primarily pragmatic interests of the participants are involved. . . . Unification is at the minimal level where coalitions are formed for the purpose of defence. Alliance, then for each particular group reflects the most minimal expression of the desire for self-preservation (pp. 148–9).

(For recent reviews on coalition formation see Murnighan 1978; Kahan and Rapaport 1984, McGrath 1984 and Wilke 1985).

The research of Sherif on realistic conflict theory

LeVine and Campbell (1972) consider Sherif (1966), an American social psychologist, as one of the most important representatives of the realistic group conflict theory. Sherif (1966) emphasized the importance of studying intergroup relations at 'the appropriate levels of analysis':

> Our claim is the study of relations between groups and intergroup attitudes of their respective members. We therefore must consider both the properties of the groups themselves and the consequence of membership on individuals. Otherwise, whatever we are studying, we are not studying intergroup problems (p. 62).

Thus, for Sherif, the study of intergroup relations does not only involve the interaction of social groups as organized social units but also the interaction of individuals as members (and representatives) of these groups.

His main hypothesis is that hostile behaviour between groups is a function of competitive interdependence in which 'the success of one group necessarily means the failure of the other' (Sherif 1966, p. 63). It should be noted that Sherif (1966) does not restrict his theory to 'instrumental competition' i.e. realistic competition about 'objective' material goods, as is sometimes suggested (e.g. Tajfel and Turner 1986) but is also concerned about the effects of 'social or relational competition', i.e. the intergroup rivalry designed to achieve prestige, status or other 'symbolic outcomes' (Rabbie 1987; Rabbie and Schot 1989*a*). The mutual hostility induced by intergroup competition may be reduced by confronting the groups with 'superordinate goals', 'being those that have a common appeal for members of each group, but that neither group can achieve without participation of the other' (Sherif 1966, p. 89).

Sherif assumed that groups are formed when individuals have to cooperate with each other to achieve a common goal. Therefore, in the first stage of group formation in his famous summer camp experiments, a large group of about 22–24 children was split into two experimental groups which were carefully matched to each other on several attributes (Brown 1988). In the first two experiments it was arranged to have the majority of each best friends in the *out*-group.

In the first few days the boys in each group were asked to cooperate with each other on common activities but did not have much to do with the other group of boys in the camp. Very quickly the groups developed an internal leadership structure with leaders and followers, common norms, and values which regulated the attitudes and behaviour toward in-group members and out-group members. They gave each of these 'organized groups' its own name and other group symbols, e.g. a flag. As a consequence of the intragroup cooperation, the boys felt more attracted to the members of their own group than to the members of the out-group at the end of the group formation, even though more of their best friends were originally assigned to the out-group than to the in-group. Although the other group did not figure much in their thinking, it was observed that the 'we' and 'they'-groups spontaneously compared themselves with one another and in these comparisons 'the edge was given to one's own group' (Sherif 1966, p. 80). Thus, it appears that intragroup cooperation is already sufficient to arouse a slight in-group bias which is probably enhanced in a later phase by social and instrumental competition between the groups.

In a second stage, after the intragroup cooperation, the groups were brought into competition with each other. As a consequence, strong in-group cohesion and out-group hostility were observed between the two organized groups. In a third stage of the experiment the antagonism between the groups could only be harmonized by confronting them with urgent problems, manufactured by the experimenter, which induced 'superordinate goals' requiring mutual cooperation by both groups to achieve them.

These experiments were replicated by Blake and Mouton (1979). Because of their background as organizational consultants they had been impressed with the wide-ranging damaging effects of intergroup conflict in industrial firms. With the help of Sherif, they designed training exercises for managers in an effort to give them experimentally based insights into the origins and effects of intergroup conflicts in their own organizations. These exercises were closely designed after the field experiments of Sherif and Sherif (1979). About 1000 managers of different industrial firms participated in these exercises. Blake and Mouton obtained very similar results to the Sherifs. The effects of intergroup competition have been summarized as follows by Schein (1980).

1. *Within* the groups, the members close ranks and there is an increase in solidarity and cohesion; each group becomes more hierarchically organized: there is a greater willingness to accept centralized leadership; deviating opinions are barely tolerated: the group demands more loyalty and conformity of its members; much more attention is paid to the accomplishment of the task which will help them to win from the other group than on the socioemotional relationships among the members.

2. *Between* the groups negative and hostile stereotypes tend to develop, the own group and its product are considered to be superior to the other group and its product; the communication between the group decreases, preventing the correction of negative stereotypes and misunderstandings. During the intergroup negotiations members pay more attention to points of disagreement than they do to agreements. Distrust and hostility toward the other group rises, sometimes erupting into open conflicts and aggression. The tactics and strategy for winning are emphasized at the expense of concerns about the merits of the problem to be negotiated (Rabbie 1982).

3. *The effects of winning and losing* may help us to understand why 'losing groups' may show sometimes 'out-group favoritism' as Brewer (1986) and

Lewin (1948) have indicated. The winning group retains its cohesion and may become even more cohesive. The winner also tends to get more relaxed, lose his fighting spirit and becomes more complacent. There is an increase of relational over instrumental intragroup cooperation: a greater concern for member's needs and a low concern for work and task accomplishments. Because of its complacency, winning has confirmed the positive attitudes about the own group and the negative attitudes about the 'enemy' group. Therefore, there is little basis for re-evaluating perceptions, or re-examining group operations in order to learn how to improve on them.

The losing group has a strong tendency to deny or distort the reality of losing by seeking psychological escapes like 'we were unfairly treated by the jury', 'the rules of the game were not sufficiently explained to us'. If the loss is finally accepted, the groups tend to splinter; unresolved conflicts, which were suppressed during the fight, come again to the surface. The losing group is tense, ready to work harder and desperate to find someone or something to blame—the leader, the jury who decided against them, the rules of the game, all in an effort to find a cause for the loss. The losing group shows little relational intragroup cooperation since it has low concerns for members' needs, and a high concern for recouping by working harder. The losing group tends to learn a great deal about the group since positive attitudes about the in-group and negative attitudes about the other group are upset by the loss and have to be revaluated. As a result the losing group is likely to reorganize and may become in time more cohesive and effective, once the loss has been accepted realistically. These effects of winning and losing can be seen in the Arab–Israeli wars, and in the conflict about the Falklands or Malvinas between Argentina and Great Britain (Rabbie 1987, 1989*b*). These striking effects of intergroup conflicts and winning and losing are of course very apparent in the aftermath of the Gulf War between Iraq and the alliance.

The experiments of Sherif (1966) and Blake and Mouton (1979) can be considered as one of the major landmarks in the social psychology of intergroup relations (Diamond and Morton 1978). However, from a methodological point of view, these experiments show several shortcomings (Rabbie 1982; Dion 1979). One of the most important methodological problems is that intergroup cooperation (and intergroup competition) were not manipulated independently of each other, but after each other. In the absence of appropriate control groups, the 'experiments' of Sherif are based on a 'pre-experimental design' which leaves too much room for alternative interpretations (Campbell and Stanley 1966). In our own research we have tried to replicate the findings of Sherif and Blake and Mouton under more controlled laboratory conditions (for reviews see Rabbie 1982, 1990; Horwitz and Rabbie 1982, 1989; Rabbie 1990).

In-group–out-group differentiation between 'minimal' groups

In our research we have been very much influenced by the work of Sherif (1966) and Lewin (1948). As we have stated earlier (Rabbie *et al.* 1989), Lewin's conception of a group was the point of departure for our first minimal intergroup experiment (Rabbie 1966; Rabbie and Horwitz 1969). In line with his work, a conceptual distinction was made between a social group and social category. A social group is defined as a 'dynamic whole' or social system, characterized by the perceived interdependence among its members, whereas a social category is viewed as a collection of individuals who share at least one attribute in common. In our 1969 experiment the aim was to compare the degree of differentiation of social categories and (minimal) social groups with each other.

On the basis of his well-known camp studies, Sherif (1966) had argued that intergroup bias, i.e. having a more favourable attitude about the in-group than about an out-group, occurred only among 'organized groups'. He believed that given two groups with well-developed structures of roles, a leadership structure and shared norms and values, the existence of competing group goals (negative goal interdependence) will lead to intergroup hostility, while a cooperative superordinate goal (positive goal interdependence) will lead to intergroup harmony. As we have seen, stimulated by Sherif's ideas, Blake and Mouton (1962) designed training exercises which enabled participants to examine the conditions that either promote or prevent intergroup antagonism. Our own work on the minimal intergroup situation grew out of a number of experiences in working with the Blake–Mouton exercise that called into question elements of Sherif's theoretical formulations. First, we found that intergroup hostility was as readily aroused by competitive instructions to newly-formed groups of strangers as in well-developed groups. Second, we observed that intergroup hostility was also aroused by what we took to be cooperative or superordinate goal instructions. Moreover, we found spontaneous intergroup rivalry among well-developed groups even in the absence of explicit competitive instructions.

These unanticipated results led us to design an experiment the aim of which was 'to isolate the minimal conditions that are sufficient to generate discriminatory in-group–out-group attitudes' (Rabbie and Horwitz 1969, p. 270). These *ad hoc* 'minimal groups' were constructed in laboratory conditions, in an effort to minimize the effects of existing prejudices, stereotyping, and other intergroup practices which are acquired in our society by social learning processes about 'naturally' existing groups.

The experiment employed several treatments that varied the perceived interdependence of fate within and between two groups. Cartwright (see p. 3) (1968) following Lewin, has suggested that the 'external designation' of members into a 'socially defined category' imposes a 'common fate' upon them in the sense that opportunities are given or denied to them 'simply because of their membership in the category'. He argued that 'interdependence among members develops because society gives them a "common fate".' (pp. 56–7).

On the basis of these suggestions and those of Lewin (1948), an experiment was designed in which strangers, males and females, were classified at random into two 'distinct' Blue or Green groups for alleged 'administrative reasons'. They had no opportunity to interact with each other, either within or between the groups involved. In the experimental 'common fate' conditions, subjects were either privileged or deprived relative to an out-group solely as a function of their membership in the Blue or Green group (negative intergroup interdependence). The subjects were rewarded or not on the basis of chance (a flip of a coin), an authority figure (the experimenter) or, allegedly, by the actions of one of the two groups. Presenting the experiment as a study of first impressions, we asked subjects to stand up in turn, introduce themselves, and rate each other on a variety of personality traits scaled along a favourable–unfavourable dimension. In addition to the ratings of individual members, subjects also rated the traits of each group as a whole. Thus, the in-group–out-group bias was measured at an individual as well as an intergroup level of analysis.

As expected, in the experimental 'common fate' conditions a significant greater in-group–out-group bias was found in favour of the own group and its members, particularly in the chance condition, in comparison with a social category (control) condition, in which an attempt was made to create minimal or near-zero interdependence within and between the groups. In that condition, the subjects were only classified as members of Blue or Green groups and were neither privileged nor deprived as a function of their membership in these groups. Our 1969 experiment failed to detect a bias in the control conditon, but by using a greater number of subjects in follow-up experiments, subjects were found to give more favourable ratings to the in-group and its members than to the out-group and its members (Horwitz and Rabbie 1982, pp. 247–8). The ingroup bias occurred especially on social-emotional or relational attributes and much less on the instrumental or achievement-oriented trait ascriptions (see Brewer 1986). Apparently, the experimenter's interests in dividing them into two groups could have suggested to them that differential consequences might befall each group as a whole, leading to the perception of a positive (minimal) interdependence of fate within each group (Rabbie and Horwitz 1988, p. 118).

In our program of research we have studied the effects of intergroup cooperation and intergroup competition on the intra- and intergroup relations, in which the different stages of group formation were systematically varied: from the initial stage of the 'minimal' group as in the experiment of Rabbie and Horwitz (1969), to the intergroup competition between 'organized groups,' as studied by Sherif (1966). In this research it was shown that in-group cohesion and out-group hostility are not invariably negatively related to one another as Sumner (1906) has proposed. On the contrary, positive, rather than negative correlations have been found between in-group and out-group ratings but more so for groups engaged in intergroup cooperation than in intergroup competition (see also Visser, not published). Moreover, the in-group bias resulting from intragroup cooperation can be more attributed to a positive attitude toward the in-group than a negative attitude toward the outgroup (see also Brewer 1979). Intergroup competition produces negative attitudes about the outgroup. For a more complete review of this research see Rabbie (1982, 1990) and Horwitz and Rabbie (1982, 1989).

Social identity theory

The social identity theory of Tajfel (1978) and Tajfel and Turner (1986), is one of the most influential contemporary theories on intergroup relations. Tajfel has been concerned with the origins and functions of intergroup stereotypes and the development of one's own social identity. These life-long concerns are probably not accidental. Tajfel, born in Poland, was sent by his worried Jewish parents to France, prior to the outbreak of World War II. He joined the French Army, but was captured as a prisoner of war during the occupation of France and sent to Germany. After the war he worked for 6 years in various European countries for organizations which tried to rehabilitate the victims of war—children and adults. This work stimulated his interest in social psychology. After a lectureship at Oxford University, he became professor in social psychology at the University of Bristol, until his untimely death in 1982.

Social identity theory offers an interpretation of the processes involved in the in-group–out-group differentiation which has both cognitive and motivational aspects (Stroebe and Insko 1989). The basic assumption of the theory is that social categories—e.g. based on nationality, occupation, and gender-are not only employed by individuals to simplify and understand their social world but are also used as a means to enhance or maintain a 'positive social identity' or self-esteem. Tajfel and Turner (1986) suppose that individuals desire a positive self-esteem and that they are therefore motivated to seek positive distinctiveness on some valued dimension for the in-group comparison with the out-group. Thus, their hypothesis is, according to Turner (1981) 'that self-evaluative social comparisons directly produce competitive intergroup processes which motivate attitudinal biases and discriminatory actions' (p. 80).

Since Lewin (1948) it is no surprise that membership in a social group provides (or denies) the individual social status, self-esteem or a 'positive social identity' as Tajfel and Turner would have labelled it. From this point of view, it is striking that they never refer to this aspect of his work in their main formulations of their theory (Tajfel 1978). Tajfel and Turner (1986) stress the importance of an intraindividual motive like the need to maintain or enhance one's self-esteem, which to them, produces 'essentially competitive' intergroup relations (p. 40). In his work on the cognitive aspects of prejudice and stereotyping, Tajfel (1981) accentuates intraindividual cognitive processes of categorization, assimilation and the search for conceptual coherence.

A critical appraisal of social identity theory

Recently, a debate has been initiated by Turner (1982, 1985) between the social identity approach and a 'cohesion' or 'interdependence perspective' as he calls our point of view. In his self-categorization theory—an elaboration of social identity theory—he rejects the interdependence perspective in favour of his own approach (Turner *et al.* 1987). The main evidence for his position is based on the allocation behavior in the minimal group paradigm (MGP) developed by Tajfel *et al.* (1971).

Doise (1988) has noted that the experiment designed by Tajfel *et al.* (1971) was modelled with some modifications on the control condition of the Rabbie–Horwitz (1969) experiment. In their initial modification of our control condition subjects were again classified into groups for 'administrative reasons'. However, this time they were led to believe that they were divided according to similarities or differences in their individual characteristics, purportedly measured by tests of aesthetic preference (pro-Kandinski versus pro-Klee) or of estimation tendencies (overestimators versus underestimators). The dependent measures of bias were a series of matrices by means of which subjects could allocate money to *and receive money from* anonymous members of each group.

As noted by Horwitz and Rabbie (1989), Tajfel's modifications had the effect, first, of transforming the experiment from one that manipulated intergroup interdependence as the sole independent variable to one that simultaneously manipulated two independent variables: intergroup interdependence and category differentiation. A second effect, as we shall show below, was to make explicit and strengthen subjects' perceptions of their differential interdependence with ingroup and outgroup members regarding their own monetary outcomes. The experiment by Tajfel *et al.* (1971) found that subjects tended to allocate more money to in-group members than to out-group members, but that they did not depart too far from fairness, i.e. from equal allocations to both groups.

The discovery of Tajfel *et al.* (1971) that the mere categorization of a group of boys into two subgroups seemed sufficient to arouse a significant in-group favouritism among them appears to be rather damaging to the basic assumptions of realistic conflict theory of LeVine and Campbell (1972). For example, in a recent review on intergroup relations, it has been concluded that: 'This discovery challenged the idea that intergroup discrimination resulted from a real conflict of interests between the two groups.' (Messick and Mackie 1989, p. 59). This point of view is shared by many other researchers (e.g. Brewer 1979; Tajfel and Turner 1986; Brewer and Kramer 1985; Brown 1988). As we will try to show, this common view is incorrect. Realistic group conflict theory cannot be excluded as a viable explanation of the 'in-group favouritism' found in the MGP.

Tajfel and Turner (1986) are insistent in their belief that in the MGP there is no realistic conflict of interests or instrumental, competition between the group members, 'nor', as they write, 'is there any rational link between economic self-interest and the strategy of in-group favoritism' (pp. 38–9).

From an interdependence perspective we take a different position. In the standard instruction of the MGP, the subjects in the Tajfel *et al.* (1971) experiment were told that 'They would always allot money to others' (i.e. always members of the in-group or out-group). 'At the end of the task . . . each would receive the amount of money that the others had rewarded him.' (p. 156). These instructions imply that the individual outcomes of the subjects in the Tajfel *et al.* (1971) experiment are dependent upon both the allocation decisions of the in-group and out-group members, but not on their own decisions. Thus, their 'outcome dependence' (Kelley and Thibaut 1986) is two-sided: on their own group and the *other* group. Although subjects in the standard MGP of Tajfel *et al.* cannot directly allocate money to themselves, they can do it *indirectly* on the reasonable assumption that the other in-group members will do the same to them. By giving more to their in-group members than to the outgroup members, in the expectation that the other in-group members will reciprocate this cooperative action, they will increase the chances of maximizing their own outcomes. Thus, although subjects in the MGP may be seen as acting independently from each other, they *perceive* themselves to be interdependent with respect to reaching their goal to maximize their outcomes. When they act on these perceptions, a 'tacit coordination' (Schelling 1963) is achieved, which helps them to gain as much as they can in the MGP.

In a recent experiment we have found strong support for our interdependence hypothesis (Rabbie *et al.* 1989). In this study three conditions of interdependence were employed among ingroup and outgroup members. One control condition replicated the standard Tajfel situation. Subjects, categorized as favouring one of two series of paintings were informed that they would receive monetary points from an anonymous individual in the in-group and an anonymous individual in the out-group. Thus, in the control condition the interdependence was two-sided. In the two other one-sided conditions, they were informed that they would receive money from someone in the in-group alone or in the out-group alone. The results showed that in the standard control condition they displayed the usual behaviour by giving more to in-group members than to out-group members. Where they viewed their outcomes as solely dependent on the actions of in-group members, they gave much more to in-group members than in the control condition and even less to out-group members. However, where they viewed their outcomes as dependent solely on the actions of the out-group, they allocated more money to *out-group* members than to in-group members.

The clear-cut evidence is that by altering the perceived interdependence of subjects' outcomes, one can obtain either in-group or out-group favouritism in the MGP, holding category differentiation constant. Moreover, in the two-sided control condition, more (instrumental) fairness was obtained than in the two one-sided experimental conditions. Thus, in contrast to Tajfel and Turner (1986), it is shown that there *is* an instrumental link between allocation behavior in the MGP and the economic self-interests of the subjects. The allocations are not so much guided by social competition in an effort to achieve a positive social identity, as they claim, but by instrumental, utilitarian cooperation with those group members they perceive themselves to be the most dependent upon.

In their early explanation of the in-group favouritism, Tajfel *et al.* (1971) stressed the importance of a 'generic group norm' to favour one's own group. Later this group norm hypothesis was abandoned in favour of social identity theory (Tajfel 1978). We have assumed that these kinds of group norms contribute to the in-group–out-group differentiation. In an earlier paper we argued that groups expect people to give more weight to the desires of in-group members than to the desires of out-group members (Rabbie and Horwitz 1982). It is to be expected that this group norm would have stronger effects on the MGP the lower the rewards that are at stake. In a replication of the Rabbie *et al.* (1989) experiment, symbolic points rather than monetary points had to be allocated (Rabbie and Schot 1989*a*). In support of the hypothesis, more in-group favouritism was obtained in the points than in the monetary conditions even where subjects perceived themselves to be dependent upon the outgroup for achieving their outcomes. Thus, it appears that the allocation of symbolic points generate more social competition than instrumental cooperation in the MGP.

A recent experiment suggests that subjects in the standard MGP engage in (inter) individual rather than in (inter) group behaviour (Rabbie and Schot 1989b). In this study, subjects participating in a standard MGP, were urged in an individual condition 'to maximize their individual interests' and in a group condition 'to maximize their group interests'. In a control condition, a standard MGP was run in which the usual instructions were given. As expected, no differences in allocation behaviour were obtained between the individual and control conditions, but both showed significantly different allocations from those in the group condition. In the group condition more (instrumental) intragroup cooperation and more (instrumental) intergroup competition occurred than in the individual and control conditions. These results indicate that in standard MGP of Tajfel *et al.* (1971), we are dealing more with interindividual than with intergroup behaviour as Turner (1982) has claimed.

Intergroup competition and in-group–out-group differentiation

According to Sumner (1906) and the realistic group conflict theory of Sherif (1966) and LeVine and Campbell (1972), it is to be expected that intergroup competition would produce a greater ingroup cohesion than intergroup cooperation. Presumably, intergroup competition provides the group with a single operational objective: the superordinate goal of winning from the other group which should override possible animosities between the members within the group. Since winning requires a strong effort on the part of all members to cooperate with each other, intergroup competition (which requires intragroup cooperation) should lead to a strong ingroup cohesiveness or solidarity. In our laboratory research there is much less support for this popular hypothesis then one would expect. Generally, intergroup competition produces no more ingroup cohesion than intergroup cooperation, except when people believe that the intergroup conflict can be brought to a successful conclusion (Rabbie *et al.* 1974). As we concluded, 'The crucial factor in producing more cohesiveness in all these studies seem to be whether the group members view themselves as achieving the goals they have set for themselves, whether in a cooperative or a competitive relationship with another group' (Rabbie 1982, p. 133).

There is much more support for the hypothesis that intergroup competition produces more negative attitudes about the outgroup than intergroup cooperation. Mutual frustration or actual interference with goal-directed activities among competitive groups create more outroup hostility than intergroup cooperation in which the groups help each other to attain their goals (Rabbie 1982).

Outgroup hostility and angry aggression

Brown (1986) has suggestged that negative stereotypes, like ethnocentrism, are not the cause of group hostilities but they serve primarily to justify and explain 'with cartoon-like simplicity', hostilities that derive from conflicts of interests leading to unjust outcomes. In his view: '*Unfair* or unjust disadvantage or frustration is the sovereign cause of anger and aggression. Disadvantage or frustration will not alone do it; perceived injustice is critical' (Brown 1986, p. 534, italics in the original).

In two experiments we have found some evidence for his point of view (Rabbie and Horwitz 1982). In these studies a two-trial and two-party prisoner's dilemma game was used (see Boyd, this volume, pp. 473–89). In the first experiment, male individuals and dyads had to make an individual or a collective decision in the game. Each individual was paired with another single male, while a dyad was always pitted against another dyad. The other party could be seen on a TV screen in front of the subjects. Unknown to them the other (programmed) party was prerecorded on video tape. At the first (sequential) trial, the programmed party made always a cooperative choice to which the subjects could respond by making a cooperative or a defecting choice. Prior to the second trial, which both parties played simultaneously, the parties could send messages to each other. The subjects received a message of the programmed party promising that he would again make a cooperative choice on the second trial and requesting the other party to do the same. The subjects could respond with a message of their own before making their own choice on a pay-off matrix whose values were increased by about a half. In the *norm-violation condition* the other party broke his promise by making a competitive rather than a cooperative choice. In the *norm-adherence condition*, the opponent kept his promise and made a cooperative choice. After the choices were exchanged, the subjects could react to the behaviour of the other party by delivering (painful) white noise to him. The noise had been described to them as one in which they could indicate the extent of their disagreement. The other party did not have this opportunity to do so. It was expected that a breach of promise to make a cooperative choice would evoke more angry aggression than when the other dyad had kept his promise by making a cooperative choice.

According to our behavioural interaction model, intragroup interactions enhances the cognitive, emotional, motivational, and normative orientations that are already present in single individuals (Rabbie 1987, 1989*a*; Rabbie and Lodewijkx, 1985). Our hypothesis was that dyads would react with more aggression than single individuals. Aggression is defined here as the intentional injury of another interdependent party: an individual or a group. The opportunity to send painful white noise to the earphones of the other party was used as a measure of physical aggression.

As expected individuals and groups reported more feelings of anger, tension, and uncertainty when the other party broke the promise than when he did not. Moreover, subjects in the *norm-violation* condition considered the other party as more hostile, less fair, morally wrong, more cowardly, too interested in earning money, and more dishonest than subjects in the *norm-adherence* condition. All these differences were significant beyond the $p = 0.0001$ level of statistical significance, using an analysis of variance.

Consistent with our hypothesis, more painful white noise was delivered to the other in the norm-violation condition (mean = 5.3) than in the norm-adherence condition (mean = 0.0); ($F = 596.3$; d.f. 1,46; $p < 0.0001$). As expected, dyads reacted with more angry aggression (mean = 5.80) to the deceptive behavior of the other party than individuals did (mean = 4.8); ($F = 4.63$; d.f. 1,46; $p < 0.03$).

The *norm-violation* condition differed from the norm-adherence condition in at least two ways. The *norm-violator* broke his promise and played competitively, while the non-violator kept his promise and played cooperatively. It is therefore impossible to ascertain whether the difference in aggression obtained between dyads and individuals reflect differences in their reaction to the competition than to the norm violation.

In a second experiment an attempt was made to disentangle these effects of the two components by varying the breach of promise but keeping the competitive behaviour of the opponent constant. On the first trial the programmed party made again a cooperative choice to which the subjects could respond. Prior to the second trial, subjects in the *breach-of-promise* condition, but not in a *no-promise* condition, received a promise that their opponents would play cooperatively. Subjects then made a first move on that trial. In both conditions, the opponents always responded with a competitive choice. Subjects could then administer the white noise which was described as enabling them to express the degree of approval or disapproval of the behaviour of the other party.

Both individuals and dyads delivered white noise of higher intensities in the *breach-of promise* (mean = 3.8) than in the *no-promise* condition (mean = 2.3); $F = 5.38$; $p < 0.02$). Again the individuals and dyads had a more negative view of the other in the promise than in the no-promise condition.

As expected dyads reacted more aggressively (mean = 4.7) to the breach of promised than individuals (mean = 3.1; $p < 0.05$), but not differently from individuals where the opponent had made no promise at all (means = 2.3 versus 2.4). These results indicate that the greater aggression by dyads than individuals is due to the difference in how they respond to the norm violation and not to the difference in how they respond to the competitive behaviour of the opponents. Apparently, intragroup interaction enhances group aggression. In other experiments, very similar conclusions have been reached (Rabbie and Lodewijkx 1987; 1990, 1991; Rabbie and Lodewijkx 1985; Rabbie and Goldenbeld 1988; Lodewijkx 1989).

CONCLUSIONS

In addition to the conclusions mentioned in the summary, this chapter has tried to make the following points:

1. Research on ethnic groups, stimulated by ethnocentrism theory, indicates that groups will not have invariably positive attitudes about the in-group and hostile attitudes about the out-group. It will depend on the attitudes being assessed and the history and outcomes of intergroup conflict whether this pattern will occur or not.

2. In contrast to ethnocentrism theory, in-group cohesion and out-group hostility are not always negatively related to one another. On the contrary, in experimental studies positive rather than negative correlations have been obtained between in-group and outgroup ratings but more so for groups

engaged in intergroup cooperation than those involved in intergroup competition.

3. The in-group bias resulting from intragroup cooperation can be more attributed to a positive attitude toward the in-group than to a negative attitude toward the out-group.

4. The mere categorization of individuals into Blue and Green groups seems sufficient to elicit more positive ratings of the in-group and its members than of the out-group and its members. The experience of a positive 'interdependence of fate' within a group and a negative interdependence of fate between the groups appears to strengthen the in-group–out-group differentiation among 'minimal groups'.

5. It is likely that the group norm to give greater weight to the desires of in-group members than of out-group members contributes to the in-group favoritism in the MGP, especially when symbolic points rather than monetary rewards have to be allocated.

6. Contrary to the claims of social identity theory, there is some evidence that allocations in the MGP can be better interpreted as interpersonal than as intergroup behavior.

7. The greater aggression by dyads than individuals when norms are violated (i.e. promises not to compete are broken) is due to the difference in how they respond to the norm violation and not to the difference in how they react to the competitive behaviour of their opponents.

SUMMARY

1. In this chapter various social psychological theories have been reviewed which attempt to explain the origins of in-group–out-group differentiation: the pervasive tendency to favour in-groups and to derogate out-groups in attitudes as well as in discriminatory behaviour. It is argued that a multi-level analysis is needed to account for this complex phenomenon.
2. Our main thesis is that intragroup cooperation enhances in-group cohesiveness and solidarity, while intergroup competition induces out-group hostility and aggression, especially when the conflict of interests seem to lead to unfair outcomes between the groups.
3. The intergroup theories of Sumner and Lewin in the first half of the century on ethnocentrism and antisemitism concentrated on how perceptions of differences between in-groups and out-groups influenced attitudes between them.
4. Individualistic, psychodynamic oriented theories have suggested that frustration leads to scapegoating, and that authoritarian personalities are more likely than non-authoritarians to favor in-groups over out-groups.

5. At a group level of analysis, realistic group conflict theory emphasized the influence on differentiation of actual and perceived conflict for resources and status. Classic studies by Sherif and Sherif and later by Blake and Mouton showed that experimentally manipulated intergroup competition produces predictable effects on intragroup and intergroup relations, like centralization of leadership and conformity pressures within the group and on mutual hostility and stereotyping between the groups and the different ways in which winning and losing groups perceive each other.
6. Our work has demonstrated that mere categorization of individuals into different groups is sufficient to cause differentiation favouring the in-group, but that this in-group bias is considerably influenced by a positive perceived interdependence of fate, especially when realistic rather than symbolic conflicts are at stake. While Tajfel and Turner have suggested that the important variable in in-group bias is the enhancement of self-esteem by social (intergroup) competition, we have focused more on conditions that facilitate instrumental (intragroup) cooperation as an explanation for this bias.
7. Further experiments suggest that in-group cohesion is not any more influenced by intergroup competition, as earlier studies suggested, than by intergroup cooperation. Instead the increase in ingroup cohesion will depend on the beliefs of the group members that, by their own efforts, they will be able to bring the intergroup conflict to a successful conclusion. Intergroup competition seems to produce intergroup hostility rather than in-group cohesion.
8. Finally, we have shown that norm violation, e.g. breaking of promises by potential competitors produces more hostility to them from both groups and individuals than does their competitiveness *per se*. Importantly in the context of this volume, groups react more angrily than do individuals to the norm violation but not to the competitiveness.

ACKNOWLEDGEMENTS

The research reported in this paper was funded by NWO, The Netherlands Organization for the Advancement of Research, grants 57-07, 57-97, and 560-270-012.

REFERENCES

Adorno, T. W., Frenkel-Brunswick, E., Levinson, D. J., and Sanford, R. N. (1950). *The authoritarian personality*. Harper and Row, New York.
Allport, G. W. (1954). *The nature of prejudice*. Addison-Wesley, Reading, MA.
Bar-Tal, Y. (1989). Can leaders change followers' stereotypes? In *Stereotyping and*

prejudice: changing perceptions (eds D. Bar-Tal, C. F. Graumann, A. W. Kruglanski, and W. Stroebe), pp. 225–42. Springer-Verlag, Berlin.

Bar-Tal, D., Graumann, C. F., Kruglanski, A. W., and Stroebe, W. (eds) (1989). *Stereotyping and prejudice: changing perceptions*. Springer-Verlag, Berlin.

Berkowitz, L. (1962). *Aggression: a social psychological analysis*. McGraw-Hill, New York.

Blake, R. R. and Mouton, J. S. (1979). Intergroup problem solving in organizations: from theory to practice. In *The social psychology of intergroup relations* (eds W. G. Austin and S. Worchel), pp. 19–32. Brooks/Cole, Monterey.

Brewer, M. B. (1979). Ingroup bias in the minimal intergroup situation: cognitive-motivational analysis. *Psychological Bulletin*, **186**, 307–24.

Brewer, M. B. (1986). The role of ethnocentrism in intergroup conflict. In *The social psychology of intergroup relations*, 2nd edn (eds S. Worchel and W. G. Austin), pp. 88–102. Nelson-Hall, Chicago.

Brewer, M. B. and Kramer, R. M. (1985). The psychology of intergroup attitudes and behavior. *Annual Review of Psychology*, **36**, 219–243.

Brown, R. (1965). *Social psychology*. MacMillan, New York.

Brown, R. (1986). *Social psychology: the second edition*. Free Press, New York.

Brown, R. J. (1988). *Group processes: dynamics within and between groups*. Basil Blackwell, Oxford.

Byrne, D. (1965). Parental antecedents of authoritarianism. *Journal of Personality and Social Psychology*, **1**, 369–73.

Byrne, D. (1971). *The attraction paradigm*. Academic Press, New York.

Campbell, D. T. and Stanley, J. C. (1966). *Experimental and quasi-experimental designs for research*. Rand McNally, Chicago.

Cartwright, D. (1968). The nature of group cohesiveness. In *Group dynamics*, 3rd edn. (eds D. Cartwright and A. Zander), pp. 45–62. Tavistock, London.

Cherry, F., and Byrne, D. (1977). Authoritarianism. In *Personality variables in social behavior* (ed. T. Blass), pp. 109–33. Erlbaum, Hillsdale, NJ.

Christie, R. and Jahoda, M. (1954). *Studies in the scope and method of the 'authoritarian personality'*. Free Press, New York.

Condor, S. and Brown, R. (1988). Psychological processes in intergroup conflict. In *The social psychology of intergroup conflict: theory, research and application* (eds W. Stroebe, A. W. Kruglanski, D. Bar-Tal, and M. Hewstone), pp. 3–26. Springer-Verlag, Berlin.

Coser, L. (1956). *The function of social conflict*. Free Press, New York.

Deutsch, M. (1982). Interdependence and psychological orientation. In *Cooperation and helping behavior* (eds V. J. Derlega and J. Grzelak), pp. 15–42. Academic Press, New York.

Diamond, S. S. and Morton, D. R. (1978). Empirical landmarks in social psychology. *Personality and Social Psychology Bulletin*, **4**, 217–21.

Dion, K. L. (1979). Intergroup conflict and intragroup cohesiveness. In *The social psychology of intergroup relations* (eds W. G. Austin and S. Worchel), pp. 211–24. Brooks/Cole, Monterey.

Doise, W. (1988). Individual and social identities in intergroup relations. *European Journal of Social Psychology*, **18**, 99–111.

Dollard, J., Miller, N. E., Doob, L. W., Mowrer, O. H., and Sears, R. R. (1939). *Frustration and aggression*. Yale University Press, New Haven.

Downing, L. L. and Monaco, N. R. (1979). *Ingroup-outgroup bias formation as function of differential ingroup-outgroup contact and authoritarian personality: a field experiment*. Mimeo, Union College, Schenectady.

Frenkel-Brunswick, E. (1954). Further exploration by a contributor to '*The authoritarian personality*'. In *Studies in the scope and method of the 'authoritarian personality'* (ed. R. Christie and M. Jahoda). Free Press, New York.

Gergen, K. J. and Gergen, M. M. (1981). *Social psychology*. Harcourt Brace Jovanowitc, New York.

Griffitt, W. and Garcia, L. (1979). Reversing authoritarian punitiveness: The impact of verbal conditioning. *Social Psychological Quarterly*, **42**, 55–61.

Hamilton, D. L. and Sherman, S. J. (1989). Illusory correlations: implications for stereotype theory research. In *Stereotyping and prejudice: changing perceptions*. (eds D. Bar-Tal, C. F. Graumann, A. W. Kruglanski, and W. Stroebe), pp. 59–82. Springer-Verlag, Berlin.

Horwitz, M. and Rabbie, J. M. (1982). Individuality and membership in the intergroup system. In *Social identity and intergroup relations* (ed. H. Tajfel), pp. 241–74. Cambridge University Press, Cambridge.

Horwitz, M. and Rabbie, J. M. (1989). Stereotypes of groups, group members and individuals in categories: A differential analysis of different phenomena. In *Stereotyping and prejudice: Changing perceptions* (eds D. Bar-Tal, C. F. Graumann, A. W. Kruglanski, and W. Stroebe), pp. 105–29. Springer-Verlag, Berlin.

Hovland, C. I., and Sears, R. R. (1940). Minor studies in aggression: VI Correlations of lynchings with economic indices. *Journal of Psychology*, **9**, 301–10.

Hyman, H. H. (1942). The psychology of status. *Archives of Psychology*. No. 269.

Kahan, J. P. and Rapaport, A. (1984). *Theories of coalition formation*. Erlbaum, Hillsdale, NJ.

Kelley, H. H. and Thibaut, J. W. (1978). *Interpersonal relations: a theory of interdependence*. Wiley Interscience, New York.

Kohn, A. (198). *No contest: the case against competition*. Houghton Mifflin, Boston.

LeVine, R. A. and Campbell, D. T. (1972). *Ethnocentrism: Theories of conflict, ethnic attitudes and group behaviour*. Wiley, New York.

Lewin, K. (1948). *Resolving social conflicts: selected papers on group dynamics*. Harper and Row, New York.

Lodewijkx, H. (1989). Aggression between Individuals and Groups. Dissertation, University of Utrecht (in Dutch).

McGrath, J. E. (1984). *Groups: interaction and performance*. Prentice-Hall, Englewood Cliffs, NJ.

Meloen, J. D. (1983). The Authoritiarian Reaction in Times of Prosperity and Crises. Dissertation, University of Amsterdam (in Dutch).

Messick, D. M. and Mackie, D. M. (1989). Intergroup relations. *Annual Review of Psychology*, **40**, 45–81.

Miller, N E. and Bugelski, R. (1948). Minor studies in aggression: The influence of frustration imposed by the in-group on attitudes expressed toward out-groups. *Journal of Psychology*, **25**, 437–42.

Murninghan, J. K. (1978). Models of coalition formation: Game theoretic, social psychological, and political perspectives. *Psychological Bulletin*, **85**, 1130–53.

Pettigrew, T. F. (1958). Personality and sociocultural factors in intergroup attitudes: a cross-national comparison. *Journal of Conflict Resolution*, **2**, 29–42.

Pruitt, D. G. and Kimmel, M. J. (1977). Twenty years of experimental gaming: critique, synthesis, and suggestions for the future. *Annual Review of Psychology*, **28**, 363–92.

Pruitt, D. G., and Rubin, J. Z. (1986). *Social conflict: escalation, stalemate, and settlement*. Random House, New York.

Rabbie, J. M. (1966). Ingroup-outgroup differentiation under minimal social conditions. Second European Conference of Experimental Social Psychology, Sorrento. Italy, December 1966.

Rabbie, J. M. (1982). The effects of intergroup competition on intragroup and intergroup relationships. In *Cooperation and helping behavior: theories and research*. (eds V. J. Derlega and J. Grzelak), pp. 123–49. Academic Press, New York.

Rabbie, J. M. (1987). Armed conflicts: towards a behavioural interaction model. In *European psychologists for peace. Proceedings of the Congress in Helsinki, 1986.* (eds J. von Wright, K. Helkama, and A. M. Pirtilla-Backman), pp. 47–76.

Rabbie, J. M. (1989*a*). Group processes as stimulants of aggression. In *Aggression and war; their biological and social bases* (eds J. Groebel and R. A. Hinde), pp. 139–55. Cambridge University Press, Cambridge.

Rabbie, J. M. (1989*b*). A behavioural interaction model: a theoretical framework for studying terrorism. Paper presented to the International Society on Aggression 5th European Conference, Szombathely, Hungary, 25–30 June 1989.

Rabbie, J. M. (1990). Aggressive conflicts between individuals and groups. In *War: An interdisciplinary approach* (eds J. A. R. A. M. van Hooff, S. Benthem van de Bergh, and J. M. Rabbie), pp. 175–236. Stubeg, Hoogezand (in Dutch).

Rabbie, J. M. (1991). Determinants on instrumental intragroup-cooperation. In *Cooperation, Trust and Commitment* (ed. R. A. Hinde and J. Groebel). Cambridge University Press.

Rabbie, J. M. and Goldenbeld, C. (1988). The effects of modelling and accountability on aggressive intergroup behavior. Paper presented to the World Congress of ISRA, Swansea, UK, 2–6 July 1988.

Rabbie, J. M. and Horwitz, M. (1969). The arousal of ingroup-outgroup bias by a chance win or loss. *Journal of Personality and Social Psychology*, **69**, 223–8.

Rabbie, J. M. and Horwitz, M. (1982). Conflicts and aggression among individuals and groups. In *Proceedings of the XXIInd International Congress of Psychology*, Leipzig, DDR, No. 8, Social Psychology (eds H. Hiebsch, H. Brandstäter, and H. H. Kelley), pp. . North-Holland, Amsterdam.

Rabbie, J. M. and Horwitz, M. (1988). Categories versus groups as explanatory concepts in intergroup relations. *European Journal of Social Psychology*, **18**, 117–23.

Rabbie, J. M. and Lodwijkx, H. (1985). The enhancement of competition and aggression in individuals and groups. In *Social/ecological psychology and the psychology of women*. (ed. F. L. Denmark), pp. 177–87. North-Holland, Amsterdam.

Rabbie, J. M. and Lodewijkx, H. (1987). Individual and group aggression. *Current Research on Peace and Violence*, **2–3**, 91–101.

Rabbie, J. M. and Lodewijkx, H. (1991). Aggressive reactions to social injustice by individuals and groups: Towards a Behavioral Interaction Model. In *Social injustice in human relations. Vol. 1: Societal and psychological origins of justice (critical issues in social justice)* (eds R. Vermunt and H. Steensma), pp. 279–361. Plenum, New York.

Rabbie, J. M., Lodewijkx, H., and Broeze, M. (1985). Individual and group aggression under the cover of darkness. Paper presented to the symposium, *Psychology of peace*, at the third European Congress of the International Society for Research on Aggression (ISRA) devoted to Multidisciplinary Approaches to Conflict and Appeasement in Animals and Men. Parma, Italy, September 3–7, 1985.

Rabbie, J. M. and Schot, J. C. (1989*a*). Instrumental and elational behavior in the minimal group paradigm. Paper presented to the 1st Congress of Psychology, Amsterdam, 2–7 July 1989.

Rabbie, J. M. and Schot, J. C. (1989*b*). Group allocations in the minimal group paradigm: fact or fiction? Paper presented to the 1st Congress of Psychology, Amsterdam 2–7 July 1989.

Rabbie, J. M., Benoist, F., Oosterbaan, H., and Visser, L. (1974). Differential power and effects of expected competitive and cooperative intergroup interaction on intragroup and outgroup attitudes. *Journal of Personality and Social Psychology*, **30**, 46–56.

Rabbie, J. M., Schot, J. C., and Visser, L. (1989). Social identity theory: a conceptual and empirical critique from the perspective of a behavioural interaction model *European Journal of Social Psychology*, **19**, 171–202.

Raven, B. H. and Rubin, J. Z. (1983). *Social psychology*. Wiley, New York.

Schein, E. H. (1980). *Organizational psychology*, 3rd edn. Prentice-Hall, Englewood Cliffs, NJ.

Schelling, T. C. (1963). *The strategy of conflict*. Harvard University Press, Cambridge, MA.

Sherif, M. (1966). *In common predicament: social psychology of intergroup conflict and cooperation*. Houghton Miffin, Boston.

Sherif, M. and Sherif, C. W. (1953). *Groups in harmony and tension*. Harper, New York.

Sherif, M. and Sherif, C. W. (1964). *Reference groups: exploration into conformity and deviation of adolescents*. Harper and Row, New York.

Sherif, M. and Sherif, C. W. (1979). Research on intergroup relations. In *The social psychology of intergroup relations* (eds W. G. Austin and S. Worchel), pp. 7–18. Brooks/Cole, Monterey.

Simmel, G. (1955). *Conflict*. Free Press, K. M. Wolf, Glencoe, Ill.

Stephan, W. G. (1985). Intergroup relations. In *The handbook of social psychology*, Vol. 2 (ed. G. Lindzey and E. Aronson), pp. 599–658. Random House, New York.

Stroebe, W., Kruglanski, A. W., Bar-Tal, D., and Hewstone, M. (eds). (1988). *The social psychology of intergroup conflict. Theory, research and application*. Springer-Verlag, Berlin.

Stroebe, W. and Insko, Ch. A. (1989). Stereotype, prejudice, and discrimination: Changing conceptions in theory and research. In *Stereotyping and prejudice: changing perceptions* (eds D. Bar-Tal, C. F. Graumann, A. W. Kruglanski, and W. Stroebe), pp. 3–34. Springer-Verlag, Berlin.

Summer, W. G. (1906). *Folkways*. Ginn, New York.

Tajfel, H. (1978). *Differentiation between social groups: studies in the social psychology of intergroup relations*. Academic Press, London.

Tajfel, H. (1981). *Human groups and social categories*. Cambridge University Press, Cambridge.

Tajfel, H., Billig, M. G., Bundy, R. P., and Flament, C. I. (1971). Social categorization and intergroup behaviour. *European Journal of Social Psychology*, **1**, 149–78.

Tajfel, H. and Turner, J. C. (1986). An integrative theory of social conflict. In *The social psychology of intergroup relations*, 2nd edn (eds S. Worchel and W. G. Austin), pp. 7–24. Nelson-Hall, Chicago.

Turner, J. C. (1981). The experimental social psychology of intergroup behaviour. In *Intergroup behaviour* (eds J. C. Turner and H. Giles), pp. 66–101. Basil Blackwell, Oxford.

Turner, J. C. (1982). Towards a cognitive redefinition of the social group. In *Social identity and intergroup relations*. (ed. H. Tajfel), pp. 15–40. Cambridge University Press, Cambridge.

Turner, J. C. (1985). Social categorization and the self-concept: a social cognitive theory of group behaviour. In *Advances in group processes: theory and research*, vol. 2. (ed. E. J. Lawler), pp. 77–122. JAI Press, Greenwich, CT.

Turner, J. C., Hogg, M. A., Oakes, P. J., Reicher, S. D., and Wetherell, M. S. (1987). *Rediscovering the social group: selfcategorization theory*. Basil Blackwell, Oxford.

Visser, L. (1990). Competition and cooperation between individuals and groups. Unpublished manuscript, University of Utrecht, Utrecht.

Wilke, H. A. M. (ed.) (1985). *Coalition formation*. North-Holland, Amsterdam.

PART II

Cooperative strategies in the 'political' arena

Cooperative strategies in the 'political' arena: Introduction

When considering the reasons for coalition formation we need to distinguish between motivational reasons, or goals—such as the desire to acquire a resource, or to defeat a rival—and the evolutionary reasons for the existence of this type of cooperative behaviour. The present section focuses on the first type of reasons, that is, on the strategic choices and immediate effects underlying coalition formation. By immediate effects we mean perceptible pay-offs that may serve as the incentive for a behaviour. Evolutionary reasons, by contrast, involve a time-scale far removed from the animal's experience and decision-making. They relate to the behaviour's impact on survival and reproduction, and will be discussed at length in Part III.

This is not to say that evolutionary considerations will be ignored here. A theme that runs through most of the chapters is that of reciprocity of support, which obviously pertains to Trivers' (1971) theory of the evolution of reciprocal altruism. The first three chapters—each in a different manner and for different species—demonstrate a reciprocal distribution of interventions in fights. Positive correlations exist between support given by one individual to another, and *vice versa*. While this result is consistent with the view of coalition formation as a form of reciprocal altruism, it is noteworthy that the authors use this term rather sparingly.

In a recent workshop, which reviewed progress during the 15 years since the theory had been set out, participants agreed that reciprocal altruism *sensu strictu* requires that (1) the exchanged acts, while beneficial to the recipients, are costly to the performers; (2) there is a significant time delay between given and received acts; and (3) giving is contingent upon receiving. No such demanding requirements are made for mutualistic cooperation. During cooperation individuals act out of an expectancy of immediate benefits, and several individuals may benefit at the same time (Taylor and McGuire 1988; see also Connor 1986).

It is hard, using exclusively observational methods, to clarify this issue for non-human primates. While their coalitions, particularly those of a defensive nature (i.e. 'aiding-at-risk'; Bernstein and Ehardt 1985), often appear to meet the first two requirements for reciprocal altruism, both Silk and de Waal comment that their analyses cannot prove the third requirement, that is, that given support is contingent upon received support. In addition, primates frequently show offensive cooperation from which all coalition members may benefit, particularly when they join forces to claim a resource (Bercovitch 1988; Noë this volume, pp. 300–2). Most likely, therefore, we are dealing with a mixed set of cooperative strategies.

Silk (Chapter 8) presents new data on bonnet macaques, a species in which affiliative relations and aggressive cooperation among adult males are unusually frequent compared to other macaque species (such as the rhesus monkeys of Ehardt and Bernstein this volume, pp. 84–8.). Apart from the already mentioned reciprocal distribution of interventions, Silk finds that support is offered to individuals who appear to need it most; it is effective in changing the outcome of the original agonistic confrontation, and it can be costly in that interveners seem to take risks. She concludes that supportive behaviour among male bonnet macaques appears to be of an altruistic character.

Silk also observed great asymmetries in supportive relationships. At first sight, this result may appear to defy an explanation in terms of reciprocal altruism. Yet, since individuals vary considerably in their supportive abilities, asymmetrical alliances are not necessarily unexpected. If a party who can provide highly beneficial services helps a relatively ineffective party less frequently than this party helps him, the two may still end up having a well-balanced relationship in terms of costs and benefits (Boyd this volume, pp. 477–9). Silk's coalition data, however, indicate that most support is given by the more effective partner in a relationship, i.e. high-ranking males help low-ranking males more often than the reverse. This may create an important imbalance unless the dominant's support is compensated for by other services by the subordinate.

De Waal (Chapter 9) reviews research on the chimpanzee colony of the Arnhem Zoo. The observations suggest a reciprocal exchange of services in many areas of social life. One simple explanation of this reciprocity, namely that it is produced by other social symmetries (e.g. kinship, preferential association), was addressed by controlling for symmetrical relationship characteristics. Because correlations persisted after this correction 'calculated reciprocity' is likely, that is, reciprocity based on mental record-keeping of given and received behaviour. Mere correlations, however, are not sufficient to prove this possibility, and de Waal goes on to discuss chimpanzee social strategy in greater detail to strengthen the case for a system of reciprocal exchange which includes 'currencies' other than coalitions (e.g. grooming, food sharing) as well as harmful behaviour, such as delayed retaliation against parties who tend to behave negatively or selfishly. He emphasizes negotiation as a way to develop beneficial relationships (see also Noë this volume, pp. 310–14), and describes a dramatic instance of a three-year-old coalition which abruptly collapsed when the pay-off matrix appeared to have become unattractive for one of the parties.

Like Kaplan (1978) and Ehardt and Bernstein (this volume, pp 90–1), de Waal points at the tension-reducing and bond-cementing effect of alliances. This effect may explain why chimpanzee males compete over performance of the 'control role' and why tensions among high-ranking individuals often lead to aggressive alliances against low-ranking individuals. Such so-called 'scapegoating' seems to ameliorate relationships among the alliance partners at the expense of the victim. **Grammer** (Chapter 10), who documented similar ganging up behaviour against low-ranking children, interprets it differently, however. Because of the absence of correlations between this form of cooperation and rank changes, he views it as an inefficient attempt to control low-ranking children.

Grammer demonstrates that children preferentially support individuals with whom they associate most, and that supportive interventions are reciprocally distributed. Both results agree with the data for non-human primates in the preceding chapters. Reciprocity appears relatively common among friends, which leads the author to conclude that friendship is a prior condition for reciprocity. He further presents evidence that children who frequently win fights as a result of support from others are likely to improve their rank before the end of the school year.

While the first three chapters lean towards an interpretation of alliances as a form of reciprocal exchange, the fourth chapter offers a different evaluation. **Noë** (Chapter 11) asserts that the theory of reciprocal altruism does not apply to alliances among male savanna baboons—precisely the species for which Packer (1977) was thought to have confirmed it—and that the often-used prisoner's dilemma game is not a suitable paradigm for this or most other forms of animal cooperation. For further comments on this paradigm see Rothstein and Pierotti (1988), who seem to share Noë's reservations, and Boyd (1988, this volume, pp. 473–7) who believes in its applicability provided realistic assumptions are being made, and who shows that the paradigm is elastic enough to cover a broad range of conditions.

It is perhaps useful to point out that coalitions among male baboons differ from those among male bonnet macaques and chimpanzees, discussed by Silk and de Waal. First, affiliative relationships among male baboons are not particularly close (e.g. grooming is virtually absent; Smuts 1985), whereas in the other two species male coalitions are an integrated part of well-developed affiliative relationships. Second, the situation of greatest interest to most students of baboon behaviour, including Noë, is that of males cooperatively stealing a female consort from the target of the coalition, a situation which is rare in the other two species. The lack of any immediate tangible benefit from cooperation in the other two species by comparison to male baboons perhaps makes reciprocation as an explanation of their behaviour more likely than for baboons.

The idea of animals competing for suitable partners is recognizable in Seyfarth's (1977) model of grooming among female monkeys as well as in de

Waal's descriptions of separating interventions among male chimpanzees. Noë takes these ideas a significant step further by proposing a model of 'market' effects regarding the supply and demand of partners. By ranking males in terms of their estimated fighting abilities, he calculates which combinations of males are most likely to benefit from aggressive cooperation. A central assumption is that the parties provide one another with reliable information about cooperative intentions, which assumption is not too far-fetched in view of the extensive exchange of signals observed in baboons before and during coalition formation (e.g. Smuts and Watanabe 1990).

Inasmuch as the coalition strategies of primates appear to combine reciprocal exchange, mutualistic cooperation, and kin-biased support, a focus on either of these explanations will miss important aspects of the behaviour. In addition, variation in partner quality (e.g. Wasser 1982), competition over partnerships, and each individual's attempts to attain maximum influence within the system make for a rather complicated picture that is best compared to a political arena. In the final chapter, **Falger** (Chapter 12) analyses the real thing, taking us to the international level and the sort of alliances that we read about in the daily newspaper.

Falger begins with pointing out one major difference between international alliances and alliances among individuals or political parties. Nations lack a higher-order organization, such as a formal hierarchy, so that their struggle for power occurs in a permanent state of anarchy. Although many of the determinants of international alliances are unique (e.g. geographic distance; arms technology), several fundamental similarities with coalition formation among individual humans and non-human primates are recognizable.

As in primates—at least among individuals without close kinship ties—expediency of international alliances appears to take precedence over stability; opportunistic changes in military cooperation are so widespread that one analyst spoke of an 'absence of memory' among the political powers of the nineteenth century. The goal of international alliances appears to be to gain as much influence as possible either by balancing against a threatening power or by bandwagoning against a losing power. The distinction between balancing and bandwagoning obviously corresponds with the ethologists' distinction between agonistic interventions in favour of the subordinate or dominant party.

The result of the strategies summarized by Falger is often a 'balance of power'. This configuration of alliances has been recognized in male chimpanzees (de Waal 1982), and may apply to other flexible coalition systems, such as that of Noë's male baboons in which middle-ranking individuals use alliances to curb the monopoly of top-ranking males. Falger notes that refined balance of power theories hold a promise of unifying biology and the social sciences.

REFERENCES

Bercovitch, F. B. (1988). Coalitions, cooperation and reproductive tactics among adult male baboons. *Animal Behaviour*, **36**, 1198–1209.

Bernstein, I. S. and Ehardt, C. (1986). Agonistic aiding: kinship, rank, age, and sex influences. *American Journal of Primatology*, **8**, 37–52.

Boyd, R. (1988). Is the repeated prisoner's dilemma a good model of reciprocal altruism? *Ethology and Sociobiology*, **9**, 211–22.

Connor, R. C. (1986). Pseudo-reciprocity: investing in mutualism. *Animal Behaviour*, **34**, 1562–6.

Kaplan, J. (1978). Fight interference and altruism in rhesus monkeys. *American Journal of Physical Anthropology*, **49**, 241–9.

Packer, C. (1977). Reciprocal altruism in *Papio anubis*. *Nature*, **265**, 441–3.

Rothstein, S. I. and Pierotti, R. (1988). Distinctions among reciprocal altruism, kin selection, and cooperation and a model for the initial evolution of beneficient behavior. *Ethology and Sociobiology*, **9**, 189–210.

Seyfarth, R. (1977). A model of social grooming among adult female monkeys. *Journal of Theoretical Biology*, **65**, 671–98.

Smuts, B. B. (1985). *Sex and friendship in baboons*. Aldine Press, New York.

Smuts, B. B. and Watanabe, J. M. (1990). Social relationships and ritualized greetings in adult male baboons (*Papio cynocephalus anubis*). *International Journal of Primatology*, **11**, 147–72.

Taylor, C. E. and McGuire, M. T. (1988). Reciprocal altruism: 15 years later. *Ethology and Sociobiology*, **9**, 67–72.

Trivers, R. (1971). The evolution of reciprocal altruism. *Quarterly Review of Biology*, **46**, 35–59.

de Waal, F. B. M. (1982). *Chimpanzee politics*, Jonathan Cape, London.

Wasser, S. (1982). Reciprocity and the trade-off between associate quality and relatedness. *American Naturalist*, **119**, 720–31.

Two male bonnet macaques simultaneously threaten a third. The coalition partners stand side by side, and one male lays his tail over the other's rump. Their opponent raises his tail in submission. (Drawing by Kathy West.)

8

Patterns of intervention in agonistic contests among male bonnet macaques

JOAN B. SILK

INTRODUCTION

Acts of altruism and cooperation among primates often seem to involve close kin, and these acts may have been shaped by kin selection (Gouzoules and Gouzoules 1987). There are, however, a number of examples of altruistic interactions in which kinship does not seem to play a significant role. For example, unrelated male baboons (*Papio* spp.) form coalitions to disrupt other males' consortships with sexually receptive females (Bercovitch 1988; Noë 1990, this volume, pp. 295–302; Packer 1977), unrelated rhesus macaque females (*Macaca mulatta*) form coalitions against adult males (Bernstein and Ehardt this volume, pp. 86–7; Kaplan 1977, 1978), and male baboons form protective relationships with unrelated adult females and their young (Altmann 1980; Busse and Hamilton 1981; Smuts 1985; Stein 1984). These kinds of behaviours may have evolved through reciprocal altruism (Trivers 1971) or through more general contingent behavioral strategies in which altruism is limited toward reciprocating partners (Axelrod and Hamilton 1981).

There is some evidence that altruism among non-relatives is selectively directed toward reciprocating partners. Packer (1977) found that male baboons tend to solicit support from one another reciprocally, although studies of baboons at other sites have produced less consistent evidence of reciprocity among coalition partners (Bercovitch 1988; Noë 1990). Unrelated pairs of rhesus macaques, stumptail macaques (*Macaca arctoides*), and chimpanzees (*Pan troglodytes*) support each other selectively in agonistic confrontations (de Waal and Luttrell 1988). Both naturalistic and experimental evidence suggests that grooming is sometimes exchanged for support in agonistic interactions (de Waal and Luttrell 1986; Seyfarth 1977, 1983; Seyfarth and Cheney 1984, 1988).

Overall, there is considerably more evidence of altruism toward kin than of altruism toward reciprocating partners. This may accurately reflect the actual distribution of altruism in primate groups. Some species may lack the cognitive capacities that are needed to monitor sequences of interactions, evaluate cost/benefit balances, and develop contingent behavioural strategies (de Waal and Luttrell 1988). However, it is also possible that reciprocity occurs more often than it is reported. It is inherently difficult to detect reciprocity in naturalistic studies. Different kinds of altruistic behaviours may be exchanged between reciprocating partners, and there may be marked asymmetries in the costs and benefits of particular kinds of interactions for reciprocating partners (Seyfarth and Cheney 1988; Boyd this volume, pp. 477–83). Therefore, it is premature to draw firm conclusions about the relative importance of reciprocity and kinship in the distribution of altruism within primate groups.

This paper describes some of the factors that influence the formation of coalitions and the patterning of support among males in a captive group of bonnet macaques (*Macaca radiata*). While earlier analyses have indicated that such coalitions can have a substantial impact upon male dominance status (Samuels *et al.* 1984), the patterning of support among these males has not been previously evaluated.

SUBJECTS AND METHODS

Study population

The data described here were collected over a 12-month period (September 1988–August 1989) at the California Primate Research Center on the campus of the University of California, Davis. The subjects were members of a large group of bonnet macaques housed in a 0.2-ha outdoor enclosure. During my study, the group contained 16 mature males (aged >6 years), approximately 21 mature females (aged >4 years) and 33 juveniles (aged <4 years). (For more information about the history and composition of this group see Silk 1990.)

Data on males' grooming behaviour are drawn from 10-min focal samples conducted regularly on each of the 16 mature males. Each male was observed at least 160 times over the course of the year. Data on dominance interactions and coalitions are drawn from the focal samples and from *ad libitum* observations.

Males' ordinal dominance ranks were asessed from the outcomes of decided dyadic agonistic contests. A separate dominance hierarchy was produced for each month of the study. Males' monthly ranks were used in most analyses. For some analyses, it was necessary to obtain aggregate measure of males' ranks over the course of the entire study. For these analyses, cardinal indices of male rank were computed, according to the procedure described in Boyd and Silk (1983).

In these analyses, I distinguish the male who initiated the original interaction from the male who was the recipient of the original act of aggression. I refer to these two parties as the *initiator* and the *recipient* of the original interaction. When the supporter intervenes on behalf of the initiator, his target is the recipient. When the supporter intervenes on behalf of the recipient, the supporter's target is the initiator.

An act of support was recorded if one male threatened, chased, hit, grappled, or bit one of the original participants. An act of suport was also recorded if the supporter and one of the original participants made physical contact with one another, but the supporter did not direct any aggressive gestures toward the target. For example, the supporter and the recipient of support sometimes wrapped their tails around each other, embraced, or advanced toward the target while maintaining contact with one another. Only one act of support per male was scored for each coalition.

All statistical analyses were performed with SPSS-PC. Pearson's correlation coefficients are computed to assess the relationships between continuous variables; all probability levels are one-tailed.

RESULTS

The frequency of coalitions

Coalitions were relatively common events in this bonnet macaque group. Males initiated 948 contests that led to coalitions, and they were the recipients of 929 contests that led to coalitions. Males were among both the initiators and recipients of 849 of the contests that led to coalitions. These events occurred on average once every 39 minutes. Coalitions usually followed spontaneous outbreaks of aggression among males. Virtually no coalitions involved direct competition over access to receptive females or other resources.

The number of male initiators in the interactions that preceded coalitions varied from one to six, and the number of male recipients in these contests ranged from one to three (Fig. 8.1). However, there was just one male initiator and one male recipient in 78 per cent of the 849 contests in which males were involved as initiators and recipients.

The analysis that follows is restricted to the coalitions that followed encounters between two mature males. For more complex cases involving multiple initiators and/or multiple recipients, it is often difficult to determine whether the supporter is helping one or all of the members of a given faction, or directing aggression against one or all of the members of the opposing faction. Thus, it is not clear how single acts of aid to multiple initiators or to multiple recipients should be apportioned.

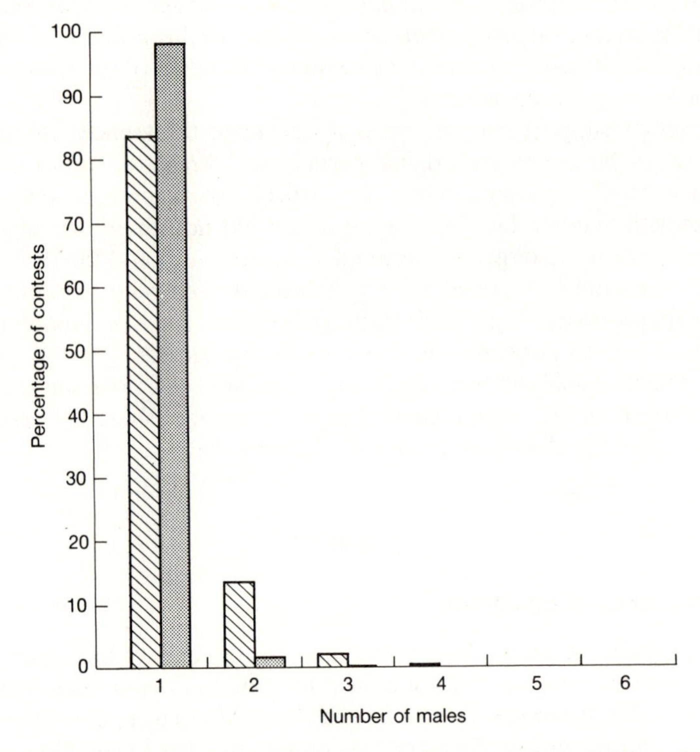

Fig. 8.1. The percentage of dyadic contests that preceded coalitions in which there were a given number of male initiators (hatched bars) and male recipients (solid bars) is illustrated. The sample is composed of all contests involving at least one male initiator and one male recipient ($N = 849$).

The form of coalitions

The initiators in dyadic contests among males received support from as many as seven individuals, while the recipients of these contests received support from as many as eight individuals (Fig. 8.2). There were 819 acts of support to the 664 initiators, and 408 acts of support to the 664 recipients. Thus, initiators were on average twice as likely to receive support as were recipients.

Is support altruistic?

If intervention is altruistic, it must be beneficial to the individual who receives support and costly to the individual who provides support. Although it is rarely possible to determine the effects of single acts upon an individual's life-

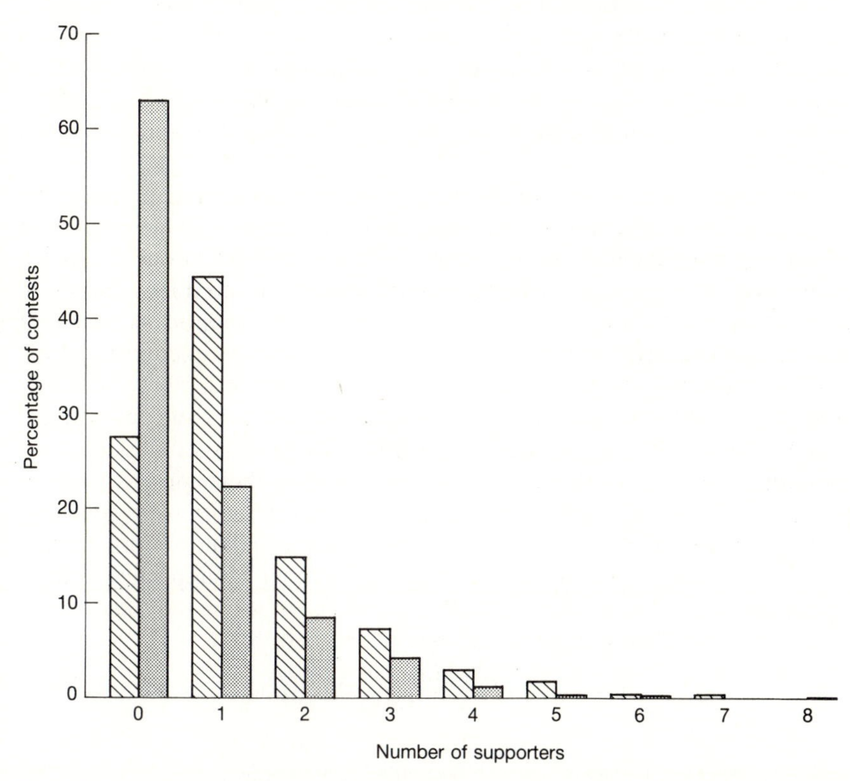

Fig. 8.2. The number of supporters who intervened on behalf of initiators (hatched bars) and recipients (solid bars) is illustrated. The sample is composed of all contests involving one male initiator and one male recipient ($N = 664$).

time fitness, we can assess the immediate impact of particular forms of behaviour upon individuals. The proximate effects of aid upon supporters and the recipients of support are evaluated in the analyses that follow.

Benefits of receiving support

Males who receive support are presumed to achieve at least short-term benefits if the support that they receive reduces the probability that they will be defeated during ongoing encounters. To assess the impact of support, I have restricted the analysis to cases in which neither the initiator nor the recipient showed any signs of submission (tail-up or grimace) during the dyadic phase of the interaction ($N = 347$ cases). For these cases, it was possible to determine whether males who received support were less likely to be defeated during coalitions than males who did not receive support. Since the effectiveness of support for a male may be influenced by the number of males who intervene on his behalf as well as the number of males who intervene on behalf of his opponent, I took the size of each males' faction into account in the analysis. Moreover, since initiators were generally higher ranking than recipients in the dyadic interactions that led to coalitions, I analysed the benefits of receiving support for initiators and recipients separately.

Differences in the level of support were measured by comparing the number of males who intervened on behalf of one party with the number of males who intervened on behalf of the other party. If support was beneficial, males would be less likely to be defeated during coalitions if they had more supporters than their opponents than if they had fewer supporters than their opponents. This prediction holds for both initiators and recipients (initiators: $\chi^2 = 48.5$, d.f. $= 2$, $p < 0.001$; recipients: $\chi^2 = 11.2$, d.f. $= 2$, $p = 0.0038$; Fig. 8.3). Although both initiators and recipients showed the same pattern, it is noteworthy that recipients were more likely to express submission than initiators when they both had more supporters than their opponents ($\chi^2 = 8.7$, $p = 0.0031$, d.f. $= 1$). Despite the fact that initiators were generally higher ranking than recipients, both recipients and initiators were equally likely to be defeated when they had the same number of supporters as their opponents ($\chi^2 = 1.2$, $p = 0.2670$, d.f. $= 1$), or when they had fewer supporters than their opponents ($\chi^2 = 0.1$, $p = 0.7638$, d.f. $= 1$).

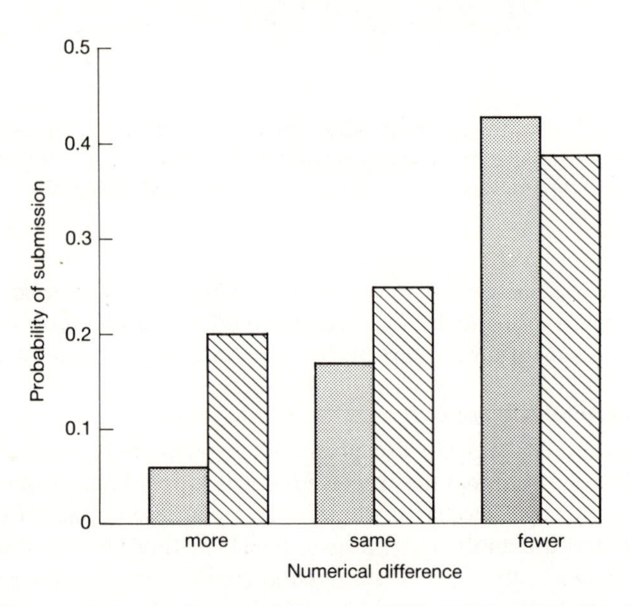

Fig. 8.3. The proportion of occasions on which initiators (solid bars) and recipients (hatched bars) expressed submission during coalitions when they had more supporters than their opponent, the same number of supporters as their opponent, or fewer supporters than their opponent is shown.

Costs of providing support

It is also important to consider the potential costs of intervention for the males who form coalitions. These analyses consider the short-term consequences of participation in coalitions, under the assumption that it is at least mildly disadvantageous for supporters to be defeated when they intervene in ongoing disputes among males.

Males who intervened in ongoing interactions expressed signs of submission approximately 8 per cent of the time ($N = 1174$ acts of support; supporter's response unknown in one case). Some males never gave submissive responses when they intervened in ongoing disputes, while one male was submissive in 35 per cent of the coalitions that he participated in.

The probability of being defeated during a coalition was influenced by the supporter's rank. Not surprisingly, low-ranking males tended to express signs of submission when they intervened on behalf of other males more often than did high-ranking males ($r = 0.74$, $p = 0.001$, $N = 16$). This may be one reason that low-raning males intervened in ongoing dyadic disputes among males less often than did higher-ranking males ($r = -0.85$, $p < 0.0001$, $N = 16$).

The rank of their targets also influenced the probability of supporters being defeated during a coalition. Taken as a whole, interventions were more likely to lead to submission by the supporter if he was lower-ranking than his target (51/295 = 17 per cent) than if he was higher-ranking than his target (44/878 = 5 per cent; $x^2 = 43.07$, d.f. = 1, $p < 0.0001$). When males are considered individually, however, the pattern is less clear. Of the 16 males, 13 intervened against both higher- and lower-ranking targets. Eight of these males showed higher levels of submission when they intervened against higher-ranking males than when they intervened against lower-ranking males, while three males showed lower levels of submission when they intervened against higher-ranking males than when they intervened against lower-ranking males and two showed no difference. This distribution is not significantly different from a random distribution ($p = 0.113$).

Although not all males were less likely to be defeated when they intervened against lower-ranking males than when they intervened against higher-ranking males, more acts of support were directed against targets who were lower-ranking than the supporters than against higher-ranking targets (supporters outrank targets: $N = 871$, targets outrank supporters, $N = 294$). Of the 16 males, 13 intervened more often against lower-ranking targets than against higher-ranking targets, and one male intervened with equal frequency against higher- and lower-ranking targets. The two remaining males intervened against lower-ranking males less often than against higher-ranking males. The probability of this distribution occurring by chance is low ($p = 0.011$).

Males may have exposed themselves to lower risks when they intervened on behalf of initiators than when they intervened on behalf of recipients. Of the 16 males, 14 intervened against lower-ranking recipients more often than against higher-ranking recipients, while two males intervened against higher and lower recipients with equal frequency ($p = 0.006$). In contrast, five of the males intervened against higher-ranking initiators more often than against lower-ranking initiators, and one male intervened against higher- and lower-ranking initiators with equal frequency. The 10 remaining males intervened against lower-ranking initiators more often than against higher-ranking initiators ($p = 0.227$).

The patterning of support among males

The results presented above suggest that support is beneficial to the recipients and costly to the donors. Thus, intervention in ongoing conflicts may be a form of altruism. If so, the patterning of intervention and the formation of coalitions is likely to have been shaped by kin selection and/or reciprocal altruism. In the analyses that follow, the effects of kinship and reciprocity upon the patterning of support among mature males are examined.

Kinship

There were five pairs of maternal kin among the 120 pairs of males. Three of these pairs were maternal siblings and the two pairs were uncle and nephew. Since paternity was not determined, the precise coefficient of relatedness among these males was not known. Moreover, the sample of 'unrelated' pairs of males, may actually include pairs of paternal kin whose kinship relationships were not recognized.

Although information about kinship is incomplete, and the number of related males is small, the data suggest that males did support their maternal relatives preferentially. Since all of the males had fewer relatives than non-relatives in the group, and therefore had more opportunities to support non-kin than to support kin, I compared the number of acts of support per related male with the number of acts of support per unrelated male for each male. All of the males who had mature male kin in the group ($N = 7$ males) supported their relatives at higher rates than they supported unrelated males ($p = 0.008$, Fig. 8.4).

Reciprocity

Although kin supported each other preferentially, kinship was not the only basis for cooperation. Only 82 of the 1165 acts of support (7 per cent) were directed toward maternal kin. All of the males in the group supported unrelated males, and at least one act of support was observed within 114 of the 115 pairs of unrelated males. Models of the evolution of cooperative or

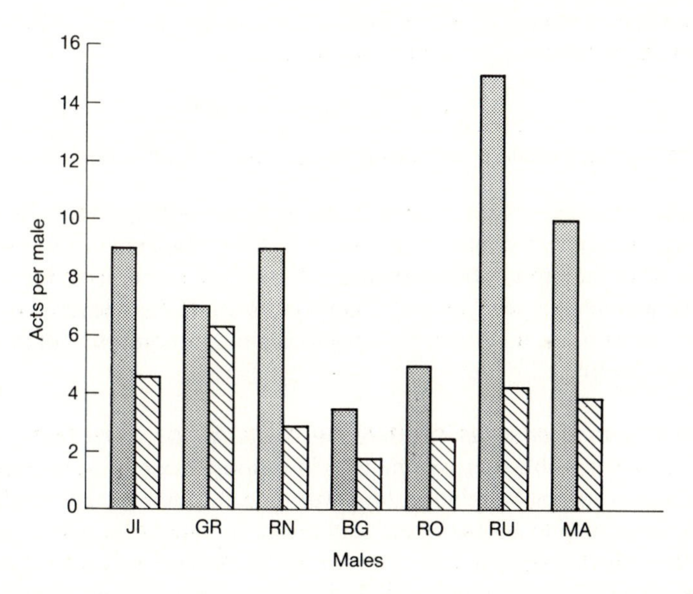

Fig. 8.4. The average rates of support given to kin (solid bars) and non-kin (hatched bars) for seven males with relatives in the group.

altruistic behaviours among non-relatives generally predict that the costs and benefits of altruism will be balanced among pairs of individuals, and that cooperative relationships will persist only as long as both partners continue to cooperate (Axelrod and Hamilton 1981). The patterning of support among unrelated males was assessed to examine these general predictions.

The number of acts of support given by one male to another was positively correlated with the number of acts of support received from the same male ($r = 0.43$, $N = 115$ pairs of males). This means that males selectively supported unrelated males from whom they had received support most often. Separate correlations were computed for each of the males. Of these correlations 15 were positive and one was negative ($p < 0.001$). Thus, there was a consistent tendency for males to selectively support unrelated males from whom they received support most often.

On the other hand, males did not consistently withhold support from unrelated males who intervened against them. There was no consistent relationship between the frequency with which males aided males and the frequency with which those males intervened against them ($r = 0.02$, $N = 115$ pairs of males). When males were considered individually, the number of positive correlations did not significantly exceed the number of negative correlations ($p = 0.151$).

Although males tended to support the males from whom they had received support most often, the number of acts of support from one male to the other was not balanced in most pairs of males. To quantify the degree of symmetry in support, the following measure was calculated:

$$\frac{\text{frequency of aid from A to B}}{\text{frequency of aid from A to B} + \text{frequency of aid from B to A}}.$$

The value of this expression ranged from 1.0, if male A was responsible for all of the support between male A and male B, to 0, when male B was responsible for all of the support between males A and B. It is clear from Fig. 8.5 that support was unbalanced among many pairs of males. Several different reasons that support might not have been balanced among pairs of males are considered below.

Did asymmetries arise as cooperative relationships were being established? It is possible that asymmetries in support arose as males attempted to establish cooperative relationships with one another. A male might have initially given help to another male, and then continued to support the other male only if he supported him in return. If this process was responsible for asymmetries in support, then we should expect to find that support was most balanced among the pairs of males who supported each other most often.

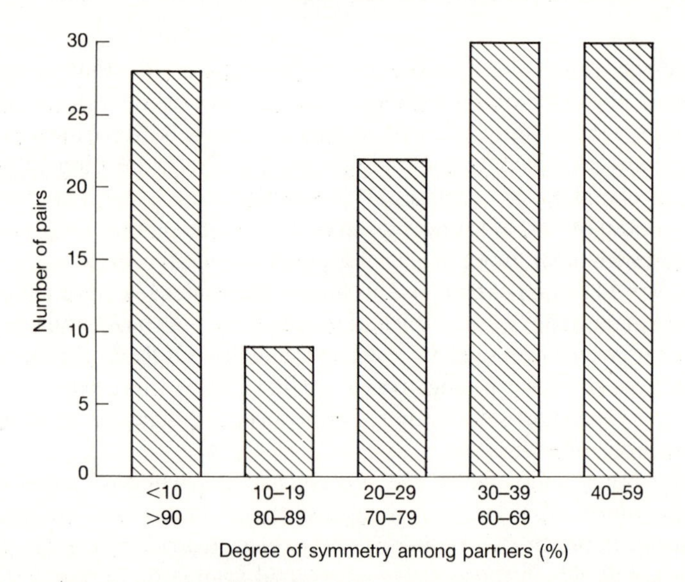

Fig. 8.5. The degree of symmetry in support among males is shown here. In the bar on the left side, one male performed at least 90 per cent of the acts of support, while his partner performed less than 10 per cent of the supports. In the next bar, one male performed 80–89 per cent of all acts of support, while his partner performed less than 20 per cent of all acts of support. The extent of symmetry in support increases from left to right.

To rest this hypothesis, I determined how much each pair's symmetry index deviated from 0.5, a perfectly balanced relationship. Then, I examined the association between these deviations and the number of acts of support exchanged within each pair of males. Support was most balanced among the pairs of unrelated males who supported each other most often ($\dot{r} = 0.25$, $N =$ 114). When this correlation is computed for each male, there are 14 positive correlations and two negative correlations ($p = 0.002$).

Although this result is consistent with the hypothesis that asymmetries arise as males establish cooperative relationships with unrelated males, it seems unlikely to provide a complete explanation of the patterning of support. Some of the most asymmetric relationships were maintained over the course of the entire study, and relationships were not necessarily terminated when one partner failed to reciprocate.

Did asymmetries reflect differences in the costs of giving aid? It is also possible that asymmetries in the extent of support reflect differences in the costs sustained by males who intervene in ongoing disputes. If support is more costly for one member of a pair to give than for the other, there may be asymmetries in the frequency with which support is given, although the costs sustained by each partner may be balanced (Boyd this volume, pp. 477–9).

Earlier, it was established that high-ranking males were somewhat less likely to express signs of submission during coalitions than are low-ranking males. If high-ranking males sustained lower costs when they intervened in ongoing disputes than did lower-ranking males, then we might expect high-ranking males to support low-ranking males more often than vice versa. Within any given pair of males, the higher-ranking of the two males typically provided more support to his partner than he received in return ($r = 0.24$, $N = 114$ pairs, one pair exchanged no acts of support). To determine whether the same pattern characterized all of the males in the group, separate correlations were computed for each male. There were 14 positive and two negative correlations ($p = 0.002$). Thus, males consistently intervened on behalf of unrelated lower-ranking partners more often than they received support from the same males.

Did asymmetries reflect differences in the effectiveness of support? It is also possible that asymmetries in the frequency of support among partners may reflect differences in the benefits associated with support from each of the partners. If altruism is balanced, males who provide the most beneficial support are expected to provide the least support. The evidence suggests, however, that the males who provide the most support are also the males whose support is most effective. We already have seen that high-ranking males intervene more often than lower-ranking males, and it seems likely that it is these males who provide the most effective support.

To evaluate the effects of male dominance rank on the outcome of coalitions, I restricted the analysis to dyadic contests among males in which neither of the original participants expressed any signs of submission before a third party intervened ($N = 347$ cases). For each case, I calculated the average rank of the initiator and his supporters. Average ranks were computed so that differences in the dominance ranks of the members of each faction would not be confounded by differences in the number of males who supported initiators and recipients. The dominance ranks of the initiators and recipients themselves were taken into account because they were important participants in these coalitions. These calculations are based upon the participation of all mature males, including kin. The exclusion of females and immature monkeys is likely to have little impact upon the results because these individuals were responsible for 2 per cent of all acts of support on behalf of initiators ($N = 451$) and 8 per cent of all acts of support on behalf of recipients ($N = 215$) in this set of coalitions.

The probability that the initiator would express signs of submission was greatly reduced if the members of his faction were on average higher-ranking than the members of the recipient's faction ($\chi^2 = 7.1$, d.f. = 1, $p = 0.0047$). For recipients, the effect of rank differences among factions was in the same direction, but not significant ($\chi^2 = 1.6$, d.f. = 1, $p = 0.2114$, see Fig. 8.6).

Did asymmetries reflect differences in males' needs for support? One reason that males may have helped lower-ranking partners more often than

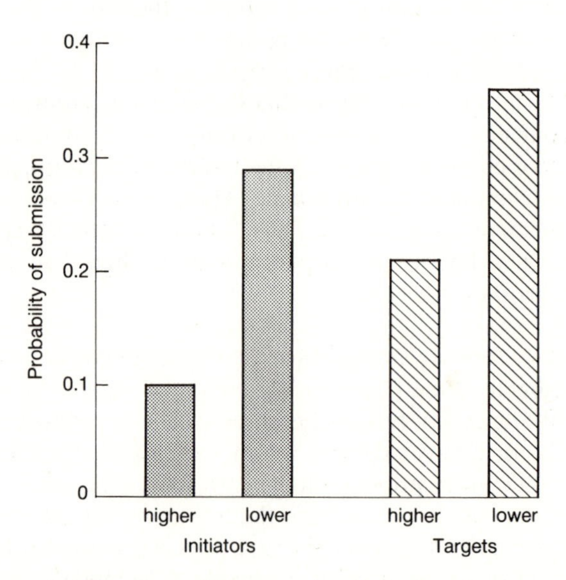

Fig. 8.6. The probability that initiators and recipients will express submission when the average rank of their faction is higher or lower than that of their opponent.

lower-ranking males helped them is that low-ranking males may have needed more help than did higher-ranking males. The males who were supported most often were those who were most often defeated ($r = 0.53$, $p = 0.017$, $N = 16$) and most often involved in undecided bouts of aggression ($r = 0.88$, $p < 0.001$, $N = 16$). However, there was no consistent relationship between the number of contests that a male won and the number of times that he received support ($r = 0.02$, $p = 0.477$, $N = 16$).

It is possible that differences among males in the need for support may help to explain the imbalance in support within pairs of males. To test this, I calculated the following measures for each pair of males:

$$\frac{\text{B aid A}}{(\text{A lose} + \text{A tie})-(\text{encounters between A and B})},$$

$$\frac{\text{A aid B}}{(\text{B lose} + \text{B tie})-(\text{encounters between A and B})}.$$

The first term in the denominator of each equation estimates how often each male needed help. Since the number of acts of support to males was correlated with the number of losses and ties, but not with the number of wins a male experienced over the course of the year, I assumed that the number of losses and ties that a male experienced was a good measure of his need for support. Encounters among male A and male B are subtracted from the totals in the denominator because it is not possible for males to intervene in ongoing contests against themselves. The value of the products of these expressions ranges from 0 to 1. If the product of the first expression equals 1 it means that male B aided male A every time that male A needed help. When the product of the second expression is 0.5 it means that male A aided male B on half of the occasions that male B needed male A's help.

If males distribute their support to unrelated males in relation to males' need for help, there should be a positive relationship between these two measures. This is, in fact, the case ($r = 0.71$, $N = 115$). The magnitude of this correlation is considerably larger than the magnitude of the correlation based upon the number of acts of support that unrelated males directed toward each other. When computed separately for each of the males, 13 were positive and three were negative ($p = 0.011$). Taken together, these results suggest that males supported the males that supported them, but gave the most support to the males who had the greatest need of their aid.

Were asymmetries in aid balanced in other altruistic currencies? Finally, it is important to consider the possibility that asymmetries in support among pairs of males were offset by altruistic behaviours that occurred in other social contexts. Social grooming is considered to be a form of altruism because individuals who groom others invest time and energy in grooming that could be invested in other activities more directly related to their own reproductive success, and grooming serves important hygienic and social functions which may enhance the recipient's health and reproductive success (Dunbar 1988). If support was exchanged for grooming, then we should expect to find that dominance rank influences the distribution of grooming, and that the frequency of support and grooming were associated within given pairs of males.

During focal observations of males, 573 grooming bouts were observed among 96 pairs of males. Low-ranking males groomed other males more often than higher-ranking males did ($r = 0.61$, $p = 0.007$, $N = 16$). Moreover, the proportion of grooming given by each male, measured as the number of grooms directed toward other males divided by the total number of grooms given and received, was positively correlated with a male's rank ($r = 0.74$, $p < 0.001$, $N = 16$). This means that low-ranking males groomed other males more often than they were groomed in return. Of all grooming bouts, 61 per cent were directed toward higher-ranking males.

Males tended to support unrelated males who groomed them most often ($r = 0.28$, $N = 115$ pairs). Of the correlations based upon individual males, 11 were positive and five were negative ($p = 0.105$). These data suggest there is only a weak association between grooming and support. More detailed analyses of the duration and quality of grooming bouts might reveal a more consistent association between grooming and support.

DISCUSSION

Intervention in ongoing encounters among males seems to be altruistic because males are less likely to be defeated when they receive support and those who provide support run some risk of being defeated during coalitions. It is difficult to assess the relative importance of kin selection and reciprocal altruism in the evolution of coalitions, but the evidence suggests that both kin selection and reciprocal altruism have shaped the evolution of these interactions.

Kinship seems to play a smaller role in the distribution of support among bonnet macaque males than it does in many other primate species. This conclusion must be treated with some caution because paternal kinship relationships may have influenced the distribution of support among males. However, some of the males who supported each other frequently could not be paternal kin because they were born in different groups. Moreover, we have no clear evidence that macaques are able to recognize paternal kinship (Gouzoules and Gouzoules 1987).

One reason that kinship may play a limited role in the distribution of support is that males may not distinguish related males from unrelated males whom they have known for many years. The males in this group have lived together longer than most males in free-ranging macaque groups. However, the fact that rates of support are higher among pairs of maternal kin than other pairs of males suggests that males can and do distinguish maternal kin from others.

The importance of kinship in the distribution of support among males might well depend upon the availability of related males. There were only five pairs of related males in this group, despite the fact that many of these males were born in the group or had lived in the group for many years. Most of these pairs of related males held relatively similar dominance ranks and were close in age. Males might have relied more heavily on their kin for support if they had been more numerous or provided more effective support.

Although males tended to support the males who most often supported them, it is not clear that these patterns conform to the technical definition of reciprocity. There is no evidence that males limited altruism to reciprocating partners, or that one male's altruism was actually contingent upon his partner's reciprocation.

Male bonnet macaques intervened in approximately 22 per cent of all dyadic aggressive interactions among males ($N = 2619$ bouts), and coalitions followed agonistic interactions among two males more than once per hour of observation. By comparison, Watanabe (1979) found that only four of 915 coalitions among Japanese macaques (*Macaca fuscata*) involved adult males. The rate of intervention among bonnet male macaques in this group is higher than the rate of intervention among free-ranging gorillas. (*Gorilla gorilla*, 10–12 per cent, Harcourt and Stewart 1987), provisioned Japanese macaques (7 per cent, Kurland 1977) or rhesus macaques (11 per cent, Kaplan 1978), and the same as the rate of intervention among female vervet monkeys (*Cercopithecus aethiops*, 22 per cent, Cheney and Seyfarth 1989). Although these figures are based upon data collected under different sampling regimes and environmental conditions, and are difficult to compare directly, they suggest that male bonnet macaques intervene in a relatively high proportion of agonistic interactions.

Why do male bonnet macaques form coalitions so often? In order to answer this question, we need to consider what function intervention might serve for the males who participate in them. Although males do not seem to use interventions to retaliate against former opponents, they might use interventions to achieve specific strategic objectives as chimpanzees do (de Waal 1982). They might also provide support to other males in order to increase the probability that they will receive support themselves in the future. Intervention by supporters confers an important advantage upon the participants in agonistic interactions because the number of supporters that a male can recruit influences the probability that he will be able to defeat his opponent. Since the theory of reciprocal altruism predicts that altruism among non-relatives will be contingent upon some degree of reciprocity, males may support others in order to maintain beneficial reciprocal relationships.

SUMMARY

1. Male bonnet macaques frequently formed coalitions in aggressive confrontations with one another.
2. Most coalitions followed spontaneous outbursts of aggression among males, and were not directly related to contests over access to receptive females or other resources.
3. Kin supported each other more often than non-kin, but coalitions were not restricted to pairs of related males.
4. Unrelated males tended to help the males who helped them most often, and support was most balanced among males who supported each other most often.
5. Support was unbalanced among many pairs of males. In such cases, males normally supported lower ranking partners more often than they were supported in return.
6. Several factors seem to contribute to rank-related asymmetries in support among males. First, support was more costly to provide for low-ranking males to provide than for high-ranking males. Second, support from lower-rank males was generally less effective than support from high-ranking males. Third, high-ranking males needed support more often than did lower-ranking males.

ACKNOWLEDGEMENTS

The research was conducted at the California Primate Research Center with support from NIH Grant RR00169. I would like to thank Robert Boyd, Dorothy Cheney, Lynn Fairbanks, Sandy Harcourt, Robert Seyfarth, and Frans de Waal for their comments upon various drafts of this manuscript.

REFERENCES

Altmann, J. (1980). *Baboon mothers and infants*. Harvard University Press, Cambridge, MA.

Axelrod, R. and Hamilton, W. D. (1981). *The evolution of cooperation*. *Science*, **211**, 1390–6.

Bercovitch, F. B. (1988). Coalitions, cooperation, and reproductive tactics among adult male baboons. *Animal Behaviour*, **36**, 1198–1209.

Boyd, R. and Silk, J. B. (1983). A method of assigning cardinal indices of dominance rank. *Animal Behaviour*, **31**, 45–58.

Busse, C. B. and Hamilton, W. J. (1981). Infant carrying by male chacma baboons. *Science*, **212**, 1281–3.

Cheney, D. L. and Seyfarth, R. M. (1989). Redirected aggression and reconciliation among vervet monkeys, *Cercopithecus aethiops*. *Behaviour*, **110**, 258–75.

Coe, C. L. and Rosenblum, L. A. (1984). Male dominance in the bonnet macaque: a malleable relationship. In *Social cohesion* (eds P. R. Barchas and S. P. Mendoza), pp. 31–63. Greenwood Press, Westport, CT.
Dunbar, R. I. M. (1988). *Primate social systems*. Cornell University Press, Ithaca.
Gouzoules, S. and Gouzoules H. (1987). Kinship. In *Primate societies* (eds B. B. Smuts, D. L. Cheney, R. M. Seyfarth, R. W. Wrangham, and T. T. Struhsaker), pp. 299–305. University of Chicago Press, Chicago.
Harcourt, A. H. and Stewart, K. J. (1987). The influence of help in contests on dominance rank in primates: hints from gorillas. *Animal Behaviour* **35**, 182–90.
Harcourt, S. A. and Stewart, K. (1989). Functions of alliances in contests within wild gorilla groups. *Behaviour*, **109**, 176–90.
Kaplan, J. R. (1977). Patterns of fight interference in free-ranging rhesus monkeys. *American Journal of Physical Anthropology*, **47**, 279–88.
Kaplan, J. R. (1978). Fight interference and altruism in rhesus monkeys. *American Journal of Physical Anthropology*, **49**, 241–50.
Kurland, J. A. (1977). *Kin selection in the Japanese monkeys*. Karger, Basel.
Noë, R. (1990). A veto game played by baboons: a challenge to the use of the Prisoner's Dilemma as a paradigm for reciprocity and cooperation. *Animal Behaviour*, **39**, 78–90.
Packer, C. (1977). Reciprocal altruism in *Papio anubis*. *Nature*, **265**, 441–5.
Samuels, A., Silk, J. B., and Rodman, P. S. (1984). Changes in the dominance rank and reproductive behavior of male bonnet macaques (*Macaca radiata*). *Animal Behaviour*, **32**, 994–1003.
Seyfarth, R. M. (1977). A model of social grooming among adult female monkeys. *Journal of Theoretical Biology*, **65**, 671–798.
Seyfarth, R. M. (1983). Grooming and social competition in primates. In *Primate social relationships: an integrated approach* (ed. R. A. Hinde), pp. 182–90. Blackwell, Oxford.
Seyfarth, R. M. and Cheney, D. L. (1984). Grooming, alliances, and reciprocal altruism in vervet monkeys. *Nature*, **308**, 541–3.
Seyfarth, R. M. and Cheney, D. L. (1988). Empirical tests of reciprocity theory: problems in assessment. *Ethology and Sociobiology*, **9**, 181–8.
Silk, J. B. (1982). Altruism among female *Macaca radiata*: explanations and analysis of patterns of grooming and coalition formation. *Behaviour*, **79**, 162–88.
Silk, J. B. (1988). Social mechanisms of population regulation in a captive group of bonnet macaques (*Macaca radiata*), *American Journal of Primatology*, **14**, 111–24.
Silk, J. B. (1990). Sources of variation in interbirth intervals among captive bonnet macaques (*Macaca radiata*). *American Journal of Physical Anthropology*, **82**, 213–30.
Smuts, B. B. (1985). *Sex and friendship in baboons*. Aldine Press, New York.
Stein, D. (1984). Ontogeny of infant–adult male relationships during the first year of life for yellow baboons (*Papio cynocephalus*). In *Primate paternalism* (ed. D. M. Taub), pp. 213–43. Van Nostrand Reinhold, New York.
Trivers, R. L. (1971). The evolution of reciprocal altruism. *Quarterly Review of Biology*, **46**, 35–57.
de Waal, F. B. M. (1982). *Chimpanzee politics*. Harper and Row, New York.
de Waal, F. B. M. and Luttrell, L. M. (1986). The similarity principle underlying social bonding among female rhesus monkeys. *Folia primatologica*, **46**, 215–234.

de Waal, F. B. M. and Luttrell, L. M. (1988). Mechanisms of social reciprocity in three primate species: symmetrical relationship characteristics or cognition? *Ethology and Sociobiology*, **9**, 101–18.

Watanabe, K. (1979). Alliance formation in a free-ranging troop of Japanese macaques. *Primates*, **20**, 459–74.

9

Coalitions as part of reciprocal relations in the Arnhem chimpanzee colony

FRANS B. M. DE WAAL

INTRODUCTION

The purpose of this chapter is to summarize and discuss the results of the coalition studies on the Arnhem chimpanzee colony. This colony, founded in 1971 at the Burgers Zoo in Arnhem (Netherlands), provides unique research opportunities into social life under semi-captive conditions. Approximately 25 chimpanzees (*Pan troglodytes*) live on an island of nearly 1 ha during 7 months of the year, and in an indoor hall during the winter.

Observations in Arnhem provide the most detailed picture to date of the chimpanzee's coalition strategies. For a complete understanding of chimpanzee behaviour, however, these observations need to be interpreted in combination with the results of fieldwork. Because wild chimpanzee females generally live dispersed over the forest, whereas males tend to travel together, the spatial conditions of captivity affect female relations probably more strongly than male relations. Opportunities for dispersal are limited even on a large island. As a result, the females in Arnhem have developed closer social relationships amongst themselves and exert greater influence on group processes than their wild counterparts. This difference is important in the present context as it manifests itself in females taking sides in male status struggles, a type of coalition that is rare or absent in the wild. Another possible difference is a greater degree of innovation in the social sphere in captive groups due to a higher interaction rate (Kummer and Goodall 1985).

De Waal (1986*a*, 1989*a*) discusses these differences as well as the many behavioural similarities between the chimpanzees in Arnhem and in the wild. Obviously, the capacity for complex social manoeuvering observed in captivity is an *evolved* capacity that must have adaptive significance for chimpanzees in their natural habitat. Studies by Nishida (e.g. 1983) and Goodall (e.g. 1986) provide evidence for similar manoeuvres in wild chimpanzees. The emerging picture is one of a high level of social reciprocity and political cognition in this species; a picture that awaits verification by means of experimental procedures. For a qualitative account the reader is referred to de Waal (1982). In the present chapter I review the quantitative evidence previously published in a series of technical reports.

RECIPROCAL COALITIONS IN CHIMPANZEES COMPARED TO MACAQUES

Chimpanzee males in a stable alliance tend to travel together, groom one another, and coordinate charging displays to a degree that makes their unity of purpose quite obvious (Fig. 9.1). Everyone in the community learns that conflict with one member of the alliance means conflict with the other. Conversely, rival males learn to take advantage of any temporary separation or disagreement within the team. The allies are well aware of this risk. De Waal (1982) observed that fights within the ruling alliance of the Arnhem colony would be broken off and hurriedly reconciled if a third male started a charging display. And Goodall (1986) describes how a new alpha male in the Gombe community would climb high in a tree and search around, giving so-called SOS screams, during a period of absence of his ally. There can be little doubt that these close alliances of male chimpanzees are reciprocal in the sense that both parties invest in and benefit from the relationship, although the nature of the investment and pay-offs may be different for each party.

This particular type of supportive relationship constitutes only a fraction of the observable aggressive cooperation in a chimpanzee community, however. There exists a much wider network of temporary coalitions in which individuals support one another for offensive reasons whenever the opportunity arises, or for defensive reasons whenever one of them comes under attack. In the Arnhem colony, for example, the screams of a distressed female may mobilize the entire female population against an aggressive male. Females also frequently interfere with power struggles among the adult males. Occasions for intervention by individual A on behalf of B may arise once per week, once per month, or even less frequently in a stable social setting. Because of the long intervals between significant events, consistent supportive relationships become apparent to the human observer only after years of exposure to these interactions. The degree of reciprocity in these relationships is impossible to judge without the assistance of statistics.

In a first attempt to determine the reciprocity in the distribution of aggressive support within the Arnhem colony, de Waal (1978) designed a support index which ranged from positive (i.e. two individuals generally help one another in confrontations with third parties) to negative (i.e. two individuals generally help third parties against one another). It was found that the number of mutually positive as well as mutually negative dyadic relations exceeded random expectation. At the time, however, the probability of this result could not be properly evaluated because of unaccounted for mathematical interdependencies. This problem affects interaction matrices in

Fig. 9.1. During their collective dominance over the colony, Nikkie and Yeroen often walked side by side, with synchronized body movements. Here, their common rival, Luit (left), throws sand in protest after the alliance (right) broke up a grooming contact between Luit and a female (from de Waal 1982).

general. The $N(N-1)$ dyadic values in a matrix are generated by interactions among only N individuals, which makes it impossible to specify the degrees of freedom. It is only recently that confirmational procedures have become available to evaluate the probability of matrix structure or of correlations between matrices. These matrix permutation or quadratic assignment procedures are discussed in detail by Dow and Cheverud (1985), Dow *et al.* (1987), Mantel (1967), and Schnell *et al.* (1985).

De Waal and Luttrell (1988) applied this technique to a database of over 2000 coalitions observed over a 5-year period in the Arnhem chimpanzee colony, as well as to a large sample of coalitions in breeding groups of rhesus and stumptail macaques (*Macaca mulatta* and *M. arctoides*) kept at the Wisconsin Regional Primate Research Center. In all studies, a coalition was defined as a third individual responding with an aggressive act against one, and only one, of the two participants in a dyadic confrontation. The aggressive intervention had to occur during or within 30 s after the end of the original incident to be scored as a response. All three parties had to engage in agonistic behaviour, as defined by particular facial expressions and vocalizations. Coalitions were recorded in a triplet format: individual A helps B against C.

This new analysis shared the aim of the earlier attempt by de Waal (1978) to cover the full spectrum of interventions by considering both supportive and harmful interventions: each choice for a particular party in a conflict is by definition a choice against the other party. I refer to these two types as *pro* and *contra* interventions, respectively. Reciprocity may occur in both domains, as reflected in 'One good turn deserves another' and its mirror image 'An eye for an eye, a tooth for a tooth'. Both expressions refer to reciprocity, but the first kind of reciprocity has received far more attention from biologists than the second (de Waal 1982).

Another aim of our analysis was to control for symmetrical relationship characteristics. This is necessary in order to determine if reciprocity is based on mental record keeping of given and received behaviour, or if it is a mere extension of the symmetry inherent in social relationships. An individual's inclination to help another may be the direct product of the amount of time it spends in association with the other. If so, helpful behaviour will be reciprocally distributed because the other individual, spending the same amount of time with the first, will be similarly inclined (Fig. 9.2, left). This mechanism of *symmetry-based reciprocity* needs to be assumed whenever reciprocity is observed in the absence of a correction for symmetrical characteristics.

If, on the other hand, reciprocity persists after symmetrical characteristics, such as association and kinship, have been taken into account, it becomes more likely that it reflects a contingency between given and received behaviour based on calculations of costs and benefits. In this case, the continuation of helpful behaviour is contingent upon benefits derived from the relationship; one possible benefit being the partner's reciprocation in kind

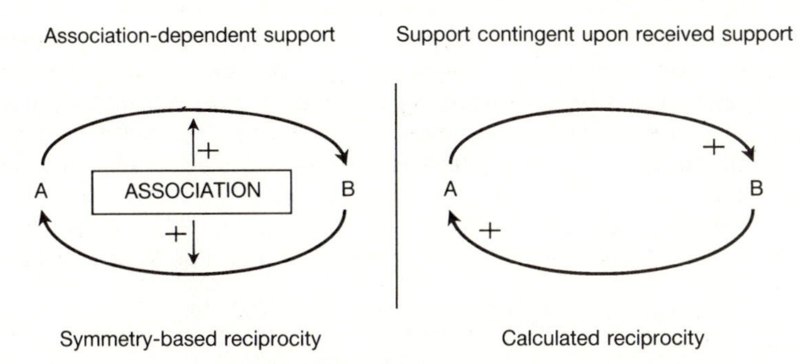

Fig. 9.2. Symmetrical relationship characteristics, such as preferential association and kinship, may cause a reciprocal distribution of beneficial behaviour if these characteristics stimulate the behaviour in both dyadic directions between individuals A and B (left). Calculated reciprocity refers to exchanges of beneficial behaviour independent of symmetrical traits, which implies that given behaviour is adjusted to received behaviour (right). The two reciprocity mechanisms are not mutually exclusive.

(Fig. 9.2, right). An intimate relationship is not required for individuals to enter into such an exchange relationship. This form of *calculated reciprocity* is expected to be most pronounced in the mentally most advanced species, which in our study was the chimpanzee. In its most complete form, calculated reciprocity manifests itself in mutual aid as well as mutual retaliation.

Our study confirmed that association between individuals affects the probability of coalition formation. All three species showed significantly positive correlations between the amount of time individuals spent in close proximity (e.g. grooming, contact-sit) and the rate of supportive interventions between them. No significant correlations existed between inter-individual association and the rate of intervention against the other. This excludes the possibility that preferential association promotes indiscriminate coalition formation through mere physical presence of the partner on the scene of the fight. Instead, the data demonstrate that frequent association between individuals *biases* interventions in favour of one another.

The next step was to investigate the reciprocity of beneficial (pro) and harmful (contra) interventions. This was done by correlating interventions in one dyadic direction with those in the other direction (i.e. interventions by individual A to B with those by B to A) for all dyads in the adult matrix. Table 9.1 presents Pearson correlations (r) as well as partial correlations (pr) after statistical removal of the effects of the following symmetrical relationship characteristics: (1) time spent in close proximity; (2) matrilineal kinship (using a binary matrix with 1's assigned to related dyads and zeros to unrelated dyads); and (3) sex-combination (using a binary matrix with 1's assigned to opposite-sex partners and zeros to same-sex partners). The

Table 9.1. Pearson correlation coefficients (*r*) between given and received agonistic interventions, for three primate species. The 'pro rate' concerns beneficial interventions, and the 'contra rate' harmful interventions (e.g. under 'pro rate' the table provides the correlation between individual A's pro rate with B vs. B's pro rate with A). Partial correlation coefficients (*pr*) are adjusted for effects of symmetrical relationship characteristics. One-tailed probability levels (*P*) concern the partial correlations, and are estimated on the basis of 200 random matrix permutations (i.e. 0.005 is the smallest *P* possible). From de Waal and Luttrell (1988).

		Reciprocity correlations		
Measure		Rhesus	Stump-tail	Chimpanzee
Pro rate	*r*	0.359	0.351	0.608
	pr	0.281	0.182	0.545
	P	(0.005)	(0.025)	(0.005)
Contra rate	*r*	−0.165	−0.225	0.334
	pr	−0.186	−0.285	0.315
	P	(0.005)	(0.005)	(0.025)

probability (*P*) of the partial correlations has been evaluated by means of matrix permutation procedures (see above).

Table 9.1 confirms a significant level of reciprocity in pro interventions among adult group members in all three species even after adjustment of the data for the effects of symmetrical relationship characteristics. The chimpanzees reach a considerably higher reciprocity correlation than the macaques. An even more important difference emerges with respect to contra interventions. These interventions are significantly reciprocal in the chimpanzee colony, but significantly antireciprocal in both macaque groups. This means that if macaque A often intervenes against B, B will rarely do so against A. On the other hand, if chimpanzee A often intervenes against B, B will do the same to A.

These results are consistent with the hypothesis that reciprocity is more complete in the mentally more advanced species. Yet there is another factor that may explain the interspecific differences: the nature of the dominance hierarchy. It has been known since Maslow (1940) that the chimpanzee's hierarchy is relatively loose and flexible, whereas the macaque's is rather rigid. Figure 9.3 illustrates how the intimidating effect of high rank may explain the absence of reciprocal contra interventions in macaques. Subordinate macaques rarely intervene against dominants, whereas chimpanzees largely ignore dominance relationships when intervening in fights.

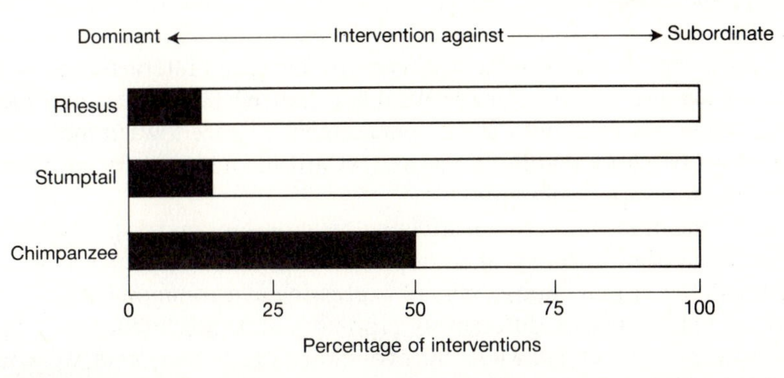

Fig. 9.3. If interventions were made regardless of dominance positions, 50 per cent would be against dominant targets and 50 per cent against subordinate targets. Based on 2053 interventions among adult rhesus macaques, 535 among adult stumptail macaques, and 1504 among adult chimpanzees, this graph shows that chimpanzee interventions are independent of dominance rank, whereas macaques mostly direct their interventions down the hierarchy (data from de Waal and Luttrell 1988).

The calculated nature of reciprocal coalitions cannot be proven without a detailed analyses of interaction sequences. It will be necessary to determine if supportive behaviour is contingent upon the partner's reciprocation. This may also be a promising area for experimental research. While the results of the above correlational analysis may not provide the final answers, they make it at least more likely than previous data suggested that primates are capable of (1) maintaining mental records of given and received support; and (2) regulating their own supportive behaviour according to a rule of reciprocity.

COALITIONS AND AFFILIATIVE RELATIONSHIPS

Affiliative behaviour and coalitions are not only correlated at the dyadic level, as discussed in the previous section; there are also indications for causal relations in time. Captive rhesus monkeys, for example, seek contact with allies following an incident in which they cooperated (de Waal and Yoshihara 1983), and free-ranging vervet monkeys (*Cercopithecus aethiops*) pay special attention to playbacks of agonistic recruitment calls by previous grooming partners (Seyfarth and Cheney 1984).

Possibly, grooming increases the probability of future support by the groomee, and vice versa. This would explain why grooming in monkey groups is predominantly directed to dominant individuals, which are undoubtedly the most valuable allies (Seyfarth 1977). In support of this hypothesis, Seyfarth (1980) reports a positive correlation in female vervet monkeys between grooming given to and support received from a partner. Yet, after confirmation of this correlation in female rhesus macaques, an even stronger correlation was found between grooming and interventions by the groomee against the groomer (de Waal and Luttrell 1986). This may mean that grooming is used both as an appeasement gesture toward individuals with a negative support attitude, and as a reward and incentive for individuals with a positive support attitude.

Even if the interplay between affiliative relationships and coalition formation can be reduced to that of an exchange of different 'currencies', we are faced with the virtual impossibility of expressing in a common currency the value of behaviours as different as grooming and interference in a fight (Seyfarth and Cheney 1988). At the evolutionary (ultimate) level, we would need to know how much each social service contributes to the survival and reproduction of the recipient, and how much it costs the performer. At the psychological (proximate) level, we would need to know how much value individuals attach to and how much satisfaction they derive from each social service. The equivalency of given and received services among animals is as yet an unresolved issue.

An entirely different (but not incompatible) perspective from which we may view the correlation between coalitions and affiliative relationships is that of coalitions as an instrument to redirect tensions that may exist between the cooperating parties. Instead of an exchange of favours, the central idea here is that the presence of a common enemy has a unifying effect. The enemy need not pose a threat; the target can be a bottom-ranking group member which serves as a 'scapegoat' (reviewed by de Waal 1986*b*). This mechanism, expressed in so-called winner-support (i.e. support given to aggressors able to win on their own), is common during increased social tension among captive primates, such as in newly established captive groups and during periods of artificial crowding (de Waal 1977, 1984; Kawai 1960). The phenomenon is by no means limited to captivity, however, and winner-support may serve to cement or confirm bonds in free-ranging monkeys as well (e.g. Cheney 1983).

Because of the low risk involved, support to a dominant individual against a defenceless victim can hardly count as an act of altruism, and we would therefore not consider these tense coalitions an exchange of valuable favours. The targets selected for this type of cooperation are usually not in a position to retaliate, and in captivity they sometimes need to be removed from the group to prevent a fatal outcome. These coalitions, however, may gradually develop into genuinely advantageous ones against more important opponents. Consequently, the interplay between affiliative behaviour and coalitions may be bidirectional, that is, low-risk coalitions may reinforce affiliative relationships, which in turn may provide a basis for high-risk coalitions. This may even occur without the mediating role of affiliative behaviour, as suggested by Noë (1990) for male baboons which occasionally engage in 'seemingly unnecessary coalitions' against weak opponents in order to confirm long-term cooperative relations. Acts of support thus become, in the words of Kaplan (1978, p. 247), 'both the mark and measure of the bond'.

Exceptions to the link between affiliative ties and coalitions can be expected in a system of calculated reciprocity in which this link is not obligatory (see previous section). In the Arnhem chimpanzee colony, coalitions of adult males became dissociated from affiliative relationships under two circumstances: (1) during instability of the rank-order; and (2) when males performed control activities (explained below). Our analysis compared the intervener's affiliative relationships with both combatants. The intervener could either support the party with which he normally spent most time in proximity or the party with which he spent least time in proximity. In case of perfect overlap between coalitions and affiliative preferences, 100 per cent of the interventions would be in favour of individuals with which the intervener associated most. If affiliative relationships had no effect, the expectation would be 50 per cent.

In agreement with previous results, almost all chimpanzees (22 out of 23) exceeded the 50 per cent limit, that is, coalitions and affiliative relationships generally corresponded. Remarkably, the individuals with the *lowest* correspondence were the three top-ranking males of the colony (de Waal 1984). These same males were found to regularly change their support preferences during periods of social upheaval: male A, which used to support male B against C, might begin to support male C against B. Such flexibility may require that existing affiliative ties be ignored, at least temporarily.

To test this hypothesis, the 5-year period of observation was divided into stable periods, in which the male hierarchy remained the same, and unstable periods, in which one or several males challenged the alpha male until either a take-over of his position or a re-establishment of previous positions had taken place. Figure 9.4 provides for each of the three males, for stable and unstable periods separately, the percentage of interventions in agreement with their affiliative preferences. The percentage exceeds the 50 per cent limit during stable periods, but drops during unstable periods. The drop is significant for each individual male (χ^2 tests, $p < 0.001$, two-tailed).

Female coalitions, on the other hand, are clearly committed to particular individuals, such as relatives and close associates. As a consequence, female coalitions are more stable and consistent over time than are male coalitions. Male coalitions appear to be linked, not to particular individuals, but to particular social goals. The above results indicate that affiliative preferences and coalitions can be disconnected during periods of social upheaval, when males need room for opportunistic manoeuvring. Because most of the coalition activity by males appears to serve the acquisition of positions of influence and high dominance status, their goals have been regarded as political (de Waal 1982).

The second exception to association-dependent intervention behaviour is

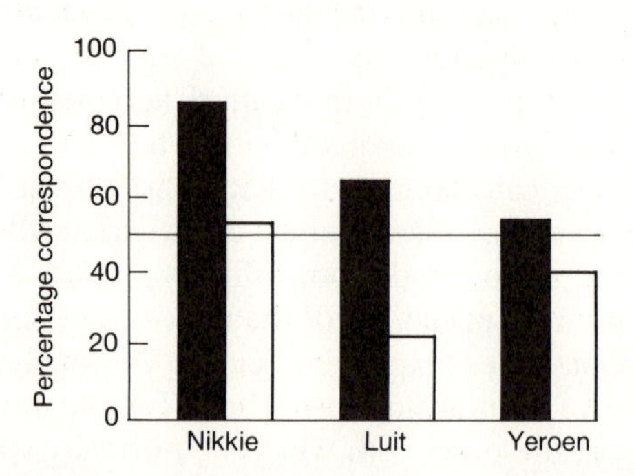

Fig. 9.4. Correspondence between interventions and affiliative relationships is said to exist if the intervener supports the party with which he associates most. Fifty per cent correspondence is expected by chance. The data for each of three adult males are divided into stable periods (solid bars) and unstable periods (open bars) in the male hierarchy (data from de Waal 1984).

the performance of control activities, which is usually done by one (and only one) high-ranking individual. This individual breaks up fights by pulling or beating the two combatants apart, after which he stands between them to prevent further aggression. If he intervenes partially, it is usually on behalf of the individual under attack. Adoption of this policing strategy may occur quite abruptly. One male in the Arnhem colony, Luit, turned from a winner-supporter into a loser-supporter in the course of a few months upon acquisition of the alpha rank (de Waal 1978). When performing control activities, males seem to place themselves *above* the conflicting parties in the sense that their interventions are not guided by affiliative preferences. The more protective the interventions by an adult male during a given time period, the less did his interventions correspond with his affiliative relationships (de Waal 1984).

At first sight this behaviour may seem unselfish; protection of the weakest group members, including unrelated ones, does not have obvious return benefits (e.g. Reinhardt *et al.* 1986). However, de Waal (1982, 1984) has speculated about possible benefits, namely that high-ranking males may stabilize their position by interfering on behalf of low-ranking individuals if these individuals return the favour by collectively supporting their protector against potential challengers. This view of control behaviour as a way to secure 'grassroots' support for one's position is based on observations of massive support to the individuals adopting this role in the Arnhem colony.

Interestingly, males compete over the 'prerogative' to perform control activites. Such competition is expected only if important advantages accrue to this behaviour. For example, attempts by low-ranking macaque males to break up fights among females are often actively inhibited by high-ranking males. As a consequence, young males learn to avoid the vicinity of fights involving females and immatures (Ehardt and Bernstein this volume, pp. 96–7; Watanabe 1979). Similarly, some of the most serious conflicts within the ruling Nikkie–Yeroen coalition of the Arnhem chimpanzee colony followed Yeroen's protests against his partner's attempts to break up fights among females. Yeroen's interference with Nikkie's interventions in female conflict did of course not help to settle the unrest within the colony, especially if the females broke off their own fight in order to support Yeroen against Nikkie. After several months of confusion, Nikkie lost interest in female disputes, and Yeroen appeared to have claimed this type of intervention for himself. Significantly, it was also Yeroen, not his higher-ranked partner, who received most female support.

These observations suggest two important characteristics of control activities, namely their exclusivity and their evaluation by others. By exclusivity I mean that performance of this type of intervention is not open to each and every group member; it appears to be subject to competition so that control behaviour is performed by one individual only (not necessarily the alpha individual). Evaluation by others is a more elusive concept; it refers to the possibility that the stability and acceptance of a male's position depends on his efficacy as arbitrator in disputes. If so, we might speak of 'norms' and 'expectations' that others have about this activity; norms and expectations are important features of the concept of role in the social sciences (Crook and Goss-Custard 1972; Hinde 1978). Further investigation into this possibility may help us determine if the term *control role* (Bernstein and Sharpe 1966) can be justifiably applied to the breaking up of intragroup disputes.

One problem is that this behaviour is not well documented for chimpanzee populations other than the one in Arnhem, or, for that matter, for other primate species (for a review see Ehardt and Bernstein this volume, pp. 84–6). Except for the tendency of male chimpanzees, particularly the alpha male, to protect recent immigrant females against resident females (Nishida 1979, 1989; Pusey 1980) there is little evidence for arbitration in conflicts among wild chimpanzees. Boehm (this volume, pp. 144–7) provides new information that may change this view, though the frequency of control behaviour appears rather low. This may be partly due to the difficulty of observing significant social events under conditions of reduced visibility, but it is also likely that the situation in Arnhem, with its constant physical proximity among the apes and the influence of females on group processes, has led to the development of a social order that differs in some fundamental respects from that among wild chimpanzees.

To summarize this section, intervention behaviour is generally biased toward close affiliates. Affiliates often are kin, but association-dependent support extends to non-kin. Possibly, the correlation between affiliative behaviour and coalitions partly reflects an exchange of different favours, such as grooming and support. There are also indications for the reverse causal connection, that is, a reinforcing effect of coalitions on social bonds.

Finally, coalitions and affiliative relationships can be disconnected, as occurs when adult male chimpanzees jockey for position or adopt a rather exclusive behavioural profile known as the control role.

COALITIONS AS SEXUAL TRANSACTIONS

A difference between the coalitions of male baboons and of male chimpanzees is that the first affect sexual competition directly, whereas the second do so indirectly via dominance status. Baboon troops are dominated by young adult males, which operate largely on the basis of individual fighting abilities. Older males may gain access to estrous females through concerted action against these younger males (e.g. Bercovitch 1988; Noë 1986, this volume, pp. 295–7; Packer 1977; Smuts 1985). In chimpanzees, on the other hand, dominance rank is not based on individual abilities alone. Both in captivity and in the wild, male dominance is normally decided by coalitions (e.g. de Waal 1982; Nishida 1983; Riss and Goodall 1977). Thus, while baboon coalitions are a strategy of subordinate males to limit the sexual monopoly of dominant males, chimpanzee coalitions modify the hierarchy itself. Although the correlation is far from perfect, high rank among male chimpanzees does seem to translate into greater access to estrous females (de Waal 1982; Nishida 1979; Tutin 1979). Because of the mediating role of dominance relations, the link between coalitions and mating success appears to be more complex in chimpanzees than in baboons.

In order to illustrate this complexity, I shall summarize the collapse of the Nikkie–Yeroen coalition in the Arnhem colony. This collapse is analysed in the light of the relation between coalitions, dominance rank, and sexual competition. For more detailed information, including the fatal consequences of this collapse, the reader is referred to de Waal (1986*c*).

Nikkie achieved the alpha position in 1978 with the help of Yeroen, and remained dependent on Yeroen in order to dominate Luit. At first, during the period immediately following the intense rivalry between Luit and Nikkie over the leadership position, Nikkie's coalition partner, Yeroen, was successful in playing off the other two males against one another. Yeroen would enlist Luit's support if Nikkie approached an estrous female, and he would enlist Nikkie's support if Luit approached a female. This changing of sides gave Yeroen a powerful key position in the sexual context. A similar strategy by an older, subordinate male chimpanzee was observed in the wild by Nishida (1983).

Yeroen's success decreased by the end of 1978 when both other males refrained from supporting him against one another. I have interpreted their sudden change in behaviour as a 'non-intervention treaty'. Over the years, this relation between Luit and Nikkie developed into a coalition. By 1979, Nikkie, though still dependent on Yeroen for dominance over Luit, co-operated with either male in the sexual context. For example, in the absence of estrous females, the Nikkie–Yeroen combination performed the majority of joint charging displays by males (i.e. 66 per cent), and the Nikkie–Luit combination a minority (i.e. 14 per cent). The latter proportion rose significantly (to 30 per cent) when estrous females were present in the colony. During these periods, it was not unusual for Nikkie and Luit to walk shoulder-to-shoulder in Yeroen's direction to separate Yeroen from a female.

Initially, Yeroen had gained considerably from supporting Nikkie. His contribution to the mating activity had increased from 9 per cent during Luit's alpha period to 39 per cent during the first part of Nikkie's alpha period, making Yeroen the most successful male. Second, Yeroen had regained the support of females, and was the male most often greeted with bows and pant-grunts by females. This is quite remarkable for a second-ranking male; usually it is the alpha male of the colony who receives most female 'respect'. Yeroen's popularity with the female population during Nikkie's alpha period has been explained above partially as a result of Yeroen's performance of control activities. Much of this changed during the second part of Nikkie's alpha period. Yeroen's mating success dropped to 19 per cent (while Nikkie's rose to 46 per cent), and the frequency of female greetings to Nikkie began to approach that to Yeroen. It is in this light that the dramatic events of 1980 need to be interpreted.

One summer day, several incidents occurred in which Nikkie and Luit together displaced Yeroen from an estrous female, as well as one incident in which Nikkie failed to support Yeroen. This incident went as follows. A sexually receptive female sat high up in a tree. When Luit started to climb towards her, Yeroen softly vocalized and looked alternately at Nikkie and Luit. In response, Luit returned to the ground, approached the other two males, and joined in their hooting chorus. After a couple of minutes, however, Luit returned to the tree and started to climb towards the female again. Now Yeroen burst out in loud screaming against Luit, while holding out his hand in a request for support to Nikkie. Nikkie, however, walked away from the scene. This led to a highly unusual surprise attack by Yeroen on Nikkie, jumping on him from behind and biting him in the back. Yeroen's violent reaction appeared to result from Nikkie's refusal to keep Luit from approaching the female.

Two days after the above incidents, an unobserved fight took place in the night-quarters of the males. In view of the injuries, this must have been the worst fighting since the establishment of the Arnhem colony. Luit showed only one superficial scratch, but both Nikkie and Yeroen had deep injuries, and swollen fingers, and Yeroen had lost one digit of a toe. Although no winner or loser could be determined on the basis of an injury count, Nikkie behaved as the loser. He had been a big and impressive alpha male, but now he looked unrecognizably small, depressed, and pitiful. Even though it did not seem that Luit had been very much involved in the physical battle, he emerged as the new dominant male. While this development may be hard to understand in dyadic terms, its explanation is simple considering the triadic structure.

There were three quantitative indications for reduced collaboration between Nikkie and Yeroen following this fight: (1) they intervened less frequently in one another's conflicts with other members of the group; (2) when intervening, they less often supported one another; and (3) they associated less frequently. On all three measures the Nikkie–Yeroen combination had scored extraordinarily high prior to the fight. The outcome of the fight, with Luit regaining control immediately following the collapse of the Nikkie–Yeroen coalition, confirmed our view that Nikkie and Yeroen had needed each other's support in order to keep Luit in check.

My interpretation of the above data is one of a broken transaction of reciprocal favours. By 1980, the advantages for Yeroen of backing up Nikkie had been reduced to almost zero, particularly in terms of access to estrous females. Nikkie's attitude towards Yeroen's sexual activities which had initially been remarkably tolerant, had become indistinguishable from Nikkie's attitude towards Luit. Sexual privileges may have constituted Yeroen's primary compensation for supporting Nikkie's successful bid for power. One would expect Yeroen to closely monitor the fulfillment of this deal. Incidents such as the one in which Nikkie, in spite of his partner's appeal, failed to interfere with their rival's sexual advances to a female may have constituted the last straw for Yeroen. When Nikkie increasingly failed to keep his end of the deal, leaning more and more toward Luit in the sexual context, Yeroen simply ended the collaboration.

TACTICS RELATED TO COALITIONS

Tactics can be divided into those temporally related to supportive interventions (i.e. occurring during such interventions, or immediately before or after), and those linked to supportive behaviour at the relationship level, without a close relation in time.

The first category includes so-called *side-directed communication*, that is, communication by the participants in an agonistic incident with individuals other than their opponent(s). In chimpanzees, this type of communication is remarkably rich in form, ranging from the seeking of reassuring contact from bystanders to the recruitment of supporters by means of begging hand gestures and agitated vocalizations (de Waal and van Hooff 1981; see also de Waal and Harcourt this volume, pp. 18–21). Side-directed behaviour often appears to reflect intense negotiation. The performer may run back and forth between its opponent and the third individual, throwing a temper tantrum if this individual walks away from the scene, yet instantly changing to a more offensive attitude toward the opponent if the third individual does take its side.

Side-directed communication is not expected when coalition formation is entirely of an instantaneous, opportunistic nature. If short-term self-interests are all that matter, there is little reason for individuals to solicit support and coordinate activities with one another; they can just enter the fray and take whichever side is most advantageous. Communication about coalitions, including the rejection of requests for support, suggests bargaining about the balance of collaboration within long-term relationships (see also Noë this volume pp. 311–13). Thus, Smuts and Watanabe (1990) view the communication gestures immediately preceding coalition formation among male baboons as telegraphic versions of sociosexual greeting ceremonies. The investigators speculate that the animals may be 'referring back' to collaborative agreements achieved earlier during routine greetings in neutral contexts.

Tactics with a possible effect on coalition formation may occur at intervals of several minutes before or after important aggressive confrontations. For example, males in the Arnhem chimpanzee colony have been observed to 'do the rounds', grooming a series of females and gently playing with their infants, before launching a challenge to a dominant male normally supported by these females. More common (or perhaps more conspicuous) than such preliminary moves are interactions in the aftermath of coalitions, such as grooming of supporters as an apparent reward and attack of the opponent's supporters as an apparent deterrent. In the Arnhem colony, Nikkie would systematically isolate and beat individual females who had assisted his rival in a previous fight (de Waal 1982), and in the Gombe community Figan's ability to withstand coalitions has been attributed to a tendency on his part to intimidate adversaries encountered alone (Goodall 1986).

At the relationship level the most important tactics related to coalitions are (1) establishment of close ties with potential allies by means of affiliative behaviour; and (2) interference with this process by third individuals. The first tactic follows logically from the previously discussed close relation between affiliative behaviour and coalitions. The second tactic manifests itself in competition over grooming and contact partners. It is quite common among primates for a dominant individual to shove the contact partner of another aside and take its place. Such interactions may represent competition over a limited number of desirable allies, as suggested by Seyfarth (1977) for competition over grooming partners among female monkeys (see also Harcourt this volume, pp. 458–60). It may also be a counter-strategy of individuals whose interests will be harmed by a close relation between others, as suggested for chimpanzees by de Waal (1978: Fig. 9.5). Most of the time, these two objectives coincide, that is A's successful prevention of the formation of a hostile alliance between B and C will also increase A's chances to establish an alliance himself with either B or C.

So-called *separating interventions*, in which one individual breaks up the contact between two others, were a highly developed aspect of the coalition structure in the Arnhem colony. Such interventions occurred on the average

Fig. 9.5. Luit's bid for the alpha position was successful only after he had isolated the established alpha male, Yeroen, from his female supporters. This process unfolded over a period of months, involving hundreds of separating interventions by Luit against contacts between Yeroen and the females. Here, Luit (centre) hoots and displays in front of Yeroen, who had been grooming a female. Yeroen (foreground; right) and the female (foreground; left) are seen leaving the scene in response to Luit's threatening behaviour. (From de Waal 1978)

once per hour during Nikkie's alpha period. By comparing the direction of 214 separating interventions with general contact frequencies in the colony, Adang (1980) could demonstrate that the most active intervener, Nikkie himself, aimed his disruptive behaviour disproportionally at contacts of his ally with his main rival. De Waal (1982) discusses this divide-and-rule strategy, and the paradox that Nikkie's ally, Yeroen, regularly provoked these interactions himself by initiating affiliative contact with their common rival, Luit. Perhaps, by seeking such contact, Yeroen 'reminded' Nikkie of his alternative political options, thus increasing his value (see Noë this volume, pp. 306–11, for a discussion of 'market' effects).

Nikkie would continue his intimidation displays for many minutes (up to 1 h) if the contact to which he responded did not cease. Evidently, he could not drag the other males apart by force. The others sometimes protested against Nikkie's disruptive behaviour; a confrontation which Nikkie lost on two occasions when they joined forces. Hence, separating interventions are not without risk. Most of the time, however, Nikkie succeeded in keeping Luit in an isolated position. Because of the option to resist the separation, each successful intervention demonstrates that one or both target males prefer a good relation with the performer of the intervention over a good relation with each other. Thus, separating interventions represent yet another area of negotiation about collaborative relations. These interventions have been confirmed for wild chimpanzees (Goodall 1986; Nishida and Hiraiwa-Hasegawa 1987).

Tactics relating to coalitions have not been studied nearly as thoroughly as the coalitions themselves. Much of the above functional interpretations are hypothetical, therefore. The study of side-directed communication, competition over potential allies, and separating interventions may become an important part of cognitive approaches to coalition formation because these tactics appear to reflect each party's perception of the power distribution and its expectation of cooperation with others.

MORALISTIC AGGRESSION AND RETALIATORY RECIPROCITY

In Trivers' (1971) model of reciprocal 'altruism' negative behaviour plays a role insofar as it helps regulate positive exchanges. Moralistic aggression is proposed as a protective mechanism against 'cheaters', that is, against individuals who try to give less than they receive. This form of aggression serves to educate unreciprocating individuals by frightening or punishing them. A possible example in the Arnhem chimpanzees is an aggressive incident in which a female, Puist, supported Luit against Nikkie. A little afterwards, Nikkie threatened Puist, who turned to Luit for protection, holding out her hand to him. Luit, however, did nothing to defend her against Nikkie's attack. Immediately Puist turned on Luit, barking furiously, chasing him across the enclosure. If her hostility towards Luit was in fact the result of his failure to help her after she had helped him, it would fit Trivers' definition of moralistic aggression (de Waal 1982, p. 207).

The first systematic evidence for this mechanism in nonhuman primates comes from a study of food-sharing in a captive group of chimpanzees at the Yerkes Regional Primate Research Center (de Waal 1989*c*). Food transfers among adults were reciprocally distributed. Aggressive behaviour, rather than being a function of dominance relationships, appeared to be an integrated part of the system of reciprocity. Individuals who were reluctant to share (i.e. showed a low rate of food distribution) had a higher probability than generous individuals of encountering aggression when they themselves requested food from others. This correlation is consistent with the view of negative sanctions against 'stinginess'.

The role of negative behaviour, however, goes well beyond that of the prevention of cheating. Negative behaviour is of equal weight to positive behaviour in reciprocity relationships, and the entire balance within a relationship may turn negative. A good example is the previously discussed reciprocity of contra interventions among chimpanzees (Table 9.1). This is interpreted as a *revenge system*, that is, negative actions by individual A against B are answered with negative actions by B against A. The difference with direct retaliation is that, instead of risking injury by fighting back against dominant aggressors, subordinate chimpanzees may wait for opportunities to intervene against dominants engaged in conflict with others. Since such interventions are known to influence status relations in the Arnhem colony, delayed retaliation provides subordinates with a powerful instrument to put pressure on dominants and to establish rules of conduct within their community.

The absence of such delayed retaliation in macaques has been attributed to the strictness of their dominance hierarchy, which inhibits actions by subordinates against dominants. The result is a less 'democratic' social organization than in chimpanzees, that is, subordinate macaques exert considerably less influence on the behaviour of dominants. The squaring of accounts in the negative domain observed in chimpanzees may represent a precursor of human systems of justice, which systems can be partially viewed as a transformation of the urge of revenge—euphemized as retribution—in order to control and regulate behaviour (e.g. Jacoby 1983).

It may seem logical in this context to distinguish between positive reciprocity (i.e. the exchange of favours) and negative reciprocity (i.e. the exchange of harmful acts). One problem is that the term 'negative reciprocity' has already been proposed to denote something quite different, namely reciprocal gift-exchanges and barter with a high degree of cheating and advantage-taking (Sahlins 1965). A more fundamental problem is that the two forms of reciprocity are not sharply distinguished in real life. They grade into one another; it is not unusual to observe elements of both in a single relationship. For example, a female chimpanzee may support a particular male in his status competition with other males, yet interfere against this male's attacks on females or juveniles. Depending on which type of interaction predominates during a given period, the male may act positively or negatively towards this female.

In conclusion, a wide range of behaviours enters into the reciprocity equation of social relationships. The focus in biology on reciprocal 'altruism' may be understandable from an evolutionary perspective (because of the special challenge to explain altruistic behaviour), but from a social psychological perspective this focus is rather narrow. A full understanding of the psychological mechanisms underlying reciprocity requires study of the entire spectrum, from helpful to retaliatory behaviour.

COMPARISONS WITH HUMAN BEHAVIOUR

Besides the general similarities between reciprocity relations in chimpanzees and humans, two specific similarities are worth further discussion. One concerns gender differences in coalition formation; the other the degree to which exchanges are centralized through a single individual or institution. To begin with the first similarity, let me summarize the difference in coalition formation by male and female chimpanzees in the Arnhem colony (de Waal 1984, in press).

Male coalitions change over time. One year, male A may support male B in his quest for dominance, only to ally the next year with male C against B if such cooperation offers better prospects for status enhancement. Female coalitions, in contrast, are committed to a limited number of female friends, and to the offspring of both the female herself and these friends. Stability of these coalitions has been documented in Arnhem for nearly two decades.

Females tend to fragment into small subgroups with high internal solidarity. Their coalitions strongly correlate with other measures of positive behaviour, such as association and grooming. Sympathies and antipathies appear to be the main determinants of coalition formation among females, therefore. Males, on the other hand, tend to form one large community with high solidarity in the face of external threats (e.g. Goodall 1986; Boehm this volume pp. 138–9) but changeable internal alliances. Rather than depending on social bonds and personal preferences, males take sides in fights dependent on strategic considerations, especially during periods of status struggle (Fig. 9.4). According to data from both the Arnhem colony and the natural habitat, the dissociation between political choices and affiliative behaviour is most pronounced in males past their physical prime. These males may build powerful positions by playing off ambitious younger and stronger males against one another. So, male opportunism may increase with age and experience (de Waal 1982; Nishida 1983).

Several strong parallels exist between these observations of chimpanzee behaviour and human behaviour. Masters' (1989) study of responses to televised displays of political figures showed that prior attitudes have a stronger influence on women (i.e. person-dependent attitudes), whereas men respond more to the specific competitive configuration of which a political leader is part (i.e. context-dependent attitudes). Similarly, experiments on human coalition formation by social psychologists suggest that the maximization models popular in this field (e.g. Murnighan 1978) do not apply well to female behaviour. Whereas men select partners on the basis of the power distribution and the prospects of winning, women appear to do so predominantly on the basis of personal attraction (Bond and Vinacke 1961; Nacci and Tedeschi 1976).

The second similarity between chimpanzees and humans is of a more speculative nature. It concerns the pooling and subsequent redistribution of power and material resources by a single individual or institution. This is common in our own species. As noted by Malinowski (1937; quoted in Sahlins 1965, p. 143): 'The chief everywhere, acts as a tribal banker, collecting food, storing it, and protecting it, and then using it for the benefit of the whole community.' The obligation of generosity associated with chieftainship, known as the *noblesse oblige* principle, may be viewed as a centralized arrangement of reciprocity relations (Sahlins 1965). Even if the redistribution of food or favours is uneven (e.g. if the chief favours his own kin), the arrangement may still benefit most members of the community.

In comparing the centralization of human exchange systems with the social organization of non-human primates, I am not thinking in the first place of the distribution of material goods, although there are indications that high-ranking chimpanzees are disproportionally involved as both collectors and (re)distributors during food-sharing (e.g. de Waal 1989*c*; Moore 1984; Teleki 1973). Instead, the comparison concerns immaterial exchanges related to control activities. We have discussed several unique features of this behaviour, such as the exclusivity of the control role; its independency of affiliative relations, which lends an element of objectivity to the arbitration, and the respect and support the performer may receive in return for his protection. This has led to the picture, sketched by de Waal (1982), of support flowing to a central individual who uses the prestige derived from it to provide social security, possibly undermining his own position if he fails to redistribute the support received. Our studies thus reveal a potential in the chimpanzee that, if shared by our own species, may have served as a blueprint for the elaborate centralized reciprocity systems of human societies.

EVOLUTIONARY CONSIDERATIONS

With regards to the evolution of the two reciprocity mechanisms discussed in this paper, the origin of symmetry-based reciprocity probably is kin-selection (Hamilton 1964). Genetic relatedness is a symmetrical trait, as is the lifelong preferential association often found among kin. The flow of help in these relationships may show periodical assymmetries (e.g. between mother and dependent offspring), but will often balance out over a lifetime, as argued by Dunbar (1980) for female gelada baboons (*Theropithecus gelada*). It seems a small step to extend this mechanism to closely associated non-relatives, especially if animals are capable of assessing the costs and benefits of collaboration. With the growing influence of these assessments, the dependency of reciprocity on kinship and association may have been reduced. The resulting calculated reciprocity, which does not require close affiliative ties, corresponds with Trivers' (1971) reciprocal altruism. Hence, the distinction at the proximate level between symmetry-based and calculated reciprocity is perhaps less sharp than appears, and agrees with the two main theories explaining the evolution of altruistic behaviour. These theories themselves are not considered mutually exclusive either (Rothstein and Pierotti 1988; Wasser 1982; Wilkinson 1988).

The female strategy of committed support to close associates most likely evolved through kin selection, because affiliative relations and kinship often overlap. In nature, female chimpanzees live dispersed over the forest, associating mainly with their offspring and females in adjacent home ranges (Halperin 1979; Wrangham 1979). The male strategy of relatively flexible partnership suggests that it is not particular individuals that are most important, but particular social goals. The apparent goals of males are to secure a large territory with many reproductive females in competition with neighbouring groups, as well as access to these females in competition with males of their own group (e.g. Goodall 1986; Nishida and Hiraiwa-Hasegawa 1987; Wrangham 1979). The ambiguity between the need for a macro-coalition of all males in intergroup conflict, and the need for each male to participate in smaller coalitions in intragroup conflict, may have stimulated the evolution of the remarkable strategic manoeuverability of this species as it requires a balancing, at two different levels, of the pros and cons of competition and cooperation (see also Boehm this volume pp. 159–61).

Bygott (1979) refers to this ambiguity when speculating that overly selfish or despotic behaviour by an alpha male chimpanzee could be penalized through the non-cooperation of other males in assisting with the defence of the community range, which would jeopardize alpha's reproductive success. However, such behaviour would harm the interests of the subordinate males as well. Nishida and Hiraiwa-Hasegawa (1987) observe that these males have little choice but to cooperate with a selfish alpha male in intergroup (and perhaps even intragroup) encounters, but may be ready to collectively overthrow him as soon as his power is diminished. If these calculations sound familiar, it is because our societies, too, are characterized by segmented coalitions and a distinction between coalitions of necessity and coalitions of choice.

SUMMARY

1. Chimpanzees show reciprocity of supportive interventions in aggressive encounters: the rate of intervention by individual A on behalf of B correlates with the rate by B on behalf of A. Reciprocity persists after adjustment of the data for symmetrical relationship characteristics, such as kinship and interindividual association. This makes it likely that the reciprocity is based on evaluations of given and received behaviour.
2. Chimpanzees also show delayed retaliation: A's interventions against B correlate with B's interventions against A. This reciprocity is discussed as a precursor of human systems of justice and morality. Two captive macaque groups did not show such reciprocity with respect to harmful interventions.
3. Supportive behaviour is generally biased toward close affiliates, but may be disconnected from affiliative preferences when adult male chimpanzees jockey for position or perform control activities.
4. A sex difference in coalition formation, with males being more opportunistic and females more committed to particular individuals, resembles one reported for humans.
5. Alliances appear to be of a transactional nature, in the sense that both parties monitor the contributions of the other, and are prepared to discontinue the collaboration if a major imbalance arises. These transactions may include 'currencies' other than coalition formation (e.g. sexual privileges, grooming). Special forms of communication and other tactics relating to coalitions serve to negotiate balanced long-term relationships.

ACKNOWLEDGEMENTS

I thank the many students who helped collect data in Arnhem, and Mary Schatz and Jackie Kinney for typing the manuscript. Writing was supported by grant No. RR00167 of the National Institutes of Health to the Wisconsin Regional Primate Research Center. This is publication No. 29-011 of the WRPRC.

REFERENCES

Adang, O. (1980). Scheidende interventies door volwassen mannelijke chimpansees. Unpublished report, University of Utrecht.

Bercovitch, F. (1988). Coalitions, cooperation, and reproductive tactics among adult male baboons. *Animal Behaviour*, **36**, 1198–1209.

Bernstein, I. and Sharpe, L. (1966). Social roles in a rhesus monkey group. *Behaviour*, **26**, 91–103.

Bond J. and Vinacke, W. (1961). Coalitions in mixed-sex triads. *Sociometry*, **24**, 61–75.

Bygott, J. D. (1979). Agonistic behavior, dominance, and social structure in wild chimpanzees of the Gombe National Park. In *The great apes* (eds D. Hamburg and E. McCown, pp. 405–27. Benjamin/Cummings, Menlo Park, CA.

Cheney, D. (1983). Extrafamilial alliances among vervet monkeys. In *Primate social relationships* (ed. R. A. Hinde), pp. 278–85. Sinauer, Sunderland, MA.

Crook, J. and Goss-Custard, J. (1972). Social ethology. *Annual Review of Psychology*, **23**, 277–312.

Dow, M. and Cheverud, J. (1985). Comparison of distance matrices in studies of population structure and genetic microdifferentiation: quadratic assignment. *American Journal of Physical Anthropology*, **68**, 367–73.

Dow, M., Cheverud, J., and Friedlaender, J. (1987). Partial correlation of distance matrices in studies of population structure. *American Journal of Physical Anthropology*, **72**, 343–52.

Dunbar, R. (1980). Determinants and evolutionary consequences of dominance among female gelada baboons. *Behavioural Ecology and Sociobiology*, **7**, 253–65.

Goodall, J. (1986). *The chimpanzees of Gombe*, Belknap, Cambridge, MA.

Halperin, S. D. (1979). Temporary association patterns in free-ranging chimpanzees. In *The great apes* (eds D. Hamburg and E. McCown), pp. 491–9. Benjamin/Cummings, Menlo Park, CA.

Hamilton W. (1964). The genetic evolution of social behaviour, I, II. *Journal of Theoretical Biology*, **7**, 1–52.

Hinde, R. A. (1978). Dominance and role—two concepts with dual meanings. *Journal of Social and Biological Structures*, **1**, 27–38.

Jacoby, S. (1983). *Wild justice: the evolution of revenge*. Harper and Row, New York.

Kaplan, J. (1978). Fight interference and altruism in rhesus monkeys. *American Journal of Physical Anthropology*, **49**, 241–50.

Kawai, M. (1960). A field experiment on the process of group formation in the Japanese monkey. *Primates*, **2**, 181–253.

Kummer, H. and Goodall, J. (1985). Conditions of innovative behaviour in primates. *Philosophical Transactions of the Royal Society of London B*, **308**, 203–14.

Mantel, N. (1967). The detection of disease clustering and a generalized regression approach. *Cancer Research*, **27**, 209–20.

Maslow, A. (1940). Dominance quality and social behavior in infra-human primates. *Journal of Social Psychology*, **11**, 313–24.

Masters, R. D. (1989). Gender and political cognition: integrating evolutionary biology and political science. *Politics and the Life Sciences*, **8**, 3–39.

Moore, J. (1984). The evolution of reciprocal sharing. *Ethology and Sociobiology*, **5**, 5–14.

Murnighan, J. (1978). Models of coalitions behavior: game theoretic, social psychological, and political perspectives. *Psychology Bulletin*, **85**, 1130–53.

Nacci, P. and Tedeschi, J. (1976). Liking and power as factors affecting coalition choices in the triad. *Social Behavior Personality*, **4**, 27–32.

Nishida, T. (1979). The social structure of chimpanzees at the Mahale Mountains. In *The great apes* (eds D. Hamburg and E. McCown), pp. 73–121. Benjamin/Cummings, Menlo Park, CA.

Nishida, T. (1983). Alpha status and agonistic alliance in wild chimpanzees. *Primates*, **24**, 16–34.

Nishida, T. (1989). Social interactions between resident and immigrant female chimpanzees. In *Understanding chimpanzees* (eds P. Heltne and L. Marquardt), pp. 68–89. Harvard University Press, Cambridge, MA.

Nishida, T. and Hiraiwa-Hasegawa, M. (1987). Chimpanzees and bonobos: co-operative relationships among males. In *Primate societies* (eds B. Smuts *et al.*), pp. 165–77. University of Chicago Press, Chicago.

Noë, R. (1986). Lasting alliances among adult male savannah baboons. In *Primate ontogeny, cognition and social behaviour* (eds J. Else and P. Lee), pp. 381–92. Cambridge University Press, Cambridge.

Noë, R. (1990). A veto game played by baboons: a challenge to the use of the prisoner's dilemma as a paradigm for reciprocity and cooperation. *Animal Behaviour*, **39**, 78–90.

Packer, C. (1977). Reciprocal altruism in *Papio anubis. Nature*, **265**, 441–3.

Pusey, A. E. (1980). Inbreeding avoidance in chimpanzees. *Animal Behaviour*, **28**, 543–52.

Reinhardt, V., Dodsworth, R., and Scanlan, J. (1986). Altruistic interference shown by the alpha-female of a captive troop of rhesus monkeys. *Folia primatologica*, **46**, 44–50.

Riss, D. and Goodall, J. (1977). The recent rise to the alpha-rank in a population of free-living chimpanzees. *Folia primatologica*, **27**, 134–51.

Rothstein, S. I. and Pierotti, R. (1988). Distinctions among reciprocal altruism, kin selection, and cooperation and a model for the initial evolution of beneficent behavior. *Ethology and Sociobiology*, **9**, 189–210.

Sahlins, M. (1965). On the sociology of primitive exchange. In *The relevance of models for social anthropology* (ed. M. Banton), pp. 139–236. Tavistock, London.

Schnell, G., Watt, D., and Douglas, M. (1985). Statistical comparison of proximity matrices: Applications in animal behaviour. *Animal Behaviour*, **33**, 239–53.

Seyfarth, R. (1977). A model of social grooming among adult female monkeys. *Journal of Theoretical Biology*, **65**, 671–98.

Seyfarth, R. M. (1980). The distribution of grooming and related behaviours among adult female vervet monkeys. *Animal Behaviour*, **28**, 798–813.

Seyfarth, R. M. and Cheney, D. L. (1984). Grooming, alliances and reciprocal altruism in vervet monkeys. *Nature*, **308**, 541–3.

Seyfarth, R. M. and Cheney, D. L. (1988). Empirical tests of reciprocity theory: problems in assessment. *Ethology and Sociobiology*, **9**, 181–8.

Smuts, B. (1985). *Sex and friendship in baboons*. Aldine Press, Hawthorne.

Smuts, B. and J. M. Watanabe (1990). Social relationships and ritualized greetings in adult male baboons (*Papio cynocephalus anubis*). *International Journal of Primatology*, **11**, 147–72.

Teleki, G. (1973). *The predatory behavior of wild chimpanzees*. Bucknell University Press, Lewisburg.

Trivers, R. (1971). The evolution of reciprocal altruism. *Quarterly Review of Biology*, **46**, 35–59.

Tutin, C. E. G. (1979). Mating patterns and reproductive strategies in a community of wild chimpanzees. *Behavioral Ecology and Sociobiology*, **6**, 39–48.

de Waal, F. (1977). The organization of agonistic relationships within two captive groups of Java-monkeys (*Macaca fascicularis*). *Zeitschrift für Tierpsychologie*, **44**, 225–82.

de Waal, F. (1978). Exploitative and familiarity-dependent support strategies in a colony of semi-free-living chimpanzees. *Behaviour*, **66**, 268–312.

de Waal, F. (1982). *Chimpanzee politics*. Jonathan Cape, London.

de Waal, F. (1984). Sex-differences in the formation of coalitions among chimpanzees. *Ethology and Sociobiology*, **5**, 239–55.

de Waal, F. (1986*a*). The integration of dominance and social bonding in primates. *Quarterly Review of Biology*, **61**, 459–79.

de Waal, F. (1986*b*). Prügelknaben bei Primaten und ein tödlicher Kampf in der Arnheimer Schimpansenkolonie. In *Ablehnung- Meidung- Ausschlusz* (eds M. Gruter and M. Rehbinder), pp. 129–45. Duncker and Humblot, Berlin.

de Waal, F. (1986*c*). The brutal elimination of a rival among captive male chimpanzees. *Ethology and Sociobiology*, **7**, 237–51.

de Waal, F. (1989*a*). Dominance 'style' and primate social organization. In *Comparative socioecology: the behavioural ecology of humans and other mammals* (eds V. Standen and R. Foley), pp. 243–63. Blackwell, Oxford.

de Waal, F. (1989*b*). *Peacemaking among primates*. Harvard University Press, Cambridge, MA.

de Waal, F. (1989*c*). Food sharing and reciprocal obligations among chimpanzees. *Journal of Human Evolution*, **18**, 433–59.

de Waal, F. (in press). Sex differences in chimpanzee (and human) behavior: a matter of social values? In *Towards a Scientific Analysis of Values* (eds L. Cooper, M. Hechter, and R. Michod).

de Waal, F. and J. van Hooff (1981). Side-directed communication and agonistic interactions in chimpanzees. *Behaviour*, **77**, 164–98.

de Waal, F. and Luttrell, L. (1986). The similarity principle underlying social bonding among female rhesus monkeys. *Folia primatologica*, **46**, 215–34.

de Waal, F. and Luttrell, L. (1988). Mechanisms of social reciprocity in three primate species: symmetrical relationship characteristics or cognition? *Ethology and Sociobiology*, **9**, 101–18.

de Waal, F. and Yoshihara, D. (1983). Reconciliation and re-directed affection in rhesus monkeys. *Behaviour*, **85**, 224–41.

Wasser, S. (1982). Reciprocity and the trade-off between associate quality and relatedness. *American Naturalist*, **119**, 720–31.

Watanabe, K. (1979). Alliance formation in a free-ranging troop of Japanese macaques. *Primates*, **20**, 459–74.

Wilkinson, G. S. (1988). Reciprocal altruists in bats and other mammals. *Ethology and Sociobiology*, **9**, 85–100.

Wrangham, R. (1979). Sex differences in chimpanzee dispersion. In *The great apes* (eds D. Hamburg and E. McCown), pp. 481–90. Benjamin/Cummings, Menlo Park, CA.

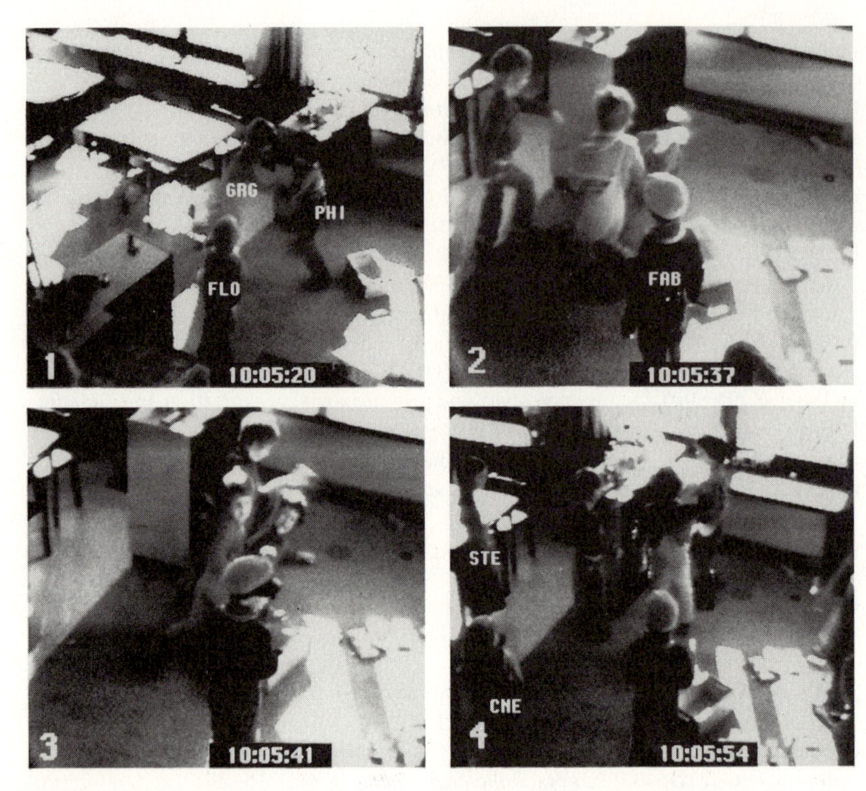

Grg and Flo take Ben's car: Ben informs Phi of this, who then threatens Grg. Ben attacks Flo, Phi attacks Grg. (1) After a melée involving all four children, Phi and Grg wrestle, watched by Flo. Phi falls, is hit by Grg and (2), (3) is attacked by Flo. (The order of attention rank is Ben, Phi, Grg, and Flo).

10

Intervention in conflicts among children: contexts and consequences

KARL GRAMMER

SUPPORTING STRATEGIES

A functional approach

Interventions in conflicts among children by peers can be observed quite often, yet intervention has generally not been a topic for research in either child ethology or developmental psychology. This may be due to numerous problems involved in the methodology of studying interventions and disagreement about the function of intervention. Some researchers describe intervention as altruism while others, who accept the existence of 'biosocial norms', see it as adherence to norm-oriented behaviour or even as goal-oriented behaviour, with the goal being the manipulation of social relationships (Grammer 1988). Indeed, most of what we know about intervention in conflicts among humans comes from a handful of observations of triadic conflicts and from laboratory experiments (pre-schoolchildren, Grammer 1982; Strayer and Noel 1986; schoolchildren, Ginsburg and Miller 1981; Loots 1985).

Although literature on supporting is rare, developmental psychology provides at least some observations in a natural setting on 'altruistic' or 'prosocial' behaviour, such as cooperation, sharing, and helping. The main question in most of the studies is whether these behaviours increase with age and experience, because the behaviours are highly prized and their development has been seen as a major goal of socialization (Rushton 1980). In a review of the literature on this topic Radke-Yarrow *et al.* (1983) conclude that this is not the case. Eisenberg *et al.* (1984) come up with still more variables:

> The factors determining an individual's choice of prosocial behaviour are both internal and external; the individual's preferences, personal characteristics, and values as well as situational contingencies and cues are potentially important determinants of whether individuals assist in a specific situation (p. 114).

Finally, Marcus (1986) reviews a total of 18 studies of pre-schoolchildren in natural settings. He concludes that empathy is the most important facilitator of prosocial behaviour in general (cf. Wispe 1972), yet the results are not consistent. As can be seen, research on prosocial behaviour mainly considers characteristics of the individual and proximate explanations. Functional considerations of this type of behaviour are rare.

In this paper I will deal with one particular aspect of intervention—supporting in conflicts among children, concentrating on the question 'What is the function of supporting?' The function of a behaviour can be elucidated by the context in which a certain kind of behaviour occurs (contextual evidence) and by its consequences (consequential evidence) (Hinde 1975).

Finding a function on the individual level is difficult if no obvious immediate benefits are present, especially as alternative proximate explanations are possible, for example empathic reactions (Wispe 1972) or the establishment of behavioural norms. Ginsburg (1980) observed that interventions and supporting among schoolchildren always took place when the loser of the conflict showed signs of submission, but the aggressor did not stop fighting. The aggressor thus was violating the norm of stopping aggression in presence of submissive signs. Of course, the biosocial norms could be induced by functional foresight; 'What the aggressor does now, he could do to me in the future' (Kohlberg 1969).

Another possibility is that there are indeed direct benefits, such as gaining status or bonding, which outweigh the costs of possible retaliation. Supporting is then a triadic interaction in which the supporter should try to augment his benefits and minimize costs by carefully choosing whom and against whom to support. Research in non-human primates has shown that supporting is closely related to both status-gain in the group hierarchy and bonding (Chapais this volume pp. 259–83; Cheney 1977; Seyfarth 1980; de Waal 1982). The benefits of supporting thus could be twofold—the successful supporter gains in status and the supported individual becomes his or her friend. In turn, on the ultimate level, high status, stable friendships, and number of friendship relations could raise the fitness of the individual (Dunbar 1984; Harcourt 1987). If this is so, individuals should adopt a strategy to support those with whom they would like to be friends, against those who could threaten their own rank positions.

Supporting can thus be a functional and attractive social strategy which may be used to manipulate relationships between individuals, if it can be used repeatedly in association with the same individuals. This view is in contrast to traditional research on prosocial behaviour. Up till now it has been neglected that prosocial behaviour might serve specific selfish functions that determine its use by individuals.

Costs of supporting

If supporting is indeed a social strategy, the individuals have to consider costs and benefits before they decide to intervene in a conflict. Possible benefits, as said, include status gain and friendship. Possible costs are:

1. The effort of fighting.
2. The amount of retaliation which can be expected, which depends on, (a) the relationships between supporter and opponent and between supported child and opponent (relative power differences—or resource holding potential (Maynard Smith and Parker 1973)—in conflicts can result in different costs, i.e. higher-ranking individuals might win conflicts more easily than others); (b) the likelihood that the opponent himself can recruit allies.
3. The outcome of possible long-term changes in relationships, which are detrimental to the intentions of the supporter, i.e. the loss of status if the opponent wins.
4. The possible gain in status by the supported individual. As soon as the supported individual also gains status, the supporter creates involuntarily possible rivals in the hierarchy of the group.

We see that as soon as there are possible benefits on a level other than immediate level, cost/benefit considerations become quite complex. We should thus expect to find different strategies of supporting, which try to augment benefit by reducing costs on different levels.

Strategies of support

Here I shall try to outline some of the strategies which are possible when individuals are mainly interested in support for personal benefit. I concentrate on three strategies: ganging-up-on, despot, and reciprocation. For the outline of the strategies, I used the costs of fighting efforts (low, moderate, and high) compared to the costs of retaliation and the costs of possible changes in relationships. The value of the strategy then is outlined in terms of minimum possible costs for the individual.

The first strategy might be called the 'ganging-up-on strategy', i.e. supporting the winner of a conflict. In this strategy, all three types of costs are probably low. The strategy primarily takes into consideration the costs of the fighting efforts, which are low. In this strategy, supporting against high-ranking children would be avoided because high-ranking individuals are socially attractive to other group members and thus have the potential to recruit supporters for retaliation. In contrast, supporting against low-ranking children would occur more frequently because they are less likely to retaliate. In the ganging-up-on strategy, unexpected changes in relationships are avoided because the supported child usually does not gain in status by the supporting act—because the child has already settled the conflict by him or herself. The main benefit of this strategy would be the maintenance of status, because it would prevent in advance low-ranking group members from rising in the hierarchy (cf. Chapais this volume pp. 29–59).

A second strategy might be called the 'despot strategy'. Here an individual supports against anybody trying to rise in the dominance hierarchy. The individual thus supports and supports against many different low-ranking individuals. The costs in terms of fighting effort are moderate when supporting against low-ranking group members, and no additional costs occur in respect to unexpected changes of relationships, because possible benefits for others are scattered throughout the group. But the despot also does not benefit from the possible bonding effect of supporting.

In the above-mentioned strategies, supporting itself can bring immediate benefit to the supporter. Thus supporting cannot be described solely under the concept of 'altruism'. Only when there is a delay in gaining a benefit from the act is the support altruistic.

A third promising strategy is the 'reciprocal supporting' strategy, where supporting is restricted to those individuals who give support in return. The strategy has the potential to increase both friendship and status gain simultaneously. Its only serious drawbacks are (1) that it can lead to status gain in the supported; (2) that at some point in the relationship cheating may occur and support not be reciprocated; and (3) that reciprocity has to be established in trial and failure.

The question then arises as to whether these strategies coexist or if they are mutually exclusive, and which strategy should be adopted if children try to maximize their benefits in supporting? Before we can address these problems, we must deal with two methodological questions. (1) Is it possible to assess probable costs and benefits for supporting? (2) Is it possible to observe the consequences of supporting for the relationships between the children involved?

Before these questions can be answered it is necessary to describe different forms of supporting.

THE DESCRIPTION OF SUPPORTING

Frequences of supporting

Compared to the total occurrence of conflicts in groups of children, triadic conflicts occur with a relatively low frequency. In a project, where 20 children (11 girls; 9 boys; 39–78 months of age) were filmed randomly for two 5-minute intervals per week during an entire pre-school year (Grammer 1982, 1988) we observed 2450 conflicts in the resulting 71 hours of video observation. In only 374 cases (15.2 per cent, 5.2 triadic conflicts per hour) was the conflict a triadic one. Comparable frequencies have been found by Strayer and Noel (1986) who recorded 4.1 triadic conflicts per hour (video observation) in a group of 17 pre-school children for an entire academic year.

When we look at the involvement of individual children in triadic conflicts we find that only some children become frequently involved in such conflicts. In our study, all boys in the group ($N = 9$) and only two girls were involved in 92.5 per cent of all triadic conflicts. Thus out of 20 children 11 were responsible for nearly all supporting, suggesting that the circle of supporters in conflicts is quite narrow. Additionally, being a supporter correlates highly with being supported (Rank Spearman, $r = 0.84$, $p < 0.001$) and with being the target—that is, being threatened by the supporter ($r = 0.73$; $p = 0.0006$).

Ginsburgs and Miller's (1981) findings are similar. In their study of school children they found that a small number of children gave aid successfully to a large number of other children. In their study a total of 14 children acted as successful aid-givers for 43 different children in 67 episodes.

Forms of supporting

Different researchers use quite different categories for the observation of triadic conflicts. For instance, Loots (1985) divided the events she observed on a school playground into rough-and-tumble interventions and aggressive interventions. In an aggressive intervention a child could take sides or the child could intervene and remain neutral or intervene and try to settle the conflict by talking non-aggressively to the participants (pacifying interventions). Strayer and Noel (1986) used definitions based on Chase's work (Chase 1974). They divided their observed events into aggressive coalitions and redirected aggression. Aggressive coalitions were further subdivided into defence (the child supports on behalf of the victim) and alliance (the child supports the aggressor). A third method of categorization concerns the functions of support and combines outcome of the initial conflict between two individuals and the time the intervention takes place (Walters 1980). According to Walters, functions may differ if the intervention is a joint threat (both individuals starting at the same time against a third individual), an event of 'aid' (starting the intervention at the point where the conflict still is open), defence (interfering on behalf of the loser), or even ganging-up-on (interfering on behalf of a winner), as pointed out above.

A further dimension is introduced by the observation of the outcome of an event of support. An episode of conflict (Grammer 1988) can be seen as a child pursuing a goal (possession or movement of an object, claiming spaces or spatial access, denying or commanding activities, or explorative goals) and another child opposing this goal. The supporter enables one of the children who is involved to reach his/her goal. Thus the event of support can be classified as effective when the supported child can reach the goal in the conflict. This distinction can be scored with a high degree of reliability. In order to demonstrate this, 15 randomly selected conflict episodes were presented to 12 students. The percentage of agreement (the number of agreements between the observers, divided by the number of agreements plus the number of disagreements) ranged between 0.92–0.97 for the goals and 0.88–0.92 for the success.

In our study we used Walter's definitions in combination with the distinction between effective and ineffective interventions.

The most common category of interventions was 'aid' (34.4 per cent), in which a child supports one of the contestants in an ongoing conflict. Defence, i.e. support of loser (29.4 per cent) and ganging-up-on, i.e. support of winner (28.0 per cent) were next in frequency. Joint threats were rare (8.3 per cent).

Strayer and Noel (1986) also compared ganging-up-on and defence. They observed in three groups of pre-school children comparable frequencies of acts of winner-support (23 per cent) and acts of defence (26 per cent).

In our study the overall effectiveness of the supporters was high: 70 per cent of all supporting episodes end with the supported child winning the conflict. Effectiveness varied to a high degree for individual children from 80 per cent to 0 per cent. Comparable results have been reported for schoolchildren by Ginsburg (1980), although he defined success differently: in 29 of 36 observed cases the benefactor could escape successfully without retaliation after the support.

There seems to be no effect of learning, in the sense that while supporting frequently might be expected to enhance the skills in the use of different fighting tactics, yet frequency of support was not related to effectiveness (Rank Spearman coefficient: $r_S = 0.28$; n.s.). In fact effectiveness in supporting seems to depend solely on the tactics which are used, with non-verbal threat having the highest effectiveness. The efficiency of tactics depends on the relative ranks of the participants: the higher-ranking the opponent is, the more escalative potential the tactics must have in order to be successful (Grammer 1982, 1988).

CONTEXT OF SUPPORTING

Age and sex of supporter, recipient of support, and target

We find significant positive correlations between the frequency of supporting with both age and body size (see Table 10.1). In other words, the older and larger the child is, the more often it is involved in episodes of supporting. But again, effectiveness is not related directly to age or body size.

Episodes in which support is usually not necessary for success, i.e. winner-support, do not show either of the above-mentioned correlations (see Table

Table 10.1. Spearman rank correlations of frequency of participation in acts of support and efficiency of supporting acts with age, body size, attention rank at the beginning of the year (RB), attention rank at the time of supporting act (RS), and attention rank at the end of the year (RE) (for definitions of attention rankings see p. 267).

	Age	Size	RB	RS	RE
Supporting	+0.53*	+0.48*	+0.38	+0.47*	+0.43
Being supported	+0.37	−0.33	+0.51*	+0.54*	+0.40
Being supported against	+0.36	+0.30	−0.55	+0.62*	+0.30
Efficiency	+0.30	+0.11	+0.21	+0.07	+0.23

$n = 18$; * $p < 0.05$, two-tailed.

10.2). These apply only to potentially high-cost supporting like aid, defence, and joint threat. In addition we find highly significant sex differences: boys appear more often in all roles of supporting (Mann–Whitney `U-test: supporter $p = 0.03$, supported child $p = 0.07$, target of aggression $p = 0.0042$, two-tailed).

As has been pointed out by several researchers, we find a clear preference for supporting the same sex. In mixed-sex conflicts, boys tended to take sides with boys and girls with girls. In the 18 cases in which girls intervened in mixed sex conflicts, in 14 they took sides of the girl: in only 4 did they side with the boy. Among boys the situation is similar. Boys supported in 40 mixed-sex conflicts; in 24 cases they sided with boys and in 14 they sided with girls. Similar results were found by Loots (1985) and by Leimbach and Hartup (1981) in a laboratory experiment. This sex-specificity of supporting will be clarified when we look at children's relationships.

Table 10.2. Spearman rank correlations of frequency of participation in episodes of ganging-up-on with age, body size, and attention rank at the beginning of the year (RB), at the time of supporting act (RS), and at the end of the year (RE).

	Age	Size	RB	RS	RE
Ganging-up-on	+0.28	+0.28	−0.25	+0.47*	+0.09
Being supported	+0.16	−0.16	+0.29	+0.54*	+0.25
Being ganged up on	+0.26	−0.19	−0.40	−0.62*	+0.32

$n = 18$; * $p < 0.05$, two-tailed.

Types of conflicts in which supporting occurs

The initial conflicts which led to an act of supporting can be divided into five different categories in terms of the desired goal. The most prominent types of conflicts had exploratory goals (45.5 per cent). This means that the conflict occurs without any apparent reason, and is carried out with a swift attack. Next most frequent were conflicts about space (22.5 per cent), objects (16.9 per cent), denying or commanding activities, which were conflicts in which one child tried to control another child's actions verbally (4.8 per cent) and previous supporting actions (10.3 per cent).

RELATIONSHIPS AND SUPPORTING

Friendship

When children are asked about their concept of friendship, pre-schoolchildren usually answer, 'because he/she plays with me' (Hayes *et al.* 1980). The concept of friendship means 'common activity' for pre-schoolchildren (Bigelow 1977; Youniss 1980). It seems reasonable, therefore, to take the frequencies of playing together as evidence for friendship (Grammer 1988). Indeed, it has been shown by Roper and Hinde (1978) that these frequencies are relatively stable over time. Thus this criterion meets the conditions of Hartup (1975) who demands that descriptions of friendship should reflect a constant and clear choice of one social partner, paired with a specific activity in the dyad, which is play in our case.

The play-partner frequencies were assessed by calculating the percentage of each child's total play that it spent playing with each other individual child. This procedure was carried out for four 2–3-month terms. As shown in Fig. 10.1, most of the dyads never play together and few play together very frequently. We divided the relationships into three types: children who never play together are 'non-friends'; children who play together occasionally (10–20 per cent) are 'friends'; and children who play together 30 per cent or more of their time when they are both present in the group are 'best friends'. Only a few (2–4 out of 18) of the relationships are reciprocal (A plays most of his time with B, and B plays most of his time with A).

An analysis of who is chosen as a friend or best friend shows clear sex preferences: girls play with girls and boys play preferably with boys. For the different terms, out of 18 relationships 12 to 15 are same-sex friend relationships (Fisher test, two-tailed; p ranges between 0.014 and 0.08 for the four terms).

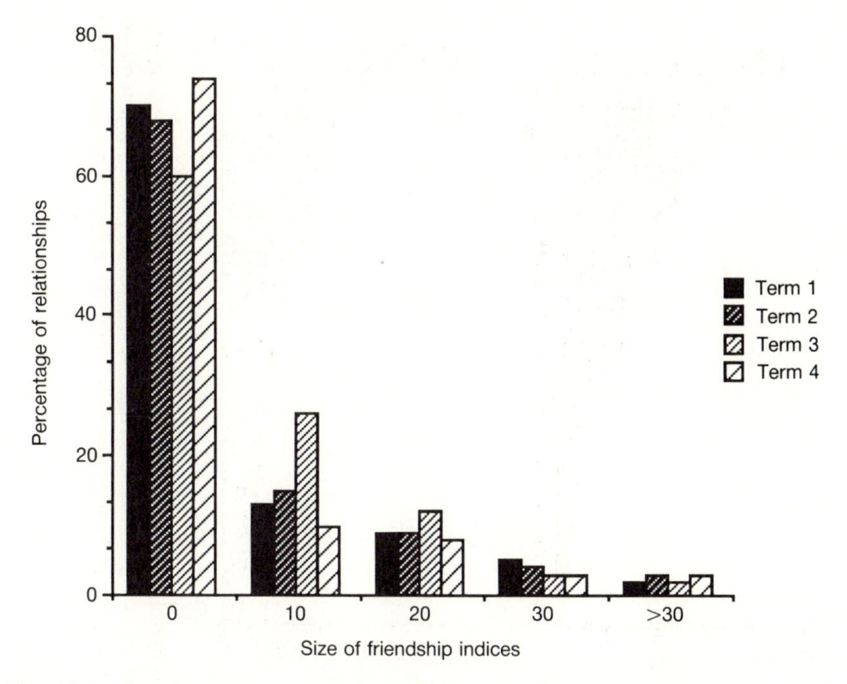

Fig. 10.1. The distribution of friendship indices shows the low proportion of dyads who play together more than 30 per cent of the time they are together in the group. Note that the distribution is not always the same. In term 3 the number of dyads who play occasionally together (10 per cent) is higher than in all other terms. Friendship relationships thus show only relative stability during the year.

Rank

For the description of status and the resulting hierarchies we used attention-structure (Hold 1976). We observed how often a child was looked at by three other children simultaneously (Fig. 10.2). Attention-structure thus reflects the amount of (visual) regard a child has in the group. It has been shown that children ranking high according to visual regard are those children who direct the group's activities (Hold 1976; Vaughn and Waters 1980). The advantage of hierarchies derived from visual regard is that they are multi-dimensional: they correlate with access to resources (Abramovitch 1976; Kalbermatten 1979), and various measures of social competence and even with hierarchies assessed by different dominance criteria (Charlesworth and LaFreniere 1983).

If we rank the children according to the amount of visual regard they recieve, we find that in all four periods this measure correlates negatively with the time the children are socially inactive (playing alone or watching

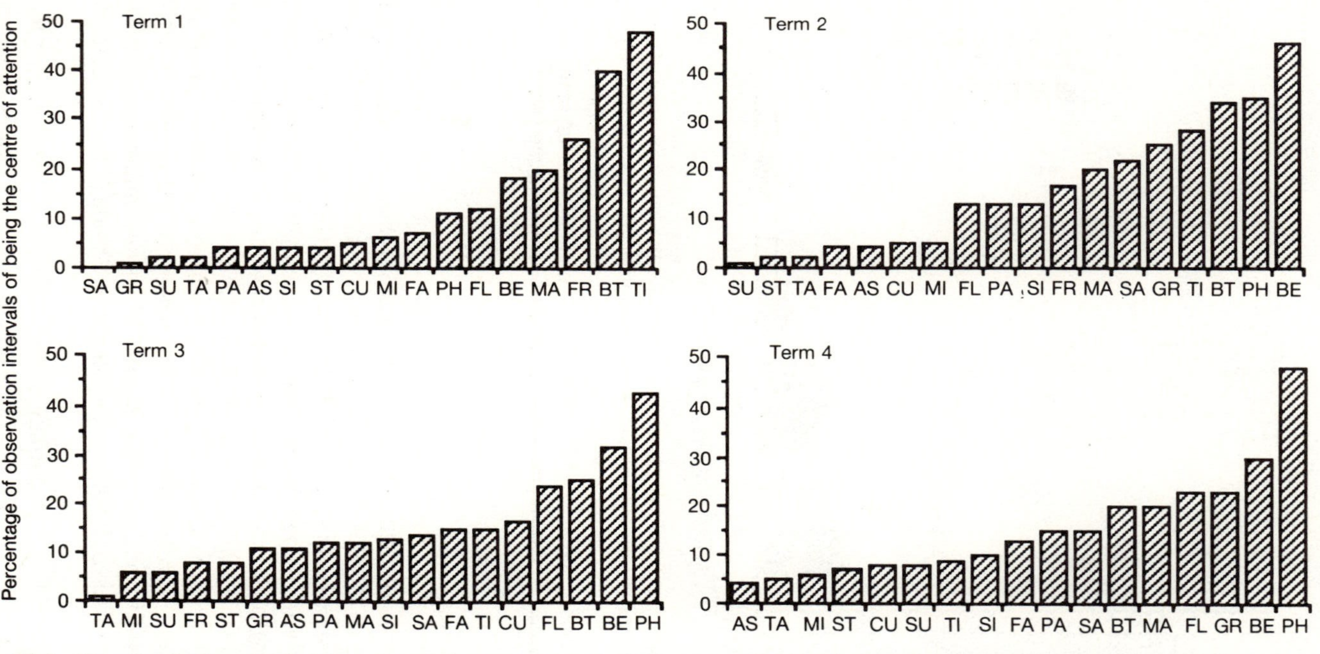

Fig. 10.2. The distribution of attention shows the percentage of 30-s intervals a child was in the focus of attention of the whole time the child was observed. Note that the form of the distribution is different in four terms: in term 3 the distribution is flat, which means that attention is not centred on some children. But not only does the quality of the hierarchy change throughout the year, single children can also move (see text).

others; Rank Spearman coefficient; first term to fourth term: first term $r_S = -0.49$, $p < 0.05$; second term $r_S = -0.55$, $p < 0.05$; third term $r_S = -0.41$, $p = 0.08$; fourth term $r_S = -0.75$, $p < 0.05$, $N = 18$; all two-tailed). Thus rank-orders generated from attention-structure accurately reflect the amount of social participation. There also is a direct relation between supporting and attention. In 72 per cent of all cases of supporting there were more than two spectators, thus supporting can result in visual regard. Nevertheless, the episodes of supporting make up only 9 per cent of all instances of being in the centre of attention. Thus both data sets can be viewed as being independent.

Neither sex received more attention than the other, but in two terms we found a positive correlation between attention rank and age (Rank Spearman correlation: $r_S = 0.48$, $p < 0.04$; $r_S = 0.47$, $p < 0.05$, $N = 18$; two-tailed).

How stable are the resulting systems? High correlations between attention rank at the beginning and at the end of the year in our work suggest relatively overall high stability for the whole observed year ($r_S = 0.75$, $p < 0.001$, two-tailed). Nevertheless, if we look at single children, they can move through the hierarchies, and take quite different places during the year. As Fig. 10.2 shows, the boy GR moves from the second lowest position to the third position in the attention hierarchy. If we have an additional look at the attention rank orders and count the changes of children's rank which are more than one rank from one month to another, we find a decline of changes from autumn 1979 to summer 1980: 9–11–9–7–7–12–4–4. The high number of changes in the antepenultimate interval could be due to long Easter holidays. If this interval is dropped, the decline is significant ($r_S = 0.92$, $p < 0.02$, two-tailed). Thus we can argue that the rank order indeed becomes stabilized at the end of the year.

Rank and friendship

The relation between attention rank and friendship was analysed by Grammer (1988) for 47 children from three different groups ($N = 19$, mean age 52 months; $N = 23$, mean age 63 months, and $N = 18$ children from the group described here). The problem in such an analysis is that the possible rank differences between each child and all other children depend on group-size and position of the child in the hierarchy. For instance, in a group of 18 children the highest-ranking child in an attention-structure has a maximum possible rank distance from his friend of -17 places whereas a middle-ranking child has only ± 8 places of possible maximum differences. In order to compare the children we determined the maximum possible status-difference for each child and calculated the percentage of the observed status-difference between the child and his best friend.

It turns out that in 31 cases out of 47, the median status-difference between friends, 37.8 per cent, is smaller than the 50 per cent we would expect by chance (sign test, $p = 0.04$, two-tailed). This difference becomes higher for second-best friends (39.6 per cent) and yet higher for third-best friends (47.6 per cent). This evidence suggests that friends tend to be similar in rank, i.e. to receive the same amount of attention. In addition, if we divide the rank orders of the groups into three categories: high-, middle-, and low-ranking, high-ranking children are chosen by 20 per cent of the respective group members, whereas low-ranking children are chosen as best friends in only 12.5 per cent of all cases ($\chi^2 = 4.91$, d.f. $= 1$, $p = 0.04$, two-tailed). Furthermore 66 per cent of the high-ranking children have reciprocal friendship relations, whereas in only 33 per cent of the cases the relations of low-ranking children are reciprocal ($\chi^2 = 5.1$, d.f. $= 1$, $p = 0.04$). These results indicate that friends tend to be of similar attention ranks and high-ranking children are more often chosen as friends than low-ranking children. This underlines the social attractiveness of high-ranking individuals.

Supporting and friendship

For friendship relationships we find that friends support (in all types of strategies) much more in favour of their friends than against them (Wilcoxon; $N = 18$ children; two-tailed $p = 0.034$) (Fig. 10.3). Furthermore they support much more against non-friends than they do in favour of non-friends (Wilcoxon; $N = 18$; two-tailed $p = 0.04$) (Fig. 10.3). If we look at actual

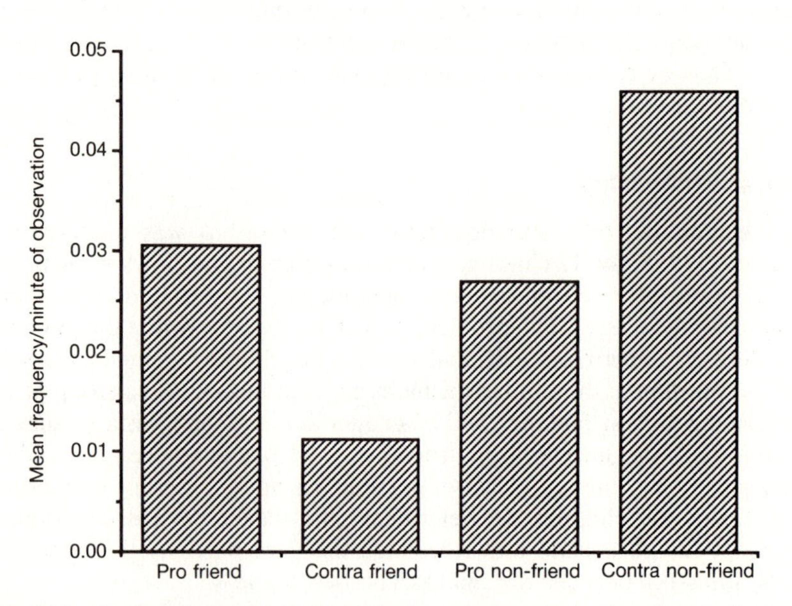

Fig. 10.3. Rate of supporting for different categories of friendships. Children support more often in favour of their friends, but less often against them, and they support more often against non-friends than in favour of them.

decisions, we find that in 120 out of 181 cases the child who was supported played more often with the supporter.

This result replicates Loots' (1985) work among school children. Loots (1985), who used the criterion of playing together in subgroups, shows that in 'rough-and-tumble interventions' 86 choices were for and 44 against friends, and for 'aggressive interventions' 38 were in favour of friends and 14 against them. Strayer and Noel (1986) did not get this result for defence (support of victims) among pre-schoolchildren, although friends engaged in joint threats in their study. They used as a criterion for friendship the frequencies of what they call 'affiliative behaviour'. This measure might not measure the selection of a social partner friendship: for example, Green (1933) reported that friends quarrel more often with each other than with non-friends.

Children show a strong tendency to support their friends. If we now compare the results on sex-differences in the choice of friends, together with the results on supporting, we can explain same sex preferences in supporting—this seems to be a by-product of supporting friends.

Supporting and rank

Relations between supporting and rank in hierarchies or dominance structures were also observed among preschool children. Strayer and Noel (1986) observed children of 4–6 years of age (for explanation of categories of support see above). They analysed dominance relationships between children on the basis of winning conflicts, and found that in defence (support of victims) the intervener was dominant over both, but did not interfere on behalf of their friend more often than expected by chance. Friends supported one another as winners and usually both were dominant over the opponent. In cases when the intervener threatened both original contestants,they were dominant over both contestants, who were their friends. Turning to a consideration of rank in the group's dominance hierarchy, as opposed to relative ranks in the interactions, Strayer and Noel (1986) found that 'In defence [support of the victim] the bystander [the intervener] tended to be a high-status group member. Similarly the original aggressor in interventions in both alliance [the aggressor was supported] and generalization [both contestants were threatened] seemed also to be high ranking in the group dominance hierarchy (p. 126). Nevertheless these results do not shed light on the question of which strategies children use.

Directly comparable with our results are the findings of Ginsburg and Miller (1981) who established an attention-structure among the children in their playground studies. They found that individuals of the upper strata of the attention hierarchy constantly give aid to the children ranked lower in the attention hierarchy.

These results all are congruent with those in our study (see Table 10.1). The rank of the supporter at the time of the supporting correlates positively with the number of support interventions given; the correlation becomes significant for the number of supports received; and there is also a positive correlation with the number of times the child was the target. Thus frequent involvement in supporting interactions seems to be a trait of high rank.

The situation is different for ganging-up on other children (see Table 10.2). Here we find that the lower-ranking a child is, the more often it is ganged up on by others, whereas the higher-ranking children are, the more often they gang up on others. In addition, the higher-ranking children are those who gang up on children who have lost the conflict. Thus the strategy of ganging-up-on, supporting winners, appears to be used by high-ranking children in order to control low-ranking children.

In sum, we find that there is a close relation between supporting, rank, and friendship independent from the type of measurement which was used by different researchers for rank or friendship. The question which then remains is, what is the history of these children in the rank order?

Consequences of supporting for rank and friendship

In order to assess possible relation between attention rank and the frequency of supporting we used a simple correlative technique. We correlated the frequencies of supporting with rank at the beginning of the year, rank at the time of supporting, and rank at the end of the year. Although this technique does not show any causal relationship between supporting and attention rank, it gives a hint concerning possible relations between the two factors.

The results (Table 10.1) show that the more often a child supports another child, the higher ranking it is at the time of supporting. Receiving support correlates with rank at the beginning of the year and at the time of support. Thus frequent supporting and being supported frequently could contribute to reaching and stabilizing rank.

Being supported against correlates negatively with rank at the beginning of the year, but positively at the time of support. This could mean that children who were low ranking at the beginning of the year and then rose in rank are likely to become a target in an episode of support. This correlation disappears at the end of the year. Thus we could argue that at least for some children, being a target has a negative effect on rank. At the end of the year these children are low-ranking again.

In summary, it seems that supporting nearly always occurs in favour of high-ranking children, and perhaps against those children who rose in rank but lose in rank again. Both supporter and supported child could benefit from episodes of support in terms of rank and rank stabilization.

Thus supporting might be one means to achieve and maintain rank-positions. The results of ganging-up-on indicate that it has a different function (Table 10.2), one that does not seem to have a permanent effect on attention rank. The main significant results for ganging-up-on occur at the time of supporting. Frequent gangers and those children on whose behalf the ganging takes place are high ranking. However, there does not seem to be any relation between ganging-up-on and rank at the beginning or end of the year. Thus ganging-up-on could be interpreted as an inefficient attempt to control the low-ranking children in a group (but see Ehardt and Bernstein this volume pp. 83–111, and de Waal this volume pp. 233–57, for alternative interpretations of 'scapegoating').

If supporting has an effect on attention rank, the gain in attention rank should correlate with the number of supporting acts the child has won. This also holds for the supported child and for the opponent if he wins. The results are shown in Table 10.3. In the first half of the year we find that the supported child and the (high-ranking) opponent (if he is able to defeat the supporters) gain in rank throughout the year (RD_a). All winners from the first half of the year gained in rank throughout the first half of the year, with the exception of the supported child (RD_h). At the end of the year the children who received a lot of effective support are high-ranking. The frequency of effective supporting in the second half of the year shows comparable results, although the amount of additional rank gain in the second half of the year no longer correlates with effectiveness in supporting. Thus we could conclude that not only does the frequency of supporting correlate with rank, even the amount of rank gain can correlate positively with effective supporting. This also suggests that there is indeed a cost effect for the supporter: the supported child also gains in rank.

The effect of supporting on 'bonding' has been described in Grammer (1988). There seems to be no such effect—children who support each other

Table 10.3. Spearman rank correlations between the number of coalitionary contests the child has won as supporter, recipient of support, or target, with attention rank at the beginning of the year (RB), change in attention rank in the whole year (RD_a) and in respective half of the year (RD_h), and attention rank at the end of the year (RE).

	RB	RD_a	RD_h	RE
Winning in the first half of the year				
Supporter	+0.06	+0.43	+0.52*	+0.43
Supported child	+0.17	+0.62*	+0.45	+0.58*
Target	+0.60*	+0.70*	+0.83*	+0.32
Winning in the second half of the year				
Supporter	+0.07	+0.45	+0.39	+0.63*
Supported child	+0.18	+0.25	+0.16	+0.48*
Target	+0.36	+0.65*	+0.37	+0.49*

$n = 18$; * $p < 0.05$, two-tailed.

do not become better friends and play more often with each other in the future, although supporting could maintain bonds. Thus up to this point, rank seems to be the sole benefit from giving and receiving support.

Reciprocity—the hallmark of rank and friendship?

Various studies have tried to establish whether reciprocal relationships in supporting among children exist. Ginsburg and Miller (1981) found little reciprocity in their direct observations. Only five episodes of 82 were reciprocal and these five were ineffective. From this result Ginsburg and Miller (1981) conclude that reciprocity might be situationally independent.

In contrast to the above-mentioned work, we found that there is reciprocity in supporting—although there is a considerable asymmetry. First, the more a child supports, the more often this child is supported, as indicated by the high correlation between giving and receiving support ($r_S = 0.85$; $p < 0.001$, $N = 18$, two-tailed). The correlation also indicates that some children are more often involved in supporting acts than others. In a group of 20 children it would be possible to observe 380 dyads of supporting. In fact, only 88 dyads were observed. Figure 10.4 shows the supporting relationships among these 88 dyads. (In this figure child 2 always was the child who supported less, whereas child 1 is the child who supported more often). Asymmetry is fairly high: a child gives a mean of 2.11 (s.d. = 2.55) supports to others, but receives only 0.65 (s.d. = 1.3) supports from those he helped. This means that investment in supporting is higher than returns. Nevertheless, the more one child supports another, the more the other supports it.

If we take a closer look at the dyads, we find that nearly 55.7 per cent were unidirectional (i.e. support was not reciprocated) and occurred only once. If we look at all unidirectional dyads, in which support was not reciprocated, we find that 69.3 per cent fall into this category. However, if we look at those dyads where at least one child supported more than four times, we find that in these cases reciprocation is much more prominent. The ratio between supporting and being supported is 9.13 (s.d. = 3.5) to 4.25 (s.d. = 1.49), or 2.15: 1, compared to the overall ratio of 3.25:1.

In order to test if reciprocity is present, I used the method developed by Mantel (1967) as suggested by Schnell *et al.* (1985). This method tests if a matrix (in this case one of giving and receiving support) is symmetric or not. The result of the test yielded a significance of $p < 0.001$ (d.f. $= \infty$; $t = 101,21$) for symmetry of the supporting matrix, meaning that children reciprocate more support than expected by chance alone. Thus we can conclude that supporting and non-supporting is reciprocal in our sample.

Therefore it seems reasonable to talk of two different types of supporting relationships: quasi-reciprocal (i.e. dyads where at least one support was coming back from the supported child) and non-reciprocal relationships (i.e. dyads where no support was coming back).

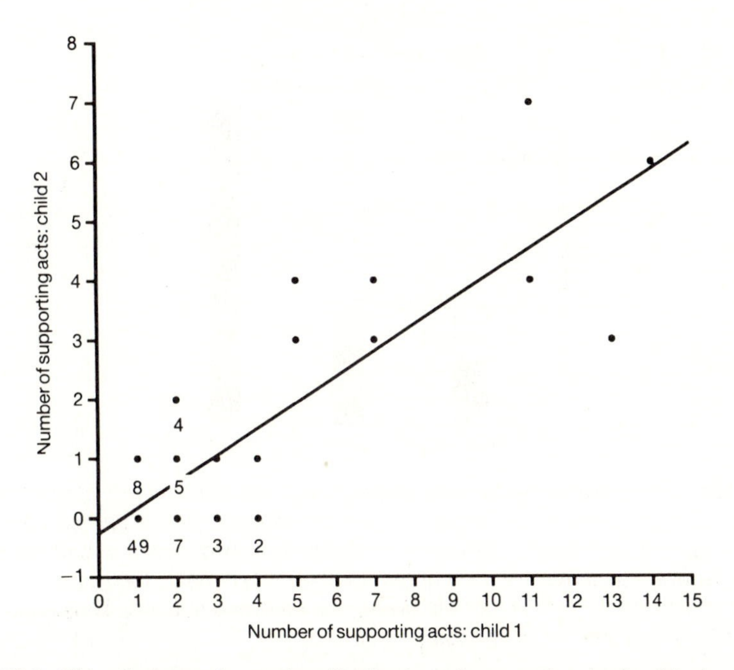

Fig. 10.4. This diagram shows the distribution of supporting in the 88 observed supporting dyads. Child 1 is the child who supported more often than child 2. We find 49 dyads where child 1 supported child 2 and child 2 did not support child 1. We also find that there are no cases where supporting becomes completely asymmetric: child 1 did not support a child 2 more than four times if child 2 did not support child 1 (two cases). On the other hand, we find 'quasi-reciprocal' dyads, where the ratio of supports given by the partners was about 2:1.

If we subdivide the 88 observed dyads into best-friend dyads, friend dyads and non-friend dyads (Fig. 10.5) we are able to cross-tabulate the unique dyads in respect to friendship. It becomes clear that in the above sense quasi-reciprocity occurs most among best friends ($N = 88$; d.f. $= 2$, $\chi^2 = 14.08$; $p = 0.0018$, two-tailed).

The situation for attention rank is not so clear. Giving support in non-reciprocal relationships shows only a weak correlation with rank at the beginning of the year ($r_S = 0.45$; $p = 0.08$, $N = 18$, two-tailed) while receiving support in reciprocal relationships correlates weakly with rank at the end of the year ($r_S = 0.46$; $p = 0.07$, $N = 18$, two-tailed). The number of partners that were recruited during the year shows a similar non-significant correlation with rank at the beginning, and rank at the end of the year ($r_S = -0.42$; $p = 0.08$ and $r_S = -0.44$; $p = 0.08$, $N = 18$, two-tailed). Now the crucial question becomes, what is better, having a few stable allies or being able to recruit many different coalition partners from the group? There is a tendency

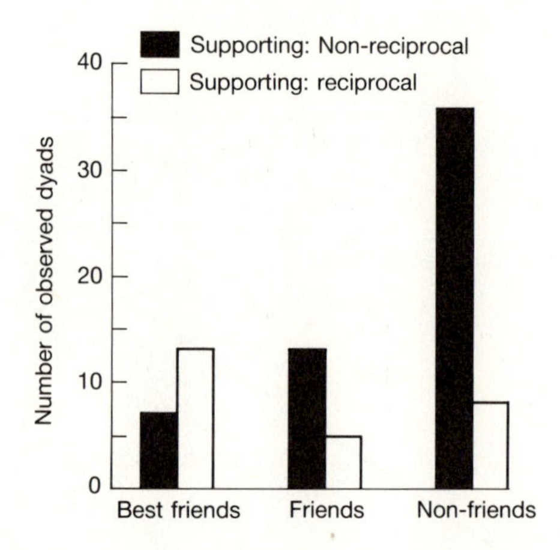

Fig. 10.5. This diagram shows the distribution of the 88 observed dyads in which supporting occurred in relation to the level of friendship of the partners. We find that in reciprocal relationships (open bars) the number of best friend dyads is the highest, whereas in non-reciprocal relationships (solid bars) the number of non-friend dyads is the highest for both giving and receiving support. Thus reciprocity occurs mainly among best friends.

for those children who rank high at the end of the year to have received support from a few stable allies. The ratio between received supports/number of different allies correlates non-significantly with rank then (r_S = 0.33; $p = 0.18$; two-tailed). This ratio does not correlate with rank at the beginning of the year either.

Thus, the fact that Ginsburg and Miller (1981) did not find reciprocity could be due to their short observation periods. Still, reciprocity appears to be poorly developed in this age group.

SUPPORTING AND SOCIAL MANIPULATION

Among non-human primates, alliances with high-ranking animals can have the effect of strengthening and perpetuating the existing hierarchy (Chapais 1983, this volume pp. 29–59; Cheney 1977; Datta 1983; Seyfarth 1980; de Waal 1982). Similarly, all the literature we found on children at different age stages shows high correlations between rank, supporting, and friendship (Ginsburg and Miller, 1981; Loots, 1985; Strayer and Noel, 1986). Furthermore we find that rank at the end of the year is associated with the frequency of receiving and giving support. Thus at least one proximate function of supporting in non-human primates and in humans might be bound to rank.

This proximate function of rank acquisition and maintenance then demands elaborate supporting systems which emerge out of cost/benefit considerations. We find that quasi-reciprocity is one possible strategy which is used by children and that effective supporting might bring rank gain. Costs and benefits of supporting are restricted to a small circle of children in the upper strata of the hierarchy, possibly because the additional costs of recruiting allies is the same for them all. Lower-ranking children should not support others, or even intervene in conflicts among high-ranking children, because the latter can recruit more allies. Quasi-reciprocal supporting can have high costs, which primarily lie in the necessary establishment of friendship as a framework for reciprocity.

The other possible strategy is ganging-up-on (supporting a winner) but this strategy is not as effective as reciprocal supporting. The investment in ganging-up-on is kept low by the children, and so apparently are the benefits in the resulting rank position.

The crux of all considerations on supporting is that there has to be a relation between winning and losing of a conflict and the establishment of rank. Various researchers have shown that it is possible to create dominance hierarchies out of these criteria (Kalbermatten 1979), and that the frequencies of winning and losing correlate with attention structure. I showed that effective supporting correlates highly with gain in rank—even if the opponent wins against the two supporters.

What then is the ultimate function of being high ranking? On this level we have to consider dominance relationships and rank separately. Dominance relationships create predictability between two individuals because a dominance relationship rests on the asymmetry of the probability of who will win a conflict. But the effect of supporting on this level is unclear (see above). Ranks are an extension of dominance relationships over the group. Dominance relationships then represent explicit power in a group. But the idea that ranks are established in struggles by everybody against everybody still prevails. The emergence of ranks through this mechanism seems to be unlikely (Chase 1974). Another mechanism could be that by observing others (visual regard) these dominance relationships are transformed into a system of implicit power. And it is those children who have implicit power in a group who are engaged in supporting. But what good is it to have implicit power? Unfortunately there are many more speculations on this topic than real data, with the exception of the work of Charlesworth and LaFreniere (1983), who show asymmetries in access to scarce resources for high- and low-ranking children. These asymmetries are bound to time of access, not to the access itself. High-ranking children have the right to the first access. But low-ranking children also have their benefits: they have much more time to play, and they can also use the predictability of the high-ranking individuals for the development of their own social strategies.

Besides access to scarce resources, reduction of aggression in a group is another ultimate function which has been proposed for hierarchies (Bernstein 1981). In addition to conceptual difficulties which invalidate the idea of a function of a group phenomenon, Hold-Cavell (1985) found that while frequencies of physical aggression declined over the year the initial phase of group formation is filled with mostly physical aggression, but in the following months the lower-ranking children become physically aggressive. Thus rank position does not only reduce aggression, it also induces aggression.

Given that conflict and support do not always concern immediate access to resources what then determines participation in such aggressive interactions, if we remember that most of the conflicts are of the 'explorative' type? The proximate cause may simply be that 'winning' conflicts could be a preoccupation of children, i.e. 'winning' is a goal in itself. The consequence of winning can be acquisition of power and supporting is an attempt to control the distribution of explicit power among different individuals.

This hypothesis would imply that the children, between 3 and 6 years of age, who are engaged in supporting are social manipulators with the preoccupation of their own payoffs. Supporting is a special case—because of its inherent possibilities that relationship parameters between the participants are affected. Thus the reduction of possible costs should play a main role. In our example it seems that two cost-reducing strategies are used, the establishment of reciprocity and ganging-up-on. This view implies knowledge about relationships and the function of supporting. However, when Strayer *et al.* (1980) asked 4-year-old children about their opinions of the social organization of a group, the dominance hierarchies which were then constructed out of the children's answers did not correlate with the observed hierarchies. This result was replicated by Sluckin and Smith (1977). On the other hand, children of this age are also not always able to tell which children are friends (Leupold 1979). Nevertheless, the impossibility of articulating a certain fact might not be correlated with not knowing it.

In contrast to these results, children from 8 to 10 years old are quite capable of telling whom they will support and who will support them (Loots 1985), which corresponds to observed frequencies of giving and receiving support.

The results suggest that supporting strategies may be at least in part the result of cost/benefit considerations relating to rank status, in which some strategies do better than others. They seem to be a demonstration of explicit power—we find that most of the conflicts are of the 'explorative' type, i.e. they have no obvious goal.

How then can such systems be established? Children could be following a benefit-optimization strategy, and we indeed see that the higher the investment is in terms of frequency of supporting, the higher the benefit, in terms of rank, in this case the resulting position in the rank order. We found that nearly 70 per cent of all supporting acts consisted of dyads which occurred only once. Whether these attempts correspond to the despot strategy or to unsuccessful efforts to establish reciprocal relationships is not clear.

Quasi-reciprocal supporting relationships correlate with friendship. Whilst supporting apparently has no enhancing effect on the strength of the bond, since frequent supporters do not play more often together after supporting, we do not know if it has an effect on bond formation. Thus we could speculate that friendship is the prior condition for reciprocity in supporting. Supporting friends has other benefits: status gain by a friend can be beneficial if support is reciprocated and the supporter himself is a high-ranking friend. Presumably a high-ranking friend is a better friend than a low-ranking friend. Thus the best decision for a child would be to form stable friendships, support each other and then rise together in status. When costs are reduced through reciprocation this strategy produces maximum benefit.

The second strategy, ganging-up-on, does not bring status gain apparently, although children apparently try to minimize costs by supporting against low-ranking children. One possible function of ganging-up-on could be control of the low-ranking children, because ganging-up-on takes place in favour of a high-ranking child against a low-ranking child. However, we did not find this effect in our data, as evident from the lack of correlation between being ganged-up-on and rank at the end of the year.

A final point is that high-ranking children are not more effective than low-ranking, and effectiveness (the percentage of effective support of all supporting acts of a child) is not linked to rank at the end of the year (see Table 10.1). We can speculate that the tactics which can be used in fighting are therefore open to all children and that effectiveness in supporting can be reached by nearly anybody. High variances in effectiveness could thus be the result of applying the appropriate tactic against the opponent (Grammer 1988). This means that the chance of winning does not depend on rank.

I have argued here that children can be viewed as 'social engineers' who constantly try to manipulate and influence others (children and adults). This being the case, supporting is a political decision in a group of children, and neither a specific personal trait of a child nor a question of positive affect. Indeed Schropp (1986) has shown that children give objects much more freely to those children they want to be friends with than to current friends or to non-friends. These aspects of behaviour regarded as prosocial or altruistic have been neglected up to now: prosocial behaviour can have selfish benefits.

The ultimate effects of adaptation in childhood are not only 'learning for life', for the risk of death in this period from weaning to adolescence is high. Mortality risk is highest just after weaning in the age-span from 2 to 6 years (Ebrahim 1983). Thus we could argue that most of the adpatations we find in this period could be a result of the strong immediate selection pressure. Friendship and rank could help a child to survive in an unpredictable environment of other children who are possible rivals.

SUMMARY

Supporting and interventions in conflicts among humans are often described as instances of 'altruism' or 'prosocial behaviour'. In contrast to this view I put forward the hypothesis that, although altruistic tendencies might be present, supporting in the first line is a tactical tool of manipulation of other individuals. In a review of observations of supporting among preschool children, together with additional data from a longitudinal study in a natural setting, the following points become clear:

1. The frequency of supporting depends on the rank position of the child in a group hierarchy: high-ranking children support more often and receive more support than low-ranking children.
2. Who is supported depends on the relationship between the children: friends are supported more often than non-friends.
3. Frequent winning in supporting pays in terms of status gain, for the supporter and also for the supported child.
4. Children try to optimize these benefits through cost/benefit considerations which lead them to select appropriate supporting strategies and to establish reciprocal supporting relationships.
5. As a result the function of supporting can be seen as serving for the establishment and the maintanenace of high rank and thus as a specific adaptation to the critical period between weaning and adolescence. Acting 'selfishly altruistic' in order to manipulate others is predominant over prosocial tendencies at this age stage.

ACKNOWLEDGEMENTS

Many thanks to P. Wiessner, A. Harcourt and F. de Waal for helpful comments on the study and the manuscript. The data presented here come from a longitudinal study on pre-schoolchildren carried out from 1978 to 1983 together with B. Hold-Cavell and H. Shibasaka.

REFERENCES

Abramovitch, R. (1976). The relation of attention and proximity to dominance in preschool children. In *The social structure of attention* (ed. M. R. A. Chance), pp. 153–76. Wiley, London.

Bernstein, I. S. (1981). Dominance: the baby and the bathwater. *Behavioral and Brain Sciences*, **4**, 419–57.

Bigelow, B. (1977). Children's friendship expectations: a cognitive developmental study. *Child Development*, **48**, 246–53.

Chapais, B. 1983. Dominance, relatedness and the structure of female relationships in rhesus monkeys. In *Primate social relationships* (ed. R. A. Hinde), pp. 208–19. Blackwell, Oxford.

Charlesworth, W. R. and LaFreniere, P. (1983). Dominance, friendship, and resource utilization in preschool children's groups. *Ethology and Sociobiology*, **4**, 175–86.

Chase, I. D. (1974). Models of hierarchy formation in animal societies. *Behavioral Science*, **19**, 374–82.

Cheney, D. L. (1977). The acquisition of rank and the development of reciprocal alliances among free-ranging immature baboons. *Behavioral Ecology and Sociobiology*, **2**, 303–18.

Datta, S. B. (1983). Patterns of agonistic interference. In *Primate social relationships: an integrated approach* (ed. R. A. Hinde), pp. 289–97. Blackwell, Oxford.

Dunbar, R. I. M. (1984). *Reproductive decisions. An economic analysis of gelada baboon social strategies*. Princeton University Press, Princeton.

Eisenberg, N., Cameron, E., and Tyron, K. (1985). Prosocial behavior in the preschool years: methodological and conceptual issues. In *Development and maintenance of prosocial behavior* (eds E. Staub, D. Bar-Tal, J. Karylowski, and J. Reykowski) pp. 101–14. Plenum Press, New York.

Ebrahim, G. J. (1983). *Nutrition in mother and child health*. Macmillan, London.

Ginsburg, H. J. (1980). Playground as laboratory: naturalistic studies of appeasement, altruism and the omega child. In *Dominance relations* (eds D. R. Omark, F. F. Strayer, and D. Freedman), pp. 341–57. Garland STPM, New York.

Ginsburg, H. J. and Miller S. M. (1981). Altruism in children: a naturalistic study of reciprocation and an examination of the relationship between social dominance and aid-giving behaviour. *Ethology and Sociobiology*, **2**, 75–83.

Grammer, K. (1982). Wettbewerb und Kooperation: Strategien des Eingriffs in Konflikte unter Kindern einer Kindergartengruppe. Ph.D. thesis, University of Munich.

Grammer, K. (1988). *Biologische Grundlagen des Sozialverhaltens*. Wissenschaftliche Buchgesellschaft, Darmstadt.

Green, E. H. (1933). Group play and quarreling among preschool children. *Child Development*, **4**, 302–7.

Harcourt, A. H. (1987). Dominance and fertility in female primates. *Journal of Zoology*, **213**, 471–87.

Hartup, W. W. (1975). The origins of friendship. In *Friendship and peer relations* (eds M. Lewis and L. A. Rosenblum), pp. 1–26. Wiley, New York.

Hayes, D. S., Gershman, E., and Bolin, L. (1980). Friends and enemies: cognitive bases for preschool children's unilateral and reciprocal relationships. *Child Development*, **51**, 1276–9.

Hinde, R. A. (1975). The concept of function. In *Function and evolution in behaviour* (eds G. P. Baerends, C. de Beer, and A. Manning), pp. 3–15. Clarendon Press, Oxford.

Hold, B. C. L. (1976). Attention-structure and rankspecific behaviour in preschool children. In *The social structure of attention* (eds M. R. A. Chance and R. R. Larsen), pp. 177–201. Wiley, London.

Hold-Cavell, B. C. L. (1985). Showing-off and aggression in young children. *Aggressive Behavior*, **11**, 303–14.

Kalbermatten, U. (1979). Handlung: Theorie, Methode, Ergebnisse. Ph.D. thesis, University of Berne, Switzerland.

Kohlberg, L. (1969). Stage and sequence. The cognitive approach to socialization. In *Handbook of socialization theory and research*. (ed. D. A. Goslin), pp. 347–480. Rand McNally, Chicago.

Leimbach, M. P. and Hartup, W. W. (1981). Forming cooperative coalitions during competitive games in same-sex and mixed-sex triads. *Journal of Genetic Psychology*, **130**, 165–71.

Leupold, K. H. (1979). Aggression und Aggressionsbeschwichtigung im Kindergarten. Diplomarbeit im Fachbereich Biologie, Universität München.

Loots, G. M. P. (1985). Social Relationships in Groups of Children: an Observational Study. Dissertation at the Vrije Universiteit de Amsterdam.

Mantel, N. (1967). The detection of disease: clustering and a generalized regression approach. *Cancer Research*, **27**, 209–20.

Marcus, R. F. (1986). Cooperation, helping, sharing: empathy and affect. In *Altruism and aggression, biological and social origins*. (eds D. Waxler, M. Cummings, and R. Iannotti), pp. 256–79. Cambridge University Press, Cambridge.

Maynard Smith, J. and Parker, G. A. (1973). The logic of asymmetric contest. *Animal Behaviour*, **24**, 159–75.

Radke-Yarrow, M., Zahn-Waxler, C., and Chapman, M. (1983). Children's prosocial dispositions and behaviour. In *Carmichael's manual of child psychology, 4th edn*, vol. IV (ed. P. Mussen), pp. 469–545. Wiley, New York.

Roper, R. and Hinde, R. A. (1978). Social behavior in a play group: consistency and complexity. *Child Development*, **49**, 570–9.

Rushton, J. P. (1980). *Altruism, socialization and society*. Prentice Hall, Englewood Cliffs, NJ.

Schnell, G. D., Watt, D. J., and Douglas, M. E. (1985). Statistical comparison of proximity matrices: applications in animal behaviour. *Animal Behaviour*, **33**, 239–53.

Schropp, R. (1986). Interaction 'objectified'. In *Ethology and psychology* (eds J. LeCamus and J. Cosnier), pp. 77–88. Université Paul Sabatier, Toulouse.

Seyfarth, R. M. (1980). The distribution of grooming and related behaviours among adult female vervet monkeys. *Animal Behaviour*, **28**, 798–813.

Sluckin, A. M. and Smith, P. K. (1977). Two approaches to the concept of dominance in preschool children. *Child Development*, **48**, 917–23.

Strayer, F. F. and Noel, J. M. (1986). The prosocial and antisocial function of preschool aggression: an ethological study of triadic conflict among young children. In *Altruism and aggression, biological and social origins* (eds C. Zahn-Waxler, E. M. Cummings, and R. Iannotti), pp. 107–34. Cambridge University Press, Cambridge.

Strayer, F. F., Strayer, J., and Chapeskie, T. R. (1980). The perception of social power relations among preschool children. In *Dominance relations* (eds D. R. Omark, F. F. Strayer, and D. Freedman), pp. 191–203. Garland STPM, New York.

Vaughn, B. E. and Waters, E. (1980). Social organization among preschool peers: dominance, attention, and sociometric correlates. In *Dominance relations* (eds D. R. Omark, F. F. Strayer, and D. Freedman), pp. 359–80. Garland STPM, New York.

de Waal, F. B. M. (1978). Exploitative and familiarity-dependent support strategies in a colony of semi-free living chimpanzees. *Behaviour*, **66**, 268–313.

de Waal, F. B. M. (1982). *Chimpanzee politics*. Cape, London.
Walters, J. (1980). Interventions and the development of dominance relationships in female baboons. *Folia primatologica*, **34**, 61–89.
Wispe, L. G. (1972). Positive forms of social behavior. An overview. *Journal of Social Issues*, **28**, 179.
Youniss, J. (1980). *Parents and peers in social development*. Chicago University Press, Chicago.

A coalition of two adult males against a third adult male. The immediate cause of this conflict was meat (note the Thompson gazelle fawn carried by the right most male).

11

Alliance formation among male baboons: shopping for profitable partners

RONALD NOË

INTRODUCTION

After Packer (1977) confirmed an earlier suggestion of Trivers (1971) that baboon allies are reciprocal altruists, this view gained the status of textbook truth. Packer's conclusions were not challenged for more than a decade. Two recent studies, however, have questioned the validity of the reciprocal altruism theory for baboon alliances. I start this review with a synopsis of this critique. Thereafter I list the phenomena that should be accounted for by a model of alliance formation. This list is based on observations of coalitions and alliances made in a large number of baboon groups at various study sites. Finally I give an overview of my own efforts to formulate such a model. I compare coalition formation to interactions on a marketplace, where supply and demand determine which deals are closed and how profits are allocated. Paradigms for this kind of processes are provided by coalition games, *N*-player games in which bargaining among the participants plays an essential role. My first concern, however, is to identify the males that qualify as players in the game. I approach this problem by asking two questions: which males would gain by coalition formation, and which males would be able to form successful coalitions?

The study of baboon coalitions: a historical overview

The first extensive descriptions of alliances among male savanna baboons were given by Hall and DeVore (1965). They described the existence of what they called 'central hierarchies' in two groups of olive baboons (*Papio cynocephalus anubis*). Hall and DeVore wrote about one of their study groups:

It became clear that certain of the adult males constantly associated with each other and tended to support each other in aggressive interactions with other males. Some of these males associated so closely that they were scarcely ever observed acting independently in such episodes, and on this basis three of them came to be designated a 'central hierarchy'. (Hall and DeVore 1965, pp. 59–60).

Following Hall and DeVore's publications (see de Waal and Harcourt this volume pp. 9–27 for more detail), data on coalitions and alliances among the adult male members of the several subspecies of the savanna baboon (*Papio cynocephalus*) have been gathered at a large number of sites in an even larger number of groups. These observations taken together now form the largest data set available on alliance formation among unrelated primates.

Hall and DeVore's pioneering work also influenced the subsequent study of alliance formation in a more indirect way. Inspired by a visit to their study site, Trivers mentioned baboon coalitions as a possible example of 'reciprocal altruism' in his seminal paper published in 1971. Alliances among male baboons became one of the textbook examples of reciprocal altruism after Packer (1977) confirmed Trivers' conjecture on the basis of field data. The power and appeal of Trivers' theory apparently led to a ready acceptance of Packer's conclusions, although one crucial step leading to these conclusions was made on the basis of six observations only (see Bercovitch 1988; Noë 1990). Further data on coalition formation were presented in studies of various aspects of the social organization and mating system of savanna baboons (Bercovitch 1985, 1986, 1987; Collins 1981; Noë 1986; Packer 1979*a*, *b*; Rasmussen 1980; Smuts 1985). The reciprocal altruism interpretation of baboon alliances was challenged in none of these publications. Then, in a paper on the use of coalitions as a tactic to improve reproductive success published in 1988, Bercovitch cast doubt on the applicability of the reciprocal altruism model. Finally, after a study especially devoted to coalition formation, I too concluded that the reciprocal altruism model was not valid and proposed an alternative approach (Noë 1989, 1990).

PROBLEMS WITH RECIPROCAL ALTRUISM AND THE PRISONER'S DILEMMA

In view of the close connection between the prisoner's dilemma model and the reciprocal altruism theory the statement that baboons are reciprocal altruists almost automatically implies that the prisoner's dilemma is a valid paradigm for baboon coalition formation. Casting doubt therefore on the validity of reciprocal altruism also casts doubt on the use of the prisoner's dilemma. Since there has been some discussion about the validity of the prisoner's dilemma paradigm in the case of reciprocal altruism, I briefly deal with this problem. My conclusion is that the prisoner's dilemma model is consistent with reciprocal altruism in most important features. Since the prisoner's dilemma model is more explicit than the original reciprocal altruism theory, I phrase my criticism as much as possible in terms of the former. One assumption specific to the reciprocal altruism approach is that coalitions between male baboons are interactions in which one male acts as an altruist. I discuss this assumption separately.

Although a general discussion of the prisoner's dilemma model is beyond the scope of this review, I devote a separate section to its general relevance.

Is one of the participants in a coalition an altruist?

Few people will have doubts about the collaborative character of baboon coalitions, i.e. the total benefit to both participants combined will on average be larger than the total cost they both suffer. The question is whether the coalitions are altruistic in character, which according to the biological meaning of the term implies that in each interaction one participant gains, while the other loses. The alternative is that coalitions are cooperative in character, i.e. both participants normally obtain a net gain. It is also possible that alliances are a mosaic of both types of interaction.

As mentioned above, Trivers (1971) suggested and Packer (1977) appeared to confirm that coalitions are altruistic acts. On what grounds did they think so? Packer concentrated on coalitions formed to obtain access to females. Since females are indivisible entities, only one participant can end up with the female. Thus, when both invest in coalitions, such coalitions will inevitably be altruistic in character. This fact is not enough, however, to make these coalitions an interesting case for the study of the evolution of altruism. For that purpose each participant should *a priori* know which role he is going to play, in other words, which male is going to end up consorting the female, when the coalition proves to be successful. The theory of reciprocal altruism says something about the *decision* to act altruistically. I refer to a strategic decision here, of course, not to a conscious decision. The decision to help an unrelated individual, so the theory goes, is taken on the grounds of the chances that aid will be received from the beneficiary in the future. The theory of reciprocal altruism is not needed to explain the occurrence of coalitions, when both participants stand a fair chance (which does not have to be an equal chance) of gaining the female after each successful interaction, and when each of the males has a net gain over a longer series of interactions (see also Bercovitch 1988). Whether or not a male will have a net gain over a series of coalitions with a particular partner will depend on (1) the cost of taking part in a coalition; (2) the success rate of the coalitions; (3) the chance to obtain the consort after a successful coalition; and (4) the benefit of obtaining a consort.

It is thus crucial to the reciprocal altruism argument that it is clear from the onset of the formation of each coalition, which of the two males will gain access to the female. This is what Packer (1977) argued to be the case. He found that in all cases he observed only one of the males asked for support (using a behavioural element called 'head-flagging') and that this male always obtained the consort. Unfortunately Packer saw this connection between head-flagging and obtaining the consort in only six cases. Bercovitch (1988) used data presented in Rasmussen (1980) and his own data to show that the initiator of the coalition was as likely as his partner to become the new consort male. He therefore concluded that baboon allies are not reciprocal altruists. I found that in some cases it was not even clear which of the males took the initiative to form the coalition, and that often both showed the head-flagging simultaneously (Noë 1989). Neither Bercovitch (1988) nor I could observe an indication at the start of a coalition as to who would profit from a successful take-over. On the contrary, in most cases it seemed to be a matter of chance. Once the close proximity between the female and the consorting male is broken, a race between the partners towards the female may take place. Reaching the female first normally suffices to make a male the undisputed owner for some time (Bercovitch 1988; own observations). This phenomenon could be based on a respect for 'ownership' reminiscent of hamadryas baboons (cf. Kummer *et al.* 1974). The new owner may even be a male that did not take part in the battle over the female. Often the male that lost the female will go on fighting one of his opponents for some time, while the other sneaks out to the female. One could say, therefore, that in such cases the previous owner largely determines which coalition partner gets the female.

Bercovitch (1988) concluded that both partners in a coalition have a chance to obtain the resource, and thus are in fact cooperating. Other coalitions, notably interferences against strong opponents, may nevertheless be altruistic interactions. It is, however, also plausible that such interferences at least partly serve the interest of the interferer: he may seize the opportunity to settle his own affairs with the target of the coalition, or his main goal may be to prevent damage to his partner, who is instrumental in getting access to certain resources (cf. Kummer 1979).

Once could argue that the question of whether altruism or cooperation is involved is bound to remain academic, since we are not, and probably never will be, able to measure the cost/benefit budgets involved. How important is it to be able to tell the difference? Historically the theory of reciprocal altruism was an answer to the challenge to explain the occurrence of altruism among unrelated animals. At the time altruism was a central problem in the debate about individual selection vs. group selection. Strong emphasis was laid on the significance of the time lag between investment and return in reciprocal altruism. The time lag would make it difficult to recognize 'cheating' by a partner. This emphasis probably exaggerated the difference between cooperation and reciprocity. At a later stage Axelrod and Hamilton (1981) proposed the iterated prisoner's dilemma as a paradigm for both cooperation and reciprocal altruism among unrelated animals and implicitly pointed out the resemblance between cooperation and reciprocity: the individuals involved have to guard themselves against exploitation by the partners in both cases. Axelrod and Hamilton's proposal implies the validity of the iterated prisoner's dilemma game as a paradigm for coalition formation among baboons. In the following sections I examine the usefulness of this approach.

Is the prisoner's dilemma game a valid paradigm for reciprocal altruism?

There has been some discussion whether the prisoner's dilemma model indeed applies to reciprocal altruism (see Packer 1986). In an iterated prisoner's dilemma each player makes a choice between 'cooperation' and 'defection' in each round. Two reciprocal altruists make these decisions one after the other. If one were to take two such decisions, one by each player, to be comparable to a single round in the game, then there would be one player in each such round that would not be faced with the dilemma so typical for the game, since he would already know the other player's decision. It is essential, however, that reciprocal altruists are not informed about the *next* decision of their partner. Therefore it is likely that a game can be constructed for reciprocal altruism, which resembles the prisoner's dilemma in essential features. Boyd (1988) indicated how the prisoner's dilemma can be adjusted to make it compatible with reciprocal altruism.

Is the prisoner's dilemma a valid paradigm for coalition formation in baboons?

In order to qualify as an example supporting the prisoner's dilemma model the following statements about baboon coalition formation should hold:

1. The formation of coalitions depends on the behaviour of two individuals only, since the model is based on the two-player iterated version of the game.
2. The (potential) costs and benefits concur with the pay-off configuration of the prisoner's dilemma.
3. Communication between the partners does not influence the decision to participate in coalitions.

The confirmation of the following statement would support the prisoner's dilemma model, although its refutation would not necessarily imply the rejection of the model:

4. The behaviour of baboon allies is consistent with playing a tit-for-tat strategy.

1. Is a two-player game suited as a paradigm for baboon alliance formation?

The two-player prisoner's dilemma is not suited as the basis of a model that takes into account the possibility to weigh one potential partner against another. The use of the two-player version of the game implies that each individual regulates its collaboration with each of its partners separately. Strategic decisions in interactions with a certain partner are supposed to be influenced only by past interactions and/or expectations about future interactions with that same partner.

Male baboons switch over to another partner, not so much because the old partner refuses collaboration, but rather because the new partner is a more effective member of the coalition (Noë 1986, 1989) (see Fig. 11.1(a) (b)). This makes sense, because a male that can reach a goal only through collaboration with one other male, and that has several potential partners, should select the best to do the job. This can be the most effective partner in one case and the least demanding in another. Two phenomena are thus crucial to the understanding of this kind of collaboration: (1) partnerships form through a process of selection from several potential partners; (2) a possibility of switching partners remains after partnerships have formed. This coin has another side as well: when the number of potential partners is limited, competition over valuable partners can be expected.

A game theoretic model that accounts for this kind of dynamics should be based on an N-player game. For those not familiar with game theory it might seem that the N-player variant of the prisoner's dilemma, best known from Hardin's (1968) explanation of the 'tragedy of the commons', could be the proper base for such a model. This is not true, however. The pay-offs for players in an N-players prisoner's dilemma game depend on the behaviour of *all* players involved, e.g. all members of a social unit. Collaboration does not pay if not all players, or at least a large proportion, decide to 'cooperate'. In an N-player prisoner's dilemma all others, defectors and fellow cooperators alike, profit from the decision of a player to cooperate. The paradigm can be applied to cases of collective collaboration, e.g. to the problem of vigilance in flocks of birds. The type of collaboration considered here is distinctly different. Individuals can increase their fitness by forming small collaborating subunits, with a neutral or disadvantageous effect on the fitness of other individuals involved.

2. Is the cost/benefit budget of coalition formation in agreement with the prisoner's dilemma's pay-off configuration?

I have already mentioned that gaining access to receptive females is an important goal of coalitions among male baboons. In these, and other coalitions that yield indivisible resources, only one partner can gain a positive benefit. Thus one possible outcome of the prisoner's dilemma, the one in

Fig. 11.1. Two scenes of a single conflict. The males at the right threaten (raised eye brow) the screaming male on the left. The conflict is unusually tense, possibly because it was one of the first conflicts in which the male on the foreground/right sided with the male in the background instead of with the other male. Both males were his allies throughout the study period, but he changed preference between them coinciding with a drop in rank of the disfavoured ally.

which both players gain the 'reward for cooperation', is fundamentally impossible. This problem is analogous to the problem of applying the prisoner's dilemma model to reciprocal altruism discussed in the previous section, and can be solved in the same way. An apparent solution would be to assume that there always is a small benefit for the less lucky partner. This would imply the assumption that the decision to take part in the coalition is based on the chance to obtain this small benefit. By invoking such undemonstrable benefits one could make all cost/benefit budgets consistent with any model; a rather fruitless procedure.

A more serious problem is that the 'temptation to defect' (T) in baboon coalition formation is not likely to be greater than the 'reward for cooperation' (R) (see PD-matrix in Boyd, this volume pp. 473–89). A high value of T implies that a 'defecting' baboon could reap the benefits of the effort of a 'cooperating' partner; in other words, the goals collaborating baboon males strive for could in principe be reached through the effort of one male alone. In most types of coalition this is not likely to be the case. Moreover, the 'cheated' partner would be able to react to the defection by breaking off the interaction at the start of the formation of the coalition in most cases, although it may be difficult to react to subtle forms of cheating in that manner. I thus consider baboon coalitions to be a form of 'synergistic cooperation' (Maynard Smith 1983) for which the trust game, in which $R > T$, would be a more appropriate paradigm than the prisoner's dilemma (see Liebrand 1983 for a classification of the social dilemma games). One could try to evade this problem and state that one of the partners could 'defect' by appropriating the resource gained after the successful collaboration. This option would, however, only be open to the strongest or most dominant individual, which is at variance with the assumption of the prisoner's dilemma model that both partners must have the option to defect unilaterally.

I mentioned above the possibility that baboon males would interfere in agonistic interactions to the advantage of their partners in order to prevent damage, which could make the partner less useful in the future. Lima's (1989) analysis shows that the pay-off configuration of coalitions would no longer be consistent with the pay-offs used in Axelrod and Hamilton's (1981) model, if indeed the future value of the partner plays a role in the decision to form coalitions.

3. Is the decision to participate in coalitions independent of communication between the males involved?

The prisoner's dilemma model is based on non-cooperative game theory, and thus explicitly excludes the possibility that negotiations influence the decision to collaborate. The argument is as follows: communication about intentions cannot be honest, because it would pay to signal willingness to cooperate and to defect as soon as the other decided to cooperate. Given the prisoner's dilemma's pay-off configuration this is a correct conclusion; not so, however, when coalition formation is indeed a form of synergistic co-operation. When both participants stand to gain by cooperation, while neither can reap direct benefits from defection, honest communication and negotiation become realistic possibilities.

Before and during coalitions baboons show extensive communication (Noë 1989). It would be rather unlikely for such behaviour to evolve if it did not have a positive influence on the behaviour of the partner. I refer to this communication more extensively in my discussion of 'bargaining', which forms an essential part of my alternative model.

4. Do baboon males play tit-for-tat?

On the basis of a number of computer tournaments, Axelrod and Hamilton concluded that tit-for-tat is an evolutionary stable strategy (ESS) of the iterated prisoner's dilemma (Axelrod 1984; Axelrod and Hamilton 1981). Although this conclusion had to be weakened in the light of later work (Boyd and Lorberbaum 1987; Selten and Hammerstein 1984) one can still assume that tit-for-tat (or strategies with similar properties, see Axelrod 1984; Boyd this volume pp. 473–89) are likely to be played in all situations for which the iterated prisoner's dilemma is a valid paradigm. It is thus worthwhile to examine whether baboons play tit-for-tat or a similar strategy. It should be noted, however, that refuting tit-for-tat does not necessarily invalidate the iterated prisoner's dilemma as a paradigm. On the other hand observing tit-for-tat does not prove that an iterated prisoner's dilemma is being played, since this strategy can also be a good strategy in other games. For example the strategy 'retaliator' in the hawk–dove game (Maynard Smith and Price 1973; this game is also known as the 'game of chicken') is identical to tit-for-tat (Rapoport 1975).

Two observations make it unlikely that baboon males play tit-for-tat. Firstly, some males participated in coalitions over consorts again and again, although they almost never gained the main prize, the oestrous female (Collins 1981; Noë 1990). Secondly a number of alliances did not fall apart, even after one ally repeatedly refused to collaborate by not responding to solicitations to do so (Bercovitch 1988; Collins 1981; Noë 1989).

Tit-for-tat is not a likely strategy, when only few potential partners are available. Individuals using this rule basically select their partners by trial-and-error, rejecting those that do not collaborate properly. The tit-for-tat rule in a strict sense is very sensitive to the occurrence of an occasional mistake, misunderstanding or inability to effectuate a decision to cooperate. Such events can very easily lead to an echo effect of repeated defections. With few partners available this can soon lead to the rejection of all potential partners. Boyd (1989) proposed a solution for the problem of mistakes, based on a modified version of tit-for-tat. Boyd's solution assumes, however, that (1) the defector-by-mistake would 'know' that he had made a mistake and that (2) animals on occasion would also make the reverse mistake and cooperate when they should have defected (but see Boyd this volume pp. 473–89). I do not know how to translate the latter assumption in terms of baboon behaviour. It therefore remains to be shown that tit-for-tat-like rules can survive in populations split up in relatively small social units when mistakes occur regularly.

Is the prisoner's dilemma a useful paradigm for other forms of collaboration?

My conclusion is that the prisoner's dilemma model does not apply when animals have a choice among partners. In those cases the dynamics of coalition formation play a role. These dynamics cannot be modelled using the prisoner's dilemma (cf. Colman 1982). This excludes the use of the prisoner's dilemma in virtually all cases of collaboration between members of the same social unit, except for those forms of collaborative behaviour which benefit the whole unit. In the latter case the N-player version of the game could be useful. Coalition formation does not play a role in social units, when for some reason each individual that could gain by collaboration has only one possible partner. In that case the proper paradigm is the trust game ($R > T > P > S$), when the collaboration is synergistic, and the prisoner's dilemma ($T > R > P > S$), when the collaboration is inessential (see section above, question 2).

This should not be taken to mean that there are no useful applications of the prisoner's dilemma paradigm. Good examples are conflicts between 'dear enemies' (Getty 1987; Whitehead 1987) and gamete exchanges in some simultaneous hermaphrodites (Fischer 1980, 1988; Leonard and Lukowiak 1984; Sella 1985). In both cases the animals involved are condemned to collaborate with the given partner and a choice among partners plays no role. Dear enemies are basically competitors that can reduce the costs of competition by some form of collaboration. The hermaphrodites figuring in the studies mentioned normally have no simultaneous choice between partners, while finding an alternative partner incurs a high cost. A mechanism to guarantee this high cost can be part of the strategies played (see Fischer 1980).

I know of one game-theoretic development that could lead to a compromise between the prisoner's dilemma model and my desire to involve the dynamics of coalition formation: the possibility of forming 'implicit coalitions' in three-player prisoner's dilemmas as studied by Fader and Hauser (1988). The problem is, however, that in Fader and Hauser's version of the game the players can opt for actions that are relatively fine-tuned intermediates between the prisoner's dilemma's 'cooperation' and 'defection'. In the case of baboon coalition formation I do not see the real world equivalent of such a continuous range of options.

The need for a new model of coalition formation in baboons

The conclusion from the above is that the prisoner's dilemma model, and thus implicitly the reciprocal altruism theory, is not satisfactory in the case of coalition formation in male baboons. The model is at variance with a number of observations and falls short in capturing important features of coalition formation: (1) the possibility to switch between partners, or at least to threaten with such a switch; (2) the competition over preferred partners; (3) the role played by communication. I am convinced that these shortcomings are not confined to the case of baboon coalition formation, but are typical for all cases in which members of social units can form collaborative subsets of varying composition (this view is further developed in Noë *et al.* 1991). I limit myself here to an alternative model of alliance formation among adult male baboons, but before presenting this model I give an overview of the relevant features of the coalitions and alliances formed by these males.

THE BIOLOGY OF ALLIANCES AMONG MALE BABOONS

A short description of coalitions

Coalitions can start in several ways: a male may interfere in an ongoing conflict; two males may attack a third individual simultaneously, or two males may defend themselves against aggression directed at both of them. In many cases the formation of a coalition is preceded by some spatial manoeuvering, by the exchange of side-directed behaviour (term coined by de Waal and Van Hooff 1981), like 'head flagging' and 'staccato grunting', and by pseudo-sexual behaviour between the participants.

Coalitions among males are formed for a variety of (apparent) reasons and against opponents of virtually all age–sex classes, apart from very young infants. The probably most frequent, and certainly best-studied, immediate cause is competition over receptive females. Typically, two or more adult males challenge a male in consort with an estrous female. Almost invariably the challenged male is dominant to each of the coalition partners. Baboon consorts are exclusive in character, i.e. a male cannot normally mate with an estrous female that is in consort with another male. The consorts can thus be seen as the entities over which males compete. Roughly one-quarter to one-third of all coalitions formed by adult males are formed to obtain or defend a consort (Noë 1989). In other words, competition over consorts is by no means the only reason to form coalitions. Other recognizable immediate causes are: (1) disputes over food (notably meat); (2) the defence of third parties (notably infants) against aggression of other group members; and (3) the redirection of mutual tension on bystanders. A considerable number of coalitions, however, have no recognizable immediate cause.

Coalition partners and their targets

Important clues to the mechanism of coalition formation are given by the attributes of coalition partners and their targets. Table 11.1 summarizes the findings of a number of studies. Note that my own study of three different groups, is the only study that concentrated on all coalitions formed. Most other studies give data on coalitions formed in the context of sexual competition only. General conclusions can therefore be drawn only for the latter type of coalitions. In all alliances I observed that a minority of the coalitions were over consorts, but this type of coalition certainly formed an important segment. If a general pattern can be recognized for the formation of coalitions over consorts, this pattern may be valid for other types of coalitions as well.

Which general pattern can be recognized? From Table 11.1 one can see that the targets of coalitions are almost invariably young, high-ranking immigrants that are relatively new to the group. This is fairly easy to understand. Rank, age, and period-of-residence show a strong correlation in adult male baboons (Collins 1981; Packer 1979*a*; Noë and Sluijter 1990; Smuts 1985). As in many vertebrates, age and vigour show an inverted U-shaped relationship in baboons (Packer 1979*a*). Individual vigour is the main factor determining the agonistic rank of a male baboon, at least in situations of dyadic conflict. Alliances have no obvious permanent effect on rank. Many males migrate for the first time after a sudden growth spurt, when they are young and strong. Such males as a rule obtain a high rank soon after their immigration in the new group. With the passing of time the immigrants drop in rank as new males arrive. This situation is reflected in a simplified form in Fig. 11.2.

Apart from idiosyncratic cases, there are a number of systematic exceptions to this pattern. (1) Males that migrate at a later time in life, usually for the second or third time, will be in the group for a shorter period than other males of comparable rank and age. (2) Some males stay in their natal group after they became sexually active adults. My impression is that such is the case especially in large groups, such as those studied in Gilgil (Bercovitch 1985, 1986, 1987; Noë 1989; Smuts 1985). My speculation is that while natal males in small groups are excluded from mating with most females due to incest-avoidance mechanisms, natal males in large groups have enough potential mates in spite of such mechanisms. (3) A few males migrate as immatures before their growth spurt and only later rise in rank in their new group.

Young, high-ranking males have an advantage in locations where conflicts tend to remain dyadic, like sleeping trees or sleeping cliffs. These males are usually in consort in the early morning as a result. Later in the day, however, the circumstances enhance the chances of a coalition (Bercovitch 1988; Noë and Sluijter 1990; Popp 1978; Rasmussen 1980; Smuts 1985).

Because of the close correlation between the main characteristics of males,

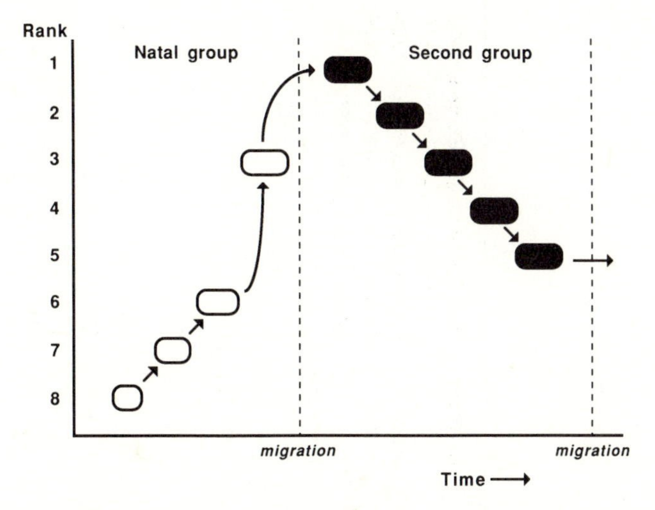

Fig. 11.2. The career of a male savanna baboon. A typical male grows up in his natal group, rises fast in rank after a growth spurt and then migrates for the first time. Young adult immigrants often obtain a high rank in their new group and drop in rank as time passes and new males immigrate.

age, rank, and residence, it is not immediately apparent whether one factor is more important than others. One obvious hypothesis is that males have to know each other for a long time, before they are able to form successful coalitions (Collins 1981; Smuts 1985). The exceptions to the rule of close correlation provide, however, a possibility to recognize the key characteristic for coalition formation. Both Bercovitch (1988) and I (Noë 1989) concluded that rank is likely to be that key factor.

The impact of coalitions on the mating success of males

Table 11.2 gives an impression of the importance of coalitions for the reproductive success of males. The data presented are likely to be biased in favour of coalitionary take-overs, since other forms of consort acquisition are more likely to occur outside the usual observation hours (see above). Nevertheless the conclusion can be that coalitions have a notable impact on mating success in general. The importance of coalitions for individual males can vary greatly, however. Figure 11.3 shows that some males obtained or defended a considerable proportion with the help of others. Two males, with a proportion of over 50 per cent of consorts obtained through coalitions, ranked first and third respectively in the rank order for frequency of consorts obtained in their group. Data presented by Rasmussen (1980), Collins (1981), and Bercovitch (1988) show comparable patterns.

Table 11.1. Attributes of coalition partners and their targets.

Sub-species	Site	Period	Group	Group size	Number of adult males	Number of coalitions observed		
						Attempted consort take-overs	Successful consort take-overs	Other coalitions
anubis	Nairobi NP Kenya	3.5 months (1959)	[a]SR	28	6	—	—	—
		6 months (1959)	[b]SV	40	5	—	—	—
	Gombe NP Tanzania	8 months+ 12 months (1972–74)	[c]A, B, and C	26–51	3–11	20	6	77
	Gilgil Kenya	20 months (1979–81)	[d]Pumphouse	78–90	4–8	55	35	—
		3 months (1983)	[e]Eburru Cliffs	~120	12	—	13	—
		9 months (1984)	[f]Eburru Cliffs	95–114	14–17	72	14[4]	105
cynocephalus	Mikumi NP Tanzania	10 months (1976)	[g]Viramba	102–120	13–15	34	23	—
	Ruaha NP	5 months	[h]Msembe	70–72	7–8	20	11	211
	Amboseli NP Kenya	12 months (1981–82)	[i]Alto	50–55	6–9	37	26	144
			[j]Hook	33–41	6–7	47	21[5]	138
ursinus	Honnet NR South Africa	6 months (1968)	[k]W	77	3	—	—	14

Long-lasting alliances?	Attributes of males frequently involved in coalitions			Attributes of targets	Source
	Rank	Age	Residence status		
[a]yes	all	late prime, old, very old	?	high rank	Hall and Devore (1965)[1]
[b]yes	high	young, very old	?	high rank	(idem)
[c]no?	middle, low	'older'	newcomer, resident	high rank	Packer (1977, 1979*a*, *b*)[2]
[d]yes	middle, low	'older'	?	young and high rank	Bercovitch (1988)
[e]yes	low	'older'	newcomer, long time resident	young and high rank	Smuts (1985) 'study 2'[3]
[f]yes	high, middle	young, prime	immigrant of long residence	all	Noë (1989)
[g]?	middle, low	'older'	natal, newcomer, long time resident	young and high rank	Rasmussen (1980)
[h]yes?	middle, low	prime, aging, aged	long time resident	young newcomers of high rank	Collins (1981)
[i]yes	middle, low	'older'	immigrant of long residence	young newcomers of high rank	Noë (1989)
[j]yes	middle, low	'older'	immigrant of long residence	young newcomers of high rank	(idem)
[k]yes	middle, low	prime, old	?	prime and high rank	Saayman (1971)

[1] Ranking not based on naturally occurring dyadic conflicts and thus not comparable to other studies.
[2] Data of 12 different males in three different groups combined.
[3] One alliance started in 1977 (Smuts' Study 1) and lasted most likely till the death of one ally during Noë's study in 1984.
[4] In addition consorts were defended successfully 14 times.
[5] In addition consorts were defended successfully five times.

Table 11.2. Proportion of consort change-overs effectuated through coalitions. The table attempts to give an impression of the order of magnitude of the impact of coalition formation on male reproductive success. The criteria for consorts, agonism, and coalitions varied considerably between studies. Note that the data are biased in favour of coalitionary take-overs (explanation in text). The data of Noë and Sluijter differ from the others in that the mode of defence of a coalition (alone or with help of others) was substituted when the start of the consort was not seen. This method reduces the bias in favour of coalitionary take-overs.

Source	Observed consort take-overs	Observed agonistic take-overs	Percentage of all take-overs	Percentage of agonistic take-overs
Rasmussen (1980) (Fig. 7.13)	—	23	—	91.3
Collins (1981) (Table 8.VIII)	69	16	23.2	64.0
Bercovitch (1988) (Fig. 1)	200	87	16.5	37.9
Smuts (1985) (p. 135ff.)	—	21	—	61.9
Noë and Sluijter (1990) (Table XII)				
Group A	128	80	20.3	32.5
Group H	125	83	16.8	25.3

The impact of the mating success of males that lost their consorts due to coalitions formed by others is harder to estimate. In the first place the number of consorts they had was not affected and in the second place it is impossible to know how long they would have had their consorts without take-overs through coalitions. The proportion of consorts thus lost can be considerable: in a number of groups studied the high-ranking males lost one-third to over one-half of their consorts due to coalitions (Collins 1981; Noë and Sluijter unpublished data).

The division of pay-offs

Little information exists about the division of pay-offs between allies. One only gets an idea who benefits when tangible resources are at stake, notably food and receptive females. Other potential benefits, like improvement of status, prevention of injury, defence of kin, etc. are almost impossible to quantify, let alone the question of whether or not receiving benefit A can be

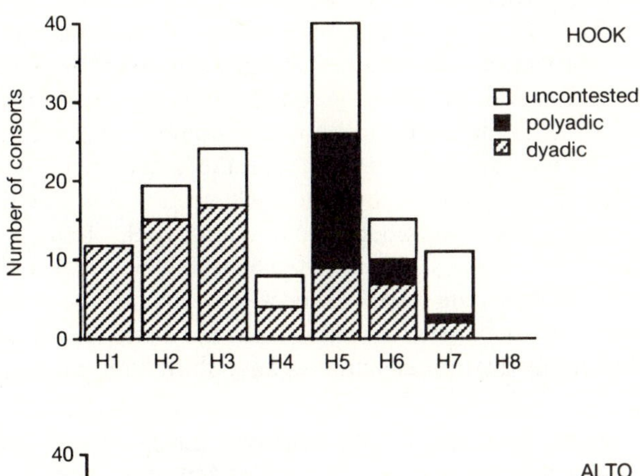

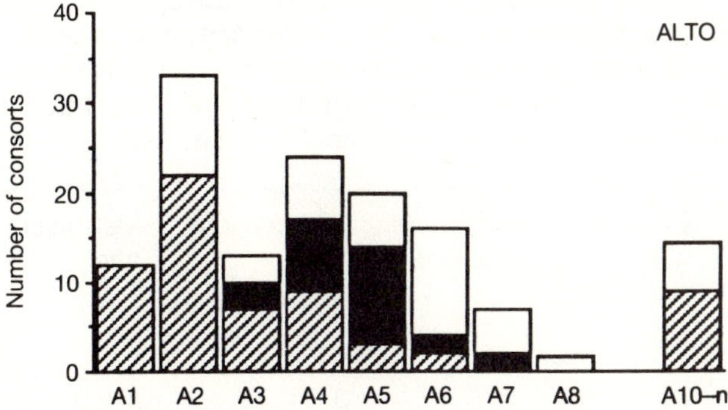

Fig. 11.3. The use and success of various ways to obtain consorts. Males can obtain consorts in three ways: (1) in a dyadic conflict, (2) with the help of others in a polyadic conflict, or (3) by forming a consort with a female that is momentarily not claimed by other males. High ranking males most frequently obtained females in dyadic fights, while middle ranking males frequently profited from the efforts of other males. A10-n was a natal male, who rose to the fourth rank at the end of the study. (From Noë and Sluijter 1990.)

compensated by receiving benefit B. A statistical analysis of the division of pay-offs can only be attempted for the few alliances in which the successful take-over of at least six tangible resources of comparable value (read consorts) has been observed. One such alliance was observed by Bercovitch (1988). Of 20 successful coalitions four involved a third male. In the remaining 16 cases the consorts were fairly symmetrically divided (7–9). In Amboseli we observed one pair of males with eight such successful coalitions; one male obtained the female in all eight cases (Noë 1990). The asymmetry in this alliance is even more remarkable, when the three-male coalitions and 'provocations' in which these males were involved are considered too (Table 11.3). A comparable asymmetry was observed by Collins (1981) in the most successful alliance in his group: one of the males obtained the female after seven of eight successful take-overs. For the model presented below it is important to know the number of potential partners each ally had. An indication for this number is the total number of adult males present during coalition formation in each group. For the three cases mentioned above these numbers were: Bercovitch: four adult males (personal communication); Collins: 7–8 males; Noë: 6–7 males. One should keep in mind that these data reflect observations on one type of benefit from coalitions only. Other benefits accruing from the same alliances may have been very differently distributed and thus have shifted the balance.

Table 11.3. Results of conflicts with coalitions (bold) or provocations over consorts between alliances of low-ranking males and single high-ranking males. H5, H6, and H7 were the three lowest-ranking of seven adult males in Hook's Group. 'Provocations' are interactions in which the allies provoke a conflict with a consorting male, without forming a coalition (from Noë 1990).

Alliance	Conflicts with coalitions	Successful coalitions	Provocations	Male in consort after conflict		
				H5	H6	H7
H5-H6	**12**	**8**	3	**8** + 3	**0** + 0	—
H5-H7	**5**	**4**	3	**4** + 2	—	**0** + 1
H6-H7	**6**	**1**	1	—	**1** + 1	**0** + 0
H5-H6-H7	**5**	**5**	2	**5** + 2	**0** + 0	**0** + 0
Total	**28**	**18**	9	**17** + 7	**1** + 1	**0** + 1

Phenomena to be explained

It is now possible to summarize the features that should be accounted for by a model of alliance formation.

1. Coalitions are formed by a limited set of males. Rank largely determines which males combine. In small groups these are middle- and lower-ranking males with a relatively strong participation of the middle-ranking males.
2. All kinds of divisions of benefits can be found, from quite symmetrical to extremely asymmetrical.
3. Some alliances are quite stable and coalitions continue to occur, despite occasional lack of response to the partner's appeals to form coalitions.

MODELS OF ALLIANCE FORMATION

I present three separate *post hoc* models that form the building blocks of a theory of alliance formation. The purpose is to explain why certain males form alliances and others do not, and why they arrive at certain divisions of pay-offs. The first model makes clear that coalitions are only formed by males below a certain rank, because only these males gain by doing so. With the help of the second model I show that the number of males likely to be involved in coalitions is further limited, because not all pairs of males can form coalitions strong enough to be successful. What is then left to be explained in the third model is why only some of the potential combinations actually develop into alliances and why some of these alliances can be remarkably stable, even when the pay-offs are asymmetrically divided between the allies. A major problem that will arise in this part is the question whether such animals as baboons are able to bargain about the formation of alliances and about pay-off distributions. A fourth section is devoted to this question.

Who would benefit from coalition formation?

I first approach the problem of who will form coalitions by asking who would gain a net profit by doing so. As discussed before, costs and benefits remain elusive quantities for the most part. We have some more grip, however, on the coalitions over consorts. Noë and Sluijter (1990) calculate at what rank a male should switch over from trying to obtain consorts on his own to trying to obtain consorts with the help of others. I give a brief account of our findings here.

Our starting point was S. A. Altmann's (1962) priority-of-access model as applied to baboons by Hausfater (1975). According to that model the mating success of a male depends on his rank and on the number of receptive females available. The male highest in rank obtains the first female, the second in rank the next and so on, till the number of available females is exhausted. It is assumed that the males can be ranked in a linear rank order, an assumption that proved to be correct in the majority of baboon studies (see discussion in the next section).

The average number of receptive females present in a baboon group at any time can be calculated on the basis of two parameters: the number of sexually mature females and the proportion of time these females are receptive. These parameters have been provided by Hausfater (1975) and J. Altmann (1980) for one of the Amboseli groups. The calculation can be made under the assumption that females do not synchronize and that there is no mating season. Although this assumption appeared to hold for our Amboseli study groups (Noë and Sluijter 1990) it may not be true for all groups and under more extreme conditions, like severe droughts (data on Gilgil population: Bercovitch personal communication; Noë and Sluijter unpublished). The calculation also asks for a precise indication of the days of the menstrual cycle that are of interest. Menstrual cycles are variable, however, for individual females and between individuals.

We therefore preferred to use a short-cut, which we think gives more relevant results. Since there were always more sexually mature males around than receptive females, we assumed that each female of interest to the males would be consorted. Our estimate of females available per day was thus directly based on the number of females consorted. This number follows a Poisson distribution, as shown in Fig. 11.4. On the basis of this distribution one can calculate the mating success of a male for each rank, under the assumption that the priority-of-access model is correct.

Hausfater (1975) found that the data are not in agreement with the model on one point: males had a much lower mating success than predicted by the model during the time they occupied rank 1. This result was corroborated by our findings. As explained above, the males on rank 1 are usually young, recent immigrants. Such males are often rather peripheral in the group and

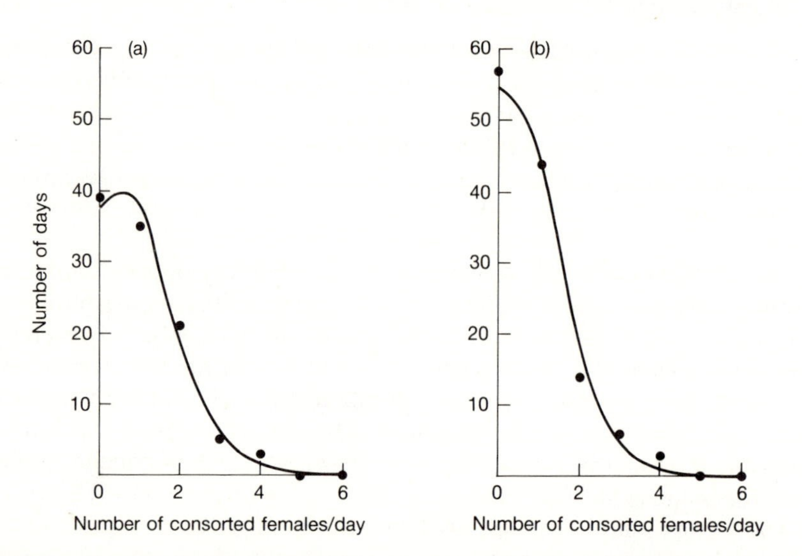

Fig. 11.4. Synchronization of menstrual cycles. The graphs show the frequency distributions of the number of females that is consorted per day. The close fit to a Poisson distribution shows that females do not influence one another, i.e. their menstrual cycles are neither synchronous nor asynchronous. (From Noë and Sluijter 1990). (a) Hook: 104 female/consort days in 103 observation days; (b) Alto: 102 female/consort days in 124 observation days. The points represent observations, and the line the expected Poisson distributon.

have difficulty forming relationships with especially the older and higher-ranking females (see also Smuts 1985). One of our groups had newcomers on rank 1 and 2 at the same time and both had a lower-than-expected success.

Hausfater (1975) did not observe the formation of coalitions and thus did not incorporate the effect of coalition formation in his calculations. We, however, tried to calculate at what rank it would benefit a male to start to form coalitions, under a number of assumptions. A first assumption is that two coalition partners divide the consorts obtained equally among themselves, and a second that every coalition is successful. From these assumptions it follows that it is worthwhile to form coalitions as soon as the success in obtaining consorts on an individual basis is less than half of the maximum number obtainable, which is the number the male on rank 1 would have according to the priority-of-access model. We did not consider the possibility of 'counter-coalitions', i.e. coalitions formed by a high-ranking male to counteract the effect of the coalitions formed by males of lower rank, since such coalitions rarely occurred. To see for which rank it would pay to form coalitions an additional factor has to be added: the number of females that become available after they have been voluntarily abandoned by the consorting male. We found that about one-third of the consorts is given up voluntarily and used this figure in our model, but even higher proportions have been reported (Bercovitch 1988).

The result of our calculations is presented in Fig. 11.5. The figure shows that in both our study groups it would pay to form coalitions for a male on rank 3 or lower. The rank thus found is of course dependent on the size of the group. Our study groups were in the medium size range (33–55, with 6–8 adult male immigrants).

Thus by taking the effect of coalitions into account the priority-of-access model can be modified to fit the data much better than Altmann's original version, but some discrepancies between data and model remain. The first, the relatively low success of high-ranking newcomers, has been discussed above. A second discrepancy we found was that both groups had a male with low mating success ranking between the successful males of high rank and the highest-ranking male that successfully used coalitions. The third deviation we found was that some middle-ranking males used coalition formation with more success than the low-ranking males. This success was the result of more frequent participation in coalitions against consorting males, of a higher success rate of these coalitions and, in some cases, of an asymmetry in the chances to obtain the consort. The second and third deviation from our modified priority-of-access model can be understood, if one considers the dynamics of coalition formation among a small number of potential participants, as I explain below. We considered all three discrepancies as systematic deviations from the basic model and incorporated them in our descriptive model as given in Fig. 11.6.

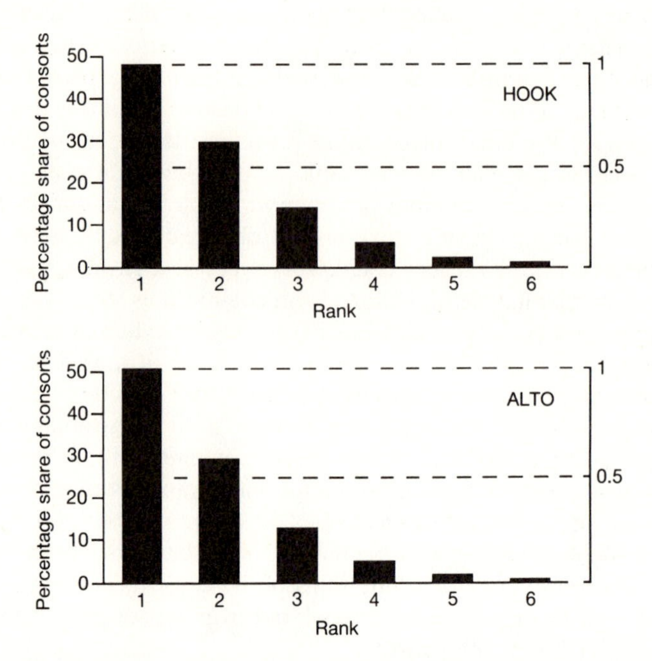

Fig. 11.5. The modified priority-of-access model. The graphs show the expected mating success for males on various ranks according to Altmann's priority-of-access model with one additional feature: it is assumed that each male abandons one-third of his consorts, which become available to the males ranking below him. Coalition formation is assumed to be of interest to males that have less than half of the maximum share. See text for further explanation.

Which males could potentially form coalitions?

There is one hidden assumption in the above model: each combination of two adult males would be able to beat each adult male member of the group. Any observer of baboons would hesitate to use this assumption. The next approach to the problem of who forms coalitions should thus be: who is able to form coalitions successfully against whom? The following model is based on differences in fighting ability between males. Perhaps I should speak of a thought experiment, rather than a model, since I have no way of estimating the most crucial parameter, (relative) fighting ability. We are not completely ignorant about the relative fighting abilities in a subgroup of adult male baboons, however.

Firstly, the rank order is usually linear at any given time. The apparent exception described by Strum (1982) is probably an artifact. Strum constructed a single rank order over a 14-month period. In the course of such a a long period many changes in the rank order can be expected (cf.

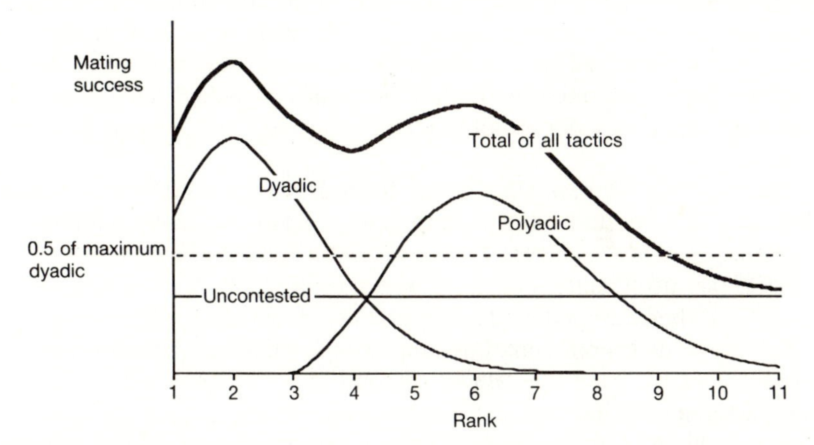

Fig. 11.6. Tactics and mating success. The graph shows a post hoc model of the contribution of three tactics used to obtain consorts to the mating success of males on various ranks. See Fig. 11.3 for the empirical data and a description of the tactics. (From Noë and Sluijter 1990.)

Bercovitch 1986; Hausfater 1975), especially in a group the size of her study group. Among the adult males, rank reversals may regularly interrupt longer periods in which rank orders are stable. When rank orders are constructed for each of these stable periods, they are likely to be linear (see for instance Bercovitch 1986). Moreover, instability in rank orders should not be confused with instability in individual dominance relationships. In the vast majority of relationships between adult males the periods of uncertain dominance are very short compared to the periods of clear-cut dominance.

Secondly, the impression is that the rank orders among adult males are rather steep, i.e. the difference in fighting ability between two males of adjacent rank can be large. This impression is based on frequent observations of high ranking males that were able to defeat several low-ranking males simultaneously. The average difference between two males of adjacent rank is less in large groups than in small groups, if the assumption is correct that rank is largely correlated with physical parameters with absolute minimal and maximal values. I see no way of quantifying relative or absolute differences, however. The 'cardinal rank' method proposed by Boyd and Silk (1983) does not work for baboons, because dominant males tend consistently to win fights, without the occasional losses needed for that method. If the relation of fighting ability to age follows roughly the same inverted U-shape (Packer 1979*a*) for all males, the estimate of relative fighting abilities can be further improved on the basis of age estimates.

For the present purpose a simplified picture will suffice. Imagine a group in which the differences in fighting ability between each pair of males of adjacent rank can be expressed as a simple ratio. An example is given in the diagonal cells of the matrix in Fig. 11.7. In this example the fighting ability of the lowest-ranking male is taken as unity and each other male is 1.3 times as strong as the male ranking below him. Such a group could result if males with equal fighting ability curves all immigrated at the same age and at regular intervals.

The possible alliances that can be formed in such a group can now be deduced, given the following assumptions: (1) Only two-male coalitions are formed. (2) A male will not form a coalition against a male weaker than himself (i.e. no 'counter-coalitions are possible). (3) A coalition will be successful when the combined fighting ability of two males exceeds the fighting ability of their opponent. I assumed simple additivity, which is perhaps more acceptable when the 'ability to contribute to a coalition' (which encompasses specific tactical skills) would be used as a parameter instead of 'fighting ability'. I suppose that the two parameters are closely related, however. (4) Coalition formation does not influence dyadic dominance relationships.

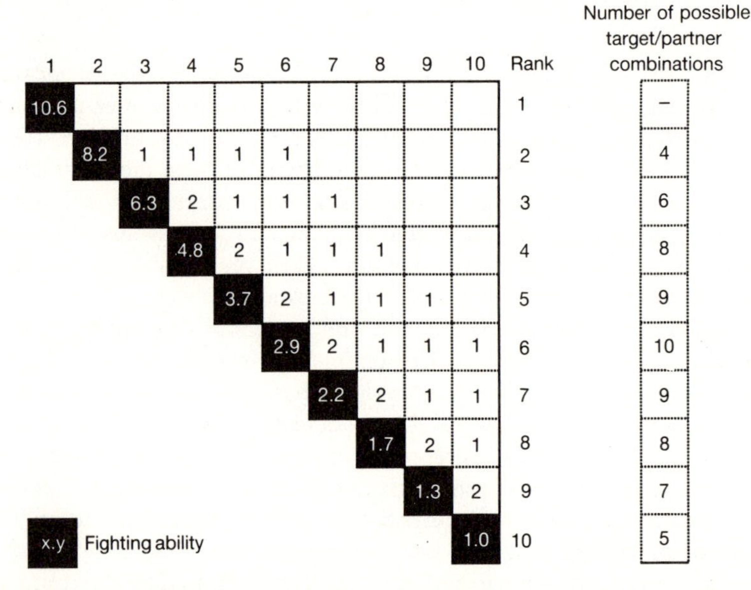

Fig. 11.7. The distribution of potential alliances. The matrix is based on an imaginary group of 10 males in which the fighting ability of the lowest ranking male is set to unity and each of the other males is 1.3 times stronger than the male ranking directly below him. The non-diagonal cells of the matrix show the number of higher ranking males that could potentially be defeated by the combination of the male at the top and the male at the right. The figures in the right-hand column give the number of options each male has, i.e. the number of combinations of one specific partner and one specific opponent.

Figure 11.7 gives an indication how many combinations of two allies with one particular target, can be formed by each pair of males, assuming a ratio of 1.3. Figure 11.8 indicates how the numbers of potential combinations vary for each rank with the ratio for relative fighting ability. A likely kind of curve for baboons would be of the shape given for the ratios 1.2 to 1.4. With a lower ratio the rank order would not be as steep as observed; with a higher ratio the formation of any coalitions would be virtually impossible.

One conclusion can be drawn from this model: the largest number of combinations can be formed by the males of middle rank. The frequency distribution predicted by the model can be expected when commodities are at stake that are fairly equally distributed over the male subgroup, in other words for cases in which the males on rank 9 and 10 have as much to gain from the male on rank 8 as the males on ranks 2 and 3 have to gain from the highest-ranking male. Commodities like consorts are not likely to be distributed this way. For such resources a disproportionate participation of the males ranking directly below those who are able to gain access on the basis of their own strength should be expected. These males will be the only ones able to form combinations that are strong enough. The lower-ranking males will have to wait, until such resources trickle further down the hierarchy.

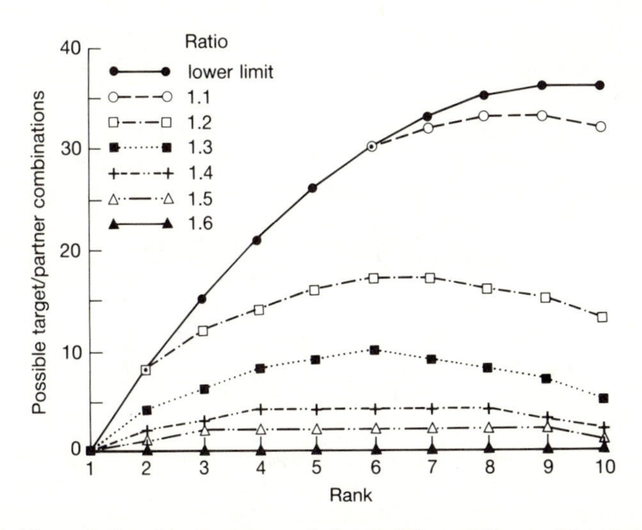

Fig. 11.8. The relationship between relative fighting ability and the distribution of potential alliances. The graph shows the relationship between the 'steepness' of the male hierarchy (expressed as the ratio between the fighting abilities of two males of adjacent rank) and the number of options each male has. The numbers for the ratio 1.3 are the same as in Fig. 11.7, right-hand column.

The dynamics of coalition formation

The assumption that two collaborating animals are equally powerful (in a game-theoretical sense) and will thus most likely divide costs and benefits equally, is implicit, or even explicit (Lombardo 1985; Whitehead 1987) in several publications on reciprocal altruism and the prisoner's dilemma model. This prediction is explicitly given in neither model, however, although Trivers (1971) hinted at it (p. 37: '... roughly equivalent benefits ... at roughly equivalent costs'). The prediction of symmetrical division also agrees with one's intuitive expectations about the outcome of a symmetrical game like the prisoner's dilemma, at least when played by two adult male baboons that strive to obtain the same resources. Some baboon allies divide the spoils rather asymmetrically, however (Collins 1981; Noë 1990), which could point at an asymmetry in power. I use the model described in the section above as a basis to show that the options of two allies are often unequal, which can explain such an asymmetry in power. How this power asymmetry is translated into an asymmetrical division of pay-offs is a problem to be discussed in the next section.

Some examples of possible configurations of the potential combinations between four males, who combine forces against two males ranking above them, are given in Fig. 11.9. Figure 11.9(a) reflects an extreme case in which one individual belongs to any successful combination that can possibly be formed. Such individuals are in a so-called *veto position*: they can demand high pay-offs, because they can play off their potential partners against each other. A simple example of a veto player is a salesman in possession of a right shoe, dealing with two colleagues who each possess a left shoe. If only a pair of shoes can be sold and bargaining is possible, then it is clear that the right-shoe salesman can ask a lion's share of the profit (see Kahan and Rapoport 1984; Murnighan 1978 for reviews of the research on veto games).

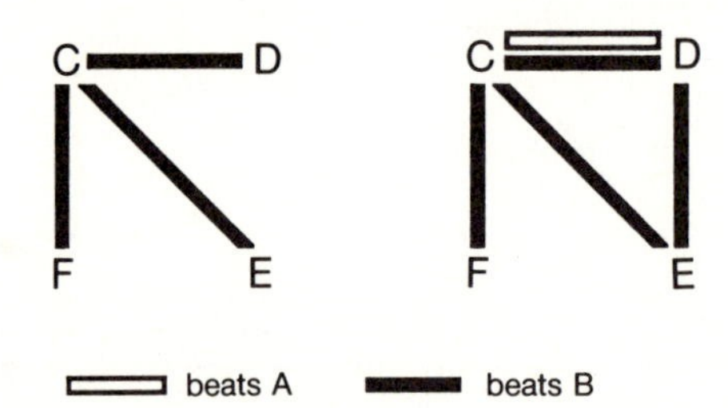

Fig. 11.9. Patterns of coalition formation. The figure shows some configurations of potential combinations of four low-ranking males (C–F) against two high ranking males (A and B). In (a), male C has a veto position: he is a member of all possible combinations against B. In (b), the males D and E have an alternative against B, but to beat A male D has no other choice but C. Since C is dependent on D too, D has much more leverage over C in configuration (b) than in configuration (a).

Although I actually found a case (Noë 1990), a veto player is bound to be rather rare among baboons seeking to form alliances. The effect of differences in options between partners, dubbed the *market effect* by Noë *et al.* (1991), may well be present in a weaker form in many alliances. No two allies are likely to have exactly the same options, which may result in a difference in leverage from the one over the other. One can expect a virtual symmetry in power, and thus in pay-offs, when both partners have a large number of options. Real symmetry is expected, if two males have only one option, namely each other. The symmetrical alliance formed by the two lowest ranking of four males observed by Bercovitch (see personal communication cited above) is probably an example. Symmetrical alliances can thus be expected in very large and very small groups. In large groups low frequencies of coalition formation per alliance can be expected, combined with frequent partner switches, in small groups high frequencies per alliance are likely. This expectation is consistent with data presently available (Bercovitch 1988 and personal communication; Noë 1989).

From power to pay-off: do baboons bargain?

The key assumption in the above story is that an asymmetry in power will be translated into an asymmetry in pay-off. I assume that some kind of bargaining will play a role in this translation process. The question whether bargaining plays a role in the collaboration among animals like baboons, can be split into four parts.

1. Are baboons able to convey information about their desire to collaborate?
2. Are baboons able to convey information about the level of costs and benefits they are willing to accept?
3. Is collaboration on the basis of communication possible in spite of temptations to convey 'dishonest' information?
4. Are baboons able to keep each other to agreements?

Are baboons able to signal their desire to collaborate?

The answer to this question is relatively straightforward: yes, they can. At first individuals willing to form a coalition have to show their desire to do so, point out with whom they would like to form a coalition, and agree about the timing of the event. Offers to form a coalition and the acceptance of such offers is made clear by baboons through special gestures and vocalizations, like 'head-flagging' and 'staccato grunting'. These behaviour patterns with their typical rhythmic and repeated character are addressed at the (potential) partner before and during coalitions. Synchronization can be achieved through the use of external cues, which will often be the behaviour of the opponent, through synchronization of locomotion, or with the help of pseudo-sexual greeting rituals (see description in Smuts and Watanabe 1990).

Are baboons able to convey information about the desired pay-off division?

In many cases the division of benefits poses no problem, because it is given by the nature of the goal reached by forming a successful alliance: e.g. status improvement, the support for an infant, etc. Cooperation that takes the form of an exchange of (indivisible) commodities likewise poses no problems. An example from a rather different species is the exchange of shells between hermit crabs (Hazlett 1983). Problems arise when costs and/or benefits can be divided in many different ways. I do not expect baboons to show exchanges such as 'I settle for 60 per cent, but not for 55 per cent'. It can be much simpler than that, provided the interaction is repeated. Suppose two animals repeatedly have to split up a divisible food source they obtain through cooperation. They can start out by any division, e.g. determined by speed of consumption, or by splitting in halves. Each animal can show discontent with the original division. A strong show of discontent is a refusal to collaborate on the next occasion. As mentioned above, observations show that blunt refusals do not necessarily lead to an end of the alliance. An adjustment of the pay-off distribution may also be reached by using some signal that conveys a threat to refuse in the near future. The behavioural repertoire of baboons contains enough signs of discontent, in a graded form if necessary, that can be used to this end.

Is the information conveyed dishonest?

Clearly, when animals communicate about levels of costs and benefits they are willing to accept, there is a temptation to ask for more than the minimum. This temptation may endanger the collaboration. Game theorists studying the theory of animal conflicts have encountered a similar problem. A number of authors concluded that animals would not convey honest information about their strength and intentions during combat (Caryl 1979; Dawkins and Krebs 1978; Maynard Smith 1982*a*, *b*). Empirical studies seem to suggest that animals in conflict tend to give honest information (see Hinde 1981). Various suggestions have been made about the circumstances in which animals would be expected to give honest signals during conflicts (Bond 1989; Gardner and Morris 1989; van Rhijn 1980; van Rhijn and Vodegel 1980). Bond (1989) argued that animals interested in winning a conflict, but at the same time interested in avoiding escalation, are likely to show some bluff (dishonest information), but that most of their message is likely to be honest.

Markl (1985) makes clear that the existence of honest information is much less of a problem in the case of cooperation. If two animals both stand to gain by the cooperation, and neither can obtain an immediate gain by refusing to cooperate (synergistic cooperation), both are likely to be honest if they signal their willingness to cooperate. In correspondence to the possibility of slight bluff in conflicts (Bond 1989), it should be possible to get away with some exaggeration of demands without jeopardizing the deal itself, when costs and/or benefits can be varied gradually. It is a prerequisite of bargaining that one asks for a bit more than strictly needed.

One can speak of a 'bargaining zone' when the costs and/or benefits can be divided in several small discrete steps or along a continuum in such a way that all divisions satisfy the minimum demands of both participants (Davies and Houston 1984; Noë *et al.* 1991). Collaboration is likely to be unstable when the bargaining zone is large, because the division of pay-offs can oscillate wildly, increasing the chance that once in a while the minimal demands of one partner are not fulfilled. Collaboration will be more stable when there is a point of attraction, or saddlepoint, on which the division of pay-offs tends to converge. Such a saddlepoint is more likely to occur when the collaborating animals differ in power than when they have equal power. For example, an animal in a veto position with two potential partners is able to push the pay-off distribution to his favour till one of the partners drops out, because his minimally acceptable pay-off is reached. In more general terms: in any case in which a threat to seek an alternative partner cannot be met with an effective counter-threat the division of pay-offs will stabilize on a level that just satisfies the minimal demand of the weakest partner but one. Likewise a saddlepoint can be expected when the maximum demand of the most powerful player falls within the bargaining zone, e.g. when he is satiated before a divisible food source is exhausted completely.

The proposition that animals bargain is likely to be met with a sceptical attitude. I should like to challenge the sceptics with the following question. Suppose one finds animals that are clearly engaged in collaboration with a variable pay-off division. Can one imagine a mechanism by which they converge on a certain division of costs and benefits other than through bargaining, and can one demonstrate the existence of this mechanism?

Do baboons conclude binding agreements?

The last problem to be discussed in this section is that of the existence of binding agreements, or contracts. Among humans the conclusion of (written) contracts about the division of costs and benefits before any party makes an investment is typical for one-off deals. In such cases each partner is kept to the contract by a threat that has little to do with the collaborative interaction itself. Such policed agreements are less necessary when the collaboration is continuous or often repeated, because then the necessary threat can be contained in the collaborative interactions themselves. It is not unusual for a dealer to send goods ordered by telephone to a regular customer without receiving a written confirmation of the order and before any payment is made.

In continuous or repeated collaboration among animals bargaining and forcing the other to accept certain pay-off divisions by threats can be one and the same process. Bargaining implies threats that the relationship will not be continued, unless a more favourable division of pay-offs is agreed upon. The other side of the same coin is that extremely high demands are not made, as long as the threat to lose future gains from it suffices to suppress the temptation to try to win high immediate gains. The analogy to Trivers' (1971) and Axelrod and Hamilton's (1981) reasoning will be obvious. Yet there is a crucial difference: I expect the 'future gains' to enter the equation as the difference between the future gains from the present collaborative relationship and the future gains from alternative relationships. The dealer in the above example has to send the goods in order to prevent the customer to do business with a competitor, who is less anxious and therefore delivers faster.

COALITION FORMATION IN BABOONS AS PARADIGM FOR OTHER FORMS OF COLLABORATION

Formation of partnerships in other social systems

The idea that competition over partners and differences in options to form partnerships determine the way in which individuals collaborate for a large part, applies to many forms of collaboration among animals. Noë *et al.* (1991) formulate this 'market' theory in a more general form, and widen its scope to collaboration among unrelated group-living animals in general. The restriction to unrelated individuals is theoretically not necessary, but it is likely that market effects will be swamped by effects of kin selection and thus will be hard to demonstrate. The restriction to collaboration among members of the same social unit is important, however. The impact of the market effect is dependent on the option to choose among potential partners, but when both partners have a virtually unlimited number of alternative partners, the impact will be minimal.

The use of coalition games as paradigms

Coalition games have been developed as paradigms for many aspects of human behaviour. Extensive theoretical and empirical work has been done in the fields of social psychology, political science, and economics. Only a small part of this vast body of knowledge is useful for biologists, however. Theories that do not assume individual rationality, i.e. the aspiration to maximize the individual pay-off, are of little use to us (cf. Parker and Hammerstein 1985). 'Social norms', 'reputation', and 'conscience' may influence the behaviour of some species, notably primates, but it would do little good to seek parallels with models from the social sciences that incorporate these phenomena. Coalition games would be easier to handle if they could be described in the normal form, i.e. using the strategic matrices biologists are familiar with through the hawk–dove game and the prisoner's dilemma. Whether the coalition games can be transformed to a matrix notation is a point of dispute (Kahan and Rapoport 1981; Michener and Potter 1981).

There are more shortcomings in the theoretical developments around coalition games. In the first place, relatively little work has been done on iterated games with the same players. In the second place, most theories about solutions for these games, i.e. predictions about the coalitions that will be formed and the pay-off configurations that will be agreed on, are deterministic in character. The theories predict a specific outcome or a set of outcomes, rather than calculating the chance of occurrence of each outcome. None of the existing theories is strongly supported by empirical data, and as a result there is no single, dominating theory (Kahan and Rapoport 1984; Parker and Hammerstein 1985). The solution could be found in the development of more stochastically oriented theories (Van der Linden and Verbeek 1985). An example of a development of such a stochastic theory for an iterated game is given by Laing and Morrison (1974).

The players and their tactics

The use of coalition games as paradigms implies that the equivalents of the players in the theoretical games are the individual animals, and not some kind of uniform vehicles for certain strategies. Publications on ESS theory and the use of the prisoner's dilemma leave the impression that the proximate tactics of animals and ESSs are one and the same thing. One could say that the strategies themselves play against each other. In such a framework it is preferable to use the mathematically more rigorously formulated form of game theory, non-cooperative theory, in which such things like bargaining and binding agreements play no role. When, however, animals are found to use communication between each other to arrive at certain decisions, one is almost forced to use the language of cooperative theory, which is generally used in the analysis of coalition games. If one does not accept that, then one is forced to postulate very complicated conditional strategies to account for the reactions of animals to the information they get from their partners. Moreover, non-cooperative game theory is ill suited to cope with phenomena like coalition formation, choice among partners, and competition over partners. The choice, unfortunately, is between realistic paradigms with sloppy theory and unrealistic theory based on sound mathematics.

Obviously players that rely on processing *ad hoc* information cannot do completely without innate capabilities and basic strategies to play the game well. Basic rules of thumb could be: (1) Break off any collaboration with a long-term negative pay-off. (2) Compare partners and seek the most profitable one(s). (3) Force any partner to yield the highest attainable pay-off. To live up to these rules of thumb players should be capable of remembering the yield of interactions with various partners and of comparing these yields.

They should also be able to signal their demands and to interpret any signals received correctly. Interesting possibilities arise with increased cognitive abilities. If animals are able to place themselves in the position of their potential partners and assess the alternatives these partners have, a much more sophisticated way of playing the game may result. This calls for one remark, although it should be superfluous: the fact that we recognize the resemblance between a certain social interaction and a certain theoretical game does not mean that the animals (or humans) involved perceive the interaction at that level.

The predictive power of models based on coalition games

Rarely will all parameters involved be so well understood that precise predictions can be made about which alliances will be formed and how pay-offs will be distributed. A comparable problem would be to predict the positions of all pieces in a game of chess after 40 moves. This is an impossible task, although one knows all the rules and the starting positions of all the pieces. It is, however, possible to predict the direction of changes in the distributions of pay-offs and the likelihood of recombinations of alliances after changes in the structure of a social unit. Such predictions also provide the opportunity to test the coalition game models.

SUMMARY

1. Coalition formation among unrelated adult male savanna baboons has been observed in a large number of groups, belonging to three different subspecies.
2. The theory of reciprocal altruism does not apply to alliances among male baboons. Although one can *a posteriori* recognize a male with a net loss and another with a net gain in several coalitions, it cannot be shown that one of the participants in a coalition *a priori* opts for an altruistic role. Moreover, alliances continue in spite of repeated refusals to collaborate by one of the allies.
3. The prisoner's dilemma game is neither a suitable paradigm for baboon alliances, nor for most other forms of collaboration among group living animals. The prisoner's dilemma model seems to apply, however, to basically competitive situations which can be mitigated by cooperation and to cooperation in situations in which alternative partners are hard to find.
4. The most frequent immediate causes for the formation of coalitions by male baboons are competition over resources, notably oestrous females and meat, defence against aggression and defence of infants. Many coalitions lack a recognizable immediate cause, however. Alliances are usually multi-purpose.

5. The most frequent participants in coalitions are males of middle and lower rank. These males are usually past their prime and resident in the group for a relatively long time. The latter findings can be explained by the correlation of age and of the period-of-residence with rank.
6. Young newcomers of high rank are frequently the targets of coalitions.
7. The division of benefits between allies can range from symmetrical to very asymmetrical.
8. A modified priority-of-access model can be used to make plausible that in a group of average size males of rank 3 or lower potentially gain from coalitions over consorts.
9. Males of middle rank have most options to form coalitions, under the assumption that relative fighting ability is the crucial parameter that determines which pairs are able to form coalitions successfully.
10. A number of features of coalition formation among baboons can be understood with the help of the theory of *N*-player coalition games. One of these games, the veto game, was shown to be a suitable paradigm in a particular case. The most relevant features of coalition games are: (1) The players compete with one another over suitable partners. (2) The relative power of the players depends on the options each of the players has to form alternative coalitions. (3) Which coalitions are formed and how pay-offs are allocated is determined by bargaining.
11. Bargaining is considered to be crucial for the translation of the power balance between allies into a corresponding distribution of pay-offs. The fact that coalitions are a form of 'synergistic cooperation' and the fact that the same individuals interact repeatedly, limits the likelihood of a disruption of the collaboration due to 'dishonest' information.
12. An agreement over a pay-off distribution is not hard to reach, if the allotment of costs and benefits is inherent to the cooperative interaction, as e.g. in an exchange of indivisible items. The distribution of variable pay-offs is relatively straightforward, if one ally is more powerful than the other.

ACKNOWLEDGEMENTS

The field study on baboons was made possible by: research permission of the Government of Kenya, access to superb field sites granted by J. Altmann, S. A. Altmann, and G. Hausfater (Amboseli) and S. Strum (Gilgil), a great investment made by my wife B. Sluijter, help received from a large number of individuals and financial support from WOTRO grant W84-195.

REFERENCES

Altmann, J. (1980). *Baboon mothers and infants*. Harvard University Press, Cambridge, MA.

Altmann, S. A. (1962). A field study of the sociobiology of the rhesus monkey, *Macaca mulatta*. *Annuals of the New York Academy of Science*, **102**, 338–435.

Axelrod, R. (1984). *The evolution of cooperation*. Basic Books, New York.

Axelrod, R. and Hamilton, W. D. (1981). The evolution of cooperation. *Science*, **211**, 1390–6.

Bercovitch, F. B. (1985). Reproductive Tactics in Adult Female and Adult Male Olive Baboons. Ph.D. thesis. University of California, Los Angeles.

Bercovitch, F. B. (1986). Male rank and reproductive activity in savanna baboons. *International Journal of Primatology*, **7**, 533–50.

Bercovitch, F. B. (1987). Reproductive success in male savanna baboons. *Behavioural Ecology and Sociobiology*, **21**, 163–72.

Bercovitch, F. B. (1988). Coalitions, cooperation, and reproductive tactics among adult male baboons. *Animal Behaviour*, **36**, 1198–209.

Bond, A. (1989). Toward a resolution of the paradox of aggressive displays: I. Optimal deceit in the communication of fighting ability. *Ethology*, **81**, 29–46.

Boyd, R. (1988). Is the repeated prisoner's dilemma a good model of reciprocal altruism? *Ethology and Sociobiology*, **9**, 211–22.

Boyd, R. (1989). Mistakes allow evolutionary stability in the repeated prisoner's dilemma game. *Journal of Theoretical Biology*, **136**, 47–56.

Boyd, R. and Lorberbaum, J. P. (1987). No pure strategy is evolutionarily stable in the repeated prisoner's dilemma game. *Nature*, **327**, 58–9.

Boyd, R. and Silk, J. B. (1983). A method for assigning cardinal dominance ranks. *Animal Behaviour*, **31**, 45–58.

Caryl, P. G. (1979). Communication by agonistic displays: what can games theory contribute to ethology? *Behaviour*, **68**, 136–69.

Collins, D. A. (1981). Social Behaviour and Patterns of Mating Among Adult Yellow Baboons (*Papio c.cynocephalus* L. 1766). Ph.D. thesis, University of Edinburgh. University Microfilms International, Ann Arbor.

Colman, A. M. (1982). Experimental games. In *Cooperation and competition in humans and animals* (ed. A. M. Colman), pp. 113–40. Van Nostrand Reinhold, Wokingham.

Davies, N. B. and Houston, A. I. (1984). Territory economics. In *Behavioural ecology: an evolutionary approach*, 2nd edn (ed. J. R. Krebs and N. B. Davies), pp. 148–69. Blackwell, Oxford.

Dawkins, R. and Krebs, J. R. (1978). Animal signals: information or manipulation. In *Behavioural ecology: an evolutionary approach* (ed. J. R. Krebs and N. B. Davies), pp. 282–309. Blackwell, Oxford.

Fader, P. S. and Hauser, J. R. (1988). Implicit coalitions in a generalized prisoner's dilemma. *Journal of Conflict Resolution*, **32**, 553–82.

Fischer, E. A. (1980). The relationship between mating system and simultaneous hermaphroditism in the coral reef fish, *Hypoplectrus nigricans* (Serranidae). *Animal Behaviour*, **28**, 620–33.

Fischer, E. A. (1988). Simultaneous hermaphroditism, tit-for-tat, and the evolutionary stability of social systems. *Ethology and Sociobiology*, **9**, 119–36.

Gardner, R. and Morris, M. R. (1989). The evolution of bluffing in animal contests: an ESS approach. *Journal of Theoretical Biology*, **137**, 235–43.

Getty, T. (1987). Dear enemies and the prisoner's dilemma: why should territorial neighbors form defensive coalitions? *The American Zoologist*, **27**, 327–36.

Hall, K. R. L. and DeVore, I. (1965). Baboon social behavior. In *Primate behavior. Field studies of monkeys and apes* (ed. I. DeVore), pp. 53–110. Holt, Rinehart and Winston, New York.

Hardin, G. (1968). The tragedy of the commons. *Science*, **162**, 1243–8.

Hausfater, G. (1975). *Dominance and reproduction in baboons*. Contributions to Primatology Series, Karger, Basel.

Hazlett, B. A. (1983). Interspecific negotiations: mutual gain in exchanges of a limiting resource. *Animal Behaviour*, **31**, 160–3.

Hinde, R. A. (1981). Animal signals: ethological and games-theory approaches are not incompatible. *Animal Behaviour*, **29**, 535–42.

Kahan, J. P. and Rapoport, A. (1981). Matrix experiments and theories of N-person games. *Journal of Conflict Resolution*, **25**, 725–32.

Kahan, J. P. and Rapoport, A. (1984). *Theories of coalition formation*. Erlbaum, Hillsdale, NJ.

Kummer, H. (1979). On the value of social relationships to nonhuman primates: a heuristic scheme. In *Human ethology: claims and limits of a new discipline* (eds M. von Cranach, K. Foppa, W. Lepenies, and D. Ploog), pp. 381–434. Cambridge University Press, Cambridge.

Kummer, H., Götz, W., and Angst, W. (1974). Triadic differentiation: an inhibitory process protecting pair bonds in baboons. *Behaviour*, **49**, 62–87.

Laing, J. D. and Morrison, R. J. (1974). Sequential games of status. *Behavioral Science*, **19**, 177–96.

Leonard, J. L. and Lukowiak, K. (1984). Male–female conflict in a simultaneous hermaphrodite resolved by sperm trading. *The American Naturalist*, **124**, 282–6.

Liebrand, W. B. G. (1983). A classification of social dilemma games. *Simulation and Games*, **14**, 123–38.

Lima, S. L. (1989). Iterated prisoner's dilemma: an approach to evolutionarily stable cooperation. *The American Naturalist*, **134**, 828–34.

Linden, W. J. van der and Verbeek, A. (1985). Coalition formation: a game-theoretic approach. In *Coalition formation* (ed. H. A. M. Wilke), pp. 29–114. North-Holland, Amsterdam.

Lombardo, M. P. (1985). Mutual restraint in tree swallows: a test of the tit for tat model of reciprocity. *Science*, **227**, 1363–5.

Markl, H. (1985). Manipulation, modulation, information, cognition: some of the riddles of communication. In *Experimental behavioural ecology and sociobiology* (ed. B. Hölldobler and M. Lindauer), pp. 163–94. G. Fischer Verlag, Stuttgart.

Maynard Smith, J. (1982*a*). *Evolution and the theory of games*. Cambridge University Press, Cambridge.

Maynard Smith, J. (1982*b*). Do animals convey information about their intentions? *Journal of theoretical Biology*, **97**, 1–5.

Maynard Smith, J. (1983). Game theory and the evolution of cooperation. In

Evolution from molecules to man (ed. D. S. Bendall), pp. 445–56. Cambridge University Press, Cambridge.

Maynard Smith, J. and Price, G. A. (1973). The logic of animal conflict. *Nature*, **246**, 15–18.

Michener, H. A. and Potter, K. (1981). Generalizability of tests in *n*-person side payment games. *Journal of Conflict Resolution*, **25**, 733–49.

Murnighan, J. K. (1978). Models of coalition behavior: game theoretic, social psychological, and political perspectives. *Psychological Bulletin*, **85**, 1130–53.

Noë, R. (1986). Lasting alliances among adult male savannah baboons. In *Primate ontogeny, cognition and social behaviour* (eds J. G. Else and P. G. Lee), pp. 381–92. Cambridge University Press, Cambridge.

Noë, R. (1989). Coalition Formation among Male Baboons. Thesis, University of Utrecht.

Noë, R. (1990). A veto game played by baboons: a challenge to the use of the prisoner's dilemma as a paradigm for reciprocity and cooperation. *Animal Behaviour*, **39**, 78–90.

Noë, R., Schaik, C. P. van and Hooff, J. A. R. A. M. van (1991). The market effect: an explanation for pay-off asymmetries among collaborating animals. *Ethology*, **87**, 97–118.

Noë, R. and Sluijter, A. A. (1990). Reproductive tactics of male savanna baboons. *Behaviour*, **113**, 117–70.

Packer, C. (1977). Reciprocal altruism in *Papio anubis. Nature*, **265**, 441–3.

Packer, C. (1979*a*). Inter-troop transfer and inbreeding avoidance in *Papio anubis. Animal Behaviour*, **27**, 1–36.

Packer, C. (1979*b*). Male dominance and reproductive activity in *Papio anubis. Animal Behaviour*, **27**, 37–45.

Packer, C. (1986). Whatever happened to reciprocal altruism? *Trends in Ecology and Evolution*, **1**, 142–3.

Parker, G. A. and Hammerstein, P. (1985). Game theory and animal behaviour. In *Evolution. Essays in honour of John Maynard Smith* (eds P. J. Greenwood, P. H. Harvey, and M. Slatkin), pp. 73–94. Cambridge University Press, Cambridge.

Popp, J. L. (1978). Male Baboons and Evolutionary Principles. Ph.D. thesis, Harvard University.

Rapoport, A. (1975). Uses of game-theoretic models in biology. *General Systems*, **20**, 49–58.

Rasmussen, K. L. R. (1980). Consort Behaviour and Mate Selection in Yellow Baboons. Ph.D. thesis, University of Cambridge.

van Rhijn, J. G. (1980). Communication by agonistic displays: a discussion. *Behaviour*, **74**, 284–93.

van Rhijn, J. G. and Vodegel, R. (1980). Being honest about one's intentions: an evolutionarily stable strategy for animal contests. *Journal of Theoretical Biology*, **85**, 623–41.

Sella, G. (1985). Reciprocal egg trading and brood care in a hermaphrodite polychaete worm. *Animal Behaviour*, **33**, 938–44.

Selten, R. and Hammerstein, P. (1984). Gaps in Harley's argument on evolutionarily stable learning rules and the logic of 'tit for tat'. *The Behavioral and Brain Sciences*, **7**, 115–16.

Smuts, B. B. (1985). *Sex and friendship in baboons*. Aldine, New York.

Smuts, B. B. and Watanabe, J. M. (1990). Social relationships and ritualized greetings in adult male baboons (*Papio cynocephalus anubis*). *International Journal of Primatology*, **11**, 147–72.

Strum, S. C. (1982). Agonistic dominance in male baboons: an alternative view. *International Journal of Primatology*, **3**, 175–202.

Trivers, R. L. (1971). The evolution of reciprocal altruism. *Quarterly Review of Biology*, **46**, 35–57.

de Waal, F. B. M. and van Hooff, J. A. R. A. M. (1981). Side-directed communication and agonistic interactions in chimpanzees. *Behaviour*, **77**, 164–98.

Whitehead, J. M. (1987). Vocally mediated reciprocity between neighbouring groups of mantled howling monkeys *Alouatta palliata palliata. Animal Behaviour*, **35**, 1615–28.

PREAMBULE DU TRAITE DE L'ATLANTIQUE NORD

Les Etats parties au présent Traité,

Réaffirmant leur foi dans les buts et les principes de la Charte des Nations Unies et leur désir de vivre en paix avec tous les peuples et tous les gouvernements, Déterminés à sauvegarder la liberté de leurs peuples, leur héritage commun et leur civilisation, fondés sur les principes de la démocratie, les libertés individuelles et le règne du droit.

Soucieux de favoriser dans la région de l'Atlantique nord le bien-être et la stabilité.

Résolus à unir leurs efforts pour leur défense collective et pour la préservation de la paix et de la sécurité,

Se sont mis d'accord sur le présent Traité de l'Atlantique Nord...

Fait a Washington le quatre avril, 1949.

For the Kingdom of Belgium:
Pour le Royaume de Belgique:

For Canada
Pour le Canada

For the Kingdom of Denmark:
Pour le Royaume du Danemark:

For France:
Pour la France:

For Iceland:
Pour l'Islande:

For Italy:
Pour l'Italie:

PREAMBLE TO THE NORTH ATLANTIC TREATY

The Parties to this Treaty reaffirm their faith in the purposes and principles of the Charter of the United Nations and their desire to live in peace with all peoples and all governments.

They are determined to safeguard the freedom, common heritage and civilization of their peoples, founded on the principles of democracy, individual liberty and the rule of law.

They seek to promote stability and well-being in the North Atlantic area.

They are resolved to unite their efforts for collective defence and for the preservation of peace and security.

They therefore agree to this North Atlantic Treaty...

Done at Washington, the fourth day of April, 1949.

For the Grand Duchy of Luxembourg:
Pour le Grand Duché de Luxembourg:

For the Kingdom of the Netherlands:
Pour le Royaume des Pays-Bas:

For the Kingdom of Norway:
Pour le Royaume de Norvége:

For Portugal:
Pour le Portugal:

For the United Kingdom of Great Britain and Northern Ireland:
Pour le Royaume-Uni de Grand-Bretagne et d'Irlande du Nord:

For the United States of America:
Pour les Etats-Unis d'Amérique:

Belgium, Canada, Denmark, France, Iceland, Italy, Luxembourg, The Netherlands, Norway, Portugal, the United Kingdom, and the United States decided, on 4 April 1949, to have a collective security arrangement against any power which might threaten their peaceful existence. This alliance has grown into the most successful and long-lasting defensive alliance in history. (Photograph by courtesy of the NATO Information Service, Brussels.)

12

Cooperation in conflict: alliances in international politics

VINCENT S. E. FALGER

Alliances are 'a natural and unexceptionable product of the working logic of the international political system' (Friedman *et al.* 1970, p. 15)

INTRODUCTION

Political science is the branch of study of human behaviour which, broadly defined, concentrates on an individual's or a social system's capacity for influencing the behaviour of other individuals or systems. This capacity, usually called power, is of course not restricted to formal institutions and formal roles. So, if one is interested in explaining relationships of influence and the extent to which institutions and individuals are successful in influencing the behaviour of other actors, there is no reason to restrict one's curiosity to 'governmental behaviour', nor to deny the existence of non-human political behaviour (DeVree 1982; de Waal 1982).

Coalition formation, as has been argued throughout this volume, is essentially to be seen as support behaviour of two or more actors against a third party. Coalition formation is a main chapter in every political science handbook, although it is usually restricted to *internal* political affairs. The formation of a cabinet in a democratic state in many instances involves different parties which have to cooperate to form a majority in parliament. But this is not the only possibility. In the United Kingdom, for example, the party that wins a majority of seats in parliament is entitled to govern and can do so without a coalition between different parties. If so, coalition formation is more informal and manifests itself within the winning party. Even in non-democratic political systems, however defined, coalition formation (resulting in party 'factions') is evidently a most important category of behaviour. *Divide et impera*—divide and rule—is a classic maxim of coalitional behaviour which applies to every political system, democratic and non-democratic, national and international.

International politics is usually seen as a special category of 'political behaviour', because the conditions of the international system, the nature of the actors and problems, and their impact are quite different from national politics. International politics, taken broadly, concerns the distribution of power on a global scale and the interaction between and among the relevant centres of comparatively major power: (governments of) national states, international organizations (ranging from the United Nations Organization and the European Community to Greenpeace, the Catholic Church, and international scientific bodies), and, incidentally, individuals (terrorists, President Gorbachev, Mrs Thatcher). Not tied to one geographically bound location are multinational corporations, organized political and religious ideologies, and even economic mechanisms such as the (world) market. The international relationships of influence are studied in the political science subdiscipline of international relations, and cover a wide range of phenomena focusing on international conflict and cooperation. Subjects such as war and peace, diplomacy and negotiation, deterrence and arms control, nation building and political integration, underdevelopment and environmental pollution, and, of course, alliances and coalitions among states fall within the field. (Standard anthologies in international relations are Rosenau 1969; Dougherty and Pfaltzgraff 1981; Viotti and Kauppi 1987; see also Waltz 1979.)

This chapter will concentrate on alliance formation in international politics. After presenting in the first section the structural context in which international alliances are to be seen, a few historical examples of alliances are described to illustrate the importance of the phenomenon. The next section is focused on a condensed review of the literature on international alliances. The penultimate section is a synthesis of alliance formation and a prospect on the disintegration of the two most important alliances are offered. The chapter will end with some conclusions and speculations.

THE CONTEXT OF ALLIANCES: THE INTERNATIONAL POLITICAL SYSTEM

The realm of international politics concerns mainly, though certainly not exclusively, the relations between sovereign states. International alliances are traditionally seen as typical outcomes of interstate behaviour, and it pays to start with this classical perspective.

The dominant theoretical framework in which international relations usually are discussed, is the so-called *'realist' analysis* of power politics (see below). In a personified way we read about it every day in the newspapers and hear it in the radio and TV news bulletins: the Soviet Union has withdrawn its troops from Czechoslovakia; the United States has quarrelled with Japan about heavy imbalances in trade between the two countries; Mrs Thatcher accuses the European Community of becoming a soviet-like bureaucratic and undemocratic institution aimed at reduction of sovereignty of the member states; the Nuclear Planning Group of NATO has today discussed the implementation of the newly adopted military strategy, and so on.

The 'realist' analysis

'Realism' in the discipline of international relations is identifiable by the following basic assumptions:

1. The state is to be seen as a unitary actor.
2. States 'behave' rationally (weighing costs and benefits).
3. States pursue national interests, of which independence and security are the most important.
4. States 'live' in an international system without a fixed hierarchical order, often called bluntly 'anarchy'. Classical realism is quite policy-oriented and emphasizes the influence of individual statesmen in the creation of balances of power (the leading realist textbook is Morgenthau 1985). In contrast to this more or less voluntaristic conception is the 'neo-realist' view on the balance of power as an attribute of the ('anarchic') system of states. Viotti and Kauppi (1987, p. 52) summarize adequately the essential neo-realist assumptions, which play an important role in this chapter:

> Given the assumptions that the state is a rational and a unitary actor that will use its capabilities to accomplish its objectives, states inevitably interact and conflict in the competitive environment of international politics. The outcome of state actions and interactions is a tendency toward equilibrium, or balance of power. Balance of power from this point of view, then, is a systemic tendency that occurs whether or not states seek to establish such a balance. Indeed, states may be motivated to improve their own positions so as to dominate others, but such attempts will likely be countered by other states similarly motivated. Thus, a balance of power inevitably occurs. Balance-of-power theory so viewed can be used to account for arms races, alliances and counter-alliances, and other forms of competitive behavior among states.

International 'anarchy'

The colloquial personified view of politics unconsciously treats states as *rational unitary actors*, represented by heads of state, ministers, and other government officials, acting in an international environment where *national interests* have to be defended against those of other states or blocs of states. National security and independence have always been considered as the highest values, as is illustrated in article 2.4 of the Charter of the United Nations: 'All Members shall refrain in their international relations from the threat or use of force against the territorial integrity or political independence of any state, or in any other manner inconsistent with the Purposes of the United Nations'. That is the ultimate ratio of the fundamental rule of international law as embodied in article 2.1 of the same Charter: 'The

Organization is based on the principle of the sovereign equality of all its Members.' Everyone knows, however, that in political *practice* these legal *principles* are not completely lived up to by all sovereign members of this world organization. Why?

The answer lies in the essential characteristic of any international system: the state is internally organized on a much higher level of complexity than the international political environment in which this state exists. Put in extreme terms, the state is a *hierarchically* organized structure existing in a non-hierarchical environment of many other independent states. These other states all try to secure their possibly conflicting national interests, resulting in enough uncertainty that they ultimately do *not* rely on formal rules (international law). International anarchy, as Thomas Hobbes called it, may sound too strong a term in modern ears, but portraying the international system as an 'anarchic order' (Waltz 1979, Chapter 6) indicates the essential difference in structure between national and international politics. The international system is ultimately one of self-help (no authoritative actor or organization can prevent a stronger state realizing its goals); the national system—the state—is not.

This is not to say that international politics is by definition violent and related to war, and national politics by definition non-violent. Many states, past and present, have internally shown or threatened much violence in order to build, save, or reconstruct the national political system. These same states may have lived in peace with their neighbours for many decades. So, if we allow ourselves to see revolution and suppression as internal war, violence and war are not exclusively international phenomena. The main point in grasping the difference between the national and international system, however, is not the absolute amount of use and/or threat of violence, but the *order–'anarchy' dichotomy*. For the remainder of this chapter it is helpful to keep in mind Robert Gilpin's characterization of international relations as continuing 'to be a recurring struggle for wealth and power among independent actors in a state of anarchy' (Gilpin 1981).

Security in the international system

However, the prospect of war and the threat to use force to settle conflicts between states is of course unacceptable to states which at least want to survive and be independent from interference of other states. This elementary wish creates the so-called *security dilemma*. Within the relatively orderless international system, states with greatly varying military and economic capabilities are trying to survive. Without a central authority to call in if necessary, all states in the very first place take care of their own security, which means that every state will be prepared to go to war to defend its own independence. This preparation itself, however, creates considerable problems. Legal protection is not enough, because this alone has never helped against states which for whatever reason did not at specific moments recognize their general obligations. That is why the League of Nations, founded in 1919, failed.

Ideally, this system of collective security would automatically turn all members against every single state which offended the obligation to settle disputes peacefully, or threatened to use force. Japan invaded Manchuria in 1931, was internationally condemned for that action, and had to accept quite severe economic sanctions from several member states, but that was not enough to stop the operation and force the troops to return to their home bases. When a European member of the League of Nations attacked Ethiopia in 1935, even less happened. Italy was condemned in Geneva, but the economic sanctions were very half-hearted, and the system of collective security simply did not work. The League proved definitively powerless when it expelled the Soviet Union after its military action against Finland in the winter of 1939.

Even the successful action against Iraq in the Gulf War of 1991 was much less a United Nations collective security response than an American restoration of the Middle Eastern *status quo ante*, legitimized by 12 UN resolutions.

Trying to solve the security dilemma by means of a *legal* solution, collective security, was an important innovation in the traditional international relations and has its roots in the first international Hague conference on peaceful settlement of disputes, in 1899. Earlier, it is possible to retrace historic examples of conference diplomacy with relatively strong tendencies of ordering the international system: the European Concert, starting with the Vienna Congress in 1815, the end of the War of Succession (Treaty of Utrecht, 1713), and the end of the great religious wars in Europe, which actually laid the foundation for the modern state system (Treaty of Westphalia, 1648). (The eurocentric accent in this historic review is undeniable, but can be defended if one accepts the premise that the international political system as we know it today is a direct consequence of the exportation of the European model all over the world since the sixteenth century. For a stimulating discussion see Kennedy 1987, Chapter 1.)

Armament and alliances

Having roughly sketched how insufficiently national security is assured by legal means in an essentially 'anarchic' world order, we now come to the two logical instruments of national self-help, armament and alliances. National states and their predecessors have always invested considerable portions of national wealth in developing, producing, or buying weapons, mostly to be used in international conflicts.

In the 1980s the United States spent on average 5.4 to 6.7 per cent of the gross domestic product on defence, the UK 5 per cent, France about 4 per cent, West Germany 3.2 per cent, the USSR 19 per cent and Israel more than

20 per cent (SIPRI 1989, pp. 188–9, 151). In 1988, the world total of military expenditure was more than 800 billion dollars (SIPRI 1989, pp. 183–7). This is not the place to discuss at length the various explanations for these huge expenditures, but the two relevant alternative standard hypotheses should be mentioned here: the military–industrial complex hypothesis and the action–reaction pattern hypothesis following from a prisoner's dilemma situation. In both hypotheses, the armament of the opponent is seen as a potential threat to one's own security.

It is not surprising that arms races and improvement of military strategy are the rule in international politics, and disarmament and arms reduction very rare (except after military defeat, but that is one-sided). The Washington Treaty of Naval Disarmament of 1922 did not work well, and the 1987 Intermediate Nuclear Forces Treaty between the United States and the USSR is still very recent and certainly not the end of the security dilemma. These historic facts are directly related to the structure of the international political system. Within all national systems it is prohibited to settle conflicts between individuals or groups directly by violent means. State monopoly of the legitimate use of force within the state's own borders is the main sanctioning mechanism. The police and the judiciary are the main embodiment of national law enforcement. Internationally, however, such an effective mechanism does not exist, and this implies self-help of sovereign states by preparing for the eventuality of armed conflicts, that is, buying or producing weapons and training huge armies, even in peace time. The difference in size and quantity of weaponry for external use illustrates the far greater perceived threat.

If the armament of states is inseparably tied up with the history of international relations, this is no less true of alliances between states. An alliance in international politics can simply be defined as

> a formal or informal relationship of security cooperation between two or more sovereign states. This definition assumes some level of commitment and an exchange of benefits for both parties; serving the relationship or failing to honor the agreement would presumably cost something, even if it were compensated in other ways. (Walt 1987, p. 1)

In the earliest written sources of history, as far as they concern international political affairs, we can easily recognize alliances. From the diplomatic archives of Amarna (Egypt, 14th century BC), Boghazkoy (Turkey, same period), and Babylon, many examples of letters and personal agreements—which we would now call treaties—between reigning kings have survived, showing that alliance formation was almost as important as waging war (e.g. Langdon and Gardiner 1920, Numelin 1950, and Helck 1971). Thucydides in his famous *History of the Peloponnesian War* described several alliances in ancient Greece. The history of international relations ever since these earliest historic records has been full of minor and major alignments, most of them now forgotten. In the next section, a few instances of modern alliances are described. So far, it is important to realize that the structure of the international system, to be described as 'anarchic order', forces sovereign states to help themselves and that alliances are obvious examples of cooperation and conflict in international politics. Cooperation in conflict may be an even better characteristic.

Alliances and the balance of power

National power, or the capacity to influence the behaviour of other states, is usually distributed unevenly among the members of an 'anarchic' international system. Faced with the security dilemma, states behave towards each other in ways that will invariably result in a temporary balance of power. Throughout history, states striving for domination over all their neighbours have met with combinations of other states which tried to counter these aspirations. Whether this tendency to counterbalance states which are too strong is seen as the willed product of statesmen, or as a structural tendency of the 'anarchic' international system itself, in either case some element of self-regulation is recognizable in the 'anarchy'. Not surprisingly, it has been noticed that the analogy of an equilibrium establishing 'invisible hand'—that central assumption of classical free-market economics—is associated with balance-of-power approaches (Holsti *et al.* 1973, p. 5).

Kenneth Waltz, a neo-realist author who favours the structural approach to international politics summarized many arguments on the origin of the balance of power very clearly: 'Balance-of-power politics prevail wherever two, and only two, requirements are met: that the order be anarchic and that it be populated by units wishing to survive.' (Waltz 1979, p. 121). This simple rule, *the* structural characteristic in international politics since the end of the Roman empire, holds in *bipolar* as well as in multipolar systems. In a bipolar system, a balance of power exists between two states (and their allies) with relatively equal power. A multipolar system less easily produces a balance of power because three or more states are engaged in checking and balancing behaviour (for an overview see Viotti and Kauppi 1987, pp. 50–5).

It will now be clear that alliance formation is an essential international political phenomenon which will remain with us until a complete world federation with an effective government, cancelling international 'anarchy', is functioning. International alliances and 'power politics' are indeed very much linked with each other, which means that alliances 'exist' only within the broad theoretical perspectives of realism and neo-realism. Within that realm, however, there is no single accepted theory of alliance formation, even if the wisdom of Thucydides, more than 2000 years old, is quoted with approval: mutual fear is the only solid basis upon which to organize an alliance (Liska 1962, p. 26; Holsti 1967, p. 111; Wight 1978, pp. 102, 138). Thucydides, in fact, has presented the nucleus of what can be seen as a 'balance-of-threat' theory of alliance formation, a useful refinement of the more current balance-of-power theory and to be discussed in the penultimate section. But before going into the details of the sometimes rather crude power political theories of realism, it may be helpful to sketch some historic examples of genuine international alliances and one borderline case.

Historic international alliances described

The purpose of this section is to illustrate historical differences in origins, purposes, and success or reasons of failure of some well-known alliances. Although the accent is, once more, on European history, this does not mean that the formation of alliances is a typically European phenomenon. But by referring to classic and more or less familiar examples, it is possible to see more quickly the essentials of the political process of alliance formation. The Quadruple and Holy Alliances (1815), the ineffective defensive alliances on the eve of World War I, the Molotov–Ribbentrop pact (1939) and the allied cooperation in World War II, NATO, and the European Community provide us with enough variation to understand the function of alliances. (See Palmer and Colton 1978, Craig and George 1983, or Kennedy 1987 for details and bibliographical references.)

The Quadruple Alliance and the Holy Alliance

A war alliance is succeeded by a postwar alliance on ideological basis, showing, however, that it is not ideology but interests that keep the partners together.

The fall of Napoleon, which began with his campaign against Russia in the summer of 1812, finally resulted from the battle of Leipzig in October 1813, the so-called 'Battle of the Nations'. The British foreign minister, Castlereagh, succeeded a few months earlier in forming a war alliance of Great Britain, Austria, Prussia, and Russia against Napoleonic France. In March 1814, the Treaty of Chaumont was concluded, the first solid coalition against France since 1792. The resulting *Quadruple Alliance* and their allied armies forced Napoleon to abdicate and to leave France for Elba, an island near the coast of Italy.

The famous Congress of Vienna 'danced' in 1814 and 1815, meanwhile laying the foundations of a new political order in Europe. English fear of Russian expansionism in the continental Europe promoted the cooperation with Austria, at that time a more important power than Prussia. Interestingly, defeated France was also present in the person of Talleyrand, once foreign minister to Louis XVIII. Although France was to pay the bill for the war, these costs were tempered by the French political balancing between the victorious powers.

All this dealing and wheeling stopped briefly when Napoleon tried to restore his authority in France after his escape from Elba. Napoleon's 'hundred days' caused new fears of revolution and war in Europe, and the armies of the Quadruple Alliance once more had to fight together against a French nation which apparently had not learnt its lesson at the battle of Leipzig. Waterloo, however, was the definitive end of Napoleon and the beginning of a firm coalition against the French threat.

In 1815 the peace negotiations led to the successful Peace of Vienna, the foundation of an unprecedentedly long period without major war in Europe. In addition, Tsar Alexander I of Russia, Emperor Franz I of Austria, and King Friedrich Wilhelm of Prussia founded a formal alliance of monarchs, inspired by Christian solidarity. The three-article alliance, which at first was not taken very seriously, promised solemnly that the three monarchs would treat each other as brothers, support each other whenever possible, and treat their subjects as fathers treat their children. They saw themselves as the representatives of God, obliged to govern the whole Christian nation in his spirit. All monarchs who could accept these principles, were invited to become a member of this exclusive *Holy Alliance*. It is a very good illustration of the postwar unity that all European monarchs signed, except—for obvious reasons—the Turkish sultan and the Pope. The British monarch, however, was forbidden to sign because parliament would not support this reactionary 'piece of sublime mysticism and nonsense' (Castlereagh).

Indeed, the significance of the Holy Alliance was less its Christian virtues than its antiliberalism and its sympathy with the restoration of the *ancien régime* in Europe. It is not surprising, then, that this personal alliance broke down when Tsar Alexander could not resist mixing in Balkan affairs where Emperor Franz I had quite different interests. However, the European conference diplomacy between 1815 and 1850, known as the *Concert of Europe*, produced a remarkable cohesion, if only temporarily, in a very much politically divided Europe. The success lasted as long as the fear of the common enemy and the interests in restoration of the old order balanced against the expected benefits of one-sidedly taking advantage of the disintegrating Turkish empire. The gradual deterioration from excellent co-operation to conflict as a result of Russia taking advantage reached its turning point in the Crimean War (1854–1856), the first major European war since Waterloo. The *tableau de la troupe*, the partnership in this war, was quite different, however.

The eve of World War I

Alliances are formed when the balance of power is threatened, but they are no guarantee against the outbreak of war.

Great Britain in 1815 was at its zenith as a world power. At the end of the nineteenth century the political conditions had changed very much for the better for Germany, which had been since 1871 a united empire of ever-growing military and economic importance in the middle of Europe. The economic success of German industry, and the wish to obtain the last remaining areas in the race for colonial territories, led Germany to manufacture a very modern fleet. The British, in diplomatic isolation since Fashoda and the Boer War, were alarmed by this; notwithstanding the close personal relationships between the respective reigning dynasties, they were forced to look for common interests with the long-time competitor in European and colonial affairs, France. In 1904 the *Entente Cordiale* between Great Britain and France laid the foundation for an alliance which later would gain support from Russia also. France and Russia had already had a formal alliance since 1893, the benefit of which for the French was a security arrangement against the Germans, and for the Russians against Austria-Hungary in the Balkans. In the same year (1904) a major war broke out between Russia and Japan, and for the first time in history a non-white people defeated a white western power, not only shocking white supremacy but also stopping Russian expansion in the Far East. This appeased the anti-Russian British, and the French tried successfully to involve Russia once more in the European political theatre. In 1907 the *Triple Entente* between Great Britain, France, and Russia was concluded, a power-political change which would certainly not have taken place without the amazing Russian defeat and the growing German threat.

The Second German Empire was the stronger partner in the alliance with Austria-Hungary. In the long series of crises in the Balkans since the beginning of the new century, Austria-Hungary opposed the Serbians and the Russians. The much disputed Austrian annexation of Bosnia and Hercegovina in 1908 would not have been possible without German tolerance. After all, it seriously weakened Russian interests and thus the Triple Entente. However, four years and two full-scale Balkan wars later, a combination of armament races (between Germany on the one side and Britain and France on the other), the loss of the defensive character of the European alliances, and a series of premature mobilizations, reached the point of no return. The new crisis was triggered by a Serbian-inspired attack on the Austrian heir to the throne, Archduke Franz Ferdinand, and his wife in Sarajevo on 28 June 1914. This resulted within one month in an Austrian declaration of war on Serbia, which was followed by a Russian mobilization.

The important point here is that Austria would not have dared to do this without full permission of her German ally, which, in its turn, thought a military confrontation in western Europe was inevitable anyway. The French did not try to calm down their Russian allies, because this new crisis meant strong pressure on Germany. The British would have preferred to keep neutral in a war on the Continent, but the German declaration of war on France on 3 August, and the demand of free passage through Belgium, forced Britain to issue an ultimatum to Germany to respect Belgian neutrality. In practice, putting up this barrier meant a British declaration of war, not so much because of their respect for international neutrality, but because the European balance of power was likely to be disturbed seriously to the detriment of Britain. The Great War had started and would only be ended after intervention—for the first time—of a non-European power, the United States (1917). The Germans called the Peace Treaty of Versailles, which was forced on them two years later, a *Diktat*, and getting rid of this dishonourable treaty was one of the main purposes of many political groups in postwar Germany.

The Molotov–Ribbentrop pact and the allied Powers in World War II

Offensive alliances of autocratic states will be held together by the prospect of gain, but overrating of capabilities by alliances leads to losses in the long run.

In many ways World War II can be seen as a continuation of the Great War now called World War I—after an interlude of 20 years. This is not to say that the course of history was predetermined and had to lead to a more definitive war (after all, Germany was the victor on the eastern front, and the war had not been fought on German soil), but the European balance of power did not work well after the Americans had withdrawn from the Paris peace conference in 1919 for national political reasons. The League of Nations, the new solution for future security problems, failed because its founding charter was part of the hated Treaty of Versailles, and it kept out guilty Germany and revolutionized Russia. Although both states became league members later on, they were not the most faithful. Shortly after coming to power in 1933, Hitler withdrew the new German *Reich* from the League and in 1936 denounced all obligations resulting from the Versailles and Locarno treaties. The Soviet Union was expelled in 1939 after having attacked Finland.

So, in the late 1930s no clear set of alliances threatened rearming Germany. The French alliance with Czechoslovakia, an important attempt to counterbalance the Germans, proved in the end not strong enough against Nazi pressure on the (Czech) Sudetenland. Although the Soviet Union too had a defensive alliance with Czechoslovakia, this power political obstacle was not enough to stand against Hitler's tactic. The French hesitated to be firm in the second crisis over Sudentenland (September 1938), because the British did not want to risk war for what was after all a part of the German nation. The Munich Conference with Hitler, Chamberlain, Daladier, and Mussolini resulted on 30 September 1938 in the well-known appeasement treaty ('Peace in our time!').

A detail which is usually forgotten is that Poland and Hungary also immediately demanded parts of Czech territory, a country that was typically a product of 'Versailles'. The Soviet Union was kept out of the Munich deal, and so it was not incidental that this country would prove pivotal in the next crises. After the German invasion of the rest of Czechoslovakia on 14 March 1939, an outright provocation to Great Britain and France which they nevertheless met with only verbal resistance, Hitler tried to seize Poland. The Poles felt very self-confident after the formal English–French guarantee of Polish neutrality, and thought that they could withstand the German army. Moreover, the Soviet foreign minister, Litvinov, discussed the possibility of a defensive alliance with Great Britain and France. But Stalin demanded a high price: Britain and France had to tolerate the stationing of Soviet troops in Poland, the Baltic states, and parts of Finland. This was especially unacceptable to the British who hoped for a German–Soviet clash in which they themselves would not have to participate.

The pro-western Litvinov was replaced by the less sophisticated Molotov, and it subsequently took only a week of secret negotiations before on 25 August 1939 the notorious *German–Soviet non-aggression pact* was signed in Moscow by the German foreign minister and his soviet colleague. Joseph Stalin kept a smiling eye on this formal part. The informal—secret—protocol which made the whole deal worthwhile, promised the Soviet Union what it had asked in vain from Britain and France: half of Poland and a free hand in the Baltic states. Japan and Italy, allied with Germany in the Anti-comintern Pact of 1936, were absolutely stunned by this diplomatic move. So were the communists in western Europe.

This aggressive alliance cleared the way for the German attack on Poland, which started on 1 September 1939. The Soviet invasion followed 16 days later. After 6 weeks Poland surrendered. Great Britain and France, although affirming their alliance with Poland and declaring war on Nazi Germany after expiration of the ultimata, hoped that the real clash might be averted and did not attack Germany in the west although this could have paralysed the advancing German army in Poland. The 'phony war' had started.

The outcome of World War II would probably have been quite different if Hitler had abstained from attacking the Soviet Union (22 June 1941). This move changed the war coalitional structure fundamentally, and proved as disastrous for the Germans as had Napoleon's attack on 22 June 1812 for the French—also a breach of alliance (with Alexander I). From now on the isolated British gained some chance of success in their struggle with the Germans. Winston Churchill had once said that he would not object to a pact with the devil himself if it was directed against Nazi Germany. Stalin fitted that role quite well, and after full American participation in the war, the structural conditions had been changed definitively. It would take time and enormous losses of lives and hardware, but a complete new world order emerged in 1945.

Alliance formation played still another important role in the duration and intensity of the war. The *Three Power Pact* of 1940 obliged the Axis powers—Germany, Italy, and Japan—to support each other in case of war. As a consequence of the American declaration of war on Japan after the attack on Pearl Harbour on 7 December 1941, Germany and Italy declared war on the United States. Churchill and Roosevelt must have thanked Hitler sincerely for this strategically very unwise step, because it was far from certain whether the US Congress would have supported an unprovoked war against Germany. Although the officially neutral Americans had gradually become more generous to the British by providing them with war equipment on very profitable financial terms (Lend Lease Act), active participation in a new European war and again having thousands of American soldiers killed was quite another matter. Hitler's *Bündnistreue*, sticking to treaties, in this case, however, had more to do with the formidable successes of the German armies in Europe and North Africa. The expected benefits must have been much higher than the costs entailed.

The NATO and EC alliances

Alliances based on nuclear deterrence tend to freeze the status quo, but military alliances as such are poor instruments of political and economic integration.

Four years after 1945 the reality proved very different from the expectations, though Hitler was right in his assessment that the war coalition against Germany would not hold. Within three years after the realization of the common purpose, unconditional surrender of the Axis powers, new patterns of conflict and cooperation restructured the European political map. What had begun as severe differences of opinion about the future of Germany and Poland in the period 1946–8 ended in the Cold War. The communist take-over in Prague (February 1948) and the beginning of the 'air lift' on Berlin (June 1948) led the war-weary western European countries to a new defensive alliance.

In March 1948, the *Brussels Treaty*, a 50-year treaty of economic, social and cultural collaboration and collective self-defence, was signed by the foreign ministers of Belgium, France, Great Britain, Luxembourg, and the Netherlands. On 4 April 1949, a more far-reaching agreement, the *North Atlantic Treaty*, was signed in Washington. In addition to all members of the Brussels Treaty, the United States, Canada, Denmark, Iceland, Italy, Norway, and Portugal participated. Within the legal framework of article 51 of the Charter of the United Nations ('Nothing in the present Charter shall impair the inherent right of individual or collective self-defence . . .'), these sovereign states allied against the danger of a possibly expansionist Soviet Union, although that name did not show up in the official documents. The following are the most important provisions of the 14-article treaty (NATO 1989: pp. 376–7):

The Parties to this Treaty are determined to safeguard the freedom, common heritage and civilisation of their peoples, founded on the principles of democracy, individual liberty and the rule of law. (. . .) They are resolved to unite their efforts for collective defence and for the preservation of peace and security. (Preamble)

> In order more effectively to achieve the objectives of this Treaty, the Parties, separately and jointly, by means of continuous and effective self-help and mutual aid, will maintain and develop their individual and collective capacity to resist armed attack. (article 3)
>
> The Parties agree that an armed attack against one or more of them in Europe or North America shall be considered an attack against them all and consequently they agree that (...) each of them (...) will assist the Party or Parties so attacked by taking forthwith, individual and in concert with the other Parties, such action as it deems necessary, including the use of armed force, to restore and maintain the security of the North Atlantic Area. (article 5)

It is interesting to see how the structural 'anarchy' of the international system is reflected here. However, equally interesting is the fact that with an absolute minimum of organizational provisions—in effect only the obligation to consult together—a formidable allied defence structure evolved which is said not only to have deterred the Soviet Union from aggression, but also effectively to have encapsulated the former enemy, now frontier-state, Germany. The *Warsaw Treaty Organization*, founded in 1955, three days after the Federal Republic of Germany became a member of NATO, is in many respects a copy of the western alliance. These two alliances formed the main institutional framework within which the conflicts in the anarchic European and world order were dealt with. Meanwhile, it has been shown that these alliances came under severe pressure when the image of each other as perverse, frightening enemies in a deadly grip of nuclear deterrence. Their dissolution will severely threaten the relative stability of the modern balance of power, as in the case of the loss of the WTO has already become evident. This problem will be discussed further in the penultimate section.

So far, only military alliances have been considered. If alliances are for whatever reason restricted to *security* cooperation between sovereign states, a wide range of highly interesting other forms of international cooperation remains out of view. In what ways do other important areas of international cooperation, in particular economic, differ from military cooperation? To conclude this section, some aspects of the world's most developed example of economic cooperation, the *European Community* (EC), are elucidated.

If we strip the recent discussion about the EC of the political rhetoric, a few clear lines can be drawn. Firstly, the idea of economic cooperation between sovereign states does not date only from the 1950s. The *German Customs Union* was founded in 1834, plans for far-reaching economic cooperation between France and Germany were discussed in the late 1920s, and the *Benelux Customs Union* between Belgium, the Netherlands, and Luxemburg (1944) worked so well that it became the main model for the construction of the EC (1957). However, the real cornerstone was laid in the *European Community for Coal and Steel* (1951), in which France and Germany, the old enemies, plus Italy and the Benelux countries agreed to place the authority over the national coal and steel production (not incidentally the essential sources of every war industry) in the hands of a supranational organ. This originally French plan gave the French at least considerable influence in a highly strategic sector of the German economy. Konrad Adenauer, who tried everything to tie the Federal Republic of Germany into the western cooperation, could live very well with this construction. It proved a very successful formula and a few years later the broad spectrum of agriculture was 'communalized'. The French interests again were well served, but now the relatively inefficient German agriculture had also to be financed.

So, considerations of national security have never been absent in the history of the EC, and it would be naive to forget them today. To put it bluntly, keeping Germany firmly within a web of international obligations of cooperation, that is, interdependent with the other western European states, is the unavowed security basis of the EC. It is no coincidence that President Mitterand of France used the EC as the first line of 'defence' against the real or imagined danger of German reunification and a disporportionate German economic influence in eastern Europe, a slumbering problem made very acute after the Berlin Wall ceased to function on 9 November 1989.

There is another reason why the EC in effect fits into the traditional perception of alliances. Although the main economic aim of the Community is to promote the economic efficiency of the member-states among each other, this has far-reaching consequences for those states which do not belong to this community. The formation of one internal, really free market (*Europe 1992*) is in many respects an act of (economic) self-defence in an international order where the anarchy of liberal capitalism does not result in the equal distribution of the benefits of international trade. Also, all kinds of protectionist arrangements to conquer each other's markets (export subsidies, quality standards, dumping, etc.) trigger counter-measures of the affected national economy. The physical wars have, at least in the western hemisphere, been replaced with 'chicken and chips wars', costing not lives but jobs.

Thus, even the essentially peaceful international behaviour of trade and regional economic cooperation had produced 'alliances'. This leads to strange results: the American and European partners in security cooperation are each others' competitors in economic respects. The, relatively speaking, declining economic and military power of the United States as the leader of the NATO alliance creates ever-increasing instability in an international system of still sovereign states. This effect is even strengthened by the far greater decline of the Soviet Union as a leading power. The times of hegemonic stability are over, and this will no doubt give rise to new patterns of alliance formation in international politics (see Gilpin 1987 for the growing importance of economic security in international politics).

INTERNATIONAL ALLIANCE FORMATION ANALYSED

When in the 1960s studies of international relations were affected by the 'behavioural revolution' in the social sciences, wars as well as international alliances became promising objects for the application of quantitative research techniques. Singer and Small, the first to break the general lack of empirical rigour in alliance studies, gained fame with their finding that whether measuring amount by the number of wars, the number of months that a nation was involved in war, or the number of battle deaths incurred, 'alliance aggregation and bipolarity predicts strongly away from [the onset of] war in the nineteenth century and even more strongly toward it in the twentieth.' (Singer and Small 1968, p. 283). In another article, Singer and Small (1969) concluded on the basis of correlational analysis of 112 formal alliances in the period 1818 to 1945 that war involvement is strongly and positively correlated with alliance commitment. Much depends on alliance membership duration, and it is hardly surprising that this correlation is most clear with alliances concluded just before war broke out. They clearly realized the limit of their own findings. The authors say that it is 'obvious that correlation and causality are rather different things, and that correlation at a high level is *necessary* to the establishment of a causal relationship, but not at all *sufficient*.' (Singer and Small 1968, 283).

This line of correlational research inspired many others and it led to various efforts to operationalize and test one or more basic propositions of the traditional balance-of-power theory. The results, however, are far from coherent, because of the use of different methodology and parameters. McGowan and Rood (1975), for example, hypothesized that in 'balance of power international systems' the occurrence of alliances will be randomly distributed, the time intervals between alliances will also be distributed randomly, and a decline in the rate of alliance formation precedes system-changing events, such as general wars. Tested on the alliances in Europe in the period 1814 to 1914, these hypotheses were confirmed.

Levy (1981) and Siverson and Sullivan (1984) agree that alliances have not generally played an important part in the process leading to most wars involving great powers, a conclusion that puts Singer and Small's findings in another perspective. Duncan and Siverson (1982), and Kegley and Raymond (1982) would ascribe this weak correlation to a rather great flexibility of alliance partner choice among the major powers, especially in the nineteenth century. Midlarsky (1983) draws attention to an apparent absence of memory in the nineteenth century international system. In an age when public opinion had no influence on actual policy, it did not matter much whether yesterday's enemy became today's ally.

In the postwar nuclear period flexibility is gone, but mutual nuclear deterrence between the superpowers, according to Weede (1983), seems so far to have extended between the *de facto* members of the respective blocs (the wars that did not occur between Greece and Turkey are an illustration). Intrabloc conventional wars, however, have not always been successfully deterred, as was shown in the limited war between China and Vietnam (1979) and the Falklands War (1982). The alliance structure in these historical cases was functioning quite differently. Weede's conclusion would support Waltz's argument that bipolarity in international politics is more stable than multipolarity. Wayman (1984) researched this classic issue in international relations with new statistical methods and arrived at the conclusion that a combination of power bipolarity (USA–USSR) and alignment multipolarity (causing all kinds of interdependency)—along with discouragement of massive transfers and proliferation of weapon technology that could destabilize the power bipolarity—is the formula for general stability in the nuclear age international system. Siverson and Tennefoss (1984) would probably agree, in particular with regard to minor powers which in conflict situations are sensible to the deterring effect of being allied with a major power.

The empirical research in alliance formation presented so far was tied up with confirmation or criticism of the balance-of-power theory. The study of alliances is also fed from two other traditions, here presented as useful additions to the balance-of-power theory. There is an important school of coalition researchers, starting with Riker (1962). In his *Theory of Political Coalitions*, no clear distinction between coalitions and alliances is made, which is quite right from the perspective of the mechanism of the processes themselves. Riker formulated a 'size principle', which essentially implies that actors will not create larger coalitions than is necessary to achieve the objective (e.g. contending a growing or threatening adversary). This is known as the rule of formation of minimal winning coalitions. In international politics this means that the decline of empires, as described by Kennedy (1987) for example, can be explained in terms of the size principle and financial overspending on allies and followers, which in the end weaken the coalition leader. Kennedy has labeled this graphically 'imperial overstretch'. The size principle figures implicitly in several of the previously discussed applications of the balance-of-power theory. In many discussions of coalition theory in *national* political contexts, implicit balance-of-power ideas are not lacking either (see for example de Swaan 1973, and Tsebelis 1987). The important dimension of cooperation is given more attention in the literature on coalitions than in the average, conflict-oriented international relations literature.

Another category of empirically oriented research adding to our understanding of alliances is the so-called collective goods approach. This economic theory of alliances tries to explain 'how much a nation acting on the basis of its national interest would contribute to an international organization such as an alliance. When provided through an alliance, security takes on the attributes of a public or collective good.' (Thies 1987, p. 300). This approach dates back to Olson and Zeckhauser (1966) and focuses on free-rider behaviour and burden sharing in a condition of alliance: the United States provide the western world with security via NATO, and cannot prevent non-members such as Finland and Austria sharing it. Besides, security will not decline if small powers like the Netherlands, Belgium, and Denmark do not pay their full agreed contribution (see also Burgess and Robinson 1969). Most and Siverson (1987) wondered whether allying would result in spending less on armament. In Europe between 1870 and 1914 no simple pattern of substituting arms for alliances was perceptible. For the smaller NATO partners in the postwar period this would certainly be demonstrable. Thies (1987) clearly illustrates in seven pre-1945 alliances that the problem of burden sharing differs depending on whether the alliance strategy is a deterrent (threatening with retaliation) or a defensive one (denying an attacker meaningful gains). The collective goods approach is impoverishing, he warns, if only post-war NATO burden sharing is researched.

Altfeld (1984), to conclude this review, uses a micro-economoic approach in his mathematical and quantitative analysis of the national decision to ally or not. He is one of the very few who sees that 'the decision not to ally is made in exactly the same way as the decision to ally, by calculating costs and benefits.' (p. 523). Securing autonomy in a tense political situation without allying will often mean the start of an armament race. Iran's consistent *Alleingang* in the war against Iraq, to take an example, enlarged the country's manoeuvring room considerably, as the use of terrorism and this Gulf War illustrated. In the end, however, it was the balance of counter-power which outweighed the costs in terms of resources and political prestige over the benefits, and forced Iran to accept an unfavourable ceasefire settlement after 8 years of exhausting war.

A SYNTHESIS

Have these different approaches, the historical-quantitative analyses, coalition theory, and theory of collective goods, provided us with more insight into the nature of international alliances and the processes of alliance formation than the classic balance-of-power theory? However interesting the findings of these researchers may be, they lack the power of a more completely elaborated theory which concentrates less on complex methodology and more on general propositions. McDonald and Rosecrance (1985) in their historical analysis of the Bismarckian system in the period 1870–90 point in an interesting direction by quoting an old Arab proverb: 'A friend of a friend is a friend; a friend of an enemy is an enemy; an enemy of a friend is an enemy; and an enemy of an enemy is a friend.' (p. 59) It is probably true that individuals are less tolerant of ambiguity and ambivalence than impersonal governments which have to decide with whom to ally, but the maxim represents a basic line in thinking on coalitions and alliances.

Formulating general propositions about alliance formation does not rule out respect for historical detail, and ideally there is an exchange of ideas between the generalizing political scientist and the particularizing historian. Baylis (1985), for example, tries to test power political alliance theory by application to the history of the 'special relationship' between the United States and Great Britain in the period from 1939 to 1984. In his theoretical introduction he quotes with approval the American political scientist K. J. Holsti who endorses the classic realist thesis, dating back to Thucydides, that common perceptions of threat are probably the most frequent source of alliance strategy (one of the most widely held and least surprising propositions, Bayliss adds). But being focused on alliance strategy (in our terms, alliance formation), this basic and apparently so self-evident proposition has not gained much more attention in this article nor elsewhere in the literature on alliances.

It is not surprising indeed if 'Thucydian' threat perceptions are presented as parts or aspects of a general balance-of-power theory, the traditional realist approach in international relations as discussed in the first section. The perspective changes, however, if subjective 'threat perceptions' rather than more or less measurable 'power' become the focus of analysis. This is precisely what the neo-realist Stephen Walt has done in his 1985 article, afterwards extended in his book *The Origins of Alliances* (1987). Here we find a good synthesis of refined theory, original historical detail (this time on alliances in the Middle East in the period between 1955 and 1979), and a stimulating discussion of modern international policy matters with a special emphasis on the role of the United States as an alliance partner.

Walt's main argument is that states ally against states that pose the greatest *threat* although the latter need not *per se* be the most powerful states in the international system. Just as national power is produced by several different components (e.g. military and economic capability, natural resources, and population), the level of threat that a state poses to others is the product of several interrelated components (in this case aggregate power, proximity, offensive capability, and perceived intentions). Whereas balance-of-power theory predicts that states will react to imbalances of power, balance-of-threat theory predicts that when there is an imbalance of threat (i.e., when one state or coalition appears especially dangerous), states will form alliances or increase their internal efforts in order to reduce their vulnerability.

The distinction is subtle but important. Balance-of-threat theory improves on balance-of-power theory by providing greater explanatory power with equal parsimony. By using balance-of-threat theory, we can understand a number of events that we cannot explain by focusing solely on the distribution of aggregate capabilities. For example, balance-of-threat theory explains why the coalitions that defeated Germany and its allies in World Wars I and II grew to be far more powerful than their opponents, in contrast to the predictions of balance-of-power theory. The answer is simple: Germany and its allies combined power, proximity, offensive capabilities, and extremely aggressive intentions. As a result, they were more threatening (though weaker) and caused others to form a more powerful coalition in response. (Walt 1987, pp. 263–4).

Stephen Walt's second important theoretical foundation is the acceptance of Kenneth Waltz's distinction between balancing and bandwagoning behaviour of states. Balancing is defined as allying with others against the prevailing threat, and bandwagoning refers to alignment with the source of danger (cf. Waltz 1979, pp. 125–6 and Walt 1987, pp. 17–21 and 28–33). Walt summarizes his explanation of alliance formation in five testable generalizations about the conditions favouring balancing or bandwagoning:

1. Balancing is more common than bandwagoning.
2. The stronger the state, the greater its tendency to balance. Weak states will balance against other weak states but may bandwagon when threatened by great powers.
3. The greater the probability of allied support, the greater the tendency to balance. When adequate allied support is certain, however, the tendency for free-riding increases.
4. The more unalterably aggressive a state is perceived to be, the greater the tendency for others to balance against it.
5. In wartime, the closer one side is to victory, the greater the tendency to bandwagon with it (Walt 1987, p. 33).

Applied to the historical alliances in Europe, mentioned in the second section, it will be clear that the balance-of-threat theory is a *refinement*, but not more than that, of the traditional balance-of-power theory. The phrase 'balance of terror' has also been used. It is rather the subjective need for uncertainty reduction, as a hardly unavoidable consequence of perceived threat, than the results of objective calculation of power positions that leads states to ally. This applies for the Holy Alliance and the EC as well as for the Molotov–Ribbentrop pact and NATO.

If this refinement of balance-of-power reasoning holds, then it is tempting to try to predict the direction of future developments in Europe. The thawing out of the Cold War bipolar balance of (power) threats will inevitably result in disintegration of the existing alliances of WTO (the WTO was dissolved in February 1991) and NATO. In fact, this process was already observable in the declining economic influence of both superpowers within their respective alliances, but since the political capabilities of the Soviet Union have been reduced substantially as *glasnost* and *perestroika* have reached the eastern European allies, new problems are to be expected. Firstly, losing the Soviet Union as 'dear enemy' as the ultimate threat will change the distribution of relative power positions within western Europe, not least the economic ones.

Secondly, the quite suddenly lost *pax sovietica* and the declining *pax americana* in Europe have meanwhile given room for conflicts hitherto hidden. The German question, fierce nationalism and ethnic problems on the Balkans and in the Soviet Union, the status of Mongolia, to mention only a few major problems, will no longer be self-evidently ruled by the superpowers. Thirdly, rising uncertainties about the future will create an acute need for a new system of more or less reliable balances. Implicit and new formal alliances will arise because the only potentially effective system available to the European Community, will be far from capable of managing the enormous national problems which became manifest so suddenly.

Even if the military component of the national power structures were for the greater part to disappear, Europe will see an economically united German nation of some 80 million people with eventually very attractive eastern European markets for 10 to 20 years of economic expansion. This perspective will dominate the European political agenda for at least 10 years, and the acute question will be: should the threat of a disporportionate German economic influence over 'Europe', however defined, be counterbalanced somehow, or should the other European countries bandwagon?

Unless a real European *political union* is formed, which would in fact make an end of the existing national state system and so of Germany, we can expect that France, the UK, and the Soviet Union will cooperate significantly more to keep the Germans in check. The '2 + 4 formula' conference on the security status of the new Germany was one clear indication (after the two old Germanies had reached agreement on when and how to reunify, the four former *Siegermächte* (United States, Soviet Union, Great Britain, and France) gave official consent to something which could not be refused any more). The broader Conference on Security and Cooperation in Europe has discussed the same matter, and even the very suddenly reopened discussion about a European political union has significance only in the perspective of serious uneasiness with the peaceful German growth among the European countries. Although nobody seriously expects the new Germany to reclaim territory lost to Poland, it is illustrative that this non-issue was blown up to absurd proportions in 1990 (even to the extent that the Dutch minister of foreign affairs publicly endorsed the Polish claim to be admitted at the '2 + 4' conference—a classic balancing act of two relatively unimportant neighbours which can only be explained by fearful expectations). The UK, however, may prefer not to commit fully, in accordance with its traditional role of balancer in continental European politics. But all that, of course, will not stop the process of further shifting relations of power to the advantage of Germany. However peacefully events may develop now—and that is to be expected—it makes a lot of (subjective) difference whether a 'European Germany', or a 'germanized Europe' will grow out of it. Even when traditional military alliances in Europe seem to have had their day, the structure of the 'anarchical' international system is still such that 'balancing' policy continues to be perceived as the most effective method of keeping the process peaceful.

The pan-European diplomacy of the historical *European Concert* has been revived and will have to 'dance' again until the European political union is a fact. A 'Brussels Empire' would be the first unrivalled new political order in Europe since the decline of the ancient Roman Empire. The magnitude of such a change can be seen as an indication of the political difficulties to be expected. Unfortunately, international politics is one of the least suitable areas of collective human behaviour in which to see fine plans realized, as the 'European' answer to the Gulf crisis of 1990–1 proved once again. Political will, vision, and a bunch of treaties are definitely not enough. Political power concentration might do, but in historical perspective this has so far always led to *de facto* counterbalancing actions elsewhere. As long as this political process goes on, there will be no end to history. . . .

CONCLUDING COMMENTS

1. Alliances are an essential part of international politics, a domain where human collective cooperation and conflict still operate under relatively 'anarchic' conditions. Self-help of national states in acute or perceived security dilemmas leads mostly to balancing against the most threatening nation or coalition. Balancing, however, is far from an assurance that violent changes in any existing distribution of power capabilities will not occur.

2. Nuclear deterrence between the United States and the Soviet Union, itself being the ultimate expression of a balancing of threats, has contributed more than any other single factor to settling of superpower conflicts without large-scale use of violence or war. The ultimate alliance, then, was the informal one between the superpowers not to use nuclear weapons, except for display. The unbalanced decline of both superpowers, accompanied by the equally unbalanced rise of new economic superpowers, destabilizes the postwar international system and will provoke new conflicts which, less than before, can be expected to be settled without violence.

3. Economic conflicts may be less costly than military ones, but as long as they are perceived as influencing the power political position of states, we can expect that alliance formation will go on.

4. An extension of the study of political alliances and coalitions with analyses of animal cooperation and contest behaviour shows that similar balancing and bandwagoning mechanisms are effective there (see for an interesting exercise in balance-of-power theorizing by biologists Alexander 1979, pp. 222–8, 258 and de Waal 1982, pp. 176–7, 187, 212; comments by Falger 1987, 1990). These mechanisms as such, then, are not merely human. After all, consciousness and rational choice are not prerequisites for alliances and coalitions, but structures allowing actors to change their

relative power positions are. Using refined balance-of-power theories in cross-species behavioural research would thus have the advantage of conceptual clarity, a broader range of empirical applicability, and more parsimony. Refined balance-of-power theories, without being super-reductionistic or anthropocentric, could serve as one of the bridges over the gap between biological studies of behaviour and the human social sciences.

SUMMARY

1. International alliances between two or more states directed at other parties are among the oldest known forms of international politics. The history of alliance formation is correspondingly long and varied. The main objective of alliances is to contribute to group or national security in a non-hierarchical international system.
2. One of the main focuses of the study of international relations is a systematic analysis of historical instances of alliance formation. Most studies concern the period between 1814 and the 1970s in the European state system, because alliances flourish in international systems not dominated permanently by one hegemonic power.
3. Historical and quantitative research has shown that balance-of-power theorizing is the most profitable approach. If attention is also paid to threat perceptions, we are equipped sufficiently to draw lines from historical situations, dominated by alliance formation, to the future which will be characterized by the loss of traditional alliances.
4. The main conclusion of this chapter is that as long as the international political system has not reached quite a high degree of hierarchical integration, comparable to the current national states, it is to be expected that alliance formation will go on. Alliances themselves certainly did serve as elements of order in 'international anarchy', but even the alliance *par excellence*, the informal non-aggression agreement between the superpowers under the aegis of nuclear deterrence, cannot guarantee future non-violence in new areas of conflict.
5. The unbalanced decline of both superpowers, accompanied by the equally unbalanced rise of new economic superpowers, destabilizes the postwar international system and will provoke new internal and external conflicts which can less than before be expected to be settled without violence. Alliances, in this perspective, are best characterized as moments of cooperation in conflict.
6. Economic conflicts may be less costly than military ones, but as long as they are perceived as influencing the power political position of states, we can expect that alliance formation will go on.
7. There is no principal reason to restrict the application of the balancing mechanism, operating so clearly in international politics, to human affairs.

The analysis of coalition and alliance formation in non-human primate groups makes most sense if the generality of this mechanism is accentuated. Besides, it may stimulate biologists and social scientists to become interested in each other's work.

REFERENCES

Alexander, R. D. (1979). *Darwinism and human affairs*. Pitman, London.

Altfeld, M. F. (1984). The decision to ally: a theory and test. *Western Political Quarterly*, **37**, 523–44.

Baylis, J. (1985). The Anglo-American relationship and alliance theory. *International Relations*, **8**, 368–79.

Burgess, Ph.M. and Robinson, J. A. (1969). Alliances and the theory of collective action: a simulation of coalitional processes. In *International politics and foreign policy* (ed. J. N. Rosenau), pp. 640–53. Free Press, New York.

Craig, G. A. and George, A. L. (1983). *Force and statecraft. Diplomatic problems of our time*. Oxford University Press, New York.

DeVree, J. K. (1982). *Foundations of social and political processes. The dynamics of human behaviour, politics and society*. Prime Press, Bilthoven.

Dougherty, J. E. and Pfaltzgraff, R. L. (1981). *Contending theories of international relations*. Harper and Row, New York.

Duncan, G. T. and Siverson, R. M. (1982). Flexibility of alliance partner choice in a multipolar system: models and tests. *International Studies Quarterly*, **26**, 511–38.

Falger, V. S. E. (1987). From xenophobia to xenobiosis? Biological aspects of the foundations of international relations. In *The sociobiology of ethnocentrism* (eds V. Reynolds, V. S. E. Falger, and I. Vine), pp. 235–50. Croom Helm, London.

Falger, V. S. E. (1990). The Arnhem zoo chimpanzee project. A political scientist's evaluation. *Social Science Information*, **29**, 33–54.

Friedman, J. R., Bladen, C. and Rosen, S. (eds) (1970). *Alliance in international politics*. Allyn and Bacon, Boston.

Gilpin, R. (1981). *War and change in world politics.* Cambridge University Press, Cambridge.

Gilpin, R. (1987). *The political economy of international relations*. Princeton University Press, Princeton.

Helck, W. (1971). *Die Beziehungen Ägyptens zu Voirderasien im 3. und 2. Jahrtausend v. Chr.* Otto Harrassowitz, Wiesbaden.

Holsti, K. J. (1967). *International politics. A framework for analysis*. Prentice-Hall, Englewood Cliffs, NJ.

Holsti, O. R. *et al.* (1973). *Unity and disintegration in international alliances: Comparative studies*. Wiley, New York.

Kegley, Ch. W. and Raymond, G. A. (1982). Alliance norms and war: a new piece in an old puzzle. *International Studies Quarterly*, **26**, 572–95.

Kennedy, P. (1987). *The rise and fall of the great powers. Economic change and military conflict from 1500 to 2000*. Random House, New York.

Langdon, S. and Gardiner, A. H. (1920). The treaty of alliance between Hattusili, king of the Hittites, and the pharaon Ramesses II of Egypt. *Journal of Egyptian Archaeology*, **6**, 179–205.

Levy, J. S. (1981). Alliance formation and war behavior: an analysis of the great powers, 1495–975. *Journal of Conflict Resolution*, **25**, 581–613.

Liska, G. (1962). *Nations in alliance. The limits of interdependence*. Johns Hopkins Press, Baltimore.

McDonald, H. B. and Rosecrance, R. (1985). Alliance and structural balance in the international system: a reinterpretation. *Journal of Conflict Resolution*, **29**, 57–82.

McGowan, P. J. and Rood, R. M. (1975). Alliance behavior in balance of power systems: applying a Poisson model to nineteenth-century Europe. *American Political Science Review*, **69**, 859–70.

Midlarski, M. I. (1983). Absence of memory in the nineteenth-century alliance system: perspectives from queuing theory and bivariate probability distributions. *American Journal of Political Science*, **27**, 762–84.

Morgenthau, H. J. (1985). *Politics among nations. The struggle for power and peace*. Knopf, New York.

Most, B. A. and Siverson, R. M. (1987). Substituting arms and alliances, 1870–1914: an exploration in comparative foreign policy. In *New directions in the study of foreign policy* (eds C. F. Hermann, C. W. Kegley, and J. N. Rosenau), pp. 131–57. Allen and Unwin, Boston.

NATO Information Service (1989). *The North Atlantic Treaty Organisation*, 11th edn. NATO, Brussels.

Numelin, R. (1950). *The beginnings of diplomacy. A sociological study of intertribal and international relations*. Oxford University Press, London.

Olson, M. and Zeckhauser, R. (1966). An economic theory of alliances. *Review of Economics and Statistics*, **48**, 266–79.

Palmer, R. R. and Colton, J. (1978). *A history of the modern world*. Knopf, New York.

Riker, W. H. (1962). *The theory of political coalitions*. Yale University Press, New Haven.

Rosenau, J. N. (ed.) (1969). *International politics and foreign policy*. Free Press, New York.

Singer, J. D. and Small, M. (1968). Alliance aggregation and the onset of war, 1815–1945. In *Quantitative international politics: insights and evidence* (ed. J. D. Singer), pp. 247–86. Free Press, New York.

Singer, J. D. and Small, M. (1969). National alliance commitments and war involvement, 1818–1945. In *International politics and foreign policy* (ed. J. N. Rosenau), pp. 513–42. Free Press, New York.

SIPRI (1989). *World armaments and disarmament 1989*. Oxford University Press, Oxford.

Siverson, R. M. and Sullivan, M. (1984). Alliances and war: a new examination of an old problem. *Conflict Management and Peace Science*, **8**, 1–16.

Siverson, R. M. and Tennefoss, M. R. (1984). Power, alliance and the escalation of international conflict, 1815–1965. *American Political Science Review*, **78**, 1057–69.

de Swaan, A. (1973). *Coalition theories and cabinet formation*. Elsevier, New York.

Thies, W. J. (1987). Alliances and collective goods: a reappraisal. *Journal of Conflict Resolution*, **31**, 298–332.

Tsebelis, G. (1987). When do allies become rivals? *Comparative Politics*, **20**, 233–40.

Viotti, P. R. and Kauppi, M. V. (1987). *International relations theory. Realism, pluralism, globalism*. Macmillan, New York.

de Waal, F. B. M. (1982). *Chimpanzee politics. Power and sex among apes*. Jonathan Cape, London.

Walt, S. M. (1985). Alliance formation and the balance of world power. *International Security*, **9** (4), 3–43.

Walt, S. M. (1987). *The origins of alliances*. Cornell University Press, Ithaca.

Waltz, K. N. (1979). *Theory of international politics*. Addison-Wesley, Reading, MA.

Wayman, F. W. (1984). Bipolarity and war: the role of capability concentration and alliance patterns among major powers, 1816–1965. *Journal of Peace Research*, **21**, 61–78.

Weede, E. (1983). Extended deterrence by superpower alliance. *Journal of Conflict Resolution*, **27**, 231–53.

Wight, M. (1978). *Power politics*. Penguin, Harmondsworth.

PART III

Coalitions and alliances: evolutionary considerations

Coalitions and alliances: evolutionary considerations: Introduction

'During fights, individuals compare the strength of their own group with that of their opponent. When they perceive that group members locally outnumber opponents, they behave aggressively, recruiting to the battle and cooperating to chase and kill intruders.' This is not a description of chimpanzee or human intercommunity strife, but Adams' (1990) summary of boundary disputes between colonies of a territorial *ant* (altered to remove the words ant, worker, and colony). Cooperation in intergroup conflict is very widespread in the animal kingdom. It occurs not only in ants (Adams 1990; Rissing and Pollock 1988; Strassman 1989), but also birds and mammals (Harcourt, this volume pp. 445–71). In all the studies, the crucial variables are the nature of the resources and the intensity of competition for them (Davies and Houston 1984).

In 1980 Wrangham argued that the advantages of cooperation in competition over valuable food sources that were both defensible and divisible was the main environmental determinant of primate social structure. Wrangham wanted to explain why primates lived in groups, and more importantly, why the groups had the structure that they did: why species differed in whether females left the group in which they were born or not, and why species in which female remained in their natal group are characterized by highly differentiated networks of social relationships within groups, based on grooming, aggression, and dominance interactions. Following Bradbury and Vehrencamp (1977) and Emlen and Oring (1977), Wrangham argued that the distribution of food influenced the distribution and behaviour of female primates, which in turn influence the distribution and behaviour of males (see also Clutton-Brock 1989). Competition for good, large, defensible food sources led to cooperative groups of female relatives; and competition between males for females, plus competitive advantages to females from the presence of males, determined the number of males per group. Concerning the effect of competition on relationships between females, Wrangham largely assumed that the effects of inter- and intragroup competition were similar.

van Hooff and van Schaik (Chapter 13) here build on and considerably extend Wrangham's model. Most importantly, they strongly distinguish between the effects of inter- and intragroup competition. Incorporating the insights of Vehrencamp's (1983) and Emlen's (1982*a*,*b*) models, which were originally produced to explain communal living of birds, they argue that while intragroup competition leads to steep, despotic dominance hierarchies of competing matrilines, intergroup conflict produces a more egalitarian outcome of competition within groups. The reason is that when groups compete, victory goes to the larger one, and therefore dominant individuals cannot afford to suppress subordinates so much that they might choose the option of leaving. Since the models apparently apply to any competing colonies also (Rissing *et al.* 1989; Strassman 1989), they clearly have extremely broad application. Not surprisingly perhaps, the situation is more complex when the effect of competition between groups of humans is considered. Nevertheless, there are some intriguing parallels relating to the effects of intergroup competition on intragroup cooperation (see Rabbie this volume pp. 175–205).

van Hooff and van Schaik also address the question of cooperation between the sexes, and cooperation between males. The story here is less clear, they argue, with the rarity of cooperation between primate males being particularly puzzling, as is the case for mammals in general (Clutton-Brock 1989). Current explanations concern intermale competition for females, and the males' role in assisting females' breeding success.

van Hooff and van Schaik considered females and males separately, because the resources of most importance to each sex's reproductive success differ. By and large, female primates (and other mammals) can best increase their reproductive output by gaining access to food, whereas males can do so by inseminating females (Trivers 1972). The contrast correlates with sometimes major differences in life histories, particularly in the case of polygynous species (e.g. Clutton-Brock *et al.* 1982). Males often grow faster, require more resources for growth, mature later, and die earlier than do females. Not surprisingly, therefore, the nature of the competition that the sexes face differs, both in immaturity and in adulthood.

With cooperation influenced by competition and so strongly influencing competitive ability, as several chapters in this volume demonstrate, it might be expected that the sexes would differ in the ways in which competitive status is acquired and maintained. **Lee and Johnson** (Chapter 14) show that they do. They emphasize, however, that coalitions and alliances are but one influence, albeit an important one, on the acquisition and maintenance of competitive ability by male and female cercopithecine primates. It is also important, they point out, to remember that immatures are not merely imperfect adults, that immaturity is not merely a stage beween birth and adulthood, but that immatures face specific problems that require solutions that sometimes differ from those appropriate in adulthood. Certainly some aspects of immatures' social behaviour can be explained on the basis of each sex's likely competitive environment as an adult. However, in at least some of the polygynous cercopithecine species that Lee and Johnson review, the sexes' different life history trajectories probably begin near birth (Altmann *et al.* 1988). So also do their competitive environments, therefore, and thus the effects of coalitions and alliances on them.

Polyadic interactions, those in which several animals take part, provide at least the potential for advantageous use of information processing ability, or intelligence: the individual that can take account, not simply of the difference in competitive ability between itself and its opponent (Maynard Smith and Parker 1976), but also that of their respective potential allies, the allies' relative proximity to the conflict, and their likelihood of actually giving support in the case of a fight, would surely be at an advantage over the individual which could cope only with its opponent's relative competitive ability? Complexity of social interactions and social relationships might therefore be expected to correlate with ability to process information. Primates have a larger information processing organ, the brain, than do most non-primates, and it has long been assumed that primates have a more complex social structure than do non-primates, though with little hard evidence. The other taxon that has a particularly large brain for its body size is the toothed whale family (Jerison 1983).

A long-term study of a wild population of one species of toothed whales that has a particularly large brain, the bottle-nosed dolphin, confirms, **Connor, Smolker and Richards** suggest in Chapter 15, the association between brain size and social complexity, as evident in the use of alliances by males. Aquatic animals are difficult for terrestrial animals to observe, and so quantitative data are few. Nevertheless, it seems clear that within a social group the bottle-nosed dolphin is rare in using two levels of alliance formation: individual males form extremely close alliances with one or two other males (identified by the most part by highly coordinated activities including herding of females); and these 'first-order' alliances then form 'second-order' alliances with one another when competing for females, and show preferences for particular first-order pairs or trios when doing so. Alliances of alliances in the face of threat from other social groups is not uncommon among animals, but Connor *et al.* argue that the fact they occur within a social group, albeit a large one, distinguishes the bottle-nosed dolphin.

Harcourt (Chapter 16) addresses the question of the potential correlation between information processing capacity and social complexity by reviewing the literature on the use of coalitions and alliances as a competitive strategy by primates compared to non-primates. He concludes that primates are indeed reported to show more complex cooperative behaviour in competition than are non-primates, dolphins perhaps excepted. However, he identifies a different sort of complexity than that emphasized by Connor *et al.* Harcourt suggests that the nature of the distinction between the taxa apparent in the literature can be explained by primates, but not non-primates, attempting to form long-term reciprocal alliances with powerful partners. It is reciprocation that Trivers (1971) originally argued required a high level of information-processing ability, because the delay in repayment of benefits opened the door to cheating. Partners and their behaviour had to be monitored closely, therefore.

Partly as a consequence, the conditions under which reciprocation can evolve are restricted (Trivers 1971). Theoretical development in ideas about the evolution of reciprocal altruism was enormously boosted by Axelrod and Hamilton's (1981) analysis of a computer tournament in which a range of strategies appropriate to the prisoners' dilemma paradigm were pitted one against the other. Part of the reason for the influence of Axelrod and Hamilton's work was surely the extreme simplicity of the winning, evolutionary stable strategy, tit-for-tat (TFT): be nice (cooperate) on the first interaction with a group member, and thereafter do what the partner does. However, the simplicity of the model's execution was accompanied by such simplicity in its assumptions that the model was often inapplicable outside of experimental situations, even with more recent modification (Axelrod and Dion 1988). Certainly, the complexities of cooperative behaviour that primatologists were seeing in social groups seemed beyond it. **Boyd** (Chapter 17) explores how relaxation of the assumptions to make them more applicable to the real world influences the models, their outcomes, and the predictions that they make about the nature of cooperative behaviour that can be expected. He shows that reciprocal cooperation can still evolve when the payoffs differ between partners and between interactions. Importantly, the pattern of exchange of cooperative acts that is evolutionarily stable might, depending on the conditions, be more complex than the simple TFT model would suggest.

REFERENCES

Adams, E. S. (1990). Boundary disputes in the territorial ant *Azteca trigona*: effects of asymmetries in colony size. *Animal Behaviour*, **39**, 321–8.

Altmann, J., Hausfater, G., and Altmann, S. A. (1988). Determinants of reproductive success in savannah baboons (*Papio cynocephalus*). In *Reproductive success* (ed. T. H. Clutton-Brock), pp. 403–18. University of Chicago Press, Chicago.

Axelrod, R. and Dion, D. (1988). The further evolution of cooperation. *Science*, **242**, 1385–90.

Axelrod, R. and Hamilton, W. D. (1981). The evolution of cooperation. *Science*, **211**, 1390–6.

Bradbury, J. W. and Vehrencamp, S. L. (1977). Social organization and foraging in emballonurid bats, III: mating systems. *Behavioral Ecology and Sociobiology*, **2**, 1–17.

Clutton-Brock, T. H. (1989). Mammalian mating systems. *Proceedings of the Royal Society of London B*, **236**, 339–72.

Clutton-Brock, T. H., Guinness, F. E., and Albon, S. D. (1982). *Red deer. Behavior and ecology of two sexes*. Edinburgh University Press, Edinburgh.

Davies, N. B. and Houston, A. I. (1984). Territory economics. In *Behavioural ecology*, 2nd edn (eds J. R. Krebs and N. B. Davies), pp. 148–69. Blackwell, Oxford.

Emlen, S. T. (1982*a*). The evolution of helping. I. An ecological constraints model. *American Naturalist*, **119**, 29–39.

Emlen, S. T. (1982*b*). The evolution of helping. II. The role of behavioral conflict. *American Naturalist*, **119**, 40–53.

Emlen, S. T. and Oring, L. W. (1977). Ecology, sexual selection, and the evolution of mating systems. *Science*, **197**, 215–23.

Jerison, H. J. (1983). The evolution of the mammalian brain as an information processing system. In *Advances in the study of mammalian behavior* (eds J. F. Eisenberg and D. G. Kleiman), pp. 113–46. The American Society of Mammalogists, Pittsburgh.

Maynard Smith, J. and Parker, G. A. (1976). The logic of asymmetric contests. *Animal Behviour*, **24**, 159–75.

Rissing, S. W. and Pollock, G. B. (1988). Pleometrosis and polygyny in ants. In *Inter-individual behavioral variability in social insects* (ed. R. L. Jeanne), pp. 179–222. Westview Press, Boulder.

Rissing, S. W., Pollock, G. B., Higgins, M. R., Hagen, R. H., and Smith, D. R. (1989). Foraging specialization without relatedness or dominance among co-founding ant queens. *Nature*, **338**, 420–2.

Strassman, J. E. (1989). Altruism and relatedness at colony foundation in social insects. *Trends in Ecology and Evolution*, **4**, 371–4.

Trivers, R. L. (1971). The evolution of reciprocal altruism. *Quarterly Review of Biology*, **46**, 35–57.

Trivers, R. L. (1972). Parental investment and sexual selection. In *Sexual selection and the descent of man* (ed. B. Campbell), pp. 136–79. Heinemann, London.

Vehrencamp, S. L. (1983). A model for the evolution of despotic versus egalitarian societies. *Animal Behaviour*, **31**, 667–82.

Wrangham, R. W. (1980). An ecological model of female-bonded primate groups. *Behaviour*, **75**, 262–300.

13

Cooperation in competition: the ecology of primate bonds

JAN A. R. A. M. VAN HOOFF AND CAREL P. VAN SCHAIK

INTRODUCTION

Primates are social animals. That is, even when solitary, they are organized into distinct societies in which animals know each other individually (e.g. Bearder 1987), and often know members of neighbouring societies as well (Dunbar 1989). Their patterns of interaction within dyads (or polyads), i.e. their social relationships (Hinde 1976), are determined by the nature (age, sex) of society members and by their previous associations and interactions (genealogy, ontogeny, and history). A nearly universal aspect of a relationship is dominance. When two animals develop a dominance relationship, escalated aggression between them is quite uncommon. Indeed, the efficiency of having such relationships may well provide an important selective pressure towards the observed stability in the membership of animal societies.

Living in a stable society of individually known animals promotes the reciprocal exchange of services (Trivers 1971). Since we are concerned here with explaining the nature of social relationships, our focus is on relationship-dependent cooperation, rather than cooperation that arises automatically from gregariousness or from anonymous exchange, which do not require bonding. In principle, the alarm calls, food calls, and contact calls of primates could exemplify anonymous exchange; but even here the nature of the audience may affect the probability that an animal produces these calls (Seyfarth 1987), making it an aspect of the caller's relationship with other group members.

Cooperation among animals may serve a variety of functions, such as building dams or nests, communal vigilance, and defence against predators or cooperative resource exploitation (e.g. communal hunting). It may also be directed at achieving social goals, such as when animals aid each other in fights or support a wounded conspecific (dolphins at sea). Primates mainly cooperate by supporting each other in agonistic conflicts (Harcourt 1989). Whenever aggression is an effective means to increase access to limiting resources, alliances may serve to boost the competitive power of the partners. Because relatives are the most reliable partners in the long run (Wrangham 1980, 1982), the need for alliance partners may affect the tendency to remain in the natal society or at least to migrate together with relatives.

Primate species vary widely in the strength of the bonds within and between the sexes, and our aim is to contribute to the analysis of the ultimate causes of this variation. Agonistic support is usually restricted to specific dyads of animals, and is sometimes reciprocated, although not necessarily at equal rates (males: e.g. de Waal this volume, pp. 233–57; females: e.g. Dunbar 1980; de Waal this volume pp. 233–57). These cooperative relationships or 'bonds' entail more, however: reciprocation may be in different behavioural currencies, such as the exchange between grooming and agonistic support (Seyfarth and Cheney 1984; Cheney *et al.* 1986; Harcourt 1989). The services exchanged may even extend into the non-social realm, such as protection against predators and the procurement of food, in particular when the commodities or abilities the animals can offer each other are very different, as in cooperation between males and females. Although bonds and their strength can be defined by the exchange of services and its intensity, they are generally recognized more easily by close proximity. For instance, in a captive group of long-tailed macaques, the mean proximity score of a dyad was correlated positively with the relative frequency with which they supported each other in agonistic conflict, and negatively with the frequency with which they had agonistic conflicts among themselves (Aureli *et al.* 1989).

We focus mainly on ecological explanations of primate bonding. We contend that the influence of ecology on the distribution of alliances across primates can be studied most profitably by considering the competitive regimes to which individuals and groups are subjected, at least for alliances between animals of the same sex. This implies, first, identifying key differences between males and females in the resources they compete for, and next, examining the nature of competition; we dissect competition into several distinct types, each expected to give rise to a different type of bond. These competitive regimes determine whether aggression, and thus coalition formation, can be effective as a means to increase fitness.

COMPETITION, AGGRESSION, AND ALLIANCES

Competition ensues whenever there is not enough of some resource to satisfy the needs of all striving to obtain it. For our present purpose, we dissect competition into two components: scramble and contest (Nicholson 1954). In the case of scramble competition (or indirect or exploitation competition) individuals cannot effectively exclude others from the resources and thus tend to share the resources equally; hence they tend to be affected equally by resource shortage. This kind of competition will lead to an ideal free distribution of animals over resource patches (Fretwell and Lucas 1970).

Whenever the distribution of resources allows it, however, animals may benefit from securing access to limiting resources by excluding others from using them, either by setting up territories or by displacing them at resources. This competition is known as *direct*, *contest*, or *interference competition* (Lazarus 1982). Resources are not shared equally. In extreme cases, the dominant animal obtains all the resources it needs, the second-dominant takes all it needs from what it left, etc. (contest in the sense of Łomnicki 1988).

In most natural situations, the competition experienced by an individual will be a mixture of both scramble and contest competition. We refer to the effects of density or group size as the *scramble component* and to those of dominance as the *contest component*. The important thing is that these two components can be estimated separately (Janson and van Schaik 1988), and that their relative importance profoundly affects social relationships.

Scramble competition is an inevitable consequence of group life. There is little an animal can do to alleviate it, apart from leaving the group or ousting some group members (which may be detrimental for other reasons to both parties involved). Where restricting the others' access to resources is impossible or too costly, dominance will not enhance reproductive success, so dominance relationships should remain faint and unstable, and coalitions should be rare. By contrast, when there is an appreciable contest component to competition, we expect to see clear-cut dominance hierarchies and alliances, aimed at improving the rank of relatives (Wrangham 1980).

The potential for contest competition should depend on the distribution and quality of the limiting resources (Davies and Houston 1984). If these are food, we expect animals to scramble when the food occurs in very small, uniformly distributed patches, not meriting defence, or in very large patches, not permitting defence. Hence, provided that there is no territorial defence, we should expect grazers and animals feeding on, cryptic insects, for example, to scramble for food, and frugivores to contest for it whereas browsers may often be intermediate.

Most primates live in groups. So far we have restricted our analysis to competition within groups, but it can obviously take place between groups as well. In theory, we can distinguish four types of competition: within-group scramble, within-group contest, between-group scramble, and between-group contest (see van Schaik and van Noordwijk 1988; van Schaik 1989). All these types of competition can be present simultaneously. The conditions giving rise to strong between-group contest are basically the same as for contest within groups: resources in short supply, and defensibility of access to them. This could either lead to territoriality between groups, or to between-group dominance in overlap areas where the resources concern especially valuable sources of water or food, or refuges.

Between-group scramble is basically an unavoidable consequence of population density, but since it cannot have direct effects on social relationships, it will not be considered further here.

When contest competition *between* groups is important, we expect group members to form a large alliance in order to improve their competitive ability. However, to keep subordinates motivated to participate in contests with neighbouring groups the dominants must relax within-group contest (Vehrencamp 1983). Otherwise subordinates might either refrain from taking any risks in between-group contests (since their gains would be small anyway) or even defect to another group. Hence, dominants should be tolerant and allow the subordinates to take a share almost as large as their own. So this is a second reason, in addition to intragroup scramble competition why societies may be egalitarian and why ranks are often not enforced in minor conflicts.

Consequently, whereas Wrangham (1980, 1982) held that between-group contest correlated with despotic hierarchies, we contend that within-group contest and between-group contest competition are forces with opposing effects on the social relationships among the group members. Between-group contest can become expressed only when the contest within groups is relaxed, or when that between groups is particularly severe.

To summarize, we expect two clear contexts in which alliances can improve animals' competitive abilities, namely significant within-group contest for limiting resources, which leads to small alliances; and between-group contest, which leads to large, group-level alliances. Before we assess whether the contexts in which primate cooperative bonds occur are in agreement with these predictions, we have to examine how these predictions differ between males and females.

SOCIAL BONDS AND THE SEXES

Due to the strong asymmetry in parental investment between males and females in mammals (Bateman 1948; Trivers 1972), the factors limiting the fitness of males and females tend to be different. As a consequence, males and females compete for different resources, and hence may have different tendencies to cooperate. In most cases females cannot increase their number of surviving offspring through obtaining extra matings. Rather they should maximize the intake of high-quality food, safety against predation and disease, both for themselves and their offspring, and safety against hostile conspecifics (especially infanticidal males; van Schaik and Dunbar 1990; Watts 1989; Wrangham 1979). The cooperative and competitive interactions that females engage in should reflect these female interests. Female bonds should therefore be used for defending access to food or safety, or for protection against hostile males.

Male strategies should be aimed at maximizing lifetime mating success, rather than at maximizing food intake (unless monogamy is imposed, in which case males should invest in their mate's reproductive condition). Females often benefit from forming an alliance with others to improve access to food, because they can usually share the food resources with their alliance partners without compromising food intake. Male alliances might be useful to improve mating access to receptive females. However, sharing copulations with an alliance partner does compromise reproductive success, and we should therefore expect male alliances to be less common.

Male–female bonds should serve the different interests of the two sexes identified above. Females are mainly expected to form alliances with males when only males can serve them with benefits in resource exploitation or protection, in return for mating privileges. This makes bonds between the sexes fundamentally different from those within one sex.

FEMALE RELATIONSHIPS

We expect that the degree of within-group contest and of between-group contest are the major forces to shape the bonds between females. In Figure 13.1 we summarize these influences (see van Schaik 1989 for details). The boundaries between the various predicted types of bonds are made fuzzy on purpose, because we expect gradual rather than discrete variations in female bonds, and because sometimes the outcome may vary when the potential for between-group contest is high. We examine this scheme by

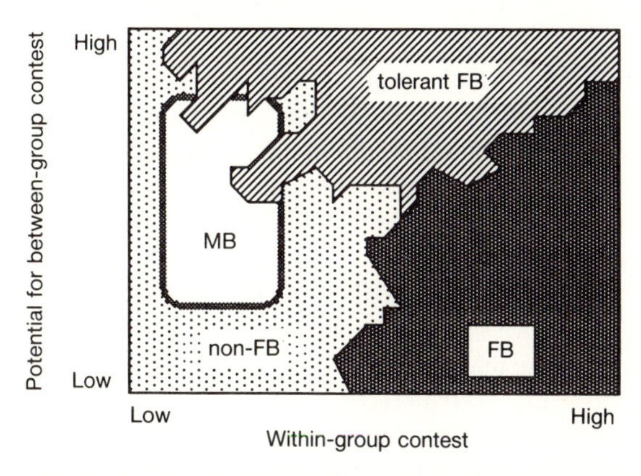

Fig. 13.1. The effect of ecological potential for between-group contest (vertical) and within-group contest on the nature of female relationships. If WGC is high and BGC is low, despotic female bonded (FB) groups will evolve. If WGC is moderate and BGC is high, FB groups will be tolerant. If WGC is low, non-FB groups will evolve, and the potential for male-bondedness exists, especially if the potential for BGC is high.

dividing the continuum according to the importance of competition within groups relative to that between groups.

Strong within-group competition

Evidence is growing that females living in compact groups do so generally (but perhaps not invariably) to reduce the risk of predation (e.g. Dunbar 1988; van Hooff 1988; van Schaik 1989; Terborgh and Janson 1986; but see Cheney and Seyfarth 1987; Robinson 1988). Living in a compact group necessarily increases within-group competition for food. The variation in within-group contest should then determine the variation in female bonds. In groups with strong within-group contest (WGC) potential, agonistic conflicts are common, and females form strong bonds, usually with relatives, characterized by grooming, proximity, and support in agonistic conflicts with third animals. As a result, female relatives tend to occupy adjacent ranks in the female hierarchy, which is usually stable and linear. Since females rely on their kin for their rank, they rarely emigrate from their natal groups (see Pusey and Packer 1987; cf. Sherman 1977; Shields 1987), unless they lack female relatives or can emigrate in the presence of alliance partners (when big groups split up, matrilines tend to remain intact: e.g. Chepko-Sade and Sade 1979). Examples include most macaques (*Macaca spp.*), most baboons (*Papio spp.*), vervets (*Cercopithecus aethiops*), and perhaps others (see Wrangham 1980; van Schaik 1989).

By contrast, in groups with weak WGC potential (scramble competition may still be strong) agonistic conflicts are rare, females do not form strong bonds and rarely if ever support others in conflicts. As a result, female hierarchies are not based on kin relations but on age, and are generally weakly developed (egalitarian) and unstable. Since reproductive success is determined mainly by group size (van Schaik 1983) and barely affected by the presence of female relatives, females are expected to move between groups when conditions in their natal groups become suboptimal. This kind of social system is found in many colobines: langurs (*Presbytis spp.*), colobus (*Colobus spp.*), proboscis monkeys (*Nasalis larvalis*) (Hrdy 1977; Struhsaker and Leland 1987; Bennett and Sebastian 1988). The lack of coalitions among colobine females is not caused by a lack of cognitive or other abilities for collaboration, because they do sometimes form coalitions against infanticidal males or other groups (Hrdy 1977). Colobines on the whole are more folivorous than the WGC species mentioned above. The social systems of non-colobine folivores tend to resemble those of colobines, the mantled howlers (*Alouatta palliata*) and mountain gorillas (*Gorilla gorilla*) being examples (Jones 1981; Harcourt and Stewart 1989). Among gorillas, adults rarely form coalitions, although support of immatures is common (Harcourt and Stewart 1989). Emigration by (single) females is also more common among folivores (Moore 1984; Pusey and Packer 1987).

The dichotomy between female-bonded (FB) and non-FB groups, first pointed out by Wrangham (1980), coincides with the distinction between a strong and a weak *contest* component to *within*-group competition. Following Wrangham (1980), we consider the low rates of aggression and agonistic support and the weakly expressed dominance hierarchy as diagnostic of non-FB groups. Note, however, that Wrangham (1980) originally saw variation in the strength of contests *between* groups as the ultimate cause of this dichotomy.

This ecological explanation for the difference between FB and non-FB groups still has several weaknesses. First, it is largely *post hoc*. The differences in social organization between colobines and the omnivorous and frugivorous cercopithecines have been well known since the 1960s (e.g. Poirier 1974). Secondly, it tends to follow taxonomic boundaries, allowing for a possible role of phylogenetic inertia. Finally, it largely coincides with a dietary distinction, allowing for alternative explanations based on the consequences of a diet in which leaves rather than fruit are a major energy source. Hence, a more direct test of the effect of food distribution on female bonding is called for. To date, only one such test is available, a comparison of two populations of squirrel monkeys (*Saimiri*).

Mitchell *et al.* (1991) compared a population of Peruvian squirrel monkeys, *S. sciureus*, in an Amazonian floodplain forest, with a population of *S. oerstedi* in a Costa Rican lowland forest. The two populations had similar group sizes, home range sizes, population density, activity budgets, and diets. There was a striking social contrast, however. Food-related aggression was about 70 times more common in the Peruvian groups. These were clearly FB, showing stable kin-based dominance hierarchies and no female emigration, whereas the Costa Rican groups were non-FB; females hardly attacked each other and had no coalitions, and routine emigration by single females was the rule. There was one corresponding ecological contrast between the two populations (Fig. 13.2). While the Peruvian monkeys fed in fruit trees of all possible sizes, up to big canopy strangling figs of over 30 m crown diameter, the Costa Rican monkeys used to feed in secondary forest with small trees (< 5 m crown diameter), which in addition tended to have very few fruits in each. The number of monkeys feeding simultaneously in a tree crown also differed dramatically: at most 3–5 in the Costa Rican groups versus an average of 17.5 in the Peruvian groups. In the latter aggressive interactions in these fruit trees frequently turned into coalitions. Confrontations between groups were not observed. These ecological contrasts confirm the part of the model illustrated in the lower half of Fig. 13.1. The similarity in diet and in phylogenetic background of these populations makes it less likely that confounding factors were involved.

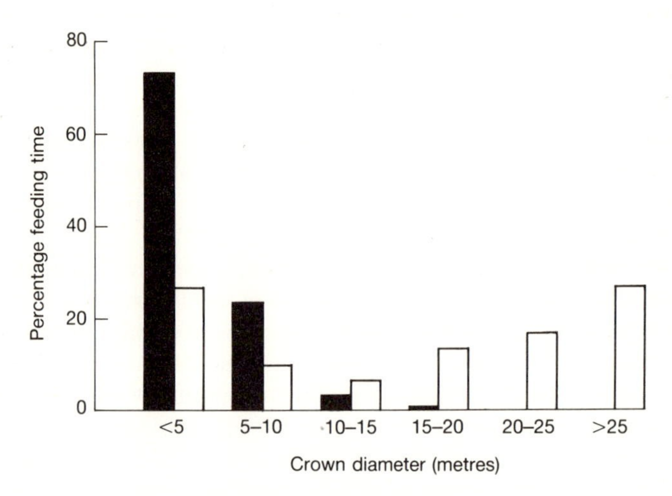

Fig. 13.2. Differences between two populations of squirrel monkeys in time spent feeding in trees of different sizes. *S. oerstedi* (solid bars) were non-FB; *S. sciureus* open bars) were FB (from Mitchell *et al*. 1991).

Relaxed within-group competition

Many primates do not live in cohesive groups. Sometimes, groups are widely dispersed during foraging, with animals being separated by tens of metres most of the time, as in mountain baboons (Byrne *et al.* 1989), patas monkeys (*Erythrocebus patas*, Chism *et al.* 1984)) or some populations of tufted capuchin monkeys (*Cebus apella*, Janson personal communication). In other societies, the members are always split into parties, of which the size is continuously adjusted to the size and density of the available food patches. Examples include bonobos (*Pan paniscus*, White 1988), forest chimpanzees (*Pan troglodytes*, Isabirye-Basuta 1988), spider monkeys (*Ateles paniscus*, van Roosmalen 1980), very large groups of long-tailed macaques (*Macaca fascicularis*, van Schaik and van Noordwijk 1988). In yet other societies, females are basically solitary, and come together only when food is temporarily superabundant: orang utans (*Pongo pygmaeus*, Sugardjito *et al.* 1987), woodland chimpanzees (Wrangham 1977).

In all these cases, the large interindividual distances or the continuous adjustment of the number of consumers to the available food supply ensure that within-group competition is relaxed. In principle, competition between groups or societies may predominate in these conditions. However, realized between-groups contest may be small, either because food is not distributed in large, defensible, high-quality patches or because the females are too dispersed to be able to form effective alliances. In these instances, we should merely expect female bonds to be weaker in these societies than in the compact FB groups discussed above and female emigratation to be more common (cf Fig. 13.1). By and large, this is correct: female emigration and undifferentiated bonds are characteristic for several fission–fusion species (Anderson 1987), and also for fissioning populations within species which elsewhere live in cohesive groups (as among a population of chacma baboons, *Papio cynocephalus*, Anderson 1987). However, since reduced predation could have made female emigration less risky, and thus more likely (Anderson 1987), a more precise test is called for, comparing the nature of food competition and female bonds in different types of groups.

At the far end of the spectrum, females are strictly solitary and antagonistic to other females. This results in monogamous systems, as in most species of gibbon (*Hylobates* spp.) and in the indri (*Indri indri*). At present, the most plausible explanation for this strict antagonism is that feeding in parties is always costly and that individual female ranges are defensible, unlike the situation in the semi-solitary fission–fusion species (van Schaik and van Hooff 1983). However, more theoretical and empirical work needs to be done on this problem.

Strong between-group contest

Between-group contest might become a shaping force in female bonds in two sets of conditions: (1) within-group contest is relaxed because groups are less compact, but rapid rallying is possible (for instance, because home ranges are small), and the potential for between-group contest for food (or water) is high; (2) special ecological circumstances reduce the competition within groups (scramble) and increase that between groups regardless of group cohesion. Female social relationships are expected to be of a more egalitarian or tolerant FB type (Fig. 13.1). The first situation may be found on oceanic islands where the absence of large predators reduces the advantage of group cohesion and that of competing species reduces home range size (van Schaik 1989). The second may be found among species specializing on exploiting large fruit trees that are scattered over huge areas. Contest between groups for these patches is conceivable, and has been observed (e.g. *Cebus albifrons*, Janson 1986). Contest within groups is relaxed in these large patches, and group size may be limited by the scramble competition experienced during the foraging trips between or around these major food sources.

There are a number of species with relaxed female hierarchies (de Waal 1989), but in most cases we do not know enough about their ecology to assess whether or not their food intake is highly dependent on contests between groups. It would be especially instructive to compare the various macaque species, among which we find usual FB species with despotic female hierarchies (such as rhesus monkeys, *Macaca mulatta*) and tolerant FB species with more egalitarian hierarchies, low rates of damaging aggression, and high rates of reconciliation (*M. tonkeana*, Thierry 1985; *M. arctoides*, de Waal 1989; both captive).

MALE COOPERATIVE BONDS

In addition to food, males compete above all over access to fertilizable females. Whether the competition over females is mainly by contest or by scramble is determined by the distribution in space and time of estrous females (Emlen and Oring 1977). Access to females can be monopolized, first, if these live in compact groups (female defence polygyny), and, second, if ranges in which females live or if resources to which females are attracted can be defended effectively (resource defence polygyny). When there is contest competition over females, intrasexual selection will favour contest vigour in males (Darwin 1871). Indeed, whereas monogamous species of primates are generally monomorphic in both body size and canine length, these characteristics are exaggerated in the males of many polygnous species (e.g. Clutton-Brock *et al.* 1977), particularly those with one-male harems (Alexander *et al.* 1979).

Female defence polygyny is very common among primates and so is sexual dimorphism associated with increased male fighting abilities. Males may attempt to obtain exclusive access to fertile females by ousting other males from the group, leading to one-male groups. If this is impossible they may either directly fight over estrous females or ensure access indirectly by fighting for a position of dominance. In the latter two cases male alliances might develop.

The potential for male contest competition over access to females may be low when females live in large and/or loose groups. It may also be reduced by female behaviour. First, all females may come into estrus simultaneously. Secondly, females may actively choose a diversity of mating partners for reasons to be discussed below (cf. van Noordwijk 1985; Hrdy and Whitten 1987). Thirdly, males may show restrained contest within groups, resulting in more equitable copulation rates, because contest would interfere with cooperation in other contexts (see below). Alternatively, in situations in which monopolization of several females is impossible, males may instead form a monogamous bond with a single female.

In these conditions, leading to what has been called 'scramble competition polygyny' (Wells 1977; Alcock 1980), natural selection may favour characteristics maximizing the chance of fertilization in sperm competition. Species of primates with a multimale or dispersed mating system often have frequent insemination and copious production of sperm (Bercovitch 1989; Møller 1988), sperm plugs, and anatomical devices to remove these, such as special penis structures (Dixson 1987). The relative testis sizes of species living in multi-male societies and assumed to experience scramble competition over females are larger than those of monogamous and single-male species (Harcourt *et al.* 1981; Møller 1988).

Where males ally in defence of females, the partners must contend with how to divide the matings. Food can often be shared in equal or unequal portions, depending on the 'bargaining power' of the contending collaborators (Noë this volume pp. 285–321), and so, in principle, can copulations or females. However, the fertilization of a single female cannot be shared, except possibly in species producing twins (e.g. in polyandrous callitrichids, perhaps). So, as a rule, the conditions for the formation of male coalitions are certainly less favourable than for female coalitions. Sacrificing a fertilization has far more serious fitness consequences than sacrificing a morsel of food.

If male relatives are available, this may well facilitate the formation of coalitions. Not only do kin make more reliable alliance partners (Wrangham 1982), but such coalitions may also raise the participants' inclusive fitness. However, sufficiently powerful male relatives may not always be around at the right time. For one thing, the females in most primate species are philopatric, for reasons outlined above. Males will, therefore, be driven to emigrate from their natal group for ultimate reasons of inbreeding avoidance (Greenwood 1980; Pusey and Packer 1987; for an opposite view see Clutton-Brock 1989*a*). The adult males in a multi-male group will not be closely related unless they have migrated together or have migrated to a group in which related males were already present (Melnick and Pearl 1987). There is evidence that migrating male macaques do exactly this (Boelkins and Wilson 1972; Sugiyama 1976, Meikle and Vessey 1981; Colvin 1983*b*; van Noordwijk and van Schaik 1985) and in vervets (Cheney and Seyfarth 1983), although, surprisingly, this has so far not been reported for baboons where the most spectacular coalitions can be observed.

Of course, male lions (*Panthera leo*) are in the same predicament as males of multi-male FB primate groups: females are philopatric. Adult males form alliances, and the size of the male alliance largely determines their success in obtaining and retaining female prides (Bygott *et al.* 1979). Consequently, they mostly migrate as a group of related peers (Packer and Pusey 1982). This illustrates the rule of female priority in philopatry patterns among mammals: even though, in this species, cooperation with kin seems to be equally important to males and females, females are the philopatric sex.

So we suggest that male philopatry becomes an option when female contest competition about resources is relaxed, and also, therefore, the pressure on females to be philopatric is low. This will facilitate the formation of male alliances, possibly consisting of related males ('fraternities'), defending access to a territory and to the females attracted to it. Hence, as with females, male alliances can occur in two main contexts: between- and within-group contests for matings.

Within-group contest competition for mates

Alliances among males in within-group contests can, of course, occur only in multi-male groups, but even here these are rarer than female alliances. They have been reported especially for baboons, e.g. savanna baboons (Packer 1977; Noë 1987, this volume pp. 285–321), where the males are mostly immigrants supposed to be non-related, and for chimpanzees (e.g. de Waal 1982; Nishida 1983) where due to male philopatry (e.g. Goodall 1986; Pusey 1979) the males are supposed to be comparatively closely related. Even so male alliances in captive chimpanzees tend to be more opportunistic and less stable over time than those between females (de Waal 1978).

Packer presented empirical data on coalitions in savanna baboons which, he believed, supported the reciprocal altruism model (Packer 1977, 1986). Later investigations have shown that the pay-off asymmetries of such alliances may be great (Bercovitch 1988; Noë 1986, 1988, 1990, this volume pp. 285–321). Noë argued that these pay-off asymmetries can be explained on the basis of differences in bargaining power of the males concerned (cf. Boyd this volume pp. 473–89).

In the above case the deal is about the division of more or less immediate-return benefits. Asymmetries in the pay-off distribution of male coalitions can also be explained as an 'investment' with delayed-return prospects. Male langurs, colobus, or guenon monkeys associating in bachelor groups, which coalesce in attempts to drive off the harem male of a bisexual single-male group, only to find that their strongest comrade subsequently claims exclusive possession of the harem (Sugiyama 1967; Oates 1977; Struhsaker 1977) may nevertheless have moved their own chances upwards by one step (Bercovitch 1988; Noë 1988).

Similarly, an increased chance to inherit a harem in due course may be the advantage to young adult males in associating with a harem in a 'follower' role (e.g. *Papio hamadryas*, Kummer 1968; Abegglen 1984; Sigg *et al.* 1982; *Theropithecus gelada*, Dunbar and Dunbar 1975; Dunbar 1984; gorillas, Harcourt 1979; Harcourt and Stewart 1981). By adopting an obviously non-provocative style and by rendering services in the protection and defence of the harem, a follower may be tolerated by the harem leader. For the latter it remains a trade-off involving risks of a take-over by the follower or a fission of the harem. These risks increase with the size of the harem (Dunbar 1984). By investing also in the goodwill of the females a follower may further pave the way for his acceptance.

Also older males that had to yield their harem may assume the 'follower' role in their former harem and be tolerated by the new leader. Their presence might add safety which is in the interest both of the new leader and of the offspring of the old leader (Stammbach 1987). Such bonds may even benefit the inclusive fitness interests of both males if these are related (Stolba 1979; Sigg *et al.* 1982).

As we noted earlier, we may expect male bonds especially when males migrate with or towards kin, or when female migratory habits allow the males to be philopatric and to stay together continuously with male relatives. The indirect fitness component to the pay-offs of tolerance and cooperation is expected to decrease the impulse for contest between males. The best-studied examples to date of species in which more permanent bonding between males has been observed, in agreement with the expectation, are the hamadryas baboon (e.g. Kummer 1968), the chimpanzee (e.g. Goodall 1986) and the red colobus, *Colobus badius* (Struhsaker and Leland 1987).

In hamadryas baboons the social organization is multi-level. Troops consist of several bands, each consisting of clans comprising a couple of harems (Stammbach 1987). Whereas females may move to different bands, males tend to stay associated in their natal clan. Clan members cooperate in harem defence and display great tolerance towards each other. Most remarkable is the 'respect' for each others' pair bond, which was experimentally demonstrated by Kummer (1968).

Between-group competition for mates

The strongest example of male bonding found to date is the chimpanzee. This species lives in societies of the fission–fusion type, adapted to a food supply that is usually low and extremely dispersed, but occasionally locally abundant (Wrangham 1986; Isabirye-Basuta 1988). In East Africa the females are comparatively solitary (Goodall 1986). They are also likely to emigrate from their local community (Pusey 1979; Goodall 1983), as may be expected where within-group competition is weak. By contrast the males, supposed to be relatives, have strong bonds. This is evident from patterns of association, grooming, and meat sharing (Goodall 1965, 1986; Teleki 1973; Ghiglieri 1984; Nishida 1979; but see Sugiyama 1988). At the same time there is a remarkable restraint in their within-group sexual competition. Although competition is certainly not absent, and males may evade such competition by the formation of 'cryptic consorts', promiscuity in a group context is often observed (Goodall 1965; Tutin 1979). Sexual tolerance seems to form an element in the bargaining about coalitions which is characteristic for the within-group power 'politics' of male chimpanzees (de Waal 1982). Their social relationships can be described as a tense affiliation (Nishida and Hiraiwa-Hasegawa 1987), with high levels of tension-reducing and reconciliatory behaviours (de Waal 1986).

This remarkable restraint in within-group competition in combination with strong between-group contest is in accordance with the conditions, theoretically deduced above, for the existence of cooperation in between-group competition. Compared with the other great apes the emphasis on *individual* contest potential (i.e. sexual dimorphism) is greatly reduced, whereas relative testis size is great (Short 1979), indicating the shift from contest for mates to sperm competition (see the woolly spider monkey, *Brachyteles*, Milton 1985, for a striking analogy).

The chimpanzee is indeed renowned for the deliberate and fierce character of its raids, in which the males of one community traverse the territory of a neighbouring community and engage in lethal fights, especially with other males, but also with older females (Goodall *et al.* 1979; Nishida *et al* 1985). As a result of such 'wars' a community may increase its territory (Goodall 1986), and, more importantly, lure females from elsewhere into its territory. The chimpanzee may thus be regarded as one of the few mammalian examples of resource defence polygyny (RDP), more precisely 'cooperative RDP'. Their excited food-calling, which can be interpreted as an advertisement of the quality and quantity of the resources they hold (Wrangham 1977), is in accordance with this.

Deliberate coordination of male actions is also obvious in another form of cooperation, namely hunting. This has now been reported in a number of populations. It occurs on a rather opportunistic basis in East African communities (Teleki 1973; McGrew 1979), and on a fairly regular basis with high degrees of purposiveness and success in at least one West African population (Boesch and Boesch 1989).

Perhaps the development of coordinated male between-group aggression has paved the way for the development of such coordinated hunting. Structured between-group aggression and the relaxation of within-group male competition is certainly not a necessary condition for the development of cooperative hunting. This behaviour has also been reported in some populations of baboons (Harding 1975; Strum 1981). However, there it never took on the deliberate, coordinated character seen in chimpanzees nor was it associated with the 'social feasting', the sharing of the spoils (Teleki 1973). The speculation is relevant to the different explanations that have been put forward for the development of intergroup conflict and primitive war in early man (see below).

It is remarkable that female emigration does not appear necessarily to lead to the formation of male within-group bonds. The chimpanzee seems rather exceptional in this respect. There are various species (e.g. many folivores) where the relaxation of female contest and of the associated female philopatry would allow males to become philopatric and to form bonds, possibly kin-based. Nevertheless, apart from hamadryas baboons (see above), red colobus (Struhsaker 1975), spider monkeys (van Roosmalen 1980; Fedigan and Baxter 1984; McFarland 1986; McFarland Symington 1988; Chapman *et al.* 1989), and woolly spider monkeys (Milton 1985) there is no explicit mention in the literature of strong and permanent male bonding. Previous suggestions that the bonobo has such bonds have recently been disputed (White 1988). All-male parties have never been observed, and previous accounts of these may be due to observational bias. Females affiliate more strongly with each other than do males (Kano 1982; Kuroda 1979;

White 1989). So their organization resembles the classical FB pattern more than had been thought, albeit in a less cohesive, fission–fusion form.

MALE–FEMALE RELATIONSHIPS

Most bonds entail more than just agonistic alliances. In several species grooming may be exchanged for agonistic support (for a review see Seyfarth 1983). Indeed, the services exchanged within a bond may extend into the non-social realm, such as protection against predators and the procurement of food. This is most obvious in cooperation between males and females, where the commodities the animals can offer are often very different. Hence a wider perspective is necessary if we study the ecology of male–female bonds.

Benefits to males and females

Males can benefit from having bonds with all or some of a group's females (or in the case of monogamy and polyandry, with a single female) mainly because it increases their access to them or because it raises the survival of their offspring (baboons, Smuts 1985, 1987*a*). Agonistic benefits to males are less obvious. Occasionally, females serve as allies in power struggles within a multimale group (e.g. rhesus monkeys, Chapais 1983*c*) or provide support against intruders who attempt to oust the leading male from the group (e.g. geladas, Dunbar 1984; long-tailed macaques, van Noordwijk and van Schaik, unpublished). The advantages to the female are more varied, but involve any conceivable combination of the following factors.

1. Males may act as allies and protectors. Although males usually refrain from playing a role in female–female conflicts, male take-overs of the top-dominant position sometimes lead to a change in the female dominance hierarchy (Jones *et al.* 1982; Marsden 1968). This explains why such events should provoke competition among the females for proximity and association with the new dominant, and why females should return to estrus sooner after a male take-over (Dunbar 1984; Stein 1984). Some experiments also suggest that females may use males as allies in their power struggles. For instance, Stynes *et al.* (1968) kept mixed-species groups of bonnet (*Macaca radiata*) and pigtailed macaque (*M. nemestrina*) females. When an adult pigtailed macaque was added to such groups, the pigtailed females became dominant over the bonnet females; but when a bonnet male was added, the opposite happened. When strangers were introduced into existing groups of (unrelated) rhesus monkeys females actually preferred males as coalition partners (Wade 1976). Female geladas without living kin may attempt to ally with the group's male (Dunbar 1984). All this suggests that, while males may not always be the preferred coalition partners where their presence in the

group if brief, females do succeed in forming alliances with males, which are sometimes effective in increasing their dominance status. Note, however, that males are not expected to offer advantages as allies in situations where females do not form clear-cut hierarchies.

2. Males may act as sentinels and defenders. Females may benefit from a close association with one or more males because of their greater vigilance (e.g. de Ruiter 1986; van Schaik and van Noordwijk 1989) and ability to mob and attack predators (e.g. Altmann and Altmann 1970).

3. In some smaller species the need for males may to act as caretakers of the young has selected for monogamy and even polyandry (Goldizen 1987).

4. Males may guarantee their females the supply of food by defending a territory or valuable food patches. However, it is unclear to what extent this defence is a by-product of male–male contest for mates rather than a direct service to the females (see Dunbar 1988; van Schaik and Dunbar 1990).

Clearly, the strength of male–female bonds among primates varies widely (see Smuts 1985). Although it has not been possible to measure bond strength in a way that allows direct comparison between species, it is our impression that the strongest bonds are found in monogamous species, in some single-male non-FB groups, and in some multi-male groups. Weak male–female bonds seem to characterize at least some single-male FB groups (Cords 1987; Rowell 1988). In some multi-male groups only a single male has a strong relationship with the females and their infants, and he turns out to be the one who was reproductively most successful in previous mating periods, or at least was dominant at the time (tufted capuchins, Janson 1984; longtail macaques, van Noordwijk and van Schaik 1988). In baboons, reproductively successful males have longer tenure in the group (Altmann *et al*. 1988), suggesting a similar pattern.

This variability should be related to the potential for male services to benefit females or their infants. Two general principles can be used to account for the preliminary generalizations presented above. First, where females have no reliable allies, they will rely more on the male for a range of benefits. Secondly, where there are more adult males in a group, female choice can in principle force males into providing services in exchange for mating opportunities (Noë *et al*. 1991).

Females without allies: strong male–female bonds

In diurnal non-human primates, continuous year-round association of both sexes is very common. Since many other gregarious mammals do not exhibit this feature (e.g. Clutton-Brock 1989*b*), the benefits of male–female bonds must be especially important and widespread among primates. One kind of male protection may be particularly important: protection against harassment by other males (Wrangham 1979). This is especially needed when females have no female allies or are solitary. Watts (1989) demonstrated the role of males in preventing a particularly damaging form of harassment, infanticide by strange males, in gorillas, a non-FB species. This may explain why male–female bonds tend to be weaker in single-male FB species, where female relatives are at hand for cooperative defence against infanticidal males, than in non-FB species.

This may also be the selective advantage of bonded monogamy in the larger primates (van Schaik and Dunbar 1990). Monogamy in these primates is generally attributed to the inability of males to defend access to more than one female, but range use data militate against that explanation; neither does the solitary female seem to derive ecological benefits from being bonded with the male. Indeed, van Schaik and Dunbar (1990) argued that primate females are more vulnerable than most other mammals to infanticide, and speculated that the need for protection against harassment by strange males, especially with regard to infanticide, has selected for male–female bonding in solitary females, leading to monogamy.

Male–female friendships

In some multi-male groups, persistent long-lasting friendships have been observed between females and certain males who were not the top dominants, e.g. in Japanese (Takahata 1982) and rhesus macaques (Chapais 1983, 1986; Hill 1990), and in baboons (Saayman 1970; Seyfarth 1976; Smuts 1985). These friendships involve frequent proximity and grooming, often on the initiative of the female; the male tends to protect the female and her infant against aggression by other group members and use them as agonistic buffers. In baboons the male friends mated the female more often than expected for their dominance rank (Smuts 1985; see also Packer 1979 and Rasmussen 1983). This may not hold for the Japanese and rhesus monkeys (Takahata 1982; Chapais 1983*a*; Hill 1990), where the long-term benefits to the male are still obscure.

When the protective male cannot be the father of the offspring already present, these friendships may be regarded as investments in reproductive prospects. In this respect these relationships differ in important ways from other friendly male–female relationships in multi-male groups where the role of friend and protector is played by the often post-prime male likely to have sired most of the group's current infants, though only after the offspring were born (e.g. van Noordwijk and van Schaik 1988).

Such investment friendships between females and non-alpha males seem to occur in a few species only. They should be expected where males render services that are diluted when shared among recipients, unlike, for instance, the benefits of vigilance. They should be expected especially in larger multi-male groups where male monopolization of all females becomes less feasible and the opportunity for female choice increases.

'Useless males' and female dominance

There are a number of primate species in which adult females can dominate all or most adult males in dyadic interactions: some lemur species (Richard 1987; Pereira *et al.* 1990), squirrel monkeys, talapoin monkeys (*Miopithecus talapoin*), patas monkeys, and others in which female coalitions can and do occasionally dominate males (Smuts 1987*b*; Mitchell 1990).

Male–female bonding in primates is expected where females derive a net benefit from associating with a male. Suppose, though, that males are not able to provide significant services, whereas they do compete for food and may pose a threat to infants. This is possible where sexual dimorphism is small, so males are not superior as allies to female relatives. In such conditions, females are expected not to develop bonds with males, and in fact should attempt to keep them away from their groups. This, then, would seem to set up the selective pressures for female dominance over males and short male tenure. This set of conditions and associated phenomena was found in Peruvian squirrel monkeys (Mitchell 1990). A similar explanation may not hold for other cases, such as the sexually dimorphic patas monkey.

DISCUSSION

Underlying principles

Reviewing present knowledge of the factors determining social evolution, Wrangham and Rubenstein (1986) recently concluded that the nature of the ecological and social pressures which operate on social relationships have been identified, but that no crisp generalizations have as yet emerged as to how these forces shape social organization. They also listed various unanswered questions, mainly concerning the nature of dominance systems and sex differences in philopatry. We hope to have shown that the answer to these questions can be brought nearer by considering the competitive regimes experienced by a species in the wild, and we propose the following set of principles which emerged from this endeavour.

1. Alliances are expected only where there is a potential for contest competition, be it within or between groups, be it for food, for safety, or for matings (cf. Wrangham 1980).

2. Alliances can be so important for the participants' fitness that they may determine philopatry (cf. Wrangham 1980).

3. Female philopatry takes priority over male philopatry, because females can refuse to mate with male relatives.

4. An alliance in which copulations are shared must bring in a greater benefit than one in which food is shared. Because of this and of point 3, male alliances are less common than female alliances.

5. The potential value to females of services provided by males in exchange for matings provides the framework to explain the relationships between the sexes.

The development of theory and of empirical knowledge has not proceeded equally in different realms of interest. In some realms rigorous deductive approaches and derived empirical predictions could be made (e.g. with respect to female–female relationships). In others the emphasis is rather on induction and *a posteriori* theorizing (e.g. with respect to female–male relationships).

Much of what we present here is, therefore, still conjectural. The detailed descriptive studies still cover too limited an array of species to allow for the framing of questions, and tests against clearly formulated alternatives have been few. Studies of intraspecific variability may offer possibilities for such tests. There is clearly much more flexibilty of social structure within genera and species than was previously thought, e.g. macaques (Thierry 1985; de Waal 1989), baboons (e.g. Byrne *et al.* 1989; cf. Stammbach 1987) and chimpanzees (Sugiyama 1988). In an isolated forest habitat at Bossou, West Africa, where chimpanzees live in a comparatively compact society, female bonds are, indeed, stronger than male bonds, and there is probably male emigration (Sugiyama 1981).

This indicates that besides phylogenetic adaptations of species-specific mechanisms of social regulation (e.g. Nagel 1973), there is a possibly wide intraspecific phenotypical variation of social adjustment behaviours. This emphasizes the need for broad descriptive studies. These can provide the comparative bases essential for understanding the extent to which competitive regime and demography shape the social relationships in primate societies, and for interpreting the causal role of the ecological context in the phylogenetic and the 'cultural' evolution of social structure. For this, field research in *undisturbed* areas is essential. Since these areas are disappearing rapidly, new studies are urgently needed, lest we lose for ever the opportunity to fully clarify these issues and, eventually, to reconstruct the social evolution that led to humans.

In addition, quantitative ethological studies are needed to evaluate the basic assumption that the frequency of food-related aggression is related to the relative rate of coalition formation, the unidirectionality of aggression within a dominance relationship, and the linearity of the hierarchy (cf. de Waal and Luttrell 1989). Since all of these variables may depend on the size of the group, its compactness, and its composition (e.g. presence of close relatives, presence of subadult and adult males), mere correlative quantitative comparisons will be insufficient and experimental approaches, such as manipulations of the food distribution, are called for.

Testing alternative explanations

As Wrangham and Rubenstein (1986) noted, there is an emerging consensus among primate socio-ecologists on the approach to the ecology of social relationships, and hence of social systems as a whole (but see Rowell 1988). This approach was pioneered in Wrangham's (1979, 1980) seminal papers, and directed by Hinde's (1976, 1979) and Kummer's (1978) emphasis on social relationships as an outcome of the interaction of individual strategies.

So far, only models for the evolution of variation in female relationships in cohesive groups have been put to the test. Wrangham (1980) considered FB-groups as grand coalitions in which females form long-term alliances in order to enhance their competitive power against other such groups. The gist of our argument, however, is that strong female bondedness is a response to within-group contest competition between females clumping together in response to other selective pressures (e.g. harassment or predation). The distinction of the relative importance of between-group contest and within-group contest is crucial. The comparison of the two populations of squirrel monkeys by Mitchell *et al.* (1991) reviewed above was the first independent test, which came out in favour of our alternative model. Moreover, while there was evidence in favour of predation avoidance as a benefit to group living, there was little evidence for the importance of between-group competition. Even such evidence as has been presented for other species (e.g. Robinson 1988) may lend itself to the alternative interpretation (van Hooff 1988).

Less well understood are the conditions under which male-bondedness evolves, in particular male-bonded polygyny, as found in some chimpanzee populations. If we accept that male philopatry becomes possible only when female philopatry is relaxed (but note that an alternative causal direction has been proposed by Pusey and Packer 1987, and by Clutton-Brock 1989*a*), the question remains when this will lead to cooperative male bonding, and, especially, to what extent male kinship is a decisive factor. The latter is assumed to play an important role in chimpanzees (e.g. Nishida and Hiraiwa-Hasegawa 1987; Gouzoules and Gouzoules 1987). However, we may wonder how often the birth intervals between male maternal siblings will allow them to be valuable allies when these are needed most. Male siblings of the same age-cohorts, i.e. paternal siblings, might be more suitable alliance partners. However, more than half of the conceptions in chimpanzees are produced under promiscuous group conditions (Goodall 1986), and juvenile males occasionally migrate with their mother (Wrangham 1986). This raises the question to what extent same-age males share fathers and whether their relatedness is at all higher than that of, say, same-age males emigrating together (cf. te Boekhorst 1991), especially if the latter stem from groups, such as macaque groups in which strong contest competition between males assures that the emigrants share the same dominant male as a father. Clearly, in order to establish whether kinship plays the crucial role attributed to it, we need further relevant demographic and genealogical information. A recent theoretical study by Hogeweg and te Boekhorst (in press) shows that kinship is not a necessary factor for the development of male associations in chimpanzees. In a 'mirror-world' computer experiment imaginary 'CHIMPS' were equipped with simple rules specifying their foraging and, for males, their mate searching strategies. Given the parameters of the distribution of 'FRUIT TREES' and 'FEMALES', similar to those of the real chimpanzee world, the 'CHIMPS' displayed association patterns strikingly similar to those found for real chimpanzees. So no specific kin-selected affiliative mechanisms had to be assumed to generate the association pattern characteristic for male and female chimpanzees.

Female emigration does not necessary lead to male within-group bonds. An alternative development might be possible and is suggested by the observations of McFarland Symington (1987) on spider monkeys. This fission–fusion species is supposed to be subjected at times to strong resource competition of the scramble type. Males, and especially, females may leave their natal groups. Both sexes will therefore, meet unrelated males, within as well as outside their natal group. This implies that the males fathering successive sons of a female need not be closely related on average. In a spider monkey community brothers and same-age males may, therefore, be less related than in a chimpanzee community.

Anyway, even more so than in chimpanzees, a male's closest 'known' relative is his mother. Assume that a son stays in the same range as his mother, and that a dominant mother can effectively protect him against the strong aggression directed especially towards males (Chapman *et al.* 1989). This set of conditions would explain why the sex ratio is strongly female-biased, with only dominant females producing sons (McFarland Symington 1987), and why a son might choose to rely on her rather than on male 'relatives'. The females could be effective supporters, because this species is rather sexually monomorphic. This is remarkable, because in the spider monkeys studied by McFarland Symington there was male competition with only the dominant males being likely to achieve an appreciable reproductive success. More data on coalition formation are obviously needed.

In this connection it is interesting to speculate whether the alleged close bond between mother and son in bonobos (Kano 1982; Kuroda 1979) reflects a similar choice. According to Furuichi (1989), adult males remaining in matrifocal units greatly benefit from their mother's assistance in dominance interactions.

Lack of pertinent data, in particular of ways to measure the net benefits, also plagues the topic of male–female relationships. There is probably appreciable variation in these, both among multi-male and single-male populations, but it still needs to be charted more, before the conditions leading to the development of 'friendships' can be evaluated. The approach stressing male services also led to speculation about the taxonomic distribution of female dominance over males in polygynous primate species. For this phenomenon, alternative hypotheses have been proposed (see Richard 1987), but no generally accepted explanation has yet emerged.

Extrapolations to humans

The principles proposed here raise the question to what extent social characteristics of our own species can also be understood in terms of these. The human species displays considerable, at least superficial variation in forms of social organization, such as patterns of exogamy, hierarchy formation, competition, and cooperation. With respect to variation in cooperative attitude, for instance, a great many cross-cultural experimental comparisons have shown that, by and large, children from rural and traditional societies behave more cooperatively in experimental game situations than those from urbanized and westernized societies (review by Bethlehem 1982). However, more specific cultural variations remain unexplained, and Bethlehem endorses the view of Mead (1937), that there is no simple relationship between, on the one hand, sociological and ecological features of small-scale societies, and, on the other hand, characteristics of competitiveness, individualism, and cooperativeness. It would be a challenge to apply the mentioned distinctions of competitive regimes to these variations. Our own species may not be the worst testing ground for some of the principles proposed, by means of comparative or experimental methods. These may also shed light on issues concerning the earliest social evolution of man (see also Ghiglieri 1990, Wrangham 1987). Thus van Hooff (1990) has argued, on the basis of the principles presented here, that it is most parsimonious to assume that also in the evolutionary line leading to man, male-bondedness, stimulated in a context of cooperative RDP, has given rise to the development of warfare and of hunting. This is in contrast to the view, held by various authors (e.g. Washburn and Lancaster 1968; Hamburg 1973; Melotti 1979), that cooperating hunting was at the roots of the development, via cannibalism and intraspecific hunting, of warfare.

SUMMARY

1. Competition for resources, such as food, shelter, safety and mates, can be of two kinds. When differential access can be achieved, leading to differential net benefits, then contest competition will occur. When limiting resources cannot efficiently and effectively be monopolized, competition is by scramble.
2. In contest competition for limiting resources, primates cooperate to improve their competitive ability; when competition is by scramble, alliances are not advantageous and hence no bonds are expected.

3. Female bonds among primates are more common than male bonds, because the resources limiting the reproductive success of females (food, safety) can be shared more easily than those of males (matings).
4. Small alliances are expected when contests for resources are mainly within groups, whereas large group-level alliances are expected when contests are mainly between groups.
5. Where alliances are vitally important, animals that rely on them are expected to be philopatric, because this allows them to stay with relatives and reap the inclusive fitness benefits of cooperation.
6. The distribution of female bonds across diurnal primates agrees with this prediction: when competition is mainly by scramble, females do not form strong bonds; when it is by within-group contest, bonds are strong, and when it is by between-group contest egalitarian bonds are common.
7. Among males, small alliances are found in multi-male groups or as leader-follower pairs in otherwise single-male groups. Their occurrence is probably determined mainly by demography.
8. Large alliances are found where males collectively defend the access to a scattered group of females. This is possible where female bonds and female philopatry are weak. In this system, sexual dimorphism in size is reduced, and males have an egalitarian system of mate competition, characterized by bargaining for sexual access and for support and by sperm competition.
9. Strong male–female bonds, a widespread trait in species without female alliances, may provide the female protection against infanticidal males, rather than the more common agonistic support in dominance struggles.
10. Special male–female relationships, or friendships, are found only in large multi-male groups, probably as a result of demographic conditions.
11. Where males cannot offer adequate protection and females derive no net benefits from having relationships with them, female dominance should evolve.

ACKNOWLEDGEMENTS

CvS was a research fellow with the Royal Netherlands Academy of the Arts and Sciences (KNAW) during the conception of this chapter. We thank Carol Mitchell for generously sharing ideas and data, D. Rubenstein for sharing ideas on philopatry, and the editors and Lynne Isbell for constructive comments.

REFERENCES

Abegglen, J. J. (1984). *On socialization in hamadryas baboons*. Associated University Presses, Cranbury, NJ.

Alcock, J. (1980). Natural selection and the mating systems of solitary bees. *American Scientist*, **68**, 146–153.

Alexander, R. D., Hoogland, J. L., Howard, R. D., Noonan, K. M., and Sherman, P. W. (1979). Sexual dimorphisms and breeding systems in pinnipeds, ungulates, primates and humans. In *Evolutionary biology and human social behaviour: an anthropological perspective* (eds N. A. Chagnon and W. A. Irons), pp. 402–603. Wadsworth, Belmont, CA.

Altmann, S. A. and Altmann, J. (1970). *Baboon ecology*. University of Chicago Press, Chicago.

Altmann, J., Hausfater, G., and Altmann, J. (1988). Determinants of reproductive success in savannah baboons. In *Reproductive success* (ed. T. H. Clutton-Brock), pp. 413–18. University of Chicago Press, Chicago.

Anderson, C. M. (1987). Female transfer in baboons. *American Journal of Physical Anthropology*, **73**, 241–50.

Aureli, F., van Schaik, C. P., and Hooff, J. A. R. A. M. van (1989). Functional aspects of reconciliation among captive long-tailed macaques (*Macaca fascicularis*). *American Journal of Primatology*, **19**, 39–51.

Baldwin, J. D. and Baldwin, J. I. (1981). The squirrel monkeys, genus *Saimiri*. In *Ecology and behavior of Neotropical primates*, vol. 1 (eds A. F. Coimbra-Filho and R. A. Mittermeier), pp. 277–330. Academia Brasileira de Ciencias, Rio de Janeiro.

Bateman, A. J. (1948). Intra-sexual selection in *Drosophila*. *Heredity*, **2**, 349–68.

Bearder, S. K. (1987). Lorises, bushbabies and tarsiers: diverse societies in solitary foragers. In *Primate societies* (eds B. B. Smuts, D L. Cheney, R. M. Seyfarth, R. W. Wrangham, and T. T. Struhsaker), pp. 11–24. University of Chicago Press, Chicago.

Bennett, E. L. and Sebastian, A. C. (1988). Social organization and ecology of proboscis monkeys (*Nasalis larvatus*) in mixed coastal forests in Sarawak. *International Journal of Primatology*, **9**, 233–55.

Bercovitch, F. B. (1988). Coalitions, cooperation and reproductive tactics among adult male baboons. *Animal Behaviour*, **36**, 1198–209.

Bercovitch, F. B. (1989). Body size, sperm competition, and determinants of reproductive success in male savanna baboons. *Evolution*, **43**, 1507–21.

Bethlehem, D. W. (1982). Anthropological and cross-cultural perspectives. In *Cooperation and competition in humans and animals* (ed. A. M. Colgan), pp. 250–68. Van Nostrand Reinhold, London.

Boekhorst, I. J. A. te (1991). The social structure of three great ape species: a study based on field work and individual oriented models. *Ph.D. Thesis*. Utrecht University.

Boelkins, R. C. and Wilson, A. P. (1972). Intergroup social dynamics of the Cayo Santiago rhesus (*Macaca mulatta*) with special reference to changes in group membership by males. *Primates*, **13**, 125–40.

Boesch, C. and Boesch, H. (1989). Hunting behaviour of wild chimpanzees in the Tai National Park. *American Journal of Physical Anthropology*, **78**, 547–73.

Bygott, J. D., Bertram, B. C. R., and Hanby, J. P. (1979). Male lions in large coalitions gain reproductive advantages. *Nature*, **282**, 838–41.

Byrne, R. W., Whiten, A., and Henzi, S. P. (1989). Social relationships of mountain baboons: leadership and affiliation in a non-female-bonded monkey. *American Journal of Primatology*, **18**, 191–207.

Chapais, B. (1983*a*). Reproductive activity in relation to male dominance and the likelihood of ovulation in rhesus monkeys. *Behavioral Ecology and Sociobiology*, **12**, 215–28.

Chapais, B. (1983*b*). Dominance, relatedness and the structure of female relationships in rhesus monkeys. In *Primate social relationships: an integrated approach* (ed. R. Hinde), pp. 208–19. Blackwell, Oxford.

Chapais, B. (1983*c*). Structure of the birth season relationship among adult male and female rhesus monkeys. In *Primate social relationships: an integrated approach* (ed. R. A. Hinde), pp. 200–8. Blackwell, Oxford.

Chapman, C. A., Fedigan, L. M., Fedigan, L., and Chapman, L. J. (1989). Post-weaning resource competition and sex ratio in spider monkeys. *Oikos*, **54**, 315–19.

Cheney, D. L. and Seyfarth, R. M. (1983). Non-random dispersal in free-ranging vervet monkeys: Social and genetic consequences. *American Naturalist*, **122**, 392–412.

Cheney, D. L. and Seyfarth, R. M. (1987). The influence of intergroup competition on the survival and reproduction of female vervet monkeys. *Behavioral Ecology and Sociobiology*, **21**, 375–86.

Cheney, D. L., Seyfarth, R. M., and Smuts, B. (1986). Social relationships and social cognition in nonhuman primates. *Science*, **234**, 1361–6.

Chepko-Sade, B. D. and Sade, D. S. (1979). Patterns of group splitting within matrilineal kinship groups: A study of social group structure in *Macaca mulatta* (Cercopithecidae: Primates). *Behavioral Ecology and Sociobiology*, **5**, 67–87.

Chism, J., Rowell, T. E., and Olson, D. K. (1984). Life history patterns of female patas monkeys. In *Female primates: studies by women primatologists* (ed. M. F. Small), pp. 47–57. Alan R. Liss, New York.

Clutton-Brock, T. H. (1989*a*). Female transfer and inbreeding avoidance in social mammals. *Nature*, **337**, 70–2.

Clutton-Brock, T. H. (1989*b*). Mammalian mating systems. *Proceedings of the Royal Society of London* B, **236**, 339–72.

Clutton-Brock, T. H., Harvey, P. H., and Rudder, B. (1977). Sexual dimorphism, socionomic sex ratio, and body weight in primates. *Nature*, **269**, 797–800.

Colvin, J. (1983*a*). Description of sibling and peer relationships among immature male rhesus monkeys. In *Primate social relationships: an integrated approach* (ed. R. A. Hinde), pp. 20–7. Blackwell, Oxford.

Colvin, J. (1983*b*). Influences of the social situation on male emigration. In *Primate social relationships. An integrated approach* (ed. R. A. Hinde), pp. 160–71. Blackwell, Oxford.

Colvin, J. (1983*c*). Rank influences rhesus male peer relationships. In *Primate social relationships. An integrated approach* (ed. R. A. Hinde), pp. 57–64. Blackwell, Oxford.

Cords, M. (1987). Forest guenons and patas monkeys: Male–male competition in one-male groups. In *Primate Societies* (eds B. B. Smuts, D. L. Cheney, R. M. Seyfarth, R. W. Wrangham, and T. T. Struksaker), pp. 98–111. University of Chicago Press, Chicago.

Darwin, C. (1871). *The descent of man and selection in relation to sex*. John Murray, London.

Davies, N. B. and Houston, A. I. (1984). Territory economics. In *Behavioural ecology, an evolutionary approach* (eds J. R. Krebs and N. B. Davies), pp. 149–69. Blackwell, Oxford.

Dixson, A. F. (1987). Observations on the evolution of genitalia and copulatory behaviour in male primates. *Journal of Zoology*, London, **213**, 423–43.

Dunbar, R. I. M. (1980). Determinants and evolutionary consequences of dominance among female gelada baboons. *Behavioral Ecology and Sociobiology*, **7**, 253–65.

Dunbar, R. I. M. (1984). *Reproductive decisions: an economic analysis of gelada baboon social strategies*. Princeton University Press, Princeton.

Dunbar, R. I. M. (1988). *Primate social systems*. Croom Helm, London.

Dunbar, R. I. M. (1989). Social systems as optimal strategy sets: the costs and benefits of sociality. In *Comparative socioecology, the behavioural ecology of humans and other mammals* (eds V. Standen and R. A. Foley), pp. 131–49. Blackwell, Oxford.

Dunbar, R. I. M. and Dunbar, E. P. (1975). *Social dynamics of gelada baboons*. Karger, Basel.

Emlen, S. T. and Oring, L. W. (1977). Ecology, sexual selection, and the evolution of mating systems. *Science*, **197**, 215–23.

Fedigan, L. and Baxter, M. J. (1984). Sex differences and social organization in free-ranging spider monkeys (*Ateles geoffroyi*). *Primates*, **25**, 279–94.

Fretwell, S. D. and Lucas, H. L. (1970). On territorial behavior and other factors influencing habitat distribution in birds. *Acta Biotheoretica*, **19**, 16–36.

Furuichi, T. (1989). Social interactions and the life history of female *Pan paniscus* in Wamba, Zaïre. *International Journal of Primatology*, **10**, 173–97.

Ghiglieri, M. P. (1984). *The chimpanzees of Kibale forest: a field study of ecology and social structure*. Columbia University Press, New York.

Ghiglieri, M. P. (1990). Hominoid sociobiology and hominid social evolution. In *Understanding chimpanzees* (eds P. G. Heltne and L. A. Marquardt), pp. 370–9. Harvard University Press, Cambridge, MA.

Goldizen, A. W. (1987). Tamarins and marmosets: communal care of offspring. In *Primate Societies* (eds B. B. Smuts, D. L. Cheney, R. M. Seyfarth, R. W. Wrangham, and T. T. Struhsaker), pp. 34–43. University of Chicago Press, Chicago.

Goldizen, A. W. (1989). Social relationships in a cooperatively polyandrous group of tamarins (*Saguinus fuscicollis*). *Behavioral Ecology and Sociobiology*, **24**, 79–89.

Goodall, J. (1965). Chimpanzees of the Gombe Stream Reserve. In *Primate behaviour* (ed. I. De Vore), pp. 425–73. Holt, Rinehart and Winston, New York.

Goodall, J. (1983). Population dynamics during a 15-year period in one communicity of free-living chimpanzees in the Gombe National Park. *Zeitschrift für Tierpsychologie*, **61**, 1–60.

Goodall, J. (1986). *The chimpanzees of Gombe: patterns of behavior*. Harvard University Press, Cambridge, MA.

Goodall, J., Bandora, A., van Zinnicq Bergmann, E., Busse, C., Matama, H., Mpongo, E., Pierce, A., and Riss, D. (1979). Intercommunity interactions in the chimpanzee population of the Gombe National Park. In *The great apes* (eds D. A. Hamburg and E. R. McCown), pp. 13–54. Benjamin/Cummings, Menlo Park, CA.

Gouzoules, S. and Gouzoules, H. (1987). Kinship. In *Primate societies* (eds B. B. Smuts, D. L. Cheney, R. M. Seyfarth, R. W. Wrangham, and T. T. Struhsaker), pp. 299–305. University of Chicago Press, Chicago.

Greenwood, P. J. (1980). Mating systems, philopatry and dispersal in birds and mammals. *Animal Behaviour*, **28**, 1140–62.

Hamburg, D. A. (1973). An evolutionary and developmental approach to human aggressiveness. *Psychoanalysis Quarterly*, **42**, 185–96.

Harcourt, A. H. (1979). Contrasts between male relationships in wild gorilla groups. *Behavioral Ecology and Sociobiology*, **5**, 39–49.

Harcourt, A. H. and Stewart, K. J. (1981). Gorilla male relationships: Can differences during immaturity lead to contrasting reproductive tactics in adulthood? *Animal Behaviour*, **29**, 206–10.

Harcourt, A. H. and Stewart, K. J. (1989). Functions of alliances in contests within wild gorilla groups. *Behaviour*, **109**, 176–89.

Harcourt, A. H., Harvey, P. H., Larson, S. G., and Short, R. V. (1981). Testis weight, body weight, and breeding system in primates. *Nature*, **293**, 55–7.

Harding, R. S. O. (1975). Meat-eating and hunting in baboons. In *Socioecology and psychology of primates* (ed. R. H. Tuttle), pp. 247–57. Mouton, The Hague.

Hill, D. A. (1990). Social relationships between adult male and female rhesus macaques: II. Non-sexual affiliative behaviour. *Primates*, **31**, 33–50.

Hinde, R. A. (1976). Interactions, relationships and social structure. *Man*, **11**, 1–17.

Hinde, R. A. (1979). *Towards Understanding Relationships.* Academic Press, London.

Hogeweg, P. and te Boekhorst, I. J. A. (in press). On the natural history of an artificial CHIMP world: group structure as a side-effect. *Behavioral Ecology and Sociobiology*.

van Hooff, J. A. R. A. M. (1988). Sociality in primates: a compromise of ecological and social adaptation strategies. In *Perspectives in the study of primates* (eds A. Tartabini and M. L. Genta), pp. 9–23. DeRose, Cosenza.

van Hooff, J. A. R. A. M. (1990). Intergroup competition and conflict in animals and man. In *Sociobiology and conflict* (eds J. M. G. van der Dennen and V. E. Falger), pp. 23–54. Chapman and Hall, London.

Hrdy, S. B. (1977). *The langurs of Abu*. Harvard University Press, Cambridge, MA.

Hrdy, S. B. and Whitten, P. L. (1987). Patterning of sexual activity. In *Primate societies* (eds B. B. Smuts, D. L. Cheney, R. M. Seyfarth, R. W. Wrangham, and T. T. Struhsaker), pp. 370–84. University of Chicago Press, Chicago.

Isabirye-Basuta, G. (1988). Food-competition among individuals in a free-ranging chimpanzee community in Kibale forest, Uganda. *Behaviour*, **105**, 135–47.

Janson, C. H. (1984). Female choice and mating system of the brown capuchin monkey *Cebus apella* (*Primates: Cebidae*). *Zeitschrift für Tierpsychologie*, **65**, 177–200.

Janson, C. H. (1986). The mating system as a determinant of evolution in capuchin monkeys. In *Primate ecology and conservation* (eds J. Else and P. C. Lee), pp. 169–79. Cambridge University Press, Cambridge.

Janson, C. H. and van Schaik, C. P. (1988). Recognizing the many faces of primate food competition: Methods. *Behaviour*, **105**, 165–86.

Jones, C. B. (1981). The evolution and socioecology of dominance in primate groups. Theoretical formulation, classification and assessment. *Primates*, **23**, 130–4.

Jones, E., Byrne, B., and Chance, M. R. A. (1982). Influence of a novel male on the social behavior of a captive group of mature female long-tailed macaques, *Macaca fascicularis*. *Laboratory Animal*, **16**, 208–14.

Kano, T. (1982). The social group of pygmy chimpanzees Pan paniscus of Wamba. *Primates*, **23**, 171–88.

Kummer, H. (1968). *Social organization of hamadryas baboons*. University of Chicago Press, Chicago.

Kummer, H. (1978). On the value of social relationships to nonhuman primates: a heuristic scheme. *Social Science Information*, **17**, 687–705.

Kuroda, S. (1979). Grouping of the pygmy chimpanzees. *Primates*, **20**, 161–83.

Lazarus, J. (1982). Competition and conflict in animals. In *Cooperation and competition in humans and animals* (ed. A. M. Colman), pp. 26–56. Van Nostrand Reinhold, London.

Łomnicki (1988). *A population ecology of individuals*. Princeton University Press, Princeton.

Marsden, H. M. (1968). Agonistic behaviour of young rhesus monkeys after changes induced in social rank of their mother. *Animal Behaviour*, **16**, 38–44.

McFarland, M. (1986). Ecological determinants of fission–fusion sociality in *Ateles* and *Pan*. In *Primate Ecology and Conservation* (eds P. C. Lee and J. Else), pp. 181–90. Cambridge University Press, Cambridge.

McFarland Symington, M. (1987). Sex ratio and maternal rank in wild spider monkeys: when daughters disperse. *Behavioral Ecology and Sociobiology*, **20**, 421–5.

McFarland Symington, M. (1988). Food competition and foraging party size in the black spider monkey (*Ateles paniscus chamek*). *Behaviour*, **105**, 117–34.

McGrew, W. C. (1979). Evolutionary implications of sex differences in chimpanzee predation and tool use. In *The great apes* (eds D. A. Hamburg and E. R. McCown), pp. 441–63. Benjamin/Cummings, Menlo Park, CA.

Mead, M. (1937). *Cooperation and Competition among Primitive Peoples*. New York, McGraw-Hill.

Meikle, D. B. and Vessey, S. H. (1981). Nepotism among rhesus monkey brothers. *Nature*, **294**, 160–1.

Melnick, D. J. and Pearl, M. C. (1987). Cercophitecines in multimale groups: genetic diversity and population structure. In *Primate societies* (eds B. B. Smuts, D. L. Cheney, R. M. Seyfarth, R. W. Wrangham, and T. T. Struhsaker), pp. 121–34. University of Chicago Press, Chicago.

Melotti, U. (1979). *L'uomo tra natura e storia: la dialettica delle origine*. Unicopoli, Milano.

Milton, K. (1985). Mating patterns of woolly spiker monkeys, *Brachyteles arachnoides*: implications for female choice. *Behavioral Ecology and Sociobiology*, **17**, 53–9.

Mitchell, C. L. (1990). The Ecological Basis for Female Social Dominance: a Behavioral Study of the Squirrel Monkey (*Saimiri sciureus*) in the wild. Ph.D. thesis, Princeton University.

Mitchell, C. M., Boinski, S., and van Schaik, C. P. (1991). Competitive regimes and female bonding in two species of squirrel monkeys (*Saimiri oerstedii* and *S. sciureus*). *Behavioral Ecology and Sociobiology*, **28**, 55–60.

Møller, A. P. (1988). Ejaculate quality, testes size and sperm competition in primates. *Journal of Human Evolution*, **17**, 479–88.

Moore, J. (1984). Female transfer in primates. *International Journal of Primatology*, **5**, 537–89.

Nagel, U. (1973). A comparison of anubis baboons, hamadryas baboons, and their hybrids at a species border in Ethiopia. *Folia Primatologica*, **19**, 104–65.

Nicholson, A. J. (1954). An outline of the dynamics of animal populations. *Australian Journal of Zoology*, **2**, 9–65.

Nishida, T. (1979). The social structure of chimpanzees of the Mahale Mountains. In

The great apes (eds D. A. Hamburg and E. R. McCown), pp. 73–121. Benjamin/ Cummings, Menlo Park, CA.

Nishida, T. (1983). Alpha status and agonistic alliance in wild chimpanzees (*Pan troglodytes schweinfurthii*), *Primates*, **24**, 318–36.

Nishida, T. and Hiraiwa-Hasegawa, M. (1987). Chimpanzees and bonobos: cooperative relationships among males. In *Primate societies* (eds B. B. Smuts, D. L. Cheney, R. M. Seyfarth, R. W. Wrangham, and T. T. Struhsaker), pp. 165–78. University of Chicago Press, Chicago.

Nishida, T, Hiraiwa-Hasegawa, M., Hasegawa, T., and Takahata, Y. (1985). Group extinction and female transfer in wild chimpanzees in the Mahale Mountains. *Zeitschrift für Tierpsychologie*, **67**, 284–301.

Noë, R. (1986). Lasting alliances among adult male savannah-baboons. In *Primate ontogeny, cognition and social behaviour* (eds J. G. Else and P. C. Lee), pp. 381–92. Cambridge University Press, Cambridge.

Noë, R. (1988). Coalition Formation Among Male Baboons. Publ. Diss., University of Utrecht.

Noë, R. (1990). A veto game played by baboons: a challenge to the use of the Prisoners' dilemma as a paradigm for reciprocity and cooperation. *Animal Behaviour*, **39**, 78–90.

Noë, R., van Schaik, C. P., and van Hooff, J. A. R. A. M. (1991). The market effect: an explanation for pay-off asymmetries among collaborating animals. *Ethology*, **87**, 97–118.

van Noordwijk, M. A. (1985). Sexual behaviour of Sumatran long-tailed macaques. *Zeitschrift für Tierpsychologie*, **70**, 277–96.

van Noordwijk, M. A. and van Schaik, C. P. (1985). Male migration and rank acquisition in wild long-tailed macaques *Macaca fascicularis*. *Animal Behaviour*, **33**, 849–61.

van Noordwijk, M. A. and van Schaik, C. P. (1988). Male careers in Sumatran long-tailed macaques. *Behaviour*, **107**, 24–43.

Oates, J. F. (1977). The social life of a black-and-white colobus monkey, *Colobus guereza*. *Zeitschrift für Tierpsychologie*, **45**, 1–60.

Packer, C. (1977). Reciprocal altruism in olive baboons. *Nature*, **265**, 441–3.

Packer, C. (1979). Male dominance and reproductive activity in *Papio anubis*. *Animal Behaviour*, **27**, 37–45.

Packer, C. (1986). Whatever happened to reciprocal altruism? *Trends in Ecology and Evolution*, **1**, 142–3.

Packer, C. and Pusey, A. E. (1982). Cooperation and competition within coalitions of male lions: Kin selection or game theory? *Nature*, **296**, 740–2.

Pereira, M. E., Kaufman, R., Kappeler, P. M., and Overdorff, D. J. (1990). Female dominance does not characterize all of the Lemuridae. *Folia Primatologica*, **55**, 96–103.

Poirier, F. E. (1974). Colobine aggression: a review. In *Primate aggression, territoriality, and xenophobia: A comparative perspective* (ed. R. L. Holloway), pp. 123–330. Academic Press, New York.

Pusey, A. E. (1979). Intercommunity transfer of chimpanzees in Gombe National Park. In *The great apes* (eds D. A. Hamburg and E. R. McCown), pp. 465–79. Benjamin/ Cummings, Menlo Park, CA.

Pusey, A. E. and Packer, C. (1987). Dispersal and philopatry. In *Primate societies* (eds

B. B. Smuts, D. L. Cheney, R. M. Seyfarth, R. W. Wrangham, and T. T. Struhsaker), pp. 250–66. University of Chicago Press, Chicago.

Rasmussen, K. L. R. (1983). Influences of affiliative preferences upon the behaviour of male and female baboons during consortships. In *Primate Social Relationships, an Integrated Approach* (ed. R. A. Hinde), pp. 200–8. Blackwell, Oxford.

Richard, A. F. (1987). Malagasy primates, female dominance. In *Primate Societies* (eds B. B. Smuts, D. L. Cheney, R. M. Seyfarth, R. W. Wrangham, and T. T. Struhsaker), pp. 25–33. University of Chicago Press, Chicago.

Robinson, J. G. (1988). Group size in wedge-capped capuchin monkeys *Cebus olivaceus* and the reproductive success of males and females. *Behavioral Ecology and Sociobiology*, **23**, 187–97.

van Roosmalen, M. G. M. (1980). *Habitat preferences, diet, feeding strategy, and social organization of the black spider monkey (Ateles p. paniscus Linnaeus 1758) in Surinam*. Rijksinstituut voor Natuurbeheer, Arnhem.

Rowell, T. E. (1988). The social system of guenons, compared with baboons, macaques and mangabeys. In *A primate radiation: evolutionary biology of the African guenons* (eds A. Gautier-Hion, F. Bourlière, J.-P. Gautier, and J. Kingdon), pp. 429–51. Cambridge University Press, Cambridge.

de Ruiter, J. R. (1986). The influence of group size on predator scanning and foraging behaviour of wedge-capped capuchin monkeys (Cebus olivaceaus). *Behavour*, **98**, 240–58.

Saayman, G. S. (1970). The menstrual cycle and sexual behaviour in a troop of free ranging chacma baboons (*Papio ursinus*). *Folia Primatologica*, **12**, 81–110.

van Schaik, C. P. (1983). Why are diurnal primates living in groups? *Behaviour*, **87**, 120–44.

van Schaik, C. P. (1989). The ecology of social relationships amongst female primates. In *Comparative socioecology, the behavioural ecology of humans and other mammals* (eds V. Standen and G. R. A. Foley), pp. 195–218. Blackwell, Oxford.

van Schaik, C. P. and Dunbar, R. I. M. (1990). The evolution of monogamy in large primates: a new hypothesis and some crucial tests. *Behaviour*, **115**, 30–62.

van Schaik, C. P. and van Hooff, J. A. R. A. M. (1983). On the ultimate causes of primate social systems. *Behaviour*, **85**, 91–117.

van Schaik, C. P. and van Noordwijk, M. A. (1988). Scramble and contest among female long-tailed macaques in a Sumatran rain forest. *Behaviour*, **105**, 77–98.

van Schaik, C. P. and van Noordwijk, M. A. (1989). The special role of male Cebus monkeys in predation avoidance, and its effect on group composition. *Behavioral Ecology and Sociobiology*, **24**, 265–76.

Seyfarth, R. M. (1976). Social relationships among adult female baboons. *Animal Behaviour*, **24**, 917–38.

Seyfarth, R. M. (1983). Grooming and social competition in primates. In *Primate social relationships: an integrated approach* (ed. R. A. Hinde), pp. 182–90. Blackwell, Oxford.

Seyfarth, R. M. (1987). Vocal communication and its relation to language. In *Primate societies* (eds B. B. Smuts, D. L. Cheney, R. M. Seyfarth, R. W. Wrangham, and T. T. Struhsaker), pp. 440–51. University of Chicago Press, Chicago.

Sherman, P. W. (1977). Nepotism and the evolution of alarm calls. *Science*, **197**, 1246–53.

Shields, W. M. (1987). Dispersal and mating systems: investigating their causal connec-

tion. In *Mammalian dispersal patterns* (eds B. D. Chepko-Sade and Z. T. Halpin), pp. 3–24. University of Chicago Press, Chicago.
Short, R. V. (1979). Sexual selection and its component parts, somatic and genital selection as illustrated by man and the great apes. *Advances in the study of behaviour*, **9**, 131–58.
Sigg, H., Stolba, A., Abeggien, J. J., and Dasser, V. (1982). Life history of hamadryas baboons: Physical development, infant mortality, reproductive parameters, and family relationships. *Primates*, **23**, 473–87.
Smuts, B. B. (1985). *Sex and friendship in baboons*. Aldine, Hawthorn, NY.
Smuts, B. B. (1987*a*). Sexual competition and mate choice. In *Primate societies* (eds B. B. Smuts, D. L. Cheney, R. M. Seyfarth, R. W. Wrangham, and T. T. Struhsaker), pp. 385–99. University of Chicago Press, Chicago.
Smuts, B. B. (1987*b*). Gender, aggression, and influence. In *Primate societies* (eds B. B. Smuts, D. L. Cheney, R. M. Seyfarth, R. W. Wrangham, and T. T. Struhsaker), pp. 400–12. University of Chicago Press, Chicago.
Stammbach, E. (1987). Desert, forest and montane baboons: multilevel-societies. In *Primate societies* (eds B. B. Smuts, D. L. Cheney, R. M. Seyfarth, R. W. Wrangham, and T. T. Struhsaker), pp. 112–20. University of Chicago Press, Chicago.
Stein, D. M. (1984). *The sociobiology of infant and adult male baboons*. Ablex, Norwood, NJ.
Stolba, A. (1979). Entscheidungsfindung in Verbänden von *Papio hamadryas*. Ph.D. dissertation, University of Zürich.
Struhsaker, T. T. (1975). *The red colobus monkey*. University of Chicago Press, Chicago.
Struhsaker, T. T. (1977). Infanticide and social organisation in the redtail monkey (*Cercophitecus ascanius schmidt*) in the Kibale forest, Uganda. *Zeitschrift für Tierpsychologie*, **45**, 75–84.
Struhsaker, T. T. and Leland, L. (1987). Colobines: Infanticide by adult males. In *Primate societies* (eds B. B. Smuts, D. L. Cheney, R. M. Seyfarth, R. W. Wrangham, and T. T. Struhsaker), pp. 83–97. University of Chicago Press, Chicago.
Strum, S. (1981). Processes and products of change: baboon predatory behaviour at Gilgil, Kenya. In *Omnivorous primates: gathering and hunting in human evolution* (eds R. S. Harding and G. Teleki), pp. 255–302. Columbia University Press, New York.
Stynes, A. J., Rosenblum, L. A., and Kaufman, I. C. (1968). The dominant male and behavior within heterospecific monkey groups. *Folia Primatologica*, **9**, 123–4.
Sugardjito, J., te Boekhorst, I. J. A., and van Hooff, J. A. R. A. M. (1987). Ecological constraints on the grouping of wild orang-utans in the Gunung Leuser National Park, Indonesia. *International Journal of Primatology*, **8**, 17–41.
Sugiyama, Y. (1967). Social organisation of hanuman langurs. In *Social communication among primates* (ed. S. A. Altmann), pp. 221–36. University of Chicago Press, Chicago.
Sugiyama, Y. (1976). Life history of male Japanese monkeys. In *Advances in the study of behaviour*, vol. 7 (eds J. S. Rosenblatt, R. A. Hinde, E. Shaw, and C. Beer), pp. 255–85. Academic Press, London.
Sugiyama, Y. (1981). Observations on the population dynamics and behavior of wild chimpanzees at Bossou, Guinea, in 1979–1980. *Primates*, **22**, 435–44.
Sugiyama, Y. (1988). Grooming interactions among adult chimpanzees at Bossou,

Guinea, with special reference to social structure. *International Journal of Primatology*, **9**, 393–407.

Tahahata, Y. (1982). Social relations between adult males and females of Japanese monkeys in the Arashiyama B troop. *Primates*, **23**, 1–23.

Teleki, G. (1973). *The predatory behavior of chimpanzees*. Bucknell University Press, Lewisburg.

Terborgh, J. and Janson, C. H. (1986). The socioecology of primate groups. *Annual Review of Ecology and Systematics*, **17**, 111–35.

Thierry, B. (1985). Patterns of agonistic interactions in three species of macaque (*Macaca mulatta, M. fascicularis, M. tonkeana*). *Aggressive Behavior*, **11**, 223–33.

Trivers, R. L. (1971). The evolution of reciprocal altruism. *Quarterly Review of Biology*, **46**, 35–57.

Trivers, R. L. (1972). Parental investment and sexual selection. In *Sexual selection and the descent of man, 1871–1971* (ed. B. Campbell), pp. 136–79. Aldine Press, Chicago.

Tutin, C. E. G. (1979). Mating patterns and reproductive strategies in a community of wild chimpanzees (*Pan troglodytes schweinfurthii*). *Behavioral Ecology and Sociobiology*, **6**, 29–38.

Vehrencamp. S. (1983). A model for the evolution of despotic versus egalitarian societies. *Animal Behaviour*, **31**, 667–82.

de Waal, F. B. M. (1978). Exploitative and familiarity-dependent support strategies in a colony of semi-free-living chimpanzees. *Behaviour*, **66**, 268–312.

de Waal, F. B. M. (1982). *Chimpanzee politics*. Jonathan Cape, London.

de Waal, F. B. M. (1986). Conflict resolution in monkeys and apes. In *Primates, the road to self-sustaining populations* (ed. K. Benirschke), pp. 341–50. Springer-Verlag, New York.

de Waal, F. B. M. (1989). Dominance 'style' and primate social organization. In *Comparative socioecology* (eds V. Standen and R. A. Foley), pp. 243–64. Blackwell, Oxford.

de Waal, F. B. M. and Luttrell, L. M. (1989). Toward a comparative socioecology of the genus *Macaca*: Different dominance styles in rhesus and stumptail monkeys. *American Journal of Primatology*, **19**, 83–109.

Wade, T. D. (1976). The effects of strangers on rhesus monkey groups. *Behaviour*, **56**, 194–214.

Washburn, S. L. and Lancaster, C. S. (1968). The evolution of hunting. In *Man the hunter* (eds R. B. Lee and I. De Vore), pp. 293–303. Aldine Press, Chicago.

Watts, C. R. and Stokes, A. W. (1971). The social order of turkeys. *Scientific American*, **224** (6), 112–18.

Watts, D. P. (1989). Infanticide in mountain gorillas: new cases and a reconsideration of the evidence. *Ethology*, **81**, 1–18.

Wells, K. (1977). The social behaviour of anuran amphibians. *Animal Behaviour*, **25**, 666–93.

White, F. J. (1988). Party composition and dynamics in *Pan paniscus*. *International Journal of Primatology*, **9**, 179–93.

White, F. J. (1989). Social organization of pygmy chimpanzees. In *Understanding chimpanzees* (eds P. G. Heltne and L. A. Marquardt), pp. 194–207. Harvard University Press, Cambridge, MA.

Wrangham, R. W. (1977). Feeding behaviour of chimpanzees in Gombe National

Park, Tanzania. In *Primate ecology* (ed. T. H. Clutton-Brock), pp. 503–38. Academic Press, New York.

Wrangham, R. W. (1979). On the evolution of ape social systems. *Social Science Information*, **18**, 335–68.

Wrangham, R. W. (1980). An ecological model of female-bonded primate groups. *Behaviour*, **75**, 262–300.

Wrangham, R. W. (1982). Mutualism, kinship, and social evolution. In *Current problems in sociobiology* (ed. King's College Sociobiology Group), pp. 269–90. Cambridge University Press, Cambridge.

Wrangham, R. W. (1986). Ecology and social relationships in two species of chimpanzees. In *Ecological aspects of social evolution* (eds D. I. Rubenstein and R. W. Wrangham), pp. 352–78. Princeton University Press, Princeton.

Wrangham, R. W. (1987). The significance of African apes for reconstructing human social evolution. In *The evolution of human behavior: primate models* (ed. W. G. Kinzey), pp. 51–71. SUNY Press, New York.

Wrangham, R. W. and Rubenstein, D. I. (1986). Social evolution in birds and mammals. In *Ecological aspects of social evolution* (eds D. I. Rubenstein and R. W. Wrangham), pp. 452–70. Princeton University Press, Princeton.

Yoshiba, K. (1968). Local and intertroop variability in ecology and social behavior of common Indian langurs. In *Primates: studies in adaptation and variability* (ed. P. C. Jay), pp. 217–42. Holt, Rinehart and Winston, New York.

14

Sex differences in alliances, and the acquisition and maintenance of dominance status among immature primates

P. C. LEE AND J. A. JOHNSON

INTRODUCTION

Reproductive success and dominance rank have been associated for males and females in a large number of cercopithecine primate species (see Harcourt 1987; Johnson 1987; Silk 1987 for reviews). Since immature primates at some point attain a rank order within their social group, answering the questions of how and when they acquire this status and of its persistence or change through time is important in attempts to relate rank to subsequent reproductive success. The process of acquisition of rank by immature primates has been of interest since Kawai (1958) initially differentiated between dependent rank (rank based on the attributes or presence of others) and basic rank (rank based on individual attributes such as age or size) in Japanese macaques (*Macaca fuscata*). Among female kin-bonded cercopithecines, young animals take a position in the hierarchy adjacent to that of their mothers as a result of support from mothers and other animals in contests with group members, as in baboons (*Papio cynocephalus*, Cheney 1977; Johnson 1984, 1987; Lee and Oliver 1979; Moore 1978; Pereira 1988; Samuels *et al.* 1987; Walters 1980), vervet monkeys (*Cercopithecus aethiops*, Horrocks and Hunte 1983*a*,*b*; Lee 1983*a*), Japanese (Chapais 1988; Kawai 1958; Koyama 1967), rhesus (*Macaca mulatta*, Chapais 1983; Datta 1983*a*,*b*; Koford 1963; Missakian 1972; Sade 1967), long-tailed (*M. fascicularis*, Netto and van Hooff 1986; de Waal 1977), and bonnet macaques (*M. radiata*, Silk *et al.* 1981). These studies have tended to concentrate on the stability of female hierarchies over generations, as exemplified by the Japanese macaques (see Chapais 1988, this volume pp. 29–59), Amboseli baboons (Hausfater *et al.* 1982; Samuels *et al.* 1987) and the Cayo Santiago rhesus (Datta 1989).

While it has long been recognized that the expression of dominance and the underlying process of status acquisition and maintenance differ between adult male and female cercopithecines, studies of rank acquisition have concentrated on immature females. It is hardly surprising that the nature of rank acquisition among immature males has been relatively neglected. Young males disperse from the natal group at around adolescence (vervet, Cheney 1983; macaque, Colvin 1986; van Noordwijk and van Schaik 1988; baboon, Manzolillo 1986; Packer 1979*a*) and thus their rank is seldom assessed for stability or consistency with that of their mother. As immatures, they tend to be lumped with female peers, and their rank relative to adult females assessed jointly (e.g. Datta 1983*a*) or ignored completely (Horrocks and Hunte 1983*a*; Walters 1980). Those studies that have examined rank acquisition among young males have suggested that different principles may operate, and that age and peer relationships are critical influences (macaque, Colvin 1983; Cords 1988; Kuester and Paul 1988; Smith and Smith 1988; baboon, Johnson 1987; Lee and Oliver 1979).

In this chapter, we examine juvenile dominance relationships with respect to differences in life histories between the sexes. Initially, the general factors affecting the process of dominance acquisition and maintenance, particularly the role of support in contests, are discussed, and then assessed for each sex of immature in relation to their specific needs. We then attempt to compare immature dominance relationships with those of adults, and ask whether the young animals are merely perfecting their final adult hierarchical position during development, or is their rank a reflection of juvenile relationships and requirements?

DOMINANCE MEASURES AND SYSTEMS OF RANKING

In most of the studies discussed below, dominance is defined either by the ability of one animal to take a resource from another, or by its ability to make another avoid or submit during an approach or aggression. These responses define the outcomes of dyadic interactions, which are then placed into a hierarchical system ranking all relevant individuals, and reflects basic or intrinsic rank. When the outcome of a dyadic interaction is influenced by aid given during coalitions (or merely by the proximity of a potential ally), the result can be considered as dependent rank. Since coalitions and alliances are thought to bear a critical influence on rank acquisition, it becomes important to separate these two types of rank, and to determine when and if they interact in the rank relations of immature animals.

It is worth noting at the outset one problem with the use of such dyadic measures. The hierarchies derived from subtle approach–retreat interactions may differ from those derived from overt high-level aggression. As such, the hierarchies determined in different studies may not be directly comparable. Ideally, the types of agonistic interactions—competitive, avoidance, or escalated aggression—should be examined separately in assessments of juvenile dominance, but the data are few at this stage.

The ranks of the two sexes of adults are influenced by different factors. As stated above, adult female rank in most cercopithecine species is relatively uninfluenced by age, but is related to the ranks of close kin (see Chapais this volume pp. 29–59). Female age-graded hierarchies do exist, for example in gelada baboons (*Theropithecus gelada*, Dunbar 1980), langurs (*Presbytis entellus*, Hrdy and Hrdy 1976), and gorillas (*Gorilla gorilla*, Harcourt and Stewart 1987). Older females tend to be dominant to younger females, with very old females declining in rank. In such systems, the status of daughters is not dependent on that of mothers and reflects basic rather than dependent rank. However, even in age-graded systems, some aspects of dependent rank can be found. The rank of ageing mothers may be stabilized by aid from daughters (e.g. Dunbar 1980) and interventions by mothers on behalf of offspring are observed in triadic interactions among gorillas (Harcourt and Stewart 1987).

Age-graded systems are typical of many hierarchies among adult males. Males tend to rank in relation to age, size, and fighting skill, with prime males dominating young and old males (e.g. baboon, Collins 1986; Hausfater 1975; Packer 1979*b*). Male ranks, unlike those of most females, tend to be fluid through time, either as males age and increase in size and skill or as they develop friendships and alliances with other males, or adult females, who will then support them during contests (e.g. baboon, Berkovitch 1988; Noe 1986; Smuts 1985; Strum 1982; vervet, Raleigh and McGuire 1989). Neither the hierarchies nor the dyadic ranks are consistent and stable in comparison to those of adult females.

Differences in the nature of hierarchies between the sexes can be related to the type of resources contested, the effect of those contests on individual survival and reproduction, and the short- and long-term ability of allies to alter the outcome of an interaction. Thus for females who live together for life, the outcome of contests over reproductively critical resources such as food will be determined by long-term knowledge of the status and allies of other females relative to the competing female. The size of matrilines, the relative ranking of matrilines as well as ranking within a matriline, and the ability to form cooperative relationships across ranks and between individuals from different matrilines, all influence the competitive relationships of females.

For males, contests occur over a variety of short-term resources, such as receptive females, and the outcome will be a complex interaction of individual strength, skill, motivation, and the presence of allies, the allies' willingness to intervene, and those allies' intrinsic strengths and abilities. All in all, males also must make complex assessments and take many factors into account when engaging in a single rank-related contest. Male decisions can be based on immediate or relatively short-term assessments, while those of the females are based on long-term knowledge and experiences within a generally stable hierarchy. For both sexes, cooperative alliances are of critical importance in determining the outcome of rank-related interactions. However, for females the outcome of a contest seldom depends on the intrinsic strength of any of the contestants, while it is more likely to do so for males.

Sex differences in the acquisition and maintenance of rank by juveniles can be examined in relation to these differences between adults in the nature of contested resources and in the factors affecting the outcome of contests. Sex differences in the development of competitive and cooperative abilities can ultimately be related to differences in life histories.

THE INFLUENCE OF MATERNAL RANK

Alliances of related females are now seen to underpin the social structure of cercopithecine primate groups and as essential in competition between grous (e.g. van Schaik 1989; van Hooff and van Schaik this volume pp. 357–89; Wrangham 1980). Within groups as well, competition between matrilines places a premium on larger alliances of adjacently ranked kin who can be depended upon for support during aggressive interactions (see Datta 1989).

Young animals, irrespective of sex, become integrated into their matrilineal alliances, and for some this process begins prior to weaning. Matriline members, especially close kin, are more likely to contact, interact with, and care for infants in a number of species (Lee 1989). Among rhesus and Japanese macaques, group members discriminate in the attention that they pay to infants and concentrate friendly interactions on infants born into their matriline (Berman 1982; Hiraiwa 1981). As a result, the friendly and supportive relationships formed by developing immatures are differentiated between family and non-family well before weaning. When large matrilines do not occur (such as with small groups or declining populations), associates of the mother ('friends') or grooming partners may function in a similar way to matriline members. These animals may be of adjacent rank, interact affiliatively and be partners in alliances even though they are not related (e.g. Cheney and Seyfarth 1986).

An infant's observations of maternal aggression and submission in the context of family and friends sets the initial conditions for acquiring rank. Knowledge about who is likely to be supportive, and who must be avoided, as the infant becomes independent of the mother, provides the background for the first indirect experiences of rank (Berman 1980). Some infants can supplant or force retreats of adult animals by the age of weaning, suggesting a very early influence of both interactions and observations of others' interactions on rank acquisition (macaque, Berman 1988: baboon, Johnson

1987; Lee and Oliver 1979: ververt, Lee 1983*a*). These infants tend to be offspring of higher-ranking females interacting with very subordinate females. The expression of such early status probably depends on a large difference in relative rank between the infant's family and that of its opponent.

The infants of each sex may be treated differently by other group members, affecting subsequent perceptions of rank and hence its acquisition. Male infants receive more attention and generally friendly contacts from other group members than do female infants (baboon, Altmann 1980; Cheney 1978), irrespective of maternal rank (vervet, Lee 1983*b*). Daughters of subordinate females are sometimes singled out as targets of aggression (baboon, Pereira 1988: macaque, Silk 1980; Silk *et al.* 1981). The effect of these differences between male and female infants in the treatment they receive from other group members may be to initiate sex differences the system of maternal rank inheritance.

Immature females

Maternal rank appears, among the cercopithecines, to be the single critical determining factor of ultimate rank for daughters. Daughters either rank immediately below their mothers, as in macaques (Datta 1983*a*; Koyama 1967) and vervets (Horrocks and Hunte 1983*a*) or above or below their mothers, as in some baboons (Johnson 1987; Moore 1978; Pereira 1988). As females produce surviving daughters, larger alliances of related females develop (Datta 1989), allowing enhanced cooperation during competitive and aggressive interactions, affecting access to those resources essential for reproductive success and infant survival. As daughters mature, their rank becomes firmly established within the female hierarchy close to that of their mother (Fig. 14.1).

The evidence for the importance of coalitionary support in affecting status of daughters is now widespread (see Datta 1989; Chapais this volume pp. 29–59). When effective assistance (e.g. Datta 1983*b*; Netto and van Hooff 1986) can be rendered by a mother to her immature daughter during contests, then the outcome of that contest can be altered from what would be predicted on the basis of the daughter's intrinsic characteristics such as size or fighting ability alone. Interventions by mothers in disputes on behalf of daughters are not, however, common in all species where maternal rank inheritance has been observed (baboon, Johnson 1987; Lee 1987; Lee and Oliver 1979; Walters 1980). Even when few alliances by mothers supporting offspring during aggression are observed, the effect of maternal rank is persistent. Maternal presence (as a potential ally) may play an important role in maintaining daughters' rank. Although orphaned baboon female infants in Kenya's Amboseli population of yellow baboons ultimately attained the rank predicted by that of their mother (Hausfater *et al.* 1982), at Gilgil in Kenya

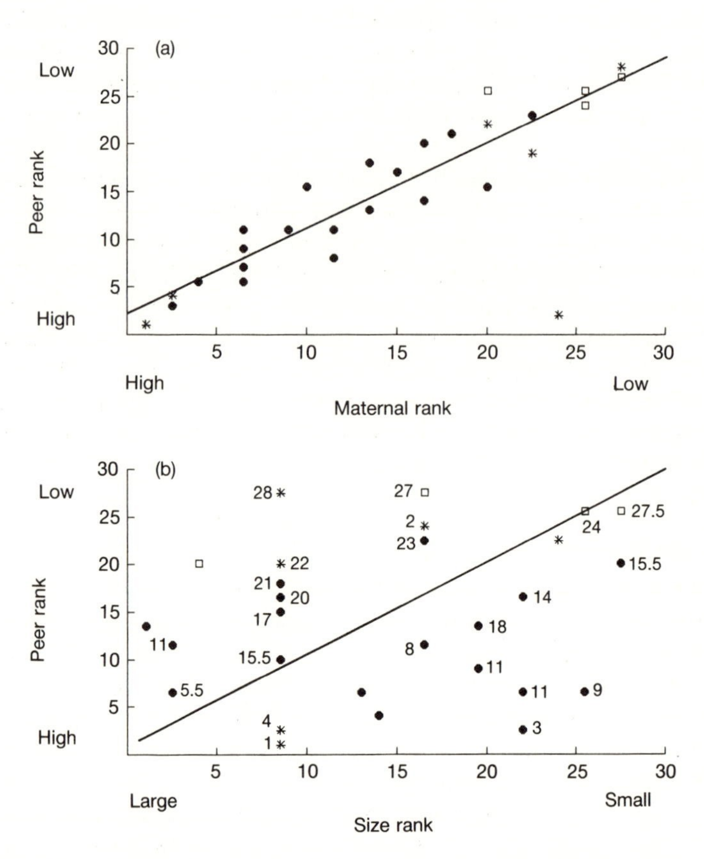

Fig. 14.1. The ranks of juvenile female baboons. (a) Ranks derived from interactions between peers compared with maternal rank of the juvenile (Spearman $r_s = 0.81$, $N = 28$, $P < 0.01$). (b) Peer ranks are compared with the relative size rank of the juvenile females ($r_s = 0.19$, $N = 28$, NS). When several females are of the same size, their relative maternal rank is given in numbers next to their symbol. Orphans are indicated by an asterisk (*) and daughters of females who fell in rank by an open square (□). (Data from Johnson 1987.)

only older juvenile olive baboons did so. Younger juveniles there fell in rank when mothers died or lost rank (Johnson 1987; Fig. 14.1).

If mothers do not form frequent alliances with their daughters, or if they are unavailable for such support as in the case of the orphans, how then can they continue to influence the ultimate rank of their daughters? As mentioned above, infants prior to weaning experience their mothers' and other animals' dominance/subordinance interactions, as well as having a history of aid. Rank acquisition depends on the outcome of a dynamic interactive process between all the participants, active or potential. There may be occasions when the mere proximity of a mother may be sufficient to deter challenges on the part of subordinate animals to the rising immature, especially if contests are over resources low in value (macaque, Datta 1983*a*: baboon, Johnson 1989; Shopland 1987: vervet, Lee, 1983*a*).

As part of this dynamic process not directly involving aggressive coalitionary support, behaviour exchanged between mothers and daughters, and with other animals, such as grooming, may play a critical role in maintaining rank. Among vervets, direction and frequency of grooming reflects dominance relations among the adult females (Seyfarth 1980), and contests over access to grooming partners result in reversals of dyadic dominance (Lee 1983*a*). Here the groomer at the focus of the contest merely grooms one of the presenting contestants, affirming its status relative to the other without any overt aggression taking place, effectively supplanting one animal from its grooming partner. Mothers will stop grooming a higher-ranking non-kin juvenile to preferentially groom their own offspring, which allows the subordinate juvenile to experience winning an encounter with a dominant (Lee 1983*a*, 1987). Among immature animals, reversals against juvenile hierarchies determined from feeding or aggressive contests are most frequent in the context of such grooming contests (Lee 1983*a*; Fig. 14.2). Grooming priorities thus may be critical in demonstrating and maintaining rank relations with non-family members (Johnson 1987, and see below).

Intrinsic aggressiveness or fighting skill can also determine whether a female will take up her predicted maternal rank order. Among the Amboseli baboons, a surge in aggressiveness coincided with puberty and reaching adult size (Walters 1980). Adolescent females intervened against specific target

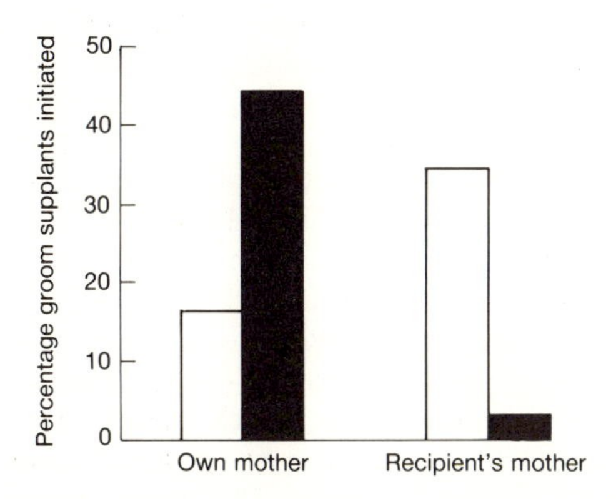

Fig. 14.2. The proportion of groom supplants by vervet monkey immatures that took place over access to a mother as a partner. These are separated into those where the dominant juvenile supplanted the subordinate (open bars, total $N = 171$) and where the subordinate supplanted the dominant (solid bars, total N 50) from grooming with either its own mother or the contestant's mother. (Data from Lee 1983a.)

females when those females were involved in aggression with others. These interventions were proposed as the mechanism by which adolescent females attained rank in the absence of frequent support from mothers. Further interventions by adult females, especially towards older juveniles, were interpreted as a mechanism for resisting reversals of relative rank (Pereira 1989). A female's past experience of aggression with others may influence her willingness to escalate, even though she is not likely to receive support during the agonistic encounter.

Immature males

The effect of maternal rank on sons has only infrequently been assessed. When the rankings of immature males have been examined separately from those of females, maternal rank influences juvenile male rank when they interact with adult females (rhesus, Koford 1963: baboons, Johnson 1987; Lee and Oliver 1979: vervets, Lee 1983*a*). While small juvenile males only infrequently challenge large adult females, among baboons they will do so when those females are subordinate to their mothers. Early experience with, or observation of, the effects of maternal rank appears to be used by males to influence the outcome of competitive and aggressive interactions with adult females. Sons of dominant females thus enter the competitive arena with the advantage of their mother's rank. As they grow in size and gain experience, the sons of subordinate mothers appear to overcome their initial rank disadvantage and before puberty are able to take resources from females irrespective of their mothers' rank (baboons, Johnson 1987; Lee and Oliver 1979: Barbary macaques, *Macaca sylvanus*, Kuester and Paul 1988).

Thus juvenile males initially experience matrilineal rank for some variable period of time, with its attendant benefits for high-born males and costs for low-born males. With age, they transcend their dependent birth rank to acquire a rank based on their own physical capacities (Fig. 14.3).

Maternal rank was found to have little or no effect on the outcome of high-level aggressive interactions between juvenile males and adult females in one study of baboons (Pereira 1988). Juvenile males received support from adult males and unrelated females. Aid from adult males may have been far more influential in determining the dependent rank of males in contests with females than help from mothers (Pereira 1989). These animals lived in a small troop with few juvenile males, which limited opportunities for the assessment of the effects of maternal rank on immature male interactions, especially within their age-sex class. The importance of demographic variation in providing opportunities for the expression of different forms of dominance relationships should not be overlooked (Datta 1989).

What few studies have yet determined is how maternal rank affects later physical capacity. It can be speculated that sons of dominant females may have an advantage that persists through life. These males should be better

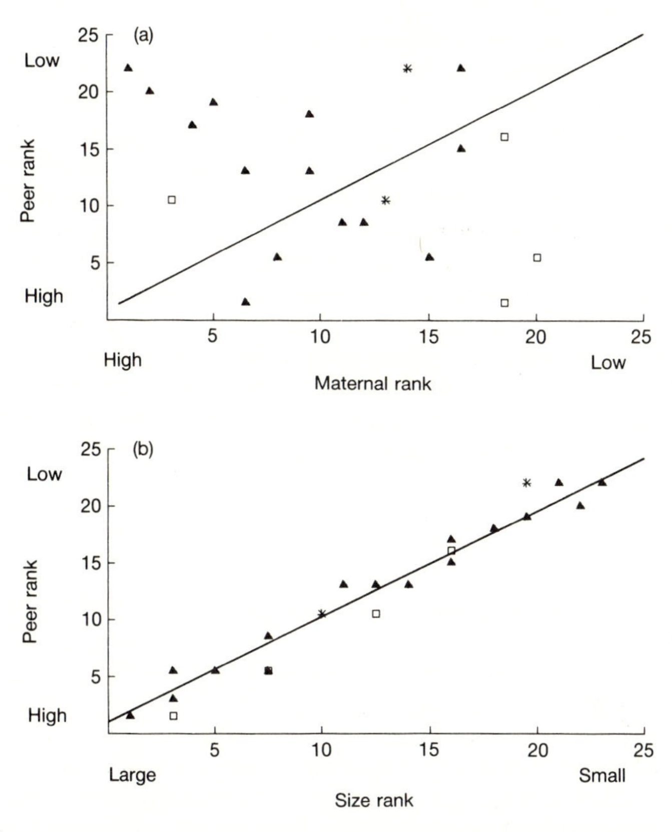

Fig. 14.3. The ranks of juvenile male baboons. (a) Ranks of males derived from interaction between male peers plotted against maternal ranks (r_s = −0.29, N = 20, NS). (b) Ranks of males with peers plotted against their relative size rankings (r_s = 0.98, N = 23, P < 0.01). Orphans and sons of females who fell in rank are indicated as in Fig. 14.1. (Data from Johnson 1987.)

nourished in infancy, and have higher survival rates and rapid growth, as observed by Mori (1979) for both sons and daughters of dominant female Japanese macaques. As juveniles, the large males could maintain this initial advantage of enhanced access to food, at least in competition with adult females (e.g. Dittus 1979). Thus, they could grow to an overall large size, mature more rapidly, and potentially transfer to begin breeding at an earlier age; all attributes which could increase the differentials between males later in life in their reproductive potential.

Male careers in baboons (Hausfater 1975) and long-tailed macaques (van Noordwijk and van Schaik 1988) suggest that not all males follow the same progression through the ranks of the male hierarchy. Some males enter their transfer groups with initially high rank and maintain this rank for long periods, with advantages for their lifetime reproductive success. At this point we can only speculate that these successful males could be the sons of dominant females. Among rhesus macaques, males who stayed longest in their natal groups were the offspring of dominant females (Colvin 1986) contrary to what would be expected if sons of dominant mothers transfer early and begin breeding sooner. However, some of these males were able to mate within their large natal group. The advantage of maternal rank here may have consisted of maintaining access to resources and mates while delaying the risks incurred in transfering between groups. A similar effect of maternal rank on reproductive performance within the natal group was found for captive rhesus (Smith and Smith 1988). Reproductive success early in the breeding lifespan was found to correlate positively with maternal rank. Interestingly, the effect was strengthened if the sons of high-ranking mothers were also the offspring of dominant males.

PEER DOMINANCE

Rank relations between peers show the complexity of the interaction between age, size, and maternal rank, i.e. intrinsic factors and allies, in determining status for both immature males and females. When the sexes are considered separately, the different patterns of rank relations seen as adults are apparent in same-sex interactions between peers.

Females tend to rank relative to peers in accordance with maternal rank, as expected from the general patterns of the establishment of a matrilineal hierarchy (Fig. 14.1). Thus, the capacity for a small weanling to supplant an adult female subordinate to her mother is mirrored in her ability to supplant a low-born, but older juvenile female (baboons, Johnson 1987; Lee and Oliver 1979; Pereira 1988). From birth, females follow a specific pattern related to the life history consequences of rank for adults.

For males, the situation is more complex. When peers are matched in size or age among cohorts of males (such as males born during the same birth season), then the rank of the mother can influence the outcome of interactions between these males of similar size (vervets, Lee 1983*a*: macaques, Kuester and Paul 1988). When there is a difference in size between the males, the larger, older male tends to win the contests irrespective of maternal rank and maternal support (macaques, Kuester and Paul 1988: baboons, Johnson 1987; Lee and Oliver 1979; Fig. 14.3). This rise in rank with age continues from immaturity through to the prime of life.

In Kuester and Paul's (1988) study of Barbary macaques, the ranks of males within peer cohorts was influenced both by maternal rank and by the presence of juvenile brothers. Males without brothers appeared to lack the crucial alliance partners, and the presence of a high-ranking mother alone was not enough to ensure that sons also became high ranking within their peer cohorts.

The influence of maternal rank on the hierarchies of immature males is likely to be most marked in those species such as macaques or vervets with seasonal births. Males born into a seasonal cohort are relatively matched in size and strength, and thus the outcome of their contest potentially can be more influenced by maternal rank.

Male peer ranks can have important consequences for social interactions. Males with low peer ranks tend to have weak affiliative relationships and a limited repertoire of friendly interactions with those peers (rhesus, Colvin 1986; Colvin and Tissier 1985). Low-ranking juvenile males may either be less interactive, or less sought as a partner for interactions. Peer rank relations tend to correlate with the direction, although not the frequency, of aggressive play among juvenile male baboons (Pereira 1989), providing differential play experiences depending on the status of the juvenile. Male rhesus who emigrate earlier than same-aged peers tend to have lower relative peer ranks, as well as being sons of low-ranking mothers (Colvin 1986). Younger emigrants among vervets tend to be those who have few or no allies in their natal groups (Cheney 1983). Thus low relative rank amongst peers may lead to long-term differences in opportunities for the acquisition of social skills which may be important in later alliances, in the age of and risks of migration, and opportunities for mating.

In contests between opposite-sexed juveniles, maternal rank generally tends to have little influence. Immature females are usually subordinate to males in most interactions. However, the outcome of an interaction can be influenced by mother's rank when the contestants are closely matched in age or size, or when the female is much older than the male. Furthermore smaller, younger males can be dominant to older females when they are the sons of dominant females, and larger females can be dominant to the sons of lower-ranking mothers (vervets, Lee 1983*a*, 1987: baboons, Johnson 1987: macaques, Kuester and Paul 1988).

SIBLING DOMINANCE

The relative rank of siblings provides information on the nature of sex differences and also on the mechanisms for the acquisition of rank. Several trends emerge for females. Among sisters close in age in rhesus and Japanese macaques, the younger tends to rise in rank above the older when the younger is 3–4 years old, largely as a result of maternal support in contests (Chapais 1988; Datta 1988, 1989; Koyama 1967; Schulman and Chapais 1980). Among vervets and baboons, some older sisters remain dominant and some fall in rank to their younger sisters (Hausfater *et al.* 1982; Horrocks and Hunte 1983*b*; Johnson 1987; Lee 1987). The age gap between sisters may again determine when reversals are likely.

Brother dyads tend to rank in straightforward relation to age, in a pattern similar to that seen in male peer ranks. Furthermore, brothers are usually dominant to sisters irrespective of age (Lee 1987).

Janus (1989) has suggested that three 'rules' operate to determine sibling and peer ranks: (1) older animals are dominant to younger; (2) males tend to dominate females; (3) members of high-ranking families tend to be dominant to those lower born. The precise rank of siblings or peers depends on the hierarchical interaction of these rules. The age and gender rules can be overridden by the family rank rule, and all rules can fail in the face of effective alliance support from outside the family (Janus 1989).

CONTEXTS OF RANK-RELATED INTERACTIONS

When do immature primates express their status, be it dependent rank or intrinsic rank? As mentioned above, one important determinant of the outcome of an aggressive interaction is the probability of that interaction becoming polyadic, with support for one or both contestants from allies. Few analyses of aggressive coalitions involving immatures have focused on sex differences. Several critical questions remain unanswered. Firstly, although mothers support offspring preferentially over non-kin (Cheney 1977; Datta 1983*a*,*b*; Lee 1987; Walters 1980) and younger offspring over older ones (Datta 1988; Lee 1987), do they discriminate between sons and daughters in frequency of support? Secondly, do mothers distribute their support depending on the rank, age, or sex of the other contestants fighting with their offspring? Thirdly, are there sex differences in the support received or aid given by juvenile males and females in contests with unrelated individuals?

These questions have yet to be addressed, possibly because of the complicated nature of such analyses. The relative ages (power) and relative rank of all participants need to be taken into account. Furthermore, group composition will also affect any observed trends, making generalizations difficult. Studies of peer and sibling rank reversals suggest that continual support is necessary in order for a reversal to be effected (Datta 1983*a*,*b*, 1988), emphasizing the importance of these alliances. In Cheney's (1977) analysis, the juvenile daughter of the dominant female baboon was given the most support by other juveniles, while low-ranking juvenile males received little or no support. However, no-low ranking females were present for comparison. An analysis of sex differences in coalition formation among immature vervets (from data collected by Lee; see Lee 1983*a*, 1987) found that juvenile males (aged 12–48 months) gave less aid than they received, but they were frequently the target of a coalition formed against them. Juvenile females were most frequently involved in coalitions as an active aider, and less frequently as the target. Infants (of both sexes) were most frequently the recipients of aid (Fig. 14.4).

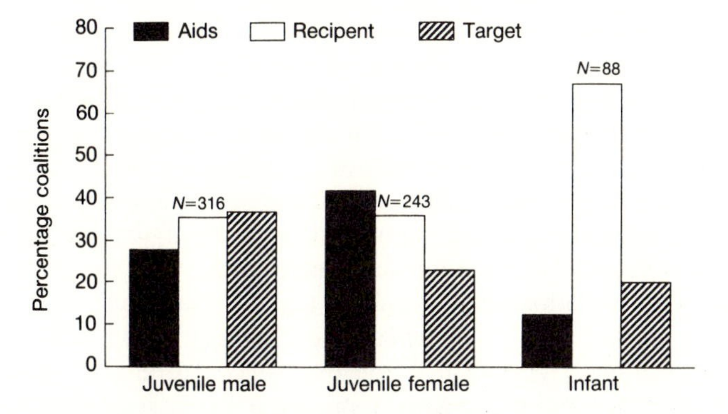

Fig. 14.4. The proportion of coalitions for each sex and age of immature vervet monkeys where the immature was the active aider, the recipient of aid, or the target of a coalition formed against it.

There were few sex differences in the partners of juveniles during coalitionary aggression. Juvenile females tended to aid other female peers and infants more than expected based on the proportion in a group, while this trend was less marked for males (Fig. 14.5*a*). Both sexes were similar in the

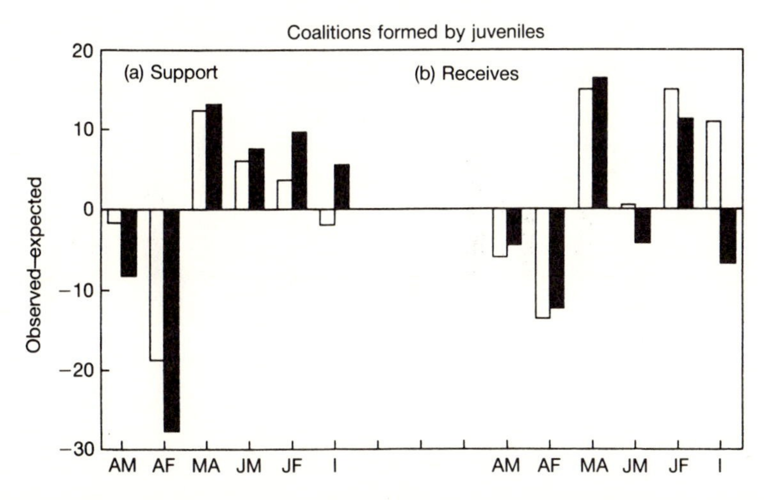

Fig. 14.5. The observed minus the expected number of coalitions. (a) Support given to each age-sex class by juvenile male ($N = 88$) or female vervet monkeys ($N = 100$); (b) Support received from each age–sex class by males ($N = 112$) or females ($N = 87$). The expected numbers are calculated from the mean proportion of each age–sex class, averaged across three groups. AM, adult males; AF, adult females; MA, mothers; JM, juvenile males; JF, juvenile females; I, infants. Open bars indicate males, solid bars females.

distribution of aids received (Fig. 14.5*b*). Juvenile females tended to be the target of a coalition formed by an unrelated adult female more than expected (Fig. 14.6*a*). When a juvenile female was the recipient of aggression from a juvenile male, she received help more than expected (Fig. 14.6*b*).

These sex differences suggest (1) that the pattern of aggressive interactions which result in coalitions differ between males and females, with females being more likely to support others while aggressive interactions involving males are more likely to elicit support for the victim; (2) that juvenile females are targeted specifically by adult females (as shown by Pereira 1988 and Walters 1980 for adolescent females); and (3) that while both sexes of juveniles give more aid than expected during aggression between peers, females appear to get more help. Again, the expected differences emerge with females more involved in the general matrilineal rank systems, while males are involved in frequent 'power struggles'. A more sophisticated analysis of sex differences in coalitions is needed to confirm these suggestive trends.

However, escalated aggressive interactions are generally less frequent than dyadic approach–retreat interactions or supplants. Supplants and avoids are unlikely to attract alliance support, but due to their relative frequency and their effect on resource acquisition, they may be extremely influential in rank acquisition and maintenance by male and female immatures. The presence of an ally in close proximity can affect both the likelihood of an approach–

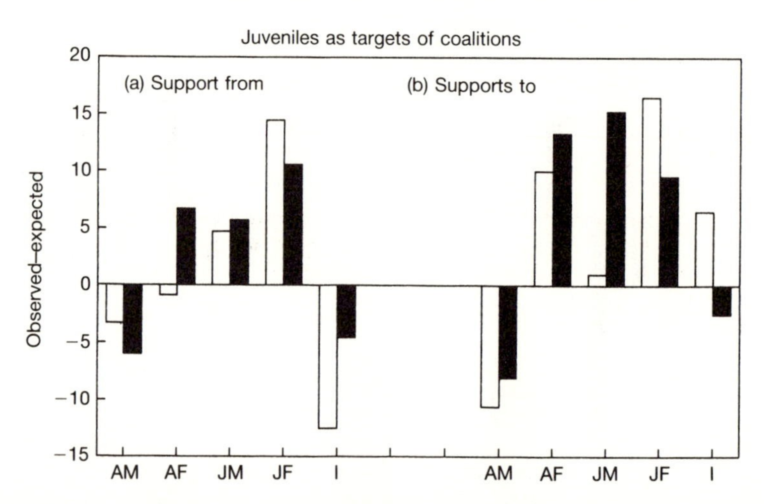

Fig. 14.6. The observed minus the expected number of coalitions where either juvenile male or female vervet monkeys were the target of the coalition. (a) Each age–sex class supports others against juveniles. (b) The different age–sex classes receive support against juveniles. Expected values calculated from mean proportions of each age–sex class available, as Fig. 14.5. Males as targets, $N = 116$; females as targets, $N = 100$. Mothers are excluded since they never targeted their own offspring. Open bars indicate males, solid bars females.

retreat interaction occurring and its outcome; should competition escalate, an alliance partner is present. Rarely occurring events could have persistent effects, and thus relatively few interventions by allies might be sufficient to maintain a rank established through early observations and interactions, especially for the immature females.

The resource under dispute during supplants may also be critical in determining the outcome of rank-related interactions. Such effects should be similar for males and females, although not necessarily over the long term. Contests over resources such as grooming partners (Lee 1983*a*, 1987) or foods of low quality and wide abundance (Johnson 1989; Shopland 1987) can produce outcomes not predicted by the juvenile hierarchies for either sex. The juveniles involved in such interactions tend to be adjacently ranked and thus with unstable ranks, and the resources appear to have a value unrelated to their short-term benefit (e.g. removal of parasites or nutritional intake). The benefits from winning such encounters lie in their effect on rank, reinforcing subtle differentials in status derived from the more visible alliances, but in a context with low costs attached to losing the resource. Both vervet and baboon immature males and females frequently supplant closely ranked rivals over grass, a relatively low-quality, widely available resource (Johnson 1989; Lee 1983*a*; Shopland 1987). For baboons, supplants by closely or ambivalently ranked immatures over the lower quality resource were more common than those over preferred foods and such contests were especially marked for young males (Fig. 14.7) (Johnson 1989).

Among vervets, reversals against the hierarchy were infrequent over food resources. When they did occur, smaller, younger, and subordinate animals were likely to attempt to supplant a dominant juvenile when the subordinate's mother was nearby. They almost never attempted to supplant a

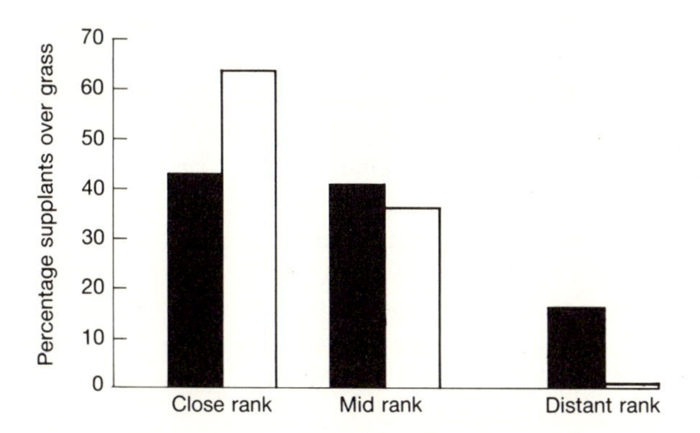

Fig. 14.7. The percentage of supplants by juvenile baboons over grass foods where the difference in rank between participants was within 4 ranks (close), between 5–10 ranks (mid) and over 10 (distant). Open bars indicate males, solid bars females. (Data from Johnson 1989.)

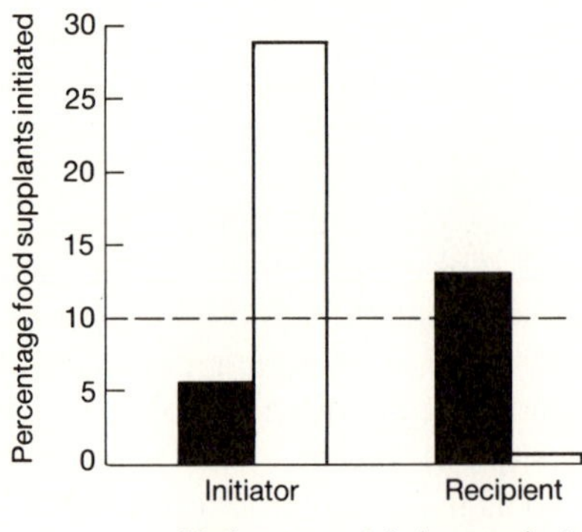

Fig. 14.8. The percentage of supplants over food initiated by dominant juvenile vervet monkeys (solid bars, total $N = 221$) or subordinates (open bars, total $N = 22$) when either the initiator or recipient's mother was within 2 m of the interaction. The median percentage of time with mothers nearby is given by the dotted line. (Data from Lee 1983a.)

dominant immature when that animal's mother was nearby (Fig. 14.8) (Lee 1983*a*; see also Chapais this volume pp. 29–59; Datta 1983*a*).

As discussed above, when a dominant animal is supplanted from a grooming partner, that partner is often related to the subordinate initiator of the supplant (Lee 1983*a*). Subordinates appear willing to initiate an encounter with a typically dominant animal when they stand to receive grooming and when the grooming partner is a reliable ally such as a relative. Such patterns are also common in contests between siblings. Supplants over grooming partners are more frequent between vervet siblings than non-kin, and the youngest sibling is most likely to succeed in gaining grooming, irrespective of sex (Lee 1987). Younger rhesus siblings are also more likely to receive maternal support during a dispute than are older siblings, encouraging the young animals to challenge their previously dominant older sibling (Datta 1988). Such contests appear to be unrelated to offspring sex.

Supplants over food or grooming partners appear to occur within the context of alliances, potential or realized. They may serve two additional functions in terms of rank acquisition: to assess and challenge the status of other immatures, which may be especially important for young males and, depending on the effectiveness of the challenge, to perpetuate existing rank, which may be critical for females.

DISCUSSION AND CONCLUSIONS

Immature primates have a status within their group that is more than a pale reflection of their ultimate rank as adults. The mechanisms of rank acquisition differ between immature male and female cercopithecines (Table 14.1). These mechanisms depend on attributes of the immature such as age and strength, the nature of contested resources, and characteristics of coalition partners and allies during competition or aggression. Differences are initiated during infancy as a result of interactions between infants and other group members, and the observations of the outcome of interactions between others. During infancy, both sexes tend to be sheltered to some extent from aggression, and they are frequently supported by mothers, siblings, and unrelated group members. With age, sex differences begin to emerge. Females challenge others in contexts where coalitionary support is likely to follow, while males appear to engage in contests where their power is an important determinant of the outcome. The difference between the types of agonistic interaction and their separate effects on rank acquisition among immature primates are only beginning to be explored.

Why should there be differences in the nature of dominance relationships between males and females prior to reproductive maturity? Adult differences are rooted in the nature of cercopithecine social structure, and in the degree and nature of competition between members of the same sex. Female alliances influence the stability of ranks, access to food, and ultimately reproductive success. Male contests may also involve cooperating with other males to enhance access to reproductive females, or with females to form stable social partnerships ensuring their acceptance within a group. Social skills are thus essential for reproductive success in both sexes and influence the outcome of contests.

However, it appears that females use their social skills for perpetuating lifetime alliances within stable groups, while males respond to relatively short-term social opportunities during their opportunistic movements between groups. Differences between the sexes of juveniles are expected, both because the functions of adult dominance relationships vary, and because the needs of the immatures themselves vary in terms of their access to resources and the establishment of long-term alliances.

For juvenile females, especially high-born daughters, establishing a position predicted by their maternal rank is critical both to the individual female and to her kin, in effect ensuring the position of a matrilineal alliance over generations. For low-born females, their prospects for defending themselves against harassments by higher-ranking females lies in acting together with matriline members. The interests of the immature females are similar to those of the adults in terms of the attainment of rank.

Young females appear to compete over social resources, for example

Table 14.1. Summary of sex differences in rank acquisition and life history variables for immature cercopithecine primates.

Developmental stage	Females	Males
Infant: dependent on mother for nutrition, transport and most interactions		
	25% of adult wt	20% of adult wt (25% of female wt)
	Harassments directed to subordinate mothers	Friendly contacts from infant handlers
	Supported by others during aggression	Supported by others during aggression
	Preferentially groomed by mother	Preferentially groomed by mother
Young juvenile: independent of mother physically, dependent socially forming cohesive family units with siblings		
	45% of adult wt	35% of adult wt (60% of female wt)
Contests with:		
Adults	Integrated into adult female hierarchy	Acquire maternal rank in contests with adult females
Cross-sex	Subordinate to males unless much older/higher born	Dominant to females unless much smaller/lower born
Same sex	Peer rank = maternal rank	Peer rank = size/age rank
Old juvenile/adolescent		
	Maintains cohesiveness within family	Peripheralization from group; maintains some contacts with mother
	65% of adult wt	50% of adult wt (80% of female wt)
Contests with:		
Adults	Targeted by females	Dominant to most females Supported by mothers against adult males prior to transfer
Cross-sex	Subordinate to males	Dominant to females
Same-sex	Established in matrilineal hierarchies for grooming and aid	Contest low value resources to reinforce peer ranks Form social subunits based on peer ranks

access to grooming partners, or high-ranking allies (Cheney 1978). This attainment of status while immature translates into the ability to compete effectively for nutritional resources as adults, when access to food can have a direct influence on reproductive success. Contests over social resources continue among adult females (Seyfarth 1980), but these occur in relation to the maintenance of alliances by reproductively active females (e.g. Cheney and Seyfarth 1986). Such a pattern is similar to the establishment of dominance between sisters (e.g. Datta 1988).

Among immature males, sons of high-ranking females also gain advantages from their mothers' rank, especially in contests with females over access to food. Sons of lower-ranking mothers may be placed at a short-term disadvantage in contests with adult females, but they literally outgrow this either before or during their pupertal growth spurt. The dominance of male juveniles over most juvenile females in contests over food also gives them an advantage during times of shortage (e.g. Dittus 1979). In contests with peers, the establishment of ranks based on individual attributes and abilities, which influences other social interactions such as play, may enable young males to develop skills necessary during contests as an adult. Young males thus appear to distribute their affiliative interactions in relation to relative peer ranks (Colvin and Tissier 1985), affecting access to social resources during immaturity.

The apparent similarities between the sexes of juveniles in the establishment of dominance lie in the social nature of the contests, in the support given and received, and the pervasive influence of maternal rank on aspects of offspring rank. The differences, then, relate to the duration of the influence of maternal rank, which is greater for daughters than for sons and which directly affects only the younger stages of the male life history. The nature of the resources contested tends to be similar—both sexes compete for food and for social partners. Females compete for social partners in the context of their alliances, matrilineal or otherwise. Males tend to concentrate their contests on those with immediate benefits, either in terms of gaining the resource or in terms of practising competitive skills.

Differences in life histories of males and females provide the basis for differences in systems of ranking, in the nature of contested resources and the outcome of contests. The importance of dependent rank—mother's rank—increases throughout development for daughters and decreases for sons. The reverse is true of basic rank—those skills and abilities intrinsic to the individual. Females are dependent on their matrilineal alliance partners throughout life for general assistance with offspring and attaining access to resources, but become increasingly so upon reaching reproductive maturity. Males are dependent on their matrilineal allies early in life, to enhance food intake and growth rates, and to protect them from harassment. With age, and especially after transfer to the reproductive group, males depend more on their own basic or acquired skills, on their ability to form short-term coalitions with other males, and affiliative relationships with reproductive females. Dependent rank among juvenile males can be related to alliances with males, brothers and adults, as well as to the rank of their mothers. The switch from a matrilineally influenced dominance rank to an intrinsic rank is based on patterns of migration, and social and reproductive opportunities specific to a novel situation in the transfer group.

Rank acquisition among immature primates is a complex and lengthy process, related to the immediate needs for access to resources in competition with other group members, and longer term needs for establishing cooperative social partnerships. The relative importance of these needs varies for males and females during development and throughout their lifetime, resulting in gender-specific systems of rank acquisition.

SUMMARY

1. In cercopithecine primates, the acquisition of dominance status within the natal group is a process of considerable duration, lasting from early infancy until reproductive maturity.
2. Cooperative alliances, especially with close kin, have considerable influence on the acquisition and maintenance of dominance rank.
3. One important factor influencing the initial rank of the immature and the nature and duration of the process of rank acquisition is that of gender: males and females acquire different types of dominance status and do so in diferent ways.
4. Since the needs of males and females differ both during development and as adults, the patterns of rank acquisition and the way in which coalitions and alliances influence these patterns can be related to the life history parameters of the two sexes. (1) Males gain experience and initial competitive benefits from a ranking close to that of their mothers in contests with adult females in their natal group as a result of support in competition. At the same time, they gradually attain a rank relative to adult and peer males that is dependent on size, strength and fighting skills, benefiting them after transfer to groups where they are reproductively active. Dependent rank changes to basic rank. (2) Females who remain in their natal groups acquire the relative rank of their mothers in a variety of interactions with females of all ages. Their rank is acquired and maintained largely as a result of support from family members. Matrilineal rank orders remain stable over the long term as long as networks of female allies persist. Juvenile females tend to rank in accordance with maternal rank in interactions with youngest males. However, as males approach adolescence and increase to approach the size of adult females, females become subordinate to males in dyadic contests, at least in part because their kin are no longer effective supporters against the larger males.

5. Juvenile dominance systems are complex reflections of the immediate needs of immatures and these can be observed in the resources contested by young animals. The rank systems develop to resemble those of the adults as the requirements of the young animals change with increasing age and maturity, and as the relative competitive abilities of the contestants and their main supporters change as the young animals mature.

ACKNOWLEDGMENTS

We thank the L.S.B. Leakey Trust for funding PCL's research and the Boise Fund and the Sigma Xi Society for funding J.A.J.'s research. We are grateful to the Government of the Republic of Kenya and the Institute of Primate Research for research permission and affiliation during field work. We thank D. A. Collins, S. B. Datta, and M. Janus for their advice and constructive criticism and the editors for their comments on the draft.

REFERENCES

Altmann, J. (1980). *Baboon mothers and infants*. University of Chicago Press, Chicago.

Berkovitch, F. B. (1988). Coalitions, cooperation and reproductive tactics among adult male baboons. *Animal Behaviour*, **38**, 1198–209.

Berman, C. M. (1980). Early agonistic experience and rank acquisition among free-ranging infant rhesus monkeys. *International Journal of Primatology*, **1**, 152–70.

Berman, C. M. (1982). The ontogeny of social relationships with group companions among free-ranging infant rhesus monkeys I. Social networks and differentiation. *Animal Behaviour*, **30**, 149–62.

Berman, C. M. (1986). Maternal lineages as tools for understanding infant social development and social structure. In *The Cayo Santiago macaques* (eds R. G. Rawlins and M. J. Kessler), pp. 73–92. State University of New York Press, Albany.

Chapais, B. (1983). Matriline membership and male rhesus reaching high ranks in their natal troop. In *Primate social relationships* (ed. R. A. Hinde), pp. 171–5. Blackwell, Oxford.

Chapais, B. (1988). Experimental matrilineal inheritance of rank in female Japanese macaques. *Animal Behaviour*, **38**, 1025–37.

Cheney, D. L. (1977). The acquisition of rank and the development of reciprocal alliances among free-ranging immature baboons. *Behavioural Ecology and Sociobiology*, **2**, 303–18.

Cheney, D. L. (1978). Interactions of immature male and female baboons with adult females. *Animal Behaviour*, **26**, 389–408.

Cheney, D. L. (1983). Proximate and ultimate factors related to the distribution of male migration. In *Primate social relationships* (ed. R. A. Hinde), pp. 241–9. Blackwell, Oxford.

Cheney, D. L. and Seyfarth, R. M. (1986). The recognition of social alliances by vervet monkeys. *Animal Behaviour*, **34**, 1722–31.

Collins, D. A. (1986). Interactions between adult male and infant yellow baboons (*Papio c. cynocephalus*) in Tanzania. *Animal Behaviour*, **34**, 430–43.

Colvin, J. (1983). Familiarity, rank and the structure of rhesus male peer networks. In *Primate social relationships* (ed. R. A. Hinde), pp. 190–200. Blackwell, Oxford.

Colvin, J. (1986). Proximate causes of male emigration at puberty in rhesus monkeys. In *The Cayo Santiago macaques* (eds R. G. Rawlins and M. J. Kessler), pp. 131–58. State University of New York Press, Albany.

Colvin, J. and Tissier, G. (1985). Affiliation and reciprocity in sibling and peer relationships among free-ranging immature male rhesus monkeys. *Animal Behaviour*, **33**, 959–77.

Cords, M. (1988). Resolution of aggressive conflicts by immature long-tailed macaques *Macaca fascicularis*. *Animal Behaviour*, **36**, 1124–35.

Datta, S. (1983*a*). Relative power and the acquisition of rank. In *Primate social relationships* (ed. R. A. Hinde), pp. 93–102. Blackwell, Oxford.

Datta, S. (1983*b*). Relative power and the maintenance of dominance. In *Primate social relationships* (ed. R. A. Hinde), pp. 103–12. Blackwell, Oxford.

Datta, S. (1988). The acquisition of dominance among free-ranging rhesus monkey siblings. *Animal Behaviour*, **36**, 754–22.

Datta, S. (1989). Demographic influences on dominance structure among female primates. In *Comparative socioecology* (eds V. Standen and R. A. Foley), pp. 265–84. Blackwell, Oxford.

Dittus, W. P. J. (1979). The evolution of behaviours regulating density and age-specific sex ratios in a primate population. *Behaviour*, **69**, 265–302.

Dunbar, R. I. M. (1980). Determinants and evolutionary consequences of dominance among female gelada baboons. *Behavioural Ecology and Sociobiology*, **7**, 253–65.

Harcourt, A. H. (1987). Dominance and fertility among female primates. *Journal of Zoology*, London, **213**, 471–87.

Harcourt, A. H. and Stewart, K. J. (1987). The influence of help in contests on dominance rank in primate: hints from gorillas. *Animal Behaviour*, **35**, 182–90.

Hausfater, G. (1975). *Dominance and reproduction in baboons*. Karger, Basel.

Hausfater, G., Altmann, J., and Altmann, S. A. (1982). Long-term consistency of dominance relations among female baboons (*Papio cynocephalus*). *Science*, **217**, 752–5.

Hiraiwa, M. (1981). Maternal and alloparental care in a troop of free-ranging Japanese monkeys. *Primates*, **22**, 309–29.

Horrocks, J. and Hunte, W. (1983*a*). Maternal rank and offspring rank in vervet monkeys: An appraisal of the mechanisms of rank acquisition. *Animal Behaviour*, **31**, 772–82.

Horrocks, J. and Hunte, W. (1983*b*). Rank relations in vervet sisters: a critique of the role of reproductive value. *American Naturalist*, **122**, 417–21.

Hrdy, S. B. and Hrdy, D. B. (1976). Hierarchical relations among female hanuman langurs (Primates: Colobinae, *Presbytis entellus*). *Science*, **193**, 913–15.

Janus, M. (1989). Social Development and Behavioural Reciprocity in Young Rhesus Monkeys with their Siblings and Non-siblings. Ph.D. thesis, University of Cambridge.

Johnson, J. A. (1984). Social Relationships of Juvenile Olive Baboons. Ph.D. thesis, University of Edinburgh.

Johnson, J. A. (1987). Dominance rank in juvenile olive baboons, *Papio anubis*: the influence of gender, size, maternal rank and orphaning. *Animal Behaviour*, **35**, 1694–708.

Johnson, J. A. (1989). Supplanting by olive baboons: dominance rank difference and resource value. *Behavioural Ecology and Sociobiology*, **24**, 277–83.

Kawai, M. (1958). On the system of social ranks in a natural group of Japanese monkeys. *Primates*, **1**, 11–48.

Koford, C. B. (1963). Ranks of mothers and sons in bands of rhesus monkeys. *Science*, **141**, 356–7.

Koyama, N. (1967). On dominance rank and kinship of a wild Japanese monkey troop in Arashiyama. *Primates*, **8**, 189–216.

Kuester, J. and Paul, A. (1988). Rank relations of juvenile and subadult natal male Barbary macaques (*Macaca sylvanus*) at Affenberg, Salem. *Folia primatologica*, **51**, 33–44.

Lee, P. C. (1983*a*). Context-specific unpredictability in dominance interactions. In *Primate social relationships* (ed. R. A. Hinde), pp. 35–44. Blackwell, Oxford.

Lee, P. C. (1983*b*). Caretaking of infants and mother-infant relationships. In *Primate social relationships* (ed. R. A. Hinde), pp. 146–51. Blackwell, Oxford.

Lee, P. C. (1987). Sibships: cooperation and competition among immature vervet-monkeys. *Primates*, **28**, 47–59.

Lee, P. C. (1989). Family structure, communal care and female reproductive effort. In *Comparative socioecology* (eds V. Standen and R. A. Foley), pp. 323–40. Blackwell, Oxford.

Lee, P. C. and Oliver, J. I. (1979). Competition, dominance and the acquisition of rank in juvenile yellow baboons (*Papio cynocephalus*). *Animal Behaviour*, **27**, 576–85.

Manzolillo, D. L. (1986). Factors affecting intertroop transfer by adult male *Papio anubis*. In *Primate ontogeny, cognition and social behaviour* (eds J. G. Else and P. C. Lee), pp. 371–80. Cambridge University Press, Cambridge.

Missakian, E. A. (1972). Genealogical and cross-genealogical dominance relations in a group of free-ranging rhesus monkeys (*Macaca mulatta*) on Cayo Santiago. *Primates*, **13**, 169–80.

Moore, J. (1978). Dominance relations among free-ranging female baboons in Gombe National Park, Tanzania. In *Recent advances in primatology* I (eds D. J. Chivers and J. Herbert), pp. 67–70. Academic Press, London.

Mori, A. (1979). Analysis of population changes by measurement of body weight in the Koshima troop of Japanese monkeys. *Primates*, **20**, 371–97.

Netto, W. J. and van Hooff, J. (1986). Conflict interference and the development of dominance relationships in immature *Macaca fascicularis*. In *Primate ontogeny, cognition and social behaviour* (eds J. G. Else and P. C. Lee), pp. 291–300. Cambridge University Press, Cambridge.

Noë, R. (1986). Lasting alliances among adult male savannah baboons. In *Primate ontogeny, cognition and social behaviour* (eds J. G. Else and P. C. Lee), pp. 381–92. Cambridge University Press, Cambridge.

van Noordwijk, M. A. and van Schaik, C. P. (1988). Male careers in Sumatran long-tailed macaques (*Macaca fascicularis*). *Behaviour*, **107**, 24–43.

Packer, C. R. (1979*a*). Inter-troop transfer and inbreeding avoidance in *Papio anubis*. *Animal Behaviour*, **27**, 1–36.

Packer, C. R. (1979*b*). Male dominance and reproductive activity in *Papio anubis*. *Animal Behaviour*, **27**, 37–45.

Pereira, M. E. (1988). Agonistic interactions of juvenile savanna baboons I. Fundamental features. *Ethology*, **79**, 195–217.

Pereira, M. E. (1989). Agonistic interactions of juvenile savanna baboons II. Agonistic support and rank acquisition. *Ethology*, **80**, 152–71.

Raleigh, M. J. and McGuire, M. T. (1989). Female influences on male dominance acquisition in captive vervet monkeys *Cercopithecus aethiops sabaeus*. *Animal Behaviour*, **38**, 59–67.

Sade, D. S. (1967). Determinants of dominance in a group of free-ranging rhesus monkeys. In *Social communication among primates* (ed. S. A. Altmann), pp. 99–114. University of Chicago Press, Chicago.

Samuels, A., Silk, J. B., and Altmann, J. (1987). Continuity and change in dominance relations among female baboons. *Animal Behaviour*, **35**, 785–93.

van Schaik, C. P. (1989). The ecology of social relationships amongst female primates. In *Comparative socioecology* (eds V. Standen and R. A. Foley), pp. 195–218. Blackwell, Oxford.

Schulman, S. R. and Chapais, B. (1980). Reproductive value and rank relations among macaque sisters. *American Naturalist*, **115**, 580–93.

Seyfarth, R. M. (1980). The distribution of grooming and related behaviour among adult female vervet monkeys. *Animal Behaviour*, **28**, 798–813.

Shopland, J. M. (1987). Food quality, spatial development, and the intensity of feeding interference in yellow baboons (*Papio cynocephalus*). *Behavioural ecology and sociobiology*, **21**, 149–56.

Silk, J. B. (1980). Kidnapping and female competition among captive female bonnet macaques. *Primates*, **21**, 100–10.

Silk, J. B. (1987). Social behavior in evolutionary perspective. In *Primate societies* (eds B. B. Smuts, D. L. Cheney, R. M. Seyfarth, R. W. Wrangham, and T. T. Struhsaker), pp. 318–29. University of Chicago Press, Chicago.

Silk, J. B., Samuels, A., and Rodman, P. S. (1981). The influence of kinship, rank, and sex on affiliation and aggression between adult female and immature bonnet macaques (*Macaca radiata*). *Behaviour*, **78**, 111–37.

Smith, D. G. and Smith, S. (1988). Parental rank and reproductive success of natal male rhesus. *Animal Behaviour*, **36**, 554–63.

Smuts, B. B. (1985). *Sex and friendship in baboons*. Aldine Press, New York.

Strum, S. C. (1982). Agonistic dominance in male baboons: an alternative view. *International Journal of Primatology*, **3**, 175–202.

de Waal, F. (1977). The organization of agonistic relations within two captive groups of Java monkeys (*Macaca fascicularis*). *Zeitschrift für Tierpsychologie*, **44**, 225–82.

Walters, J. (1980). Interventions and the development of dominance relationships in female baboons. *Folia primatologica*, **34**, 61–89.

Wrangham, R. W. (1980). An ecological model of female bonded primate groups. *Behaviour*, **75**, 262–300.

15

Dolphin alliances and coalitions

RICHARD C. CONNOR, RACHEL A. SMOLKER, AND ANDREW F. RICHARDS

INTRODUCTION

Recently, much attention has focused on a role for complex social interactions, including alliance formation, in the function and evolution of large brains and intelligence in primates (e.g. Jolly 1966; Humphrey 1976; Kummer 1982; Cheney *et al*. 1986; Alexander 1989; papers in Byrne and Whiten 1988). Based largely on captive studies, Herman (1980) and Connor and Norris (1982), suggested a link between complex social interactions and the large dolphin brain. Until recently, our knowledge of the social lives of free-ranging dolphins was scant. Now, however, long-term studies of several dolphin species are revealing highly complex societies. Here we review: (1) our study of alliances and coalitions among male bottlenose dolphins in Shark Bay, W. Australia; (2) the limited data on coalitions among females at Shark Bay; (3) evidence for alliances in other populations and species. We then compare alliance formation among male dolphins with alliances and coalitions among male and female primates, including humans.

The Monkey Mia Dolphin Study

Following a preliminary visit in 1982 (Connor and Smolker 1985), in 1984 we began a long-term study of a social network of bottlenose dolphins in a 130 km^2 area off the east side of Peron Peninsula, which bisects Shark Bay, 850 km north of Perth. By 1990 we had identified over 300 individuals in our study area. The focal point of the study area is the Monkey Mia campground where, since at least the early 1960s, people have hand-fed 6–12 dolphins in shallow waters by the beach. The combination of several provisioned dolphins that can be closely observed in shallow water at Monkey Mia, and a large number of non-provisioned dolphins in the surrounding waters, provides exceptional observational opportunities (Connor and Smolker 1985).

We conducted a 25-month study of male social relationships in 1987–9 (i.e., March–November 1987; March–December 1988; February–July

1989). Operating in small (3–3.5 m) boats we collected survey and *ad libitum* data on the composition and activity of groups encountered in the bay, and conducted over 700 h of focal individual sampling on 30 known or suspected males. Individuals selected for focal sampling were those encountered frequently during surveys. Most results reported here were obtained during focal and *ad libitum* observations on 21 males (535 focal hours). Three of these 21 males were among the provisioned dolphins which were also observed at close range in the Monkey Mia shallows.

Individual dolphins are identified by the scar patterns and shape of the dorsal fin. A photograph of each individual is kept in an identification catalogue. The sex of 16 of the 21 focal males was determined by direct observation of an erection, two were sexed by the gap (⩾2.5 cm) between the genital and anal slits and a noted lack of mammaries, and three by the gap alone (it is much larger in males). The Monkey Mia dolphins acquire ventral speckles around the time of sexual maturity which can be used as an approximate indicator of age and sexual maturity (Ross and Cockroft 1990; Smolker *et al.* unpublished data).

The Monkey Mia dolphins live in a fission–fusion society (like chimpanzees and spider monkeys) in which party size and membership is variable rather than fixed, and in which all of the members of the social group are never together in one party (Struhsaker and Leland 1979; Symington 1990). Foraging and feeding assemblages of dolphins in Shark Bay are typically ephemeral and poorly defined so we restrict our analysis of associations to resting, travelling, and socializing parties. We include as party members any individuals within 10 m of any other dolphin in the group. The modal distance between individuals in such parties is typically <2 m. The mean size of resting, travelling, and socializing parties from 1985–9 at Monkey Mia was 4.8 non-calf individuals and the range was 2–20, or 22 if calves are included (Smolker *et al.*, unpublished data). Association coefficients are calculated using the 'half-weight' method (Cairns and Schwager 1987). The association coefficient between two individuals A and B is defined as $100 \times 2N_t/(N_A + N_B)$ where N_t is the number of parties in which A and B are found together and N_A and N_B are the total number of party sightings for each individual, respectively. This formula yields numbers ranging from 0, for individuals that are never seen together, to 100, for individuals that are always together.

MALE–MALE ASSOCIATIONS

Males associate in exclusive subgroups of 2–3 individuals called 'pairs' and 'triplets', which are often stable for years. Members of pairs are each other's closest associate. An individual is included as a member of a triplet if (1) he is the second closest associate of the other two members; (2) his coefficient of association with them is within 20 points of their coefficient with each other.

Some male–male association coefficients are in the same range as those found between females and their nursing calves (81–100). Similar high levels of male–male associations in bottlenose dolphins have been reported from Sarasota Bay, Florida (Wells *et al.* 1987) and Indian River, Florida (Duffield *et al.* 1985).

During 1987–8 the 21 focal males could be assigned to eight pairs or triplets. Each pair or triplet of males preferentially associated with one or two other pairs or triplets (Fig. 15.1). Pairs and triplets often travel, forage, and socialize together. In following sections we refer to each pair and triplet by a letter code. The reader should keep in mind that each letter (A–I) represents two or three dolphins, not an individual.

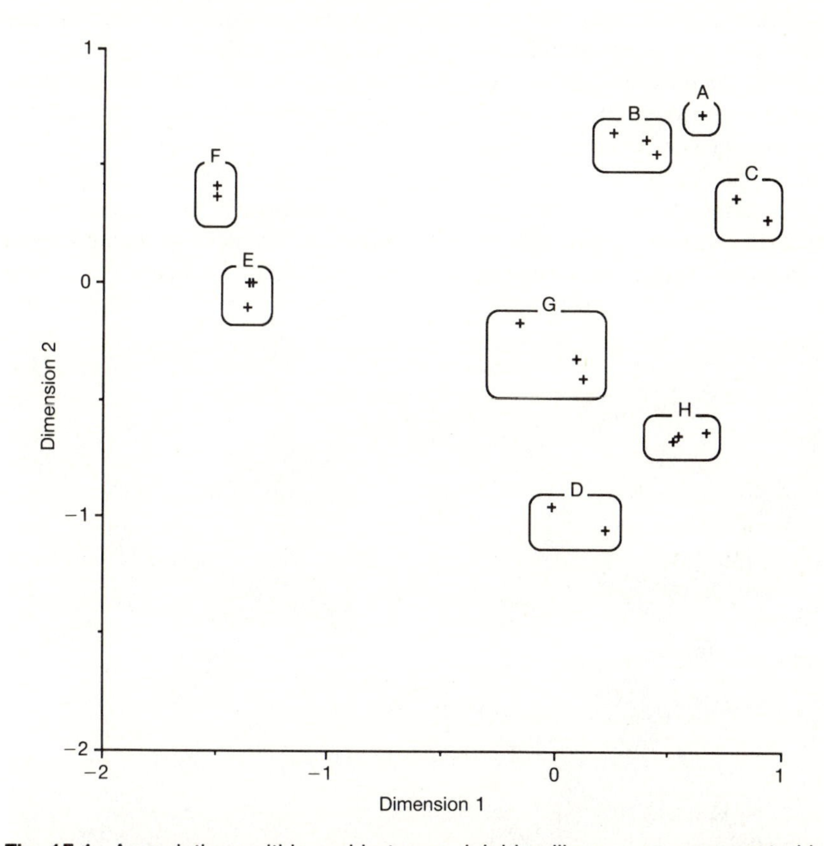

Fig. 15.1. Associations within and between dolphin alliances are represented by scaling, in two dimensions, the association matrix for the 21 focal males in all male groups (Kruscal's stress formula 1, stress = 0.048). The letters correspond to those given for alliances in the text and Table 15.1. Each point represents one individual, except for one point in A and one in C in which the points for two dolphins overlap exactly.

MALE–MALE ALLIANCES: HERDING AND THEFTS OF FEMALES

Males in pairs and triplets jointly herd females. Herding begins with the capture of a female by two or more males, and ends when the female escapes (she may or may not be pursued by the males). When travelling with a herded female, males are usually positioned on either side and just behind the female or abreast behind her (Fig. 15.2). Aggression which males direct at herded females includes chasing, hitting with the tail, 'head jerks' (sharp lateral or vertical movements of the head), charging, biting, and slamming bodily into the female. Males enforce herding partly by making a 'popping' vocalization that induces the herded female to approach. Data from 12 herded females show that females are significantly more likely to approach a popping male than the same male when he is not popping (unpublished data). If the female does not approach she may be threatened or attacked. Males often perform spectacular synchronous displays around herded females (Fig. 15.3). We refer to pairs and triplets of males as 'first-order' alliances.

The associations between first-order alliances were initially surprising; if pairs and triplets simply competed for access to females, we would expect exclusively hostile relationships between first-order alliances. We subse-

Fig. 15.2 Males traveling in 'formation' with herded female. The males remain on either side and just behind the female.

Fig. 15.3 Synchronous display. Males often engage in synchronous displays around a herded female. In this case, the two males are performing 'belly-slaps' on either side of the female. Note that the males are going in opposite directions, an example of a 'reverse-parallel' display.

quently discovered a basis for affiliative associations between some pairs and triplets: two first-order alliances may form coalitions to aggressively take a female being herded by another pair or triplet. Because the first-order alliances that form coalitions for thefts also exhibit relatively high levels of association in all-male groups (Fig. 15.1) we refer to these associations as 'second-order' alliances (Fig. 15.4).

FIRST-ORDER ALLIANCES: HERDING FEMALES

Herding by the provisioned males

We discovered aggressive herding of females while observing the three males that visit the provisioning area. Because all 30 herded dolphins we were able to sex were female (observed mammary slits or accompanied by a dependent calf), we assume here that all 48 individuals herded by the provisioned males were female. We observed 19 captures by the provisioned males, of which 17 included obvious chases and two merely involved the males rushing up to and around the female. In the most prolonged capture, chasing, hitting, and

Fig. 15.4 Second-order alliance. A pair and a triplet of males travelling together. Two first-order alliances will cooperate to take females being herded by other first-order alliances, or to defend females from theft attempts.

displaying continued for 85 min over a 7 km distance. The provisioned males, all large speckled individuals, often swam into the Monkey Mia shallows with a herded individual in tow. For unhabituated females being herded into Monkey Mia for the first time, the proximity to human activity was clearly upsetting as they often attempted to escape by 'bolting'. Females who were herded repeatedly during a season sometimes became so accustomed to the presence of humans that they began to solicit fish handouts, yet still sought to escape from the males. Herded females bolted from the provisioned males on 179 occasions, escaping in 45 (25 per cent) of the attempts. The males chased in 25 (56 per cent) of the escapes.

Only two of the three provisioned males herded together at a time; the two males herding together were said to be 'partners' and the third was referred to as the 'odd-male-out.' There were frequent 'partner changes' in which two of the three provisioned males herded together. Thus, on one occasion, the males Snubby and Sickle would herd a female together, and on a subsequent occasion the males Snubby and Bibi might herd together. The odd-male-out would sometimes herd with either member of a pair of non-provisioned males (alliance D) that formed a second-order alliance with the provisioned males. Complicating matters further, a third non-provisioned male (previously known from only two sightings) suddenly began associating (and herding) with alliance D in the Monkey Mia area in August 1988. Because the provisioned males visited the Monkey Mia shallows every day, we were able to monitor pairings and partner changes among the provisioned males on a daily basis.

A herding event begins with the capture of a female and ends when she escapes or is released by the males. However, because we could not monitor the provisioned males round the clock, we can only estimate the duration of herding events. On 13 occasions, we observed a female escape only to be herded again by the provisioned males later the same day or on the following day. We can reasonably assume, then, that females occasionally escaped and were recaptured between observation periods. Thus, if we define the length of a herding event as the number of consecutive days that a female is herded, we will underestimate the total number of events.

The vast majority of partner changes among the three provisioned males occurred when they were not herding a female. In 8 of the 13 cases in which a female escaped and was recaptured by the next day, a partner change occurred between events. Thus, partner changes can be used to detect some of the cases where a female escaped and was recaptured while we were not observing. However, incorporating partner changes into our definition of a herding event may also bias our estimation upwards to the small extent that partner changes occur while a female is being herded. Incorporating partner changes into the definition based on the number of consecutive days a female was herded increases the number of 240 herding events derived from the 'consecutive female day' method by only 15 events (6 per cent). Using the partner-change method we recorded 255 suspected cases of herding by two provisioned males or one of the provisioned males and a member of alliance D. In this paper, we will employ the 'partner change' definition for herding events.

For provisioned and non-provisioned males, we recorded a 'confirmed' herding event upon observing any of the following: (1) a capture; (2) a bolt (escape attempt); (3) a male 'popping'; (4) a theft, in which two alliances aggressively take a female from another alliance. A fifth criterion was employed for the provisioned males: aggression in the form of a charge, head-jerk, or hit directed at a non-provisioned female in the provisioning area. At least one of these behaviours was observed in 208 (82 per cent) of the 255 suspected cases and in 142 (95 per cent) of 149 events in which there was at least 20 min of continuous observation of the males with the female. These data indicate that the females were being herded in nearly all, if not all, of the 255 events.

Herding by non-provisioned males

During 1987–9 we confirmed 58 of 100 suspected cases of herding by nine non-provisioned alliances on the basis of observations of a capture, a bolt, pops, or a theft. A Mantel test (Mantel 1967; Schnell *et al.* 1985) reveals a strong association between male–male association in all-male groups and joint participation in herding for 13 non-provisioned focal males ($t = 6.65$, $p < 0.001$; Table 15.1).

Table 15.1. Association coefficients from all-male groups (upper right) and the number of times males herded together (lower left) for the 13 non-provisioned focal males who were observed herding. Blank cells have zero values. A Mantel test ($t = 6.65$; $p < 0.001$) reveals a strong association between male–male association and herding partners. The numbers above each dolphin's letter code are the number of sightings in all male groups for each individual.

	12	12	24	26	20	21	23	21	31	28	35	31	25
	A1	A2	B1	B2	B3	C1	C2	C3	D1	D2	E1	E2	E3
A1	—	100	44	37	44	24	39	24					
A2	15	—	44	37	44	24	39	24					
B1			—	92	82	13	21	13					
B2			15	—	83	8	16	8					
B3			15	16	—	14	23	14					
C1						—	91	100					
C2	1	1	1			10	—	91					
C3						10	10	—					
D1									—	88			
D2									5	—			
E1											—	93	81
E2											1	—	80
E3											1	1	—

PARTNER CHANGES: COMPETITION FOR HERDING PARTNERS

The three provisioned males were not a consistent triplet in 1987–8. First, association coefficients based on sightings in offshore groups suggest that relationships among the provisioned males during 1987–8 (coefficients of 67, 74, and 82) were not as strong as those between individuals in the three non-provisioned triplets of adult-sized, speckled males under observation (coefficients ranged from 83–91; 93–100, and 81–95). Second, while we have identified an odd-male-out in non-provisioned triplets based on association data, the phenomenon of an odd-male-out herding with a member of another first-order alliance was rare in the other triplets but common among the provisioned males, especially in 1988. We observed a total of 24 confirmed and unconfirmed cases of herding by the triplet B and 16 by the triplet C, but only two cases of herding by a pair formed from a member of each. By contrast, in 75 of 236 total cases of herding involving the provisioned males, one of the three provisioned males was paired with a member of the D alliance (we did not know the partners in 19 events that occurred early in the study, which were reported to us by the rangers, or for which we made inadequate observations).

The relationships among the provisioned males were complex and appear to have developed during 1987–88 toward a solidification of the pair Sickle–Bibi. Table 15.2 presents the number of events in which each provisioned male herded with another provisioned male or with one of the members of the *D* alliance. In 1987 Sickle and Bibi herded twice as much with each other as either did with Snubby. The proportion of Sickle-Bibi pairings increased significantly in 1988 ($G = 10.642$, $p = 0.005$); they herded together five times more than either did with Snubby.

The changes in relationships among the provisioned males in 1987–8 were accompanied by an increase in herding by the odd-male-out with members of the D alliance. In 1987 a member of the D alliance, Wave or Shave, herded

Table 15.2. Change in herding relationships among the provisioned males (Snu, Sic, Bib) and between the provisioned males and the non-provisioned D alliance: Sha, Wav, and Spu. 1988 was characterized by greater participation by Wav and Sha in herding with the provisioned males (particularly Snu) and a reduction in herding by Snu with the other provisioned males, Sic and Bib.

	1987	1988
Sic–Bib	41	55
Sic–Snu	24	9
Bib–Snu	20	12
Snu–Sha	6	29
Snu–Wav	3	13
Bib–Sha	1	1
Bib–Wav	2	3
Sic–Sha	1	4
Sic–Wav	1	9
Sic–Spu	****	2

with the off-male-out in 14 events, but in 1988 they were involved in 61 such herding events. The dramatic increase in herding by the odd-male-out in 1988 coincided with our first observations of Wave and Shave herding by themselves. We suspect that the maturation of Shave during our study played a role in this development. Shave was much smaller than the others in 1987, but grew considerably during the study period and was observed to have some ventral speckling in 1989.

The occurrence of partner changes suggests an element of competition for herding partners among the provisioned males. Partner changes could be considered non-competitive if the role of odd-male-out was taken voluntarily when a male was not interested in herding. However, herding by the odd-male-out with members of other alliances demonstrates an interest in herding. Also, on two occasions a female escaped and was recaptured on the same day, but herded the second time by a male who had been odd-male-out (and ignored her) earlier in the day. Further, an elaborate interaction called a 'circle competition' preceded two partner changes and one attempted partner change that we observed.

In each circle competition, the three provisioned males swam abreast within one metre of each other and often in contact. The males swam in tight circles 2–4 m in diameter, while the dolphin on the outside of the circle performed excited displays such as rapid flailing of his flukes or alternating rostrum and fluke slaps on the water surface. While engaged in displays, the outside male also actively petted the middle male with his pectoral fin, or simply pressed the fin against the side of the central male. The two inside males turned away from the excited or agitated outside male who appeared to be trying to establish or re-establish a herding bond with the middle dolphin. In one videotaped case, the dolphin in the inside position initiated the turns. In all three cases, the individual on the outside of the circle became the odd-male-out following the circle competition. In two cases this male had been part of the herding pair before the circling began.

We speculate that the provisioning itself may have contributed to the instability of relationships among the provisioned males. Competition for fish handouts was intense, and the herded female typically swam several metres offshore, in the opposite direction from that in which fish were offered. Females often attempted to escape during feeding periods and their motivation to escape from Monkey Mia, where they typically were unable to forage, may have been unusually high. The males sometimes took turns in remaining just offshore 'guarding' a herded female, but we suspect that the location of the female and fish in opposite directions exacerbated conflicts among the males over guarding versus feeding. It follows that the frequency and intensity of aggression directed at herded females by the provisioned males may also have been unusually high.

Instability before the mating season

Another interesting facet of partner changes is the occurrence of a striking seasonal shift in the frequency of partner changes in both 1987 and 1988. The frequency of partner changes, defined on the basis of observing a change in which two males herded together, was highest late in the austral winter (July–August) before declining sharply in the spring and early summer (September–December, Fig. 15.5). The number of partner changes per month during 1987–8 was significantly higher for the period May–August than for the months September–December (Mann–Whitney U test, $U = 0$, $p = 0.001$). September marks the beginning of the mating season in Shark Bay, as defined by the time of births (given a 12-month gestation period). The shift in partner stability from July–August to September–December may have a

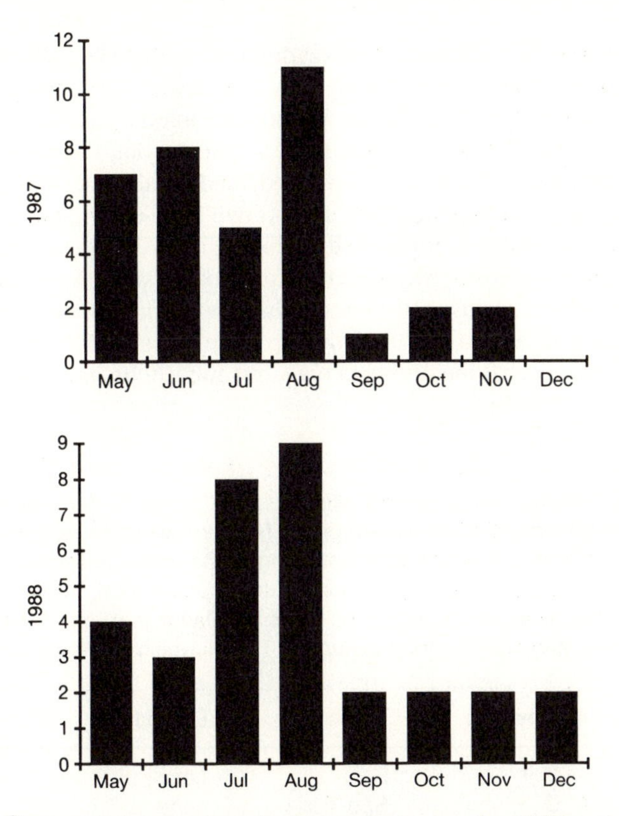

Fig. 15.5. Seasonal changes in pair change frequency for 1987 and 1988. The bars represent the number of pair changes that occurred in each month. Data are not available for December 1987. More pair changes occurred in the winter months (May–August) than in the spring and summer months (September–December) (Mann–Whitney U test, $U = 0$, $p = 0.001$).

hormonal correlate. Male *Tursiops* exhibit a marked surge in serum testosterone levels several weeks prior to the mating season. Testosterone levels then decline to low levels during the mating season, when sperm production and density are highest (Schroeder and Keller 1989).

The competitive nature of partner changes seems clear, but we do not know whether a high frequency of partner changes reflects higher or lower levels of competition. We also do not know how copulations are distributed between a pair of males with a herded female, as we have never observed an intromission. However, in 8 of 26 herding events in which we observed a male mount a herded female, we observed two males mount the female synchronously from either side, and sometimes both males had erections.

SECOND-ORDER ALLIANCES: THEFTS

We observed six thefts. To assist the reader, we have listed the salient features of each theft in Table 15.3, and shall refer to each theft by number in the text. In each theft, the attacking alliance was accompanied by another alliance. In theft 1, we were unable to see whether the accompanying alliance assisted the attacking alliance. In thefts 5 and 6 the defending alliance was assisted by another alliance resulting in a 'two against two' (five individuals against five individuals) encounter. Cooperative defence against thefts may explain why, during the mating season, two pairs or triplets often 'shadow' each other, i.e., coordinate their travel direction and turns even though they may be up to 200 m apart. The outcome of the four 'two against one' thefts were decided in less than 2 min each. One of the 'two against two' thefts (theft 6) lasted 10–

Table 15.3. Thefts. The attacking alliances, the 'victims', defenders, and the alliance that obtained the female are given. Non-provisioned alliances are represented by a single letter code corresponding to the codes in the text and Table 15.1. Three-letter codes are given for Snubby, Sickle, Bibi, Wave, Spud, and Shave. An asterisk by an aggressor indicates an alliance that already had a female before, during, and after the theft except for Theft 6 which was joined in progress.

Theft	Aggressors	Target	Defenders	Winner
1	A* C	B	none	C
2	A* B	Snu–Sic	none	B
3	A B	Sic–Bib	none	A
4	E I	Snu–Sha	none	E
5	A* B	Sic–Bib	Snu, Wav, Sha	B
6	A B*	Sic–Bib	Snu, Sha, Spu	A

20 min (it was discovered in progress); chasing and fighting in theft 5 lasted for 70 min and covered 8–9 km. Theft 5 may have been a three against two encounter, as two small males (one clearly immature) from the triplet G joined the theft in progress, but it was unclear if they actively participated.

The following sequence shows that interactions between alliances may be dependent on context. In four thefts (2, 3, 5, and 6) a pair of males A combined with triplet B to take females from the provisioned males H, (which were assisted by alliance D in thefts 5 and 6). In theft 1, A (which already had a female) accompanied C as they took a female from B, and then the next day assisted B as they took a female from the provisioned males (theft 2). A week later (theft 3), A combined with B to take a female from the provisioned males which A then herded.

Recruitment and support intervention

Alliances may recruit other alliances to participate in thefts. During 9 months of observation (1986–7) before thefts 2 and 3, alliances A and B had been sighted in the provisioning area only once each and on both occasions the provisioned males were absent. Prior to both thefts, however, B visited the provisioning area, approaching as close as 4 m to the provisioned males of alliance H with no apparent interaction. Alliance B then left the provisioning area and returned with A 150 and 105 min later, respectively, for the co-operative thefts from the H alliance. In one of these two thefts we found alliances A and B together 1.5 km north of the provisioning area 85 min prior to the theft.

In theft 4, the supporting alliance joined while the theft attempt was in progress. Triplet E began following Snubby and Shave with the female, then initiated a chase and were joined shortly by a pair of non-focal males (I) that came leaping in from a perpendicular direction. The theft was over 75 s after the pair I joined, when Snubby and Shave dropped out of the group. The female bolted three minutes after the theft and E led the chase. Upon catching the female, alliances E and I continued travelling. When they returned to the area where the theft had occurred 30 min earlier, alliance I dropped behind and left in the direction they had originally joined from, leaving E with the female.

Although our sample is small, our observations suggest that participation in second-order alliances by pairs and triplets of male dolphins cannot be accounted for by each having a greater chance of obtaining the female, as in baboons (Noë this volume pp. 295–302). In our total sample of 355 suspected herding events involving provisioned or non-provisioned males, we observed only one case in which an alliance herded two females at the same time. It thus seems highly unlikely that an alliance which was already herding a female would join in a theft attempt in order to acquire a second female for themselves. In four of the six thefts (1, 2, 5, and 6), however, one of the attacking alliances already had a female before the theft. In thefts 1, 2, and 5 the same alliance retained the female throughout the encounter and afterwards; we joined theft 6 in progress so we do not know if the same alliance had the female before or whether there was a switch during the theft. When we arrived, A had the newly taken female and B had the female that had previously been with A or B.

Our detailed observations from theft 5 (videotaped), in which A already had a female, are instructive. In the 12 min before the attack, as A and B approached D and H at Monkey Mia, A seemed hesitant. Alliance A turned away from the beach four times and B appeared to solicit A by turning repeatedly toward them, then back to the beach. Finally, with B in front of A and both alliances facing the beach, all three members of B synchronously accelerated and leapt into the beach shallows, initiating the theft. Alliance A followed slowly. For the next 55 min A typically lagged 20–40 m behind the rest of the chasing and fighting group, catching up to them on five occasions. However, A played a critical role in the theft. Some of the most intense fighting occurred during three occasions when A caught up to the group and once, as the group turned back toward Monkey Mia, A swam out to the side and front of the group and redirected it away from Monkey Mia. The final bout of fighting began when A accelerated up into the group from behind. During the ensuing fighting, alliance B emerged from the group with the female, performing breaches and synchronous displays. However, this left A, who still had their female, fighting with the three provisioned males (D had left by this time). Having travelled over 100 m from the others, B then turned around and travelled back, rejoining the engagement 10 min after they had left. Alliances A and B then jointly defeated H, and left together 70 min after the theft had begun. In this case it was clear that A was assisting B and was at some risk of losing their own female in the confrontation.

We do not know how members of different alliances are related and consequently cannot determine the role kinship might play in thefts. Either reciprocity (Trivers 1971) or pseudo-reciprocity (Connor 1986) could be involved, depending on the nature of the benefits returned to the assisting alliance. Reciprocation of helping is suggested by the observations in which B and then A acquired the female in successive thefts (2 and 3), and in theft 5, in which A assisted B who then returned to assist A after acquiring the female. If, however, A still had mating access to females herded by B (if A is dominant to B), help provided by A would be an example of pseudo-reciprocity. A's benefit would derive from B's selfish behaviour (herding the female). B's returning to help A after B had acquired the female in the two on two theft would, in this scenario, be self-serving: if A lost their female to D then A might take the female from B. Unfortunately, we have no data on dominance rankings of first-order alliances. Observations of symmetrical help by either A and B under similar circumstances would support a reciprocity interpretation. Our observations suggest that one first-order alliance will participate in a theft in order to help another alliance acquire a female, though why they do so remains unclear.

COMPETITION FOR SECOND-ORDER ALLIANCE PARTNERS?

The changes in association between first-order alliances together with a few behavioural observations, suggest that first-order alliances may compete for second-order alliance partners. We traced the associations of several first-order alliances in 1985–9. We illustrate second-order alliance dynamics by outlining the relationships of eight males who belonged to a pair and two triplets during 1987–9: males A1, A2, B1, B2, B3, C1, C2, and C3. In 1985 triplets B and C were a second-order alliance based on association coefficients, while males A1 and A2 were actually members of two different but closely associating pairs. This situation changed with the disappearance of A2's partner in July 1986. On a few occasions in June–August 1986, A1's partner was subjected to aggression from other males, including A1 and A2. In one case, six males from the A and C alliances lined up head-to-head against A1's partner, with one of the males taking his position beside the others after coming from behind A1's partner. The six males then attacked A1's partner. Thus in 1986 A1 appeared to be losing his position in his first-order alliance in the context of a newly forming second-order alliance. A1's partner had disappeared from our study area by the time we returned in March 1987, and A1 and A2 had formed a new pair. In 1987–9, a 'triangle' existed between the new A pair and the triplets B and C. During this period, A was observed with B or C but B and C were rarely seen together. These changes are reflected in the association coefficients (calculated for each year) among the males during this period. In 1985, individual members of B and C shared association coefficients of 55–71 with each other but only 10–15 with males A1 and A2. During 1987–8 coefficients for B and C were only 7–16 with each other but were 31–46 for members of B and C with the A males. Pair A was clearly 'splitting time' with the triplets B and C.

On one of the few occasions in which we saw all three alliances together, we clearly observed pair A initiate aggression with C against B. Pair A dropped behind the group, then synchronously charged into the group at a member of B who lay on his side at the surface (a behaviour we interpret as submissive). An explosive chase followed with all members of A and C leaping synchronously abreast, chasing all three members of B. One other bizarre set of observations concerned herding relationships among A, B, and C. One day during the 1987 mating season we found B herding a female and travelling with A. The next day we found C herding the same female and travelling with A. The third day we found B herding the same female again, while travelling with A. During these observations A had no female. The changing patterns of association and behaviour of A, B, and C lead us to suspect that A may have been 'playing' B and C off against each other. The ABC triangle lasted until May 1989, when two members of C disappeared. The third member of C began travelling with A as a clear 'odd-male-out'; sometimes he herded with them and sometimes he did not. When we returned in 1990 (March–July) the triplet B had disappeared, and A (with the former C member) had formed a second-order alliance with a pair of maturing males.

FEMALE COALITIONS AND ALLIANCES

Unlike males, female–female association patterns are commonly not transitive; i.e. A associates often with B, and B with C but not A with C. The result is more of a 'network' of associations in which important individual differences exist in the number of associates females have and the strength of those associations.

The study of female behaviour and social relationships in Shark Bay is in its infancy. We do not know, for example, how female choice might be expressed within or outside the herding context. We also do not know whether the relatively stable associations of some females are alliances. Nonetheless, there are several intriguing observations that suggest that females may form coalitions to protect themselves from male harassment. We briefly describe two of the most interesting cases.

Case 1

The provisioned males were inshore at the beach with a non-provisioned female that they had been herding for 5 days. The female bolted due north, straight out from the beach, and all three males chased. Seconds later, the non-provisioned female returned to the shallows alone while the males apparently searched for her about 100 m offshore. Immediately, two of the provisioned females took up positions flanking her on either side and all three swam northeast at moderate speed, away from the park and the males, who were still milling about to the north. We had never before seen a non-provisioned female bolt and then return inshore without the males. A possible interpretation is that the non-provisioned female, trapped between the males and the provisioning area (in a channel), was aided in her escape by provisioned females who are familiar with the area.

Case 2

During a follow of a male triplet, the males approached and joined another group and within 40 s began chasing an adult-sized individual from the group. The individual was never identified, but repeated small splashes about 30 m in front of the group suggested the presence of a calf with the individual being chased. The males caught up at times and were observed attempting to mount the dolphin and charging it before the chase resumed, at speeds of 16–18 km/h. Nine minutes after the chase began, we noted another large group leaping in towards the chase from the southeast. As the group from the southeast closed in, the chase suddenly veered to the east. The group from the southeast then veered east, chasing the three males who were chasing the other dolphin. The group chasing the males closed to within 10 m of the males, and both the males and the chasing group slowed down to normal speed (3–5 km/h), ending the chase 12 min after it started. The males next surfaced in the middle of the group that had chased them: six known females. For the next 5 min before the males left, they remained very tight, side-by-side among the females, and one male performed several displays in which he slapped his rostrum on the water immediately followed by a fluke slap. All six females were without calves and at least four were believed to be cycling that year as they were herded by males during the previous 4 months. Three were adult females who had calved previously, and the other three were thought to be subadult (one was known to be 9.5 years old). A possible interpretation is that the six females interfered with the attempted capture of another female. These observations suggest that females may cooperate to reduce harassment from male alliances.

EVIDENCE FOR ALLIANCES IN OTHER POPULATIONS AND SPECIES

Although they have not established whether males in pairs and triplets herd females in Sarasota, Wells *et al.* (1987) report that they have observed several instances in which 'a male pair has been observed apparently separating an individual female from a school'. In one case, two older males approached a group of six dolphins that included two pairs of young males and two females. When resighted several minutes later, only one of the females from the group of six was with the two older males, who chased the female in formation at moderate speed for 2.8 k. The chase ceased when the three dolphins swam toward an area of feeding dolphins and began to feed.

Pryor and Shallenberger (1990) observed schools of spotted dolphins (*Stenella attenuata*) while they were surrounded by tuna nets during fishing operations in the eastern tropical Pacific. Each school of spotted dolphins contained at least one adult male subgroup of 3–8 individuals that often swam side-by-side and threatened other dolphins in unison. In one net containing over 200 dolphins, they counted eight of these adult male subgroups. At this point, however, we cannot label *Stenella* with *Tursiops* as separate genera in which male alliances have been documented, as *Stenella* are apparently not a natural group. Perrin *et al.* (1981, 1987) provide evidence that *S. attenuata* and *S. frontalis* are more closely allied with *Tursiops* than with other *Stenella* species.

Do bottlenose dolphins live in closed groups?

We do not yet know if the Monkey Mia dolphin society is part of a closed group or community in which members are hostile to non-members. We have identified photographically over 300 individuals in our study area, but have not detected evidence of a boundary such as that dividing two chimpanzee communities. Wells *et al.* (1987) describe the social network of about 100 bottlenose dolphins in Sarasota Bay, Florida, as a 'relatively closed group' resembling a chimpanzee community. However, they state that in 17 per cent of sightings of Sarasota community members, members of adjacent communities were present (Wells *et al.* 1987). Wells (1986) reports that most of the groups (72 per cent) that contained dolphins from both the Sarasota and adjacent communities were in the periphery of the Sarasota community range and that Sarasota males were preferentially sighted in such mixed schools. Sarasota males were identified in 74 per cent of the mixed community schools but only 54 per cent of schools with Sarasota dolphins only. In contrast, Sarasota females were in 61 per cent of the mixed schools and 78 per cent of the schools with Sarasota dolphins only. Although Wells *et al.* have clearly identified different levels of association, they have not shown that the 100 Sarasota dolphins live in a closed society comparable to chimpanzees, which almost never mix with individuals from adjacent communities (Kawanaka 1984; Goodall 1986). Except for encounters of males with a few nulliparous or cycling females, intercommunity encounters are almost always accompanied by overt hostility in common chimpanzees (Boehm this volume pp. 138–9; Kawanaka 1984; Nishida *et al.* 1985; Goodall 1986). The lack of sharp community boundaries, combined with the stepwise pattern of association found among female dolphins in Sarasota Bay, is reminiscent of the association patterns of female elephants (Moss and Poole 1983).

LEVELS OF ALLIANCES IN PRIMATES

In our discussion of levels of alliance formation in primates, we wish to distinguish intragroup from intergroup interactions.

Competition for partners and other complexities of coalition formation emerge from triadic interactions (e.g. two against one, Kummer 1967). While triadic interactions are common within groups, interactions between 'closed' or 'semi-closed' primate groups are usually one on one encounters. Further, relationships between groups are hostile or at most tolerant. Thus, consider a potentially triadic interaction among groups, such as two neighbouring primate troops (or neighbouring territorial birds, for that matter) simultaneously threatening an intruder. Such joint action may fit a general definition of coalition formation, but affiliative interactions (not agonistic) mediate complexities in coalition formation such as competition for allies, recruitment of allies, and context-dependent relationships with potential allies. This distinction is important. Even among male savannah baboons, which have little to do with each other outside of the coalition itself, a specific set of soliciting gestures and 'greeting' behaviours accompany coalition formation (Bercovitch 1988; Noë this volume pp. 295; Packer 1977; Ransom 1981; Smuts and Watanabe 1990). Hostile or merely tolerant relationships that typify intergroup interactions preclude the kind of complex coalition formation that derives from affiliative interactions within groups. Humans, as we will see below, are an important exception.

Males

Two-level male alliances with both hostile and affiliative interactions between particular alliances appear not to have been identified in any primate except humans. Among non-human primates, hamadryas baboon groups (*Papio hamadryas*) are a classic example of a multi-level primate society (Kummer 1984; Stammbach 1987). The term 'multi-level' as applied to hamadryas baboons, however, refers to association, not alliance formation. The lowest level of male–male association is the 'clan', which is a strongly age-graded grouping of one (and rarely two) 'leader' males, each with a harem of herded females, plus subadult 'follower' males and old ex-leaders (Abegglen 1984; Kummer 1984). The next level of male–male association is the 'band' which is a semi-closed, social group equivalent to the 'troop' in savannah baboons (*Papio cynocephalus*) (Dunbar 1988; Kummer 1984). Males in savannah baboon troops regularly form first-order alliances to take estrous females from another male. An equivalent level of alliance formation in hamadryas baboons would involve male clan members acting against other clans of the same band. There is, however, no evidence that clans cooperate in defence of females and Kummer (1988) states that 'the only observed selective advantage of clans is that a subadult male obtains juvenile females mainly from his clan.' Males from clans within a band participate in aggressive encounters against males from other bands (Abegglen 1984; Kummer 1968, 1984), which is the equivalent of intertroop encounters in savannah baboons. See also Boehm's discussion of territorial encounters between 'macro-coalitions' of male chimpanzees (this volume pp. 137–73).

Nested alliances may have been an important factor in human evolution. Chagnon (1974, 1977) described several levels of male alliance formation in the Yanomamö Indians of Venezuela and Brazil. A typical Yanomamö village contains 75–80 people, and contains two and rarely three dominant male lineages. Lower limits on village size are constrained by the need to have enough adult males (10–15) to engage successfully in intervillage conflict. Within a village lineage, close male kin may form factions in order to have controlling power within their lineage. Such factions within lineages are the first-order alliances in Yanomamö society. The lineages themselves are second-order alliances. Disputes between males of different lineages eventually cause villages to split at around 150 occupants. Lineages within a village cooperating in warfare against other villages are third-order alliances. Villages also ally with each other in warfare (small villages may even fuse into one village) which represents fourth-order alliance formation. The fourth and first alliance levels of the Yanomamö yield the greatest capacity for complex triadic interactions. For example, interactions between villages are sometimes context-dependent; villages that cooperate against one village may be on opposing sides when the warfare involves a different village (Chagnon 1974, 1977).

Alliance formation between villages is interesting in relation to the distinction we make between 'within' versus 'between' group interactions in other animal societies. Villages form and strengthen bonds by reciprocal exchanges of women between villages, so in effect, a village strengthens a bond with another village by placing part of itself in that other village. The problem lies in whether groups are defined by spatial association or social relationships; in most animal societies the two closely correspond. An individual's network of relationships may extend to other groups in some non-human primates (Cheney and Seyfarth 1982, 1983; Meikle and Vessey 1981), but humans are probably unique in the degree to which they maintain relationships with kin who have 'emigrated' to different groups (Alexander and Noonan 1979; Rodseth *et al.* 1990). While non-human primates may recognize relationships between individuals (Cheney and Seyfarth 1986), it is not known whether they recognize relationships between groups. Humans, however, both recognize and attempt to manipulate relationships between groups.

Females

In female-bonded Old World monkeys such as vervets, macaques, and baboons, females acquire dominance ranks just below that of their mothers so that matrilines ranks as a unit below or above other matrilines (reviewed in Gouzoules and Gouzoules 1987; Chapais this volume pp. 29–32). Relationships between matrilines are characterized by long periods of stability punctuated by brief periods of severe aggression which result in entire matrilines changing rank as a unit (Chance *et al.* 1977; Ehardt and Bernstein 1986; Samuels and Hendrickson 1983; Samuels *et al.* 1987). These matriline 'overthrows' are usually described as 'mob' attacks on one individual at a time by members of lower ranking matrilines (Ehardt and Bernstein 1986; Samuels and Hendrickson 1987). Ehardt and Bernstein (1986) report a case where individuals in the high-ranking matriline under attack initially coalesced into a defensive unit before being routed. The nature of attacks in matriline overthrows suggests that matrilines are perceived as matrilines by other units (Ehardt and Bernstein 1986). The matriline can thus be considered a second level of alliance formation when alliances also occur within matrilines. Indeed, Ehardt and Bernstein (1986) describe an overthrow of an alpha matriline by lower-ranking matrilines that was precipitated by fighting between coalitions within the alpha matriline.

PRIMATES AND DOLPHINS: COMPARISONS ACROSS LEVELS

It is clear from our descriptions that first-order dolphin alliances interact much like individual male primates of a few species. Individual male chimpanzees and alliances of male dolphins (1) associate consistently with one another in non-agonistic contexts; (2) form alliances with the same parties with which they have agonisitic interactions on other occasions; (3) herd females temporarily (de Waal this volume pp. 244–6; Goodall 1986; Nishida 1983; Wrangham and Smuts 1980). Interestingly, the only reported case of herding by a group of male chimpanzees was in an intergroup interaction (Goodall *et al.* 1979). Like individual male chimpanzees and baboons, first-order dolphin alliances may recruit other alliances for agonistic interactions, and alliances may compete for second-order alliance partners (de Waal this volume pp. 247–9; Nishida 1983; Noë 1989, pp. 311–13). Finally, second-order alliances in male bottlenose dolphins are used to acquire access to (apparently) estrous females as are (first-order) alliances in savanna baboons, in contrast to intragroup alliances in male chimpanzees, which are used in rank-related interactions (de Waal this volume p. 244; Nishida and Hiraiwa-Hasegawa 1987).

While the emphasis in this paper, and indeed in the data itself, is on coalition formation that is congruent with the two levels of alliance, we have occasionally observed dolphin coalitions within and across each level of alliance. Earlier we described a case in which a male was set upon by his first-order alliance partners and another alliance. Here, the coalition cut across the two alliance levels, albeit in a period of instability in those alliances. Coalitions may form within first-order alliances; we have observed two males chase an 'odd-male-out' or push him away from a female. We once observed aggression between two pairs within a second-order alliance and suspect that the two provisioned males Sickle and Bibi occasionally took females from Snubby and the maturing male Shave. Thus, we might properly say that there are two levels of alliance formation, but several levels at which coalitions may form in male bottlenose dolphins.

SOCIAL COMPLEXITY AND BRAIN SIZE EVOLUTION IN CETACEA

Harcourt (1988, this volume pp. 445–7) argues that the relatively large brains and high intelligence of primates are linked to the frequency and complexity of alliance formation in primates compared to other taxa. He suggests that information-processing requirements increase geometrically as more participants are included in an interaction and that alliances require abilities not needed in dyadic encounters. First, in alliances, one needs to discriminate among other individuals without reference to oneself, such as when evaluating the dominance rank of two contestants. Second, one may also have to know about relationships involving individuals who are not involved in the interaction. For example, in their study of redirected aggression, Cheney and Seyfarth (1986) show that individuals take into account an adversary's relationships with other troop members.

We suggest that social interactions involving nested alliances within groups, as in bottlenose dolphins, demand greater information processing capacity than interactions involving only a single level of alliance formation. In nested alliances, individuals must not only take into account the effects of decisions involving potential allies and adversaries in first-order alliances, but the effect those decisions have on alliances at other levels must be considered as well. In theory, individuals might even make decisions that are costly to themselves at one level, because of a greater benefit they obtain at a different level of alliance.

Although we suggest that participation in multi-level alliances requires greater information processing abilities than single-level interactions, we are not suggesting that abilities increase in proportion to the number of levels of alliances. For example, once humans acquired the cognitive machinery to engage in multi-level interactions, additional levels might be added by 'extrapolation' without demanding greater information-processing capabilities. This may allow human groups to form context-dependent alliances with other groups irrespective of whether their kin reside in the other groups, as in interactions among nation-states (Falger this volume pp. 327–9). Also, individuals become further and further removed from decision-making processes at higher levels of interaction as the size of the participating group increases.

The brains of some delphinids, including *Tursiops*, are relatively larger than those of any non-human primate (Ridgway 1986, Table 15.4), but there is also enormous variation in brain size for dolphins of the same body size. The brains of many delphinids, including *Tursiops*, are 2–3 times larger than those of some other toothed whales (odontocetes) with the same adult body size (Ridgway and Brownson 1984; Ridgway 1986; Table 15.4). If complex social behaviour is linked to large brain size, then we can predict that social

Table 15.4. Extreme differences in brain size for apes, toothed whales, and an elephant of similar body size. *Neophocaeana* belongs to the family phocoenidae; *Pontoporia*, *Platanista*, and *Lipotes* to the plantanistidae (a questionably monophyletic group); *Ziphius* to the Ziphiidae; and *Sotalia*, *Delphinus*, *Tursiops*, and *Orcinus* are delphinids. Sources: Pilbeam and Gould (1974), von Bonin (1937), Gihr *et al*. (1979), Kamiya and Yamasaki (1974), Pilleri and Chen (1982), Pilleri and Gihr (1970, 1971, 1972), Morgane and Jacobs (1972), Ridgway and Brownson (1984) and Ridgeway (unpublished) for *Orcinus* values.

Genus	*N*	Body weight (kg)	Brain size*
Neophocaeana	4	37.3	471.3
Pontoporia	9	39.0	227.0
Sotalia	1	42.2	688.0
Pan troglodytes	—	45.0	395.0
Platanista	4	59.6	295.3
Delphinus	10	67.6	835.6
Gorilla (female)	—	70.0	460.0
Tursiops	19	167.4	1587.5
Lipotes	2	230.5†	570.0
Gorilla (male)	—	140.0	550.0
Orcinus	3	2262.0	6143.3
Ziphius	1	2273.0	2004.0
Elephas (male)	1	2547.0	5045.0

* Brain weight in grams; ape cranial capacity in cubic centimetres.

† Data from pregnant females close in body length ($\bar{x}$ = 251.5 cm) to the *Tursiops* ($\bar{x}$ = 245.7 cm).

interactions, including alliance formation, will be more complex among the large-brained forms than in small-brained genera.

What selection pressures might have favoured increasingly complex social interactions in ancestral odontocetes? In his classic essay on the social function of intellect, Humphrey (1976) linked the evolution of complex societies to advances in technical knowledge. He then writes, 'the open sea is an environment where technical knowledge can bring little benefit and thus complex societies—and high intelligence—are contraindicated (dolphins and whales provide, maybe, a remarkable and unexplained exception).'

Rather than technical knowledge, the key to the evolution of the complex social capacities of both humans and dolphins might have been the threat posed by inter- or intragroup alliances (Alexander 1979, 1989; Alexander and Tinkle 1968; Bigelow 1969; Harcourt 1988, this volume pp. 447–9). The interplay between competition with alliance members and the need to cooperate against other alliances could favour increasingly sophisticated social abilities. Alternatively, Connor and Norris (1982) suggested that sharks may have been the 'threat' driving social evolution in dolphins, and pointed out that dolphins giving birth in the three-dimensional aquatic habitat are especially vulnerable. There are, of course, other factors to consider in brain evolution. An obvious one is that large brains entail substantial energetic costs. The dolphins' energy-rich diet and relatively high metabolic rate (Hampton *et al.* 1971; Irving *et al.* 1941; Ridgway 1990; Ridgway and Patton 1971) may have paved a smoother road to large brain evolution than would be the case in other taxa.

SUMMARY

1. Male bottlenose dolphins in Shark Bay, Western Australia, associate in pairs and triplets that are often stable for years. Males in these pairs and triplets cooperate in aggressive herding of apparently estrous females. We refer to pairs and triplets as 'first-order' alliances.
2. Detailed observations on three provisioned males indicate that males compete for alliance partners.
3. First-order alliances combine to form 'second-order' alliances. Males in second-order alliances cooperate to aggressively take females being herded by other alliances or to defend females from theft attempts by other alliances. Alliances may recruit other alliances to participate in theft attempts.
4. Alliances may participate in theft attempts to help other alliances obtain a female rather than each alliance acting to increase its chance of acquiring the female. We do not know if benefits returned to members of first-order alliances for such support derive from nepotism, reciprocity, or pseudo-reciprocity.
5. Changes in association patterns over 5 years and some behavioral observations suggest that first-order alliances may compete for second-order alliance partners.
6. Preliminary observations suggest that females may cooperate to protect themselves from male harrassment.
7. Increasing the number of levels of alliance formation, particularly within groups where affiliative interactions mediate coalition formation, may require increased information processing capacity.
8. Interactions among first-order dolphin alliances resemble interactions among individual males of some primate species.
9. The large brains of humans and many delphinids may reflect convergent adaptations to selection for increasingly complex social interactions. Among odontocetes there exists 2–3 fold variation in brain size for species of the same adult body size. The social selection hypothesis for

brain evolution predicts that alliance formation will be less complex or absent in the smaller brained forms.

ACKNOWLEDGEMENTS

We thank Richard D. Alexander, Gary W. Fowler, Alexander H. Harcourt, Brian A. Hazlett, Kenneth S. Norris, Barbara B. Smuts, Frans B. M. de Waal, and Richard W. Wrangham for valuable comments and discussion. RCC was supported by grants from the National Geographic Society, an NSF Predoctoral Grant (BNS-8601475), a Fulbright Fellowship to Australia, and a Rackham Predoctoral Grant from the University of Michigan. The dolphin project is also supported by private contributions to The Dolphins of Shark Bay Research Foundation, Inc.

REFERENCES

Abegglen, J. (1984). *On socialization in hamadryas baboons*. Associated University Presses, Cranbury, NJ.

Alexander, R. D. (1979). *Darwinism and human affairs*. University of Washington Press, Seattle.

Alexander, R. D. (1989). The evolution of the human psyche. In *The human revolution: behavioural and biological perspectives on the origins of modern humans* (eds P. Sellars and C. Stringer), pp. 455–513. University of Edinburgh Press, Edinburgh.

Alexander, R. D. and Noonan, K. N. (1979). Concealment of ovulation, parental care, and human social evolution. In *Evolutionary biology and human social behavior: an anthropological perspective* (eds N. A. Chagnon and W. G. Irons), pp. 436–53. Duxbury Press, North Scituate, MA.

Alexander, R. D. and Tinkle, D. W. (1968). Review of *On aggression* by Konrad Lorenz and *The territorial imperative* by Robert Ardrey. *Bioscience*, **18**, 245–8.

Bercovitch, F. B. (1988). Coalitions, cooperation and reproductive tactics among adult male baboons. *Animal Behaviour*, **36**, 1198–209.

Bigelow, R. S. (1969). *The dawn warrior: mans's evolution towards peace*. Little, Brown, Boston.

von Bonin, G. (1937). Brain-weight and body-weight of mammals. *Journal of General Psychology*, **16–17**, 379–89.

Byrne, R. and Whiten, A. (1988). *Machiavellian Intelligence*. Clarendon Press, Oxford.

Cairns, S. J. and Schwager, S. J. (1987). A comparison of association indices. *Animal Behaviour*, **35**, 1454.

Chagnon, N. A. (1977). *Yanomamö: the fierce people*. Holt, Rinehart, and Winston, New York.

Chagnon, N. A. (1974). *Studying the Yanomamö*. Holt, Rinehart, and Winston, New York.

Chance, M. R. A., Emory, G. R., and Payne, R. G. (1977). Status referents in long-tailed macaques (*Macaca fascicularis*): precursors and effects of a female rebellion.*Primates*, **18**, 611–32.

Cheney, D. L. and Seyfarth, R. M. (1982). Recognition of individuals within and between groups of free-ranging vervet monkeys. *American Zoologist*, **22**, 519–29.

Cheney, D. L. and Seyfarth, R. M. (1983). Non-random dispersal in free-ranging vervet monkeys: social and genetic consequences. *American Naturalist*, **122**, 392–412.

Cheney, D. L. and Seyfarth, R. M. (1986). The recognition of social alliances by vervet monkeys. *Animal Behaviour*, **34**, 1722–31.

Cheney, D. L., Seyfarth, R. M., and Smuts, B. B. (1986). Social relationships and social cognition in nonhuman primates. *Science*, **234**, 1361–6.

Connor, R. C. (1986). Pseudo-reciprocity: investing in mutualism. *Animal Behaviour*, **34**, 1562–84.

Connor, R. C. and Norris, K. S. (1982). Are dolphins reciprocal altruists? *American Naturalist*, **119**, 358–72.

Connor, R. C. and Smolker, R. A. (1985). Habituated dolphins (*Tursiops* sp.) in Western Australia. *Journal of Mammalogy*, **66**, 398–400.

Connor, R. C. and Smolker, R. A. (in prep.). 'Pop' goes the dolphin: a vocalization males use to herd females.

Duffield, D. A., Asper, E. D., Odell, D. K., and Provancha, J. (1985). Movements and association patterns of wild bottlenose dolphins of different genotype in the Indian river system. *Sixth biennial conference on the biology of marine mammals, abstracts*. Vancouver, British Columbia.

Dunbar, R. I. M. (1988). *Primate social systems*. Cornell University Press, New York.

Ehardt, C. L. and Bernstein, I. S. (1986). Matrilineal overthrows in rhesus monkey groups. *International Journal of Primatology*, **7**, 157–81.

Gihr, M., Pilleri, G., and Zhou, K. (1979). Cephalization of the Chinese river dolphin *Lipotes vexillifer* (Platanistoidea, Lipotidae). *Investigations on Cetacea*, **2**, 129–44.

Goodall, J. (1986). *The chimpanzees of Gombe: patterns of behavior*. Belknap, Harvard, MA.

Goodall, J., Bandora, A., Bergmann, E., Busse, C., Matama, H., Mpongo, E. *et al.* (1979). Intercommunity interactions in the chimpanzee population of the Gombe National Park. In *The great apes* (eds D. A. Hamburg and E. R. McCown), pp. 13–53. Benjamin/Cummings, Menlo Park, California.

Gouzoules, S. and Gouzoules, H. (1987). Kinship. In *Primate societies*, (eds B. B. Smuts, D. L. Cheney, R. M. Seyfarth, R. W. Wrangham, and T. T. Struhsaker) pp. 299–305. University of Chicago Press, Chicago.

Hampton, I. F. G., Whittow, G. C., Szekercezes, J., and Rutherford, S. (1971). Heat transfer and body temperature in the Atlantic bottlenose dolphin, *Tursiops truncatus*. *International Journal of Biometeorology*, **15**, 247–53.

Harcourt, A. H. (1988). Alliances in contests and social intelligence. In *Machiavellian Intelligence* (eds R. Byrne and A. Whiten) pp. 132–52. Clarendon Press, Oxford.

Herman, L. M. (1980). Cognitive characteristics of dolphins. In *Cetacean behavior: mechanisms and functions*. (ed. L. M. Herman) pp. 363–429. Wiley, New York.

Humphrey, N. K. (1976). The social function of intellect. In *Growing points in ethology* (eds P. P. G. Bateson and R. A. Hinde). Cambridge University Press, Cambridge.

Irving, L., Scholander, P. F., and Grinnel, S. W. (1941). The respiration of the porpoise, *Tursiops truncatus*. *Journal of Cellular and Comparative Physiology*, **17**, 145–68.

Jolly, A. (1966). Lemur social behavior and primate intelligence. *Science*, **153**, 501–6.

Kamiya, T. and Yamasaki, F. (1974). Organ weights of *Pontoporia blainvillei* and *Platanista gangetica* (Platanistidae). *Scientific Reports of the Whale Research Institute, Toyko*, **26**, 265–70.

Kawanaka, K. (1984). Association, ranging, and the social unit in chimpanzees of the Mahale Mountains, Tanzania. *International Journal of Primatology*, **5**, 411–34.

Kummer, H. (1967). Tripartite relations in hamadryas baboons. In *Social communication among primates* (ed. S. A. Altmann). University of Chicago Press, Chicago.

Kummer, H. (1968). *Social organization of hamadryas baboons*. University of Chicago Press, Chicago.

Kummer, H. (1982). Social knowledge in free-ranging primates. In *Animal mind–human mind (Dahlem Konferenzen)* (ed. D. R. Griffen), pp. 113–30. Springer-Verlag, Berlin.

Kummer, H. (1984). From laboratory to desert and back: a social system of hamadryas baboons. *Animal Behaviour*, **32**, 965–71.

Kummer, H. (1988). Hamadryas baboons: definitions of social units and a sketch of their interaction with the environment. Paper presented at the 1988 International Primatalogical Congress, Brazil.

Mantel, N. (1967). The detection of disease clustering and a generalized regression approach. *Cancer Research*, **27**, 209–20.

Meikle, D. B. and Vessey, S. H. (1981). Nepotism among rhesus monkey brothers. *Nature*, **294**, 160–1.

Morgane, P. J. and Jacobs, M. S. (1972). Comparative anatomy of the cetacean nervous system. In *Functional anatomy of marine mammals, vol. 1* (ed. R. J. Harrison) pp. 117–244. Academic Press, New York.

Moss, C. J. and Poole, J. H. (1983). Relationships and social structure of African elephants. In *Primate social relationships* (ed. R. A. Hinde) pp. 315–25. Sinauer, Sunderland, MA.

Nishida, T. (1983). Alpha status and agonistic alliance in wild chimpanzees (*Pan troglodytes schweinfurthii*). *Primates*, **24**, 318–36.

Nishida, T. and Hiraiwa-Hasegawa, M. (1987). Chimpanzees and bonobos: cooperative relationships among males. In *Primate societies*, (eds B. B. Smuts, D. L. Cheney, R. M. Seyfarth, R. W. Wrangham, and T. T. Struhsaker) pp. 165–77. University of Chicago Press, Chicago.

Nishida, T., Hiraiwa-Hasegawa, M., and Hasegawa, T. (1985). Group extinction and female transfer in wild chimpanzees in the Mahale National Park, Tanzania. *Zeitschrift für Tierpsychologie*, **67**, 284–301.

Noë, R. (1989). A veto game played by baboons: a challenge to the use of the prisoner's dilemma as a paradigm for reciprocity and cooperation. *Animal Behaviour*, **39**, 78–90.

Packer, C. (1977). Reciprocal altruism in *Papio anubis*. *Nature*, **265**, 441–3.

Perrin, W. F., Mitchell, E. D., Mead, J. G., Caldwell, D. K., and van Bree, P. J. H. (1981). *Stenella clymene*, a rediscovered tropical dolphin of the Atlantic. *Journal of Mammalogy*, **62**, 583–98.

Perrin, W. F., Mitchell, E. D., Mead, J. G., Caldwell, D. K., Caldwell, M. C., van Bree, P. J. H., and Dawbin, W. H. (1987). Revision of the spotted dolphins. *Stenella* spp. *Marine Mammal Science*, **3**, 99–170.

Pilbeam, D. and Gould, S. J. (1974). Size and scaling in human evolution. *Science*, **186**, 892–901.

Pilleri, G. and Chen, P. (1982). The brain of the Chinese finless porpoise *Neophocaena asiaeurientalis* (Pilleri & Gihr, 1972): (1) Macroscopic anatomy. *Investigations on Cetacea*, **13**, 27–78.

Pilleri, G. and Gihr, M. (1970). Brain–body weight ratio of *Platanista gangetica. Investigations on Cetacea*, **2**, 79–82.

Pilleri, G. and Gihr, M. (1971). Brain–body weight ratio in *Pontoporia blainvillei. Investigations on Cetacea*, **3**, 69–73.

Pilleri, G. and Gihr, M. (1972). Contribution to the knowledge of the cetacea of Pakistan with particular reference to the genera *Neomeris, Sousa, Delphinis*, and *Tursiops* and description of a new Chinese porpoise (*Neomeris asiaeorientalis*). *Investigations on Cetacea*, **4**, 107–62.

Pryor, K. and Shallenberger, I. K. (1990). Social structure in spotted dolphins. (*Stenella attenuata*) in the tuna purse seine fishery in the eastern tropical Pacific. In *Dolphin societies* (eds K. Pryor and K. S. Norris). University of California Press, Berkeley.

Ransom, T. W. (1981). *Beach Troop of the Gombe*. Associated University Presses, Toronto.

Ridgway, S. H. (1986). Physiological observations on dolphin brains. In *Dolphin cognition and behaviour: a comparative approach* (eds R. J. Schusterman, J. A. Thomas, and F. G. Wood) pp. 31–59. Erlbaum, Hillsdale, NJ.

Ridgway, S. H. (1990). The central nervous system of the bottlenose dolphin. In *The bottlenose dolphin* (eds S. Leatherwood and R. R. Reeves) pp. 69–97. Academic Press, San Diego.

Ridgway, S. H. and Brownson, R. H. (1984). Relative brain sizes and cortical surface areas in odontocetes. *Acta Zoologica Fennica*, **172**, 149–52.

Ridgway, S. H. and Patton, G. S. (1971). Dolphin thyroid: some anatomical and physiological findings. *Zeitschrift für Vergleichende Physiologie*, **71**, 129–41.

Rodseth, L., Wrangham, R. W., Harrigan, A. M., and Smuts, B. B. (1991). The human community as a primate society. *Current Anthropology*, **32**, 221–54.

Ross, G. J. B. and Cockcroft, V. G. (1990). Comments on Australian bottlenose dolphins and the taxonomic status of *Tursiops aduncus* (Ehrenberg, 1832). In *The bottlenose dolphin* (eds S. Leatherwood and R. R. Reeves) pp. 101–28. Academic Press, San Diego.

Samuels, A. and Henrickson, R. V. (1983). Outbreak of severe aggression in captive *Macaca mulatta. American Journal of Primatology*, **5**, 277–81.

Samuels, A., Silk, J. B., and Altman, J. (1987). Continuity and change in dominance relations among female baboons. *Animal Behaviour*, **35**, 785–93.

Schnell, G. D., Watt, D. J., and Douglas, M. E. (1985). Statistical comparison of proximity matrices: applications in animal behaviour. *Animal Behaviour*, **33**, 239–53.

Schroeder, J. P. and Keller, K. V. (1989). Seasonality of serum testosterone levels and sperm density in *Tursiops truncatus. Journal of Experimental Zoology*, **249**, 316–21.

Smuts, B. B. and Watanabe, J. M. (1990). Social relationships and ritualized greetings in adult male baboons (*Papio cynocephalus anubis*). *International Journal of Primatology*, **11**, 147–72.

Stammbach, E. (1987). Desert, forest, and montane baboons: multilevel societies. In *Primate societies* (eds B. B. Smuts, D. L. Cheney, R. M. Seyfarth, R. W. Wrangham, and T. T. Struhsaker) pp. 112–20. University of Chicago Press, Chicago.

Struhsaker, T. T. and Leland, L. (1979). Socioecology of five sympatric monkey species in the Kibale forest, Uganda. *Advances in the Study of Behavior*, **9**, 159–228.

Symington, M. M. (1990). Fission–fusion social organization in *Ateles* and *Pan*. *International Journal of Primatology*, **11**, 47–61.

Trivers, R. L. (1971). The evolution of reciprocal altruism. *Quarterly Review of Biology*, **46**, 35–57.

Wells, R. S. (1986). Structural Aspects of Dolphin Societies. Ph.D. dissertation, University of California at Santa Cruz.

Wells, R. S., Scott, M. D., and A. B. Irvine. (1987). The social structure of free-ranging bottlenose dolphins. *Current Mammalogy*, **1**, 247–305.

Wrangham, R. W. and Smuts, B. B. (1980). Sex differences in the behavioural ecology of chimpanzees in the Gombe National Park, Tanzania. *Journal of Reproduction and Fertility (Supplement)*, **28**, 13–31.

Cooperation in conflict in a gorilla group. During competition over access to a dead tree, the leftmost individual, an adult female, cuffs the juvenile male to her right. This was not a sensible move. The juvenile male's mother (above the tree), grandmother (large female on the right of the tree), and aunt (smaller individual at fore right) all intervene aggressively on behalf of the juvenile male. The result was the supportive family supplanted the unrelated female, and gained sole access to the tree.

16

Coalitions and alliances: are primates more complex than non-primates?

A. H. HARCOURT

INTRODUCTION

Frank Bruno and Mike Tyson did not need to exercise their respective brains overmuch to work out which of them was going to win the title of world heavyweight boxing champion in 1989. Tyson was bigger, faster, and better than Bruno and almost everyone knew it. But what if each had been allowed a boxing partner? What if they had a range of partners available, each with their own special skills? What if they were given a day to persuade partners to join the fight, and both knew that they were competing for partners? The situation is now extremely complex. Only with the ability to process a lot of information about potential opponents, partners, and the relationships among them would accurate prediction and sure bets be possible.

In other words, coalitions potentially introduce an order of complexity into competitive interactions not present in dyadic (two-animal) contests (Kummer 1967). It seems reasonable to suggest, therefore, that the ability to process information might impose an evolutionary limit on the use of coalitions and alliances among vertebrates. The vertebrates' information-processing organ is the brain, and the brain of primates is larger than that of any other vertebrate family relative to body size. Mammals have larger brains for their body weight than do other vertebrates, and primates, along with toothed whales, have larger brains for their body weights than do other mammals (Armstrong 1983; Jerison 1983; Martin 1981, 1984). However, the brain has a high energy demand and therefore brain size might be constrained by basal metabolic rate (BMR) (Armstrong 1985; Martin 1981; but see Deacon 1990). Toothed whales (and pinnipeds) have unusually high metabolic rates in relation to their body weight, and if brain weight is plotted against body weight with the effects of BMR removed, primates alone are above the common regression line (Armstrong 1983, 1985), although still not significantly different from odontocetes. Within families, relative brain size varies, and dolphins have some of the largest recorded relative brain sizes of any non-human animal (Jerison 1983; Ridgway 1986; Connor *et al.* this volume pp. 415–43).

Of course, not all the brain is concerned with processing information of the sort that might be relevant to winning contests in a social situation in which individuals compete by forming coalitions and alliances. However, the size of the region generally considered to be concerned with higher mental functions, the neocortex, appears to scale isometrically with total brain size in primates (Armstrong 1985; Dunbar, in press; Sawaguchi and Kudo 1990); and in mammals as a whole, including dolphins, the surface area of the cerebral cortex also scales isometrically with brain weight (Jerison 1983; Ridgway 1986). The isometry occurs because the cortex is a large proportion of the brain in these taxa (Deacon 1990). Nevertheless it indicates that relative brain size might be a good measure of information-processing ability, although the dolphin cortex, which constitutes a larger proportion of the brain than is the case for primates, is clearly different from that of terrestrial mammals (Ridgway 1986).

Is there any evidence that measures of brain size do indeed correlate with measures of information processing abilities, or 'intelligence'? Testing differences between species in intelligence is extremely difficult (Essock-Vitale and Seyfarth 1987; Macphail 1982). Nevertheless, it still seems to be the case that where differences in intelligence are detected, the cleverest primates are probably more intelligent—they appear to be able to solve more quickly a greater variety of more complex problems—than the cleverest terrestrial non-primates (see Essock-Vitale and Seyfarth 1987; Macphail 1982; Passingham 1981). Therefore, if information-processing ability is important, and if relative brain size is an acceptable measure of the appropriate sort of information-processing ability (see Deacon (1990) for a contrary view), we should find that primates use coalitions and alliances as means of competing either more often than do non-primates, more efficiently, in more ways, or in more complex ways.

Influenced by Altmann (1962), Hall and DeVore (1965) and especially Kummer (1967), Wilson (1975, p. 517) suggested that a distinctive behavioural trait of primates was indeed the frequent use of alliances in a social context of individual recognition and manipulation of the 'social field'. Wrangham (1982, 1983) took up the idea about the distinctiveness of primate coalitionary and alliance behaviour, highlighting the potential complexity of such interactions and their capacity for generating further social complexity, and pointing out that alliances among adult females were common in social groups of primates, but very rare among non-primates. Harcourt (1988) went into more detail about the complexity involved in the use of alliances and coalitions and, following earlier suggestions about the number of different types of social relationships that individual primates had (e.g. Kummer 1982), suggested a wide range of partners in contests as characteristic of primates. A more thorough comparison of primate and non-primate use of coalitions and alliances led to the suggestion that only primates choose allies on the basis of their competitive ability (Harcourt 1989).

This chapter reviews in detail the information available in the literature on the use of coalitions and alliances by primates and non-primates. Its emphasis is on the consequences, the pay-offs, of coalitions to the participants, and little attention is paid to either the proximate cues or the development of coalitionary behaviour. Rather, it assumes that coalitions and alliances are competitive strategies that might benefit the individual, and asks whether and how non-primates differ from primates in the way that individuals gain the potential benefits from the use of coalitions.

PRIMATES AND NON-PRIMATES COMPARED

I begin with the simplest form of coalition and proceed to what I see as the most complex. Two developmental, and perhaps evolutionary, origins of coalitions and alliances seem likely. One is parental protection of offspring (Kummer 1967); the other is cooperative defence against predators or competing social groups. Two origins, rather than one, are proposed because the proximate cues and the benefits of the two types of coalitions involved appear so different. In a protective coalition, any benefit to the supporter is only via consanguinity with the recipient of support. In mutualistic coalitions, on the other hand, the supporter itself directly benefits by joining with others in a threat or attack. In both cases, the suggestion is that offensive uses develop during maturation or have evolved from the presumed defensive origins of the coalitions; and in both cases the coalitions, the acts of support, might or might not be part of a long-term supportive relationship between the partners, an alliance.

Mutualistic coalitions and alliances: cooperative defence

In several species of both primate and non-primate, not only is a coalition more powerful than one animal acting alone, but larger coalitions defeat smaller ones. Acorn woodpeckers (*Melanerpes formicivorus*) are a non-primate example (Fig. 16.1(a), Hannon *et al.* 1985). Similarly, larger raiding parties or colonies of ants (*Azteca trigona*) defeated smaller ones (Adams 1990), and larger families of white-fronted geese (*Anser albifrons*) defeated smaller ones at winter feeding grounds (Boyd 1953); changes in number of defenders correlated with changes in the same direction in the size of territory of both ants (Adams 1990) and purple gallinules *Porphyrula martinica* (Hunter 1985); groups of over three male lions (*Panthera leo*) retained prides for more than twice as long as did smaller coalitions or singletons (Bygott *et al.* 1979); and two thirds of male cheetahs (*Acinonyx jubatus*) with partners had territories, as against only 4 per cent of singletons (Caro and Collins 1987). Among primates, larger groups of capuchin monkeys (*Cebus apella*) won a greater proportion of their intergroup

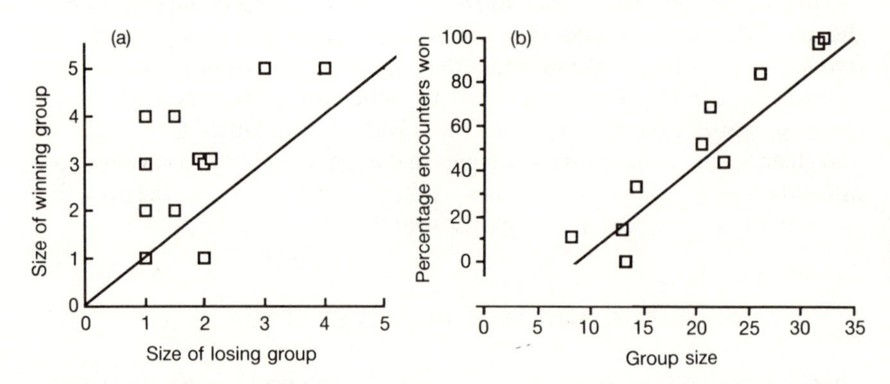

Fig. 16.1. Large coalitions defeat small ones. (a) Non-primates—acorn woodpecker: size of winning group in relation to mean size of groups it defeated. $N = 12$ contests, $P < 0.05$, binomial test. (Data from Hannon *et al.* 1985.) (b) Primates—capuchin: proportion of contests won against other groups in relation to size of winning group. $r_s = 0.90$, $P < 0.01$. (Data from Robinson 1988.)

encounters than did smaller groups (Fig. 16.1(b), Robinson 1988), and in the territorial vervet monkey (*Cercopithecus aethiops*) groups with more females had larger territories (Cheney and Seyfarth 1987).

So far, all the non-primate examples of mutualistic coalitions and alliances concern contests between social groups, not within them. Adult non-primates do support one another in contests within social groups (dolphins, Connor *et al.* this volume pp. 415–43; plains zebra *Equus burchelli*, Schilder 1990), but primates appear to do so far more commonly. For example, female vervet monkeys will join forces against males (Cheney 1983), as will female Japanese macaques (Packer and Pusey 1979). Baboon males are too large by comparison to females for coalitions of females to be often successful against them (Packer and Pusey 1979), but males join forces against others in competition over females (Bercovitch 1988; Hall and DeVore 1965; Packer 1977; Smuts 1985). It seems that middle- and low-ranking males are particularly likely to form a coalition, often against a high-ranking male whom neither could defeat alone (Bercovitch 1988; Noë 1990, this volume pp. 285–321; cf. Murnighan 1978). Coalitions between male baboons are by no means always successful, but the potential benefits, access to a fertile female and hence possibly the production of offspring, are great.

In a number of these examples of mutualistic cooperation, the partners are unrelated. Baboon males in a group are likely to be unrelated, for example, because they have immigrated from other groups (Pusey and Packer 1987; female vervets who cooperatively attack males are not necessarily close relatives (Cheney 1983); and in the case of male lions in the Serengeti plains in Tanzania, about a third of the alliances contained non-relatives (Packer and Pusey 1982). The possibility of obtaining benefits from coalitions and alliances with non-relatives enormously increases the number of potential partners available, and therefore enormously increases the potential complexity of decisions concerning the use of coalitions and alliances as a competitive strategy.

Protective coalitions

Both primates and non-primates, and even invertebrates, protect offspring and other kin in the face of danger from either potential predators, or conspecifics. Such active protection from harm is the most common form of 'coalition', if indeed it can be called such, and will not be treated further.

Protective coalitions allow improved access to resources

If the protective support not only stops the opponent from harming the recipient, but also drives it away, then a further benefit to the recipient is possible, namely access to otherwise unobtainable resources. Both non-primates and primates give support in situations that improve the recipient's access to resources.

Bewick swan parents (*Cygnus columbianus*) protected their immature offspring, intervening on their behalf in 34 per cent of contests (Scott 1980). The result was that the amount of time a cygnet spent feeding was inversely proportional to its distance from its parents: cygnets within three swan-lengths of their parents spent about 80 per cent of the time feeding, compared to about 50 per cent beyond this distance, a significant difference (Scott 1980).

Support is provided not only to offspring, or other close kin, but also to mates. The feeding success of pigeons (*Columbia livia*) was proportional to the amount of time that their mates spent threatening competitors (Fig. 16.2(a)), with the effect being stronger for females being defended by males than vice versa (Lefebvre and Henderson 1986). Jackdaws (*Corvus monedula*) won a greater proportion of contests (15 per cent) against otherwise dominant individuals when their mate was present than when it was not (7 per cent), and were four times as likely to win (27 per cent) when their mate was not only present but also supported them (Wechsler 1988).

Turning to primates, Harcourt and Stewart (1987) showed that weaned immature gorillas won more contests against dominant opponents when supported than when not supported in contests (Fig. 16.2(b)). Protection by mates has not yet been quantitatively proved to provide increased access to resources in social groups of primates, but six of eight female savannah baboons (*Papio cyncocephalus*) were supported by adult males more often when in consort than at other times; and, presumably as a result, seven of them more often ignored dominant females' threats when in consort (Seyfarth 1978).

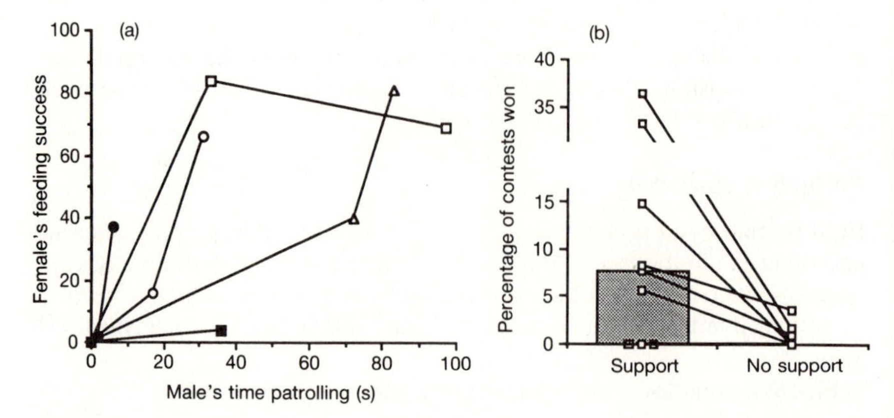

Fig. 16.2. Support in contests improves the recipient's access to resources. (a) Non-primates—pigeon: index of latency to accumulate 5 minutes of feeding time by female in relation to time spent patrolling by male in three separate trials for each of five pairs. (Data from Lefebvre and Henderson 1986). (b) Primates—gorilla: proportion of contests in which subordinate immature contestants gained access to a resource in relation to whether the immature was supported or not. Lines connect points for the same individual; histogram shows value for median individual. $N = 9$ immatures; $P < 0.05$, binomial test. (Median no. of contests with support/individual = 12; without support = 106). (Data from A. Harcourt and K. Stewart, unpublished).

Does the increased access to resources correlate with increased reproductive success? As far as coalitions in intergroup competition is concerned, the answer is yes (Brown 1987; Bygott *et al.* 1979; Emlen 1984). Within the group, the only substantive data that I know of come from Fairbanks and McGuire's (1986) study of captive vervet monkeys. Daughters whose mothers were still present were supported nearly twice as frequently as those without a mother and produced three times as many surviving offspring. (The females with mothers had the same mean age and the same mean dominance rank as those without.)

Protective coalitions influence the recipient's dominance rank

Dominance hierarchies result from individuals learning differences in competitive ability. Chickens (*Gallus gallus*) give way to those who have defeated them in the past, but are confident in the presence of those whom they defeated, and the result is a dominance hierarchy, the classic pecking order (Schjelderup-Ebbe 1935). It is perhaps not a big step from learning whom you can and cannot defeat on your own to learning your relative competitive ability in the presence of your own and your opponents' supporters? Both primates and non-primates make this step.

All 11 Bewick swan second- and third-year offspring studied by Scott (1980) were more dominant in the presence of their parents than in their absence, a phenomenon that correlates with the finding that five of six Bewick swan cygnets were threatened less often by others when near their parents than when far from them. Datta (1983*a*) reports similar findings for rhesus macaques. Ten of 12 immature rhesus macaques in the presence of their mother threatened otherwise dominant animals more often than they were threatened, whereas the reverse occurred for 10 of 11 in the absence of their mother (see also Chapais 1988*a*, this volume pp. 29–59).

The information-processing ability required in the decision of whether to fight or not according to the presence or absence of the opponent's main supporter is perhaps not great. The situation is potentially more complex when both contestants have supporters. What determines the outcome of the contest then, and on what basis should recipients base future decisions about whether to compete or not?

The answer to both questions seems to be the relative competitive abilities, the dominance ranks, of the supporters. In the same way that dominant animals are successful competitors, by definition, so they ought to be successful supporters, other things being equal. Evidence for an effect of competitive ability on quality of support is found in both non-primates and primates. In jackdaw flocks, females with dominant mates were apparently more likely to win contests against otherwise dominant birds than were females with subordinate mates (Fig. 16.3(a); Wechsler 1988), although the data are presented as acts, not individuals, and so admit the possibility that the result might have been due to a minority of the individuals. Similarly, dominant savannah baboon mothers were more successful in their support (more often brought the original contest to an end) than were subordinate ones (Fig. 16.3(b); Cheney 1977). Datta's (1983*b*) results indicate that dominant rhesus macaque (*Macaca mulatta*) mothers were also less likely to suffer a counter-attack than were subordinate mothers, in other words they risked less in supporting offspring; and Berman (1980), studying the same population of macaques as Datta, showed that dominant supporters were more confident (gave fewer fear-grimaces) when supporting than were subordinate supporters.

Perhaps because of this greater success and lower risk, dominant mothers in some primate groups support their offspring more frequently than do subordinate ones. Thus in a baboon group and a Caribbean island vervet group dominant mothers supported their offspring in about twice as many of their fights as did subordinate mothers, at least partly as a result in this latter study of mothers never supporting their juvenile offspring against opponents dominant to themselves. The medians were 13 per cent vs. 8 per cent for 10 baboons (Cheney 1977); and 5.3 per cent vs. 2.3 per cent for

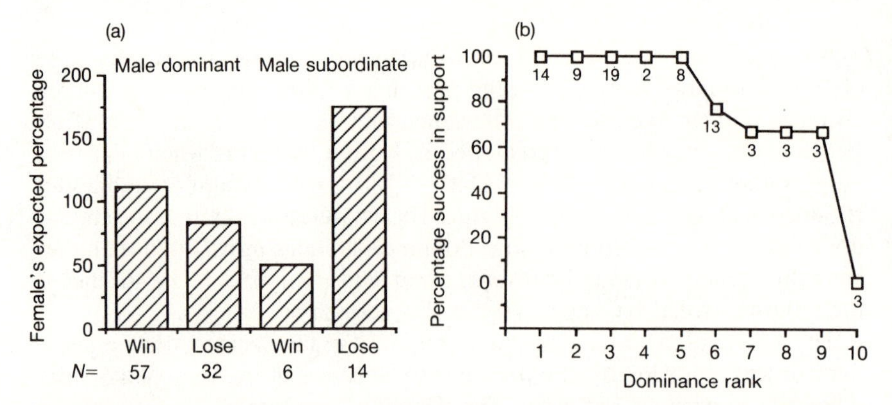

Fig. 16.3. Dominance rank of supporter influences the outcome of contests. (a) Non-primates—jackdaw: observed probability of females changing dominance rank (expressed as percentage of expected probability) in relation to dominance rank of mate relative to female's opponent's rank. *N*'s refer to number of rank changes among 13 females. (Data from Weschler 1988). (b) Primates—baboon: proportion of incidents of support by the mother that resulted in cessation of the offspring's contest in relation to rank of mother. $r_s = 0.93$, $P < 0.01$. (Data from Cheney 1977.)

eight vervets (Horrocks and Hunte 1983), a significant difference in each case.

Given that the outcome of contests can be learnt even by invertebrates (Caldwell 1985), given that quality of support differs between individuals, and given that the contestants remember one another and their respective supporters, the way is open for offspring (or mates) of dominant animals to learn that they can win contests against all whom their supporters defeat, and for offspring (or mates) of subordinate animals to learn that they should submit to those to whom their supporters defer. Associates of high-ranking animals themselves eventually acquire high rank, while associates of low-ranking animals remain low ranking. Rank is 'inherited'.

'Inheritance' of rank occurs in both non-primates and primates. In winter flocks of Bewick's swans, the index of dominance of immature birds in the absence of their parents is proportional to their parents' dominance rank (Fig. 16.4(a), Scott, 1980); subadult spotted hyaenas defeat all animals that their mother defeats (Franks 1986; see Zabel *et al.* this volume pp. 113–35, for more details); and in primates, the inheritance of rank system, first described in the fifties for Japanese macaques (*M. fuscata*) (Kawai 1958; Kawamura 1958; see de Waal and Harcourt this volume pp. 9–27) has now been demonstrated in four genera and ten species of primate (see Chapais, Datta this volume pp. 29–59, 61–82). Vervet monkeys are an example (Fig. 16.4(b), Horrocks and Hunte 1983).

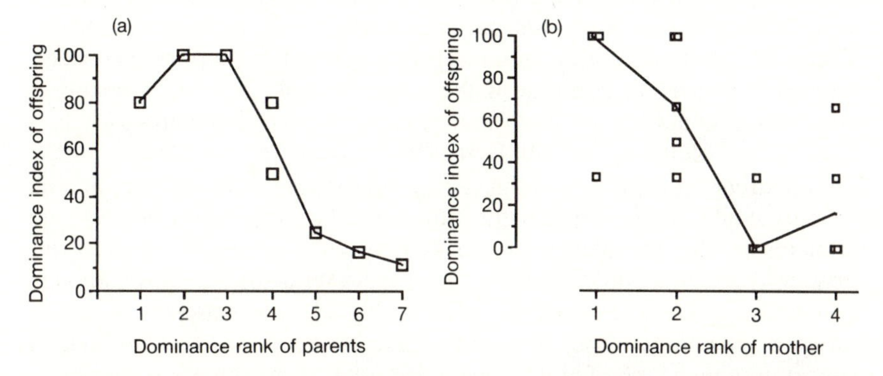

Fig. 16.4. Individuals achieve a dominance rank equivalent to that of their main supporter, i.e. they 'inherit' rank. (a) Non-primates—swan: proportion of opponents defeated by yearlings in relation to dominance rank of parents. $r_s = 0.88$, $P < 0.02$. (Data from Scott 1980.) (b) Primates—vervet: proportion of same-aged opponents defeated by immature (1–3 yrs old) in the absence of their mother in two free-ranging groups. $r_s = 0.67$, $P < 0.02$. (Data from Horrocks and Hunte 1983.)

Learning is not the only process by which rank can become inherited. Offspring of dominant barnacle geese (*Branta leucopsis*), for example, grow faster than offspring of subordinate ones, and since a correlation of rank with weight exists, they become themselves dominant at least partly as a result of being heavier (Black and Owen 1987). In hyaena clans, there is such a strong association between mother's rank and cubs' access to carcasses (Frank 1986) that in this species too, rank could be 'inherited' as much through good feeding and greater weight as directly through learning of the effects of support in contests. At least two studies of primates have also found associations between rank and weight, in one case among adult female vervets (Whitten 1983), and in the other among immature Japanese macaques in relation to the rank of their mother (Mori 1979). Furthermore, attributes that promote dominance (aggressiveness, large size) can also be inherited genetically, as indicated by cross-fostering experiments: aggressive barnacle geese produce aggressive offspring, and dominance tends to be associated with level of aggression (Black and Owen 1987).

Nevertheless, the evidence for a strong effect of support on dominance rank via learning is compelling. Offspring of dominant Japanese macaque mothers and swans might be heavier than their peers, but they dominate animals heavier than themselves; and Chapais' experiments with social groups of macaques, where he removed and reintroduced supporters, demonstrate fully that rank can be 'inherited' through behavioural processes (Chapais 1988*a*, this volume pp. 29–59).

Although subordinance is learned, few individuals permanently accept a subordinate position; dominants are continually being probed for weakness. Therefore, where rank is acquired largely as a result of support in contests, support will need to continue if the acquired rank is to be maintained, especially if coalitions of subordinates challenge the dominants (Chapais 1988*b*, this volume pp. 29–59; Datta 1983*b*; Harcourt and Stewart 1987). The apparent necessity for continued support to maintain inheritance of rank systems could partly explain the difference between non-primates and primates in the prevalance of the systems: non-primates relatively infrequently support adult offspring, it appears (Wrangham 1983), although hyaenas might be an exception (Frank 1986). The distinction, however, is surely a quantitative one, if it exists, and has more to do with lack of opportunity than lack of ability, given that adult non-primates support one another in other contexts.

Manipulating support

The healthy dog who limps in order to attract attention is known to us all, yet I have so far written as if recipients of support and partners in coalitions passively accept the support and its consequence as something over which they have no control. Here I consider evidence that the behaviour of non-primates and primates is more complex than this and that they do in fact attempt to manipulate the amount of support that they receive, and also its quality.

Manipulating the frequency of support

Three obvious ways for an animal to manipulate the probability that it will be supported in a contest suggest themselves. Animals can directly solicit support; they can coerce others to give support by punishing them if they refrain from giving it; and they can induce others to give support, or in general to cooperate, by being cooperative themselves (Axelrod 1984; Trivers 1971). Both non-primates and primates appear to use all three means.

1. *Solicitation*. When immature swans (Scott 1980), baboons (Walters 1980) and rhesus macaques (Gouzoules *et al.* 1984) enter contests with dominant, larger opponents, they give distinctive calls, which appear to stimulate support (D. Scott personal communication; Gouzoules *et al.* 1984). In the case of macaques, the calls have been acoustically analysed, and not only has their distinctiveness been confirmed, but the calls vary with the precise context and nature of the contest (Gouzoules *et al.* 1984). Male baboons have a special gesture to solicit support from other males in contests over estrous females (Smuts and Watanabe 1990). Success varies, but in one

group in Kenya, about 70 per cent of solicitations were responded to (Bercovitch 1988).

2. *Coercion*. Coercion of support itself has not been seen, as far as I know, but ornithologists have suggested that breeding birds might in effect coerce previous offspring, or others, to feed their current brood by threatening to drive them off the territory when they do not help (Ligon 1983); gorillas give threatening vocalizations to animals who have stopped grooming them, apparently forcing them to start again (personal observation); and chimpanzees, but not macaques, give support in contests against those who have previously formed coalitions against them (de Waal and Luttrell 1988; de Waal this volume, pp. 233–57).

3. *Reciprocation*. Unequivocal demonstrations of reciprocal altruism are extremely rare (Packer 1986), but one of the better demonstrations of it so far comes from a study of a non-primate (Wilkinson 1984). Well-fed vampire bats (*Desmodus rotundus*) regurgitate blood to starving roost mates at little cost to themselves but with immense benefit to the recipients; and bats recently fed by others are more likely to regurgitate later when they find their own food than are previously unfed bats, and apparently more likely to regurgitate to those individuals that fed them previously. Reciprocation of support in contests, as opposed to other forms of help, has not been reported in non-primates. Packer's (1977) observations initially indicated its occurrence among male baboons, but further studies make that interpretation less likely; they are more convincing in their suggestion that mutual support for personal benefit is involved (Bercovitch 1988; Noë 1990, this volume pp. 285–321). However, other studies have now indicated that in groups of chimpanzees and some species of macaque, individuals support often those who support them often, independently of, for example, consanguinity or time spent in proximity (de Waal and Luttrell 1988; this volume pp. 233–57).

4. *Friendly relationships*. A distinction needs to be made between reciprocal interactions, in which specific acts of help are alternately reciprocated, and friendly relationships, alliances, which involve a more or less frequent exchange of small altruistic acts of various sorts without close 'score-keeping'. A non-primate example of the establishment of such a relationship might by courtship feeding by male seabirds and birds of prey. Although interpretable as a means for the female to select males of high quality (Wiggins and Morris 1986), in the current context males can be seen as establishing a friendly relationship with the female in which later mating rights are exchanged for food now.

Among primates, grooming (picking through the fur to remove dirt and parasites) is commonly interpreted as a beneficial and friendly behaviour for several reasons: the behaviour seems so obviously useful; groomed animals give every impression of enjoying it; the most frequent grooming partners seen are mothers grooming young offspring; and among adults, grooming partners frequently spend a lot of time together, are relaxed in each other's presence, eat near one another, and so on. Grooming, then, is treated here as a behaviour that initiates and maintains friendly relationships.

For primates, Stammbach (1988) performed a beautiful experiment in which he trained only some individuals in a captive group of macaques to obtain food from an apparatus, and showed that extra grooming of these individuals by others correlated with the trained animals allowing the groomers increased access to the food. It is not necessarily the case that grooming was exchanged for food; rather, the friendly relationship established by the grooming might have prevented the subordinates being frightened away from the apparatus by the dominants. Seyfarth and Cheney (1984) found that unrelated vervet monkeys were more likely to look in the direction of the distress calls of individuals who had groomed them in the recent past than of individuals who had groomed them in the recent past than of individuals who had not. Finally, de Waal (1989) observed that a low-ranking rhesus monkey gained priority of access to water as a result of friendly association with one member of a high-ranking family which led to the whole family tolerating it at a water source. If friendly relationships are indeed important to an individual's competitiveness, it should invest in their maintenance in the face of alternative calls on its time (Dunbar 1988), and Dunbar and Dunbar (1988) have shown that as time spent feeding by lactating gelada females *Theropithecus gelada* increased above 35 per cent of the day, so time spent resting dropped, whereas it was not until feeding took up 60 per cent of the day that time spent in social activity began to drop.

Manipulating the quality of support

All of us know that if we want to advance our careers, one necessity is to cultivate not simply a number of contacts, as the geladas are apparently doing, but contacts with powerful potential supporters. In probably most populations, individuals' competitive abilities differ (Lomnicki 1988). If coalitions are used as a competitive strategy, the individual that can through friendly behaviour form an alliance with a powerful group member is going to be at an advantage. Nevertheless, it seems that only primates may attempt to manipulate the quality of future support by preferentially attempting with friendly behaviour to form alliances with powerful group members. Manipulation of quality of support in this way might be an absolute difference between primates and non-primates, according to presently available data.

In a study of three captive groups of vervet monkeys of three adult females each, Fairbanks (1980) showed that the dominant female in each group received more grooming from a greater proportion of group members than did any of the other females (Fig. 16.5). In general, where a bias in the direction of grooming has been reported, subordinates are usually found to groom

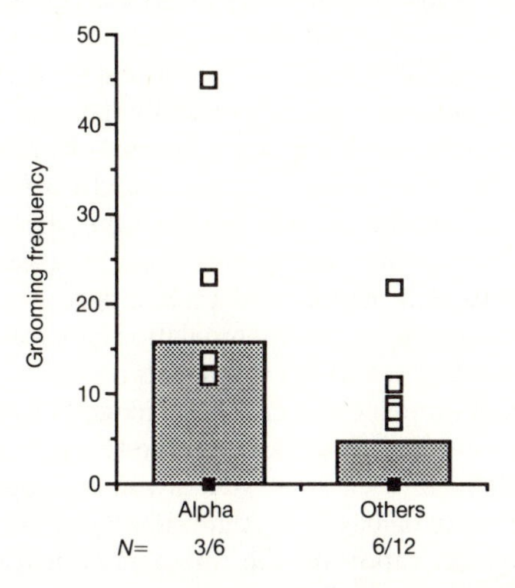

Fig. 16.5. Individuals are preferentially 'friendly' (establish alliances with) dominant group members. Primates—vervet: frequency with which the alpha adult female and two other adult females in each of three captive groups was groomed by the other two females. Points are individual females' values; histogram is mean value. *N* is the number of individuals/number of possible values (zero values not shown). (Data from Fairbanks 1980.)

dominants more than vice versa (Oki and Maeda 1973; Seyfarth 1977, 1980, 1983; Silk 1982, this volume pp. 215–31).

If grooming is concerned with establishment of alliances with powerful potential allies, there is little point in grooming low-ranking group members. Thus grooming among subordinate animals should be less frequent than among dominant ones. In three groups of wild vervet monkeys the top-ranking 50 per cent of females groomed one another for a mean of twice as long as did the low-ranking females (Seyfarth 1980).

Most reports of grooming relationships concern female cercopithecines, but the pattern is neither taxon nor sex specific. Male chimpanzees in Gombe Stream National Park in Tanzania also groomed dominant animals more than they groomed subordinate ones; moreover, high-ranking chimpanzee males groomed one another more than did low-ranking males (Simpson 1973).

With regard specifically to coalitions. Cheney (1977) observed that in a wild group of savannah baboons, 81 per cent of incidents of support from immatures to aggressive adult females were on behalf of the top three of the ten females in the group, the animals least in need of support (Cheney 1977). In a study of free-ranging adult female rhesus macaques, Chapais (1983*b*) found that in 28 coalitions involving different combinations of unrelated participants, support was always given to the dominant contestant. Chapais' observations accord to behaviour expected as the first step in a reciprocal exchange: it was given to a familiar animal who was almost certain to remain in the social group (another female) at very little cost to the donor (because the dominant contestant was going to win the fight anyway) (Trivers 1971).

It is also the case that captive immature hyaenas distribute support in this way. (Zabel *et al.* this volume pp. 113–35). Zabel *et al.* have a far simpler explanation for the observed distribution than the one offered here. They suggest that their results are simply explained by a general tendency of individual hyaenas to join attacks and threats against subordinate group members. The consequence is not manipulation of quality of support, but suppression of competitors, and the stimulus is a simple one. This could be true of primates too, in which case contrasts in cognitive capacities probably have nothing to do with the differences described here. However, primatologists could be justified in their inference of greater cognitive complexity, because more appears to be occurring than simply joining in attacks on subordinates. For instance, rarely do the interactions involve as many individuals as appears usually to be the case in the captive hyaena group; attacking the subordinates (support of dominants) is associated with preferential grooming of dominants; the attacks are not as inevitable as they seem to be among the hyaenas; and some studies demonstrate clear choice by primates of which subordinate to attack.

The classical example of reciprocal altruism concerns saving a drowning man by throwing a rope to him, an act of low cost but high benefit, which could later be reciprocated (Trivers 1971). Reciprocation from the man might be almost as likely if it was his drowning child who was saved. Cheney (1977) found that immature baboons, when supporting animals younger than themselves, helped only the offspring of animals dominant to their own mother. When supporting other immatures of similar age, however, the relative ranks of their mothers were unimportant. The younger recipients of help, being smaller and subordinate to their supporters, could hardly have been any use as allies, unlike their peers. Cheney suggested that the supporters were forming coalitions with the offspring of high-ranking females as a means of establishing an alliance with the mothers (see also Cheney 1978). That this sort of second-order calculation is within primates' cognitive capacity is indicated by observations of adult vervet monkeys being more likely to threaten and to be extra friendly with another if that other's relative had recently threatened them and also if the other's relative had recently threatened one of their own relatives (Cheney and Seyfarth 1989; see also York and Rowell 1988).

Competition for powerful allies

Dominant group members cannot ally with everybody. In a situation in which animals are attempting to improve the quality of future support by being friendly with powerful group members, competition to be the powerful animal's 'friend' is going to result. The competition can take two forms. Animals can attempt to perform more or better services than their rivals. Or they can simply prevent their rivals either from interacting with the dominants: one of Alexander Haig's complaints when he was US Secretary of Defense under Richard Nixon was that Haldeman and Erlichman continually hindered his access to the President.

Among non-human primates, competition to groom dominant animals appears to be more frequent than competition to groom subordinate ones (Fig. 16.6, Silk 1982). In vervet monkey groups in Amboseli National Park in Kenya, the frequency with which adult females prevented grooming among others was proportional to the rank of the groomed animal (Cheney and Seyfarth 1990; Seyfarth 1980). Cheney (1977) and Seyfarth (1977) modelled the effect on the distribution of grooming among adult females of a tendency to groom dominant animals coupled with competition to do so. One result was that individuals tended to groom most those close in rank to themselves. Seyfarth (1977) showed that precisely such a distribution is found in several primate groups, whether wild or captive, whether composed of related or unrelated females.

In addition to being forced by competition to groom animals of nearby rank, individuals might actively chose them as partners. They might not be the most powerful animals in the group, but because of kinship in species in which rank is inherited (Colvin 1983; Seyfarth 1977), or similarity of competitive ability (de Waal and Luttrell 1986), they might be the most likely to reciprocate support or to cooperate for mutual benefit (cf. Noë 1990, this

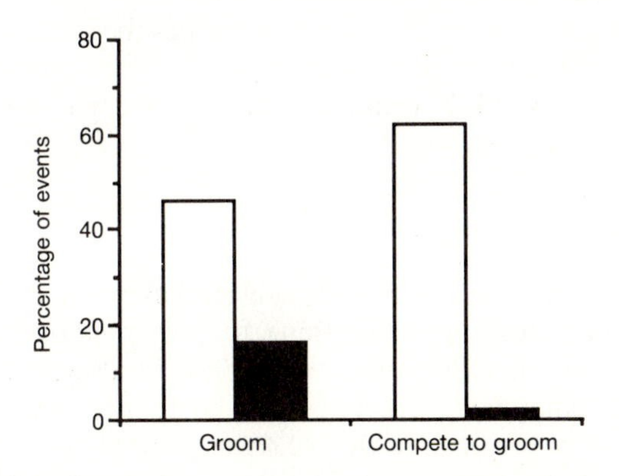

Fig. 16.6. Individuals compete to be friendly with (establish alliances with) dominant group members. Primates—bonnet macaque: percentage of events of grooming and supplanting from grooming by ten adult females that were directed at higher (open bars) or lower ranking (solid bars) adult female group members. (Histograms do not sum to 100 because the focal females groomed other group members besides adult females.) (Data from Silk 1982.)

volume pp. 285–321). Thus, of eight free-ranging immature rhesus macaque males who had a dominant non-kin and a subordinate close kin ranking adjacent to them, seven chose the subordinate kin (Colvin 1983). In this case, individuals are essentially adopting a tactic of encouraging fairly certain support from kin or those similar in rank, even if weak, rather than going for uncertain returns from powerful animals. In other words, primates could be following two tactics in their manipulation of alliances, which might or might not be mutually exclusive depending on the situation (Cheney and Seyfarth 1990). In either case, though, the behaviour is interpretable as manipulation of support.

If the power differential between group members is not too great, two subordinate animals can defeat one dominant to both. This being the case, an alliance with a subordinate is valuable, even if perhaps not quite as valuable as an alliance with a dominant animal. On the basis of the argument being presented here, grooming of subordinates by dominants is only to be expected, therefore. The concept might also explain the finding that support of one subordinate against another by unrelated adult female rhesus macaques can be frequent (Chapais 1983*b*): if alliances with subordinates are valuable, alliances among subordinates can be dangerous. de Waal and Luttrell (1986) did not find a correlation between rank of groomed female and the frequency with which others competed to groom her. However, they found that about 80 per cent of observations of macaques preventing others grooming concerned animals hindering grooming between two subordinate group members, as did Silk (1982). In both these studies, the most common combination of interactants was a dominant animal preventing grooming of a subordinate by an even more subordinate individual.

In summary, the social life of primates has become extraordinarily complex in comparison to that so far reported for non-primates, as individuals attempt to establish alliances with particularly useful partners and prevent others from doing so.

Reciprocation or mutualism?

Methodological problems exist with substantiation of the idea that the advantage to individuals in establishing friendly relationships is help of various sorts later, including support in contests (Seyfarth and Cheney 1984). From the point of view of motivation, goals of behaviours are difficult to analyse, and from the point of view of consequences, reciprocation can be very difficult to demonstrate. How does the observer detect reciprocity in a natural social group when different currencies of different value might be exchanged with time lags of days or even weeks among partners of differing power in a social milieu in which any two partners probably have several other partners with whom they are also involved in more or less unbalanced exchange of services? No wonder that reciprocation was not quantitatively proven in two studies that specifically searched for it (Fairbanks 1980; Silk 1982), although the tests might have been unnecessarily rigid (see Boyd this volume pp. 473–89). Of course, the difficulties do not mean that reciprocation is impossible to demonstrate (see Silk, de Waal this volume pp. 215–31, 233–57).

A reasonable alternative hypothesis for many of the observations reviewed above is that the supporting or other cooperative behaviour was not directed at increasing the probability of future cooperation from the partner, but rather at gaining immediate advantage from the association (Chapais 1983*a*, this volume pp. 29–59; Cheney 1977; Seyfarth 1977, 1980). Thus grooming might pacify a dominant animal and so inhibit it from chasing the groomer from a good food source (Silk 1982); however, tolerance itself could be viewed as reciprocated altruism. Alternatively, joining a dominant animal in an attack on a subordinate one could lead to the subordinate associating the supporter with the dominant and hence make the subordinate more fearful of the supporter and therefore less likely to challenge it later (Chapais 1983*a*, this volume pp. 29–59). As said before, though, subordinates do not always pacifically accept their status. Therefore if association with the dominant was not backed up by support from the dominant, power over the subordinate might not be reinforced by a coalition with the dominant animal.

Even if association with dominant group members brings immediate benefit, it is perhaps more difficult to think of a reasonable immediate benefit that could be obtained from association with a subordinate group member. Perhaps an alpha female's status is enhanced by joining a beta female in contest against a gamma female, but would not the alpha do better to suppress the beta, its more dangerous rival, by joining with the gamma? Here, reciprocity seems a more viable explanation. In the case of immature baboons supporting younger, smaller, subordinate group members, immediate gain via mutualism becomes a very difficult hypothesis to hold; Cheney's (1977) reciprocation hypothesis explains the observations better.

DISCUSSION

The data reviewed here indicate that non-primates differ from primates in their use of coalitions and alliances. It seems that only primates react to differences in competitive ability among group members by attempting to cultivate particularly useful animals, often dominant ones, as allies; that only primates compete over access to these useful allies; and that only primates prevent the formation of rival alliances. The data appear to be telling us that for primates, coalitions and alliances are an end in themselves (Kummer 1978); for non-primates they are simply a means to an end.

Two provisos are necessary. First, only a very small proportion of primate species, less than 5 per cent, have been identified as using complex coalitions and alliances of the sort described here. In generalizing about primates, therefore, I am in fact generalizing only about Old World monkeys and apes, and not even all genera of them. Secondly, I have been careful throughout the chapter to write about apparent differences between primates and non-primates, for it is not impossible that the contrast detected is a false one resulting simply because most non-primatologists, not expecting complexity, have not looked for it (Harcourt 1988; Rowell 1988). Thus Schilder (1990), who describes one of the most complex distributions of interventions in contests and friendly interactions for a non-primate, the plains zebra, was clearly influenced in his analysis by primatologists; and similarly for Zabel *et al.* (this volume pp. 113–35), whose study of coalitions among immature hyaenas perhaps comes the closest to negating the existence of the taxonomic distinction that I am suggesting.

If the differences between primates and non-primates are real, why do they exist? In this review I have concentrated on the behavioural complexity of coalitionary behaviour because of the possibility that information processing ability might constrain the use of coalitions and alliances as a competitive strategy. However, information-processing ability is not the only constraint. Environmental and societal constraints exist also (Davies and Houston 1984; Wrangham 1980, 1982; van Schaik 1989; van Schaik and van Hooff this volume pp. 357–89), as they do on the use of cooperation in general (Axelrod 1984; Brown 1987; Emlen 1984; Trivers 1971; Vehrencamp 1983).

For example, the adaptive establishment of alliances with powerful group members, and the expenditure of energy and risk in prevention of others from so doing, assumes that the benefits of the alliance arise in part through reciprocation. This being the case, the behaviour requires that associations among potential partners are stable, in order that they can reap the benefit of their alliance and retaliate against non-reciprocation (Axelrod 1984; Trivers 1971); it requires that differences in competitive abilities exist within the group; and it requires that the differences in competitive abilities are stable, in order that investment in establishing an alliance with a dominant group member is not wasted by the partner losing its status. Such characteristics will in turn be determined in part by the nature of the habitat and the animals' use of it, for social organization is influenced by ecology (Clutton-Brock and Harvey 1977; Emlen and Oring 1977; Wrangham and Rubenstein 1986). More particularly, if differences in competitive ability are to be apparent, individuals must contest access to resources, which means that the resources must be both worth defending and defensible, in other words of relatively high quality and compactly distributed (Davies and Houston 1984). For mutualistic coalitions to be advantageous, it helps if the resources are divisible (Wrangham 1980, 1982).

In sum, the social and ecological contexts provide the setting in which individuals must play their competitive games. Change either, and the value of using coalitions and alliances as a competitive strategy can change. With particular reference to the use of coalitions and alliances, female vervet monkeys (Cheney 1983) and male chimpanzees (de Waal 1984) use them in what could be interpreted as more complex ways than do their counterparts of the opposite sex, and similarly with cercopithecine primates by comparison to gorillas (Harcourt and Stewart 1989). However, in no case is the former more intelligent than the latter; rather, their socio-ecological settings differ (Harcourt and Stewart 1989). We need to consider how the relevant social and environmental factors might differ between primates and non-primates (or for that matter, between New and Old World primates), if they do.

The question for this chapter is whether more species of primates have richer, more defensible, or more divisible resources than do most or all non-primates, or whether they live in more stable groups with more obvious or more stable dominance hierarchies. In some habitats it is not impossible that many primates could have access to richer foods than other sympatric species, because the primates' dexterity might mean that food can be processed minutely to obtain the highest-quality parts. However, many non-primates live on divisible resources of high quality that they defend, for instance the acorn woodpecker (Hannon *et al.* 1985). Many non-primates also live in stable groups, for example female vampire bats (Wilkinson 1984) and female lions (Schaller 1972), and dominance hierarchies are widely reported in the non-primate literature. It is the combination of 'primate' characters that is important, though, and it could be that primate-like socio-ecological systems are rare (Wrangham 1983). Nevertheless, they do occur, the hyaena being an example (Frank 1986; Zabel *et al.* this volume pp. 113–35), yet hyaenas have not been reported to manipulate allies in the way that primates do; and while dolphins have allies, and possibly have to balance costs of competition with them in one context with cooperation in another (Connor *et al.* this volume pp. 415–43), primate-like manipulation of potential allies has not yet been reported.

In making the suggestion that information-processing abilities might be a constraint on the use of coalitions and alliances as a competitive strategy, it is important to say that despite the functional analysis of coalitions and alliances presented here, no implications of intentionality are implied. Subordinate vervet monkeys, for example, probably do not groom dominant ones in order consciously to increase the chances of future support in contests. Nevertheless, the data suggest that primates either have greater knowledge of their group companions, or make greater use of their knowledge.

Experiments on mate choice by females in fallow deer, *Dama dama* (Clutton-Brock *et al.* 1989) and on feeding competition in goldfinches, *Carduelis tristis* (Popp 1987), to name just two, indicate that non-primates are as capable as primates at distinguishing individuals and acting on the perceived differences among them. Perhaps the difference between primates and non-primates, if there is one, lies in the timing and context of the consequences of the distinction, therefore. Current literature indicates that non-primates can easily distinguish group members in order themselves to mate or differentially avoid them now; primates distinguish and select among group members to change the chances that the chosen partners will behave differentially towards themselves in the future.

Nevertheless, there is poor or even non-existent evidence that non-primates are in fact incapable of using coalitions and alliances in the way that primates do because of lack of information-processing ability. This question of the level of social cognition required to account for the sort of observations of apparent social complexity discussed here, along with questions about the relative levels of influence of social, environmental, and information-processing constraints on the use of coalitions and alliances in competition, will require detailed experimentation for their resolution (Cheney and Seyfarth 1990; Harcourt and de Waal this volume pp. 493–510; Humphrey 1976; Kummer *et al.* 1990). At the least, I hope this review may have identified where it would be valuable to look more closely for differences in social complexity between the taxa.

Whether or not primates are eventually confirmed to be more socially complex in this way than non-primates, whether or not any such dichotomy is shown to be due to differences in information-processing abilities, the fact remains that in the process of alliance formation and maintenance in primate groups, some fairly complex information about social companions is probably being processed (Cheney and Seyfarth 1990). Why primates have particularly large brains is still not understood (Bennett and Harvey 1985; Sawaguchi 1990). The social complexity itself has led to suggestions that primate, and odontocete (Connor *et al.* this volume pp. 415–43), brain size and intelligence might have evolved to cope with the complexity (see Byrne and Whiten 1988). However, correlations of relative brain size with either the social or the physical environment (Byrne and Whiten 1988; Clutton-Brock and Harvey 1980; Dunbar in press; Milton 1988; Sawaguchi 1990; Sawaguchi and Kudo 1990) say little about which environment was responsible for the enlargement in the first place, or indeed whether the selection acted on the brain or the body (Deacon 1990). Not only is brain size a very crude measure indeed of ability to process information relevant to decisions in social interactions and relationships, but information-processing abilities evolved in response to one environment could presumably be used in the other, or a third factor could be operating.

Both the physical environment and the social environment probably select for information-processing abilities (Byrne and Whiten 1988; Clutton-Brock and Harvey 1980; Dunbar in press; Humphrey 1976; Milton 1988; Sawaguchi 1990; Sawaguchi and Kudo 1990). A crucial distinction between them is that only in the social environment, or to a far greater extent in the social environment, is there positive feedback in both directions (Bateson 1988). Once one animal uses coalitions as a competitive strategy, the other members of the society have to do so too if they are to compete effectively. Once one attempts to manipulate the probability and quality of future coalitions, strong pressure operates on the others to be more effectively manipulative. The same is in principle true of the physical environment. In the case of the social environment, however, a runaway process is set in train (Humphrey 1976). Selection for manipulative abilities via information-processing ability necessarily increases the complexity of the social environment, in a way and at a rate that I cannot see happening with the physical environment. The increased complexity of the social environment then selects for greater information-processing power, and so on. Thus as soon as the process is started, it would seem to be the social environment more than the physical environment that provides the greater selection pressure, because the social environment is changed more than is the physical environment by intelligent manipulation—until we get to humans.

SUMMARY

1. Coalitions are a complex social behaviour that might require processing a lot of information about participants in present and possible future contests. Primates as a family have relatively larger information-processing organs, brains, than do other vertebrate families. Do they use coalitions in different ways than do non-primates?
2. Both primates and non-primates use coalitions for the same functional reasons: (1) to protect others, especially kin; (2) to allow others to gain access to resources; (3) to raise the dominance rank of kin; (4) to improve their own access to resources.
3. It appears that primates might form competitive coalitions with adults of the same sex within social groups more often than do non-primates, but the distinction is by no means absolute.
4. However, primates appear to be unique in the way in which they obtain some of the benefits that cooperation in contests can offer. It seems that only primates use alliances tactically for their own benefit. (1) Both primates and non-primates give mutual support in contests, but currently, only primates have been recorded to reciprocate support. (2) Perhaps more importantly, although in both non-primates and primates the quality of support in conflicts depends on the dominance rank of the supporter, only primates seem to have made use of this fact by preferentially attempting to cultivate friendly, supportive relationships, namely alliances, with dominant group members. (3) Furthermore, only primates appear to compete for such allies.
5. Fairly specific ecological and social conditions are necessary for

coalitions and alliances to be used as primates use them. A stable group membership of individuals with consistent differences among them in competitive ability is required, which in turn necessitates a rich, divisible resource, compactly distributed.

6. Primates could be unusual in the frequency with which these conditions are met, but are not unique. Nevertheless, their use of alliances as a competitive strategy is unique, according to presently available data. Such a dichotomy substantiates the suggestion that information-processing ability might be an operative constraint.
7. Once one individual in a social group starts to manipulate others for its own benefit, once one animals starts down the route of tactical alliance formation, others have to follow. A positive feedback loop is then set in process in which alliance formation selects for information-processing abilities which in turn allow further refinements in the use of alliances as a competitive strategy.

ACKNOWLEDGEMENTS

I thank the Department of Zoology at Cambridge, especially Tim Clutton-Brock, and the Department of Anthropology at Davis, especially Peter Rodman and David Smith, for hospitality and facilities. The manuscript was considerably improved by criticism from Dorothy Cheney, Thelma Rowell, Robert Seyfarth, Kelly Stewart, and Frans de Waal.

REFERENCES

Adams, E. S. (1990). Boundary disputes in the territorial ant *Azteca trigona*: effects of asymetries in colony size. *Animal Behaviour*, **39**, 321–8.

Altmann, S. A. (1962). A field study of the sociobiology of the rhesus monkey, *Macaca mulatta*. *Annals of the New York Academy of Sciences*, **102**, 338–435.

Armstrong, E. (1983). Relative brain size and metabolism in mammals. *Science*, **220**, 1302–4.

Armstrong, E. (1985). Allometric considerations of the adult mammalian brain, with special emphasis on primates. In *Size and scaling in primate biology* (ed. W. L. Jungers), pp. 115–46. Plenum, New York.

Axelrod, R. (1984). *The evolution of cooperation*. Basic Books, New York.

Bateson, P. P. G. (1988). The active role of behaviour in evolution. In *Evolutionary processes and metaphors* (eds M.-W. Ho and S. W. Fox), pp. 191–207. Wiley, London.

Bennett, P. M. and Harvey, P. H. (1985). Brain size, development and metabolism in birds and mammals. *Journal of Zoology*, **207**, 491–509.

Bercovitch, F. B. (1988). Coalitions, cooperation and reproductive tactics among adult male baboons. *Animal Behaviour*, **36**, 1198–209.

Berman, C. M. (1980). Early agonistic experience and rank acquisition among free-ranging infant rhesus monkeys. *International Journal of Primatology*, **1**, 153–70.

Black, J. M. and Owen, M. (1987). Determinants of social rank in goose flocks: acquisition of social rank in young geese. *Behaviour*, **102**, 129–46.

Boyd, H. (1953). On encounters between wild white-fronted geese in winter flocks. *Behaviour*, **37**, 291–319.

Brown, J. L. (1987). *Helping and communal breeding in birds*. Princeton University Press, Princeton.

Bygott, J. D., Bertram, B. C. R., and Hanby, J. P. (1979). Male lions in large coalitions gain reproductive advantages. *Nature*, **282**, 839–41.

Byrne, R. and Whiten, A. (eds) (1988). *Machiavellian intelligence. Social expertise and the evolution of intellect in monkeys, apes, and humans*. Clarendon Press, Oxford.

Caldwell, R. L. (1985). A test of individual recognition in the stomatopod *Gonodactylus festae*. *Animal Behaviour*, **33**, 101–6.

Caro, T. M. and Collins, D. A. (1987). Male cheetah social organization and territoriality. *Ethology*, **74**, 52–64.

Chapais, B. (1983*a*). Adaptive aspects of social relationships among adult rhesus monkeys. In *Primate social relationships* (ed. R. A. Hinde), pp. 286–9. Blackwell, Oxford.

Chapais, B. (1983*b*). Dominance, relatedness and the structure of female relationships in rhesus monkeys. In *Primate social relationships* (ed. R. A. Hinde), pp. 208–19. Blackwell, Oxford.

Chapais, B. (1988*a*). Experimental matrilineal inheritance of rank in female Japanese monkeys. *Animal Behaviour*, **36**, 1025–37.

Chapais, B. (1988*b*). Rank maintenance in female Japanese macaques: experimental evidence for social dependency. *Behaviour*, **104**, 41–59.

Cheney, D. L. (1977). The acquisition of rank and the development of reciprocal alliances among free-ranging immature baboons. *Behavioral Ecology and Sociobiology*, **2**, 303–18.

Cheney, D. L. (1978). Interactions of immature male and female baboons with adult females. *Adult Behaviour*, **26**, 389–408.

Cheney, D. L. (1983). Extrafamilial alliances among vervet monkeys. In *Primate social relationships* (ed. R. A. Hinde), pp. 278–86. Blackwell, Oxford.

Cheney, D. L. and Seyfarth, R. M. (1987). The influence of intergroup competition on the survival and reproduction of female vervet monkeys. *Behavioral Ecology and Sociobiology*, **21**, 375–86.

Cheney, D. L. and Seyfarth, R. M. (1989). Redirected aggression and reconciliation among vervet monkeys, Cercopithecus aethiops. *Behaviour*, **110**, 258–75.

Cheney, D. L. and Seyfarth, R. M. (1990). *How monkeys see the world: inside the mind of another species*. University of Chicago Press, Chicago.

Clutton-Brock, T. H. and Harvey, P. H. (1977). Primate ecology and social organisation. *Journal of Zoology*, London, **183**, 1–39.

Clutton-Brock, T. H. and Harvey, P. H. (1980). Primates, brains and ecology. *Journal of Zoology*, London, **190**, 309–23.

Clutton-Brock, T. H., Hiraiwa-Hasegawa, M., and Robertson, A. (1989). Male choice on fallow deer leks. *Nature*, **340**, 463–5.

Colvin, J. (1983). Familiarity, rank and the structure of rhesus male peer networks. In *Primate social relationships* (ed. R. A. Hinde), pp. 190–200. Blackwell, Oxford.

Datta, S. B. (1983*a*). Relative power and the acquisition of rank. In *Primate social relationships* (ed. R. A. Hinde), pp. 93–103. Blackwell, Oxford.

Datta, S. B. (1983*b*). Relative power and the maintenance of dominance. In *Primate social relationships* (ed. R. A. Hinde), pp. 103–12. Blackwell, Oxford.

Davies, N. B. and Houston, A. I. (1984). Territory economics. In *Behavioural ecology*, 2nd edn (eds J. R. Krebs and N. B. Davies), pp. 148–69. Blackwell, Oxford.

Deacon, T. W. (1990). Fallacies of progression in theories of brain-size evolution. *International Journal of Primatology*, **11**, 193–236.

Dunbar, R. I. M. (1988). *Primate social systems*. Croom Helm, London.

Dunbar, R. I. M. (in press). Neocortex size as a constraint on group size in primates. *Journal of Human Evolution*.

Dunbar, R. I. M. and Dunbar, P. (1988). Maternal time budgets of gelada baboons. *Animal Behaviour*, **36**, 970–80.

Emlen, S. T. (1984). Cooperative breeding in birds and mammals. In *Behavioural ecology* (eds J. R. Krebs and N. B. Davies), pp. 305–39. Blackwell, Oxford.

Emlen, S. T. and Oring, L. W. (1977). Ecology, sexual selection, and the evolution of mating systems. *Science*, **197**, 215–23.

Essock-Vitale, S. and Seyfarth, R. M. (1987). Intelligence and social cognition. In *Primate societies* (eds B. B. Smuts, D. L. Cheney, R. M. Seyfarth, R. W. Wrangham, and T. T. Struhsaker), pp. 452–61. University of Chicago Press, Chicago.

Fairbanks, L. A. (1980). Relationships among adult females in captive vervet monkeys: testing a model of rank-related attractiveness. *Animal Behaviour*, **28**, 853–9.

Fairbanks, L. A. and McGuire, M. T. (1986). Age, reproductive value, and dominance-related behaviour in vervet monkey females: cross-generational influences on social relationships and reproduction. *Animal Behaviour*, **34**, 1710–21.

Frank, L. G. (1986). Social organization of the spotted hyaena (*Crocuta crocuta*). II. Dominance and reproduction. *Animal Behaviour*, **34**, 1510–27.

Gouzoules, S., Gouzoules, H. and Marler, P. (1984). Rhesus monkey (*Macaca mulatta*) screams: representational signalling in the recruitment of agonistic aid. *Animal Behaviour*, **32**, 182–93.

Hall, K. R. L. and DeVore, I. (1965). Baboon social behavior. In *Primate behavior* (ed. I. DeVore), pp. 53–110, Holt, Rinehart and Winston, New York.

Hannon, S. J., Mumme, R. L., Koenig, W. D., and Pitelka, F. A. (1985). Replacement of breeders and within-group conflict in the cooperatively breeding acorn woodpecker. *Behavioral Ecology and Sociobiology*, **17**, 303–12.

Harcourt, A. H. (1988). Alliances in contests and social intelligence. In *Machiavellian intelligence. Social expertise and the evolution of intellect in monkeys, apes, and humans* (eds R. W. Byrne and A. Whiten), pp. 132–52. Clarendon Press, Oxford.

Harcourt, A. H. (1989). Social influences on competitive ability: alliances and their consequences. In *Comparative socioecology* (eds V. Standen and R. A. Foley), pp. 223–42. Blackwell, Oxford.

Harcourt, A. H. and Stewart, K. J. (1987). The influence of help in contests on dominance rank in primates: hints from gorillas. *Animal Behaviour*, **35**, 182–90.

Harcourt, A. H. and Stewart, K. J. (1989). Functions of alliances in contests within wild gorilla groups. *Behaviour*, **109**, 176–90.

Horrocks, J. A. and Hunte, W. (1983). Maternal rank and offspring rank in vervet monkeys: an appraisal of the mechanisms of rank acquisition. *Animal Behaviour*, **31**, 772–82.

Humphrey, N. K. (1976). The social function of intellect. In *Growing points in ethology* (eds P. P. G. Bateson and R. A. Hinde), pp. 303–17. Cambridge University Press, Cambridge.

Hunter, L. A. (1985). The effects of helpers in cooperatively breeding purple gallinules. *Behavioral Ecology and Sociobiology*, **18**, 147–53.

Jerison, H. J. (1983). The evolution of the mammalian brain as an information processing system. In *Advances in the study of mammalian behavior* (eds J. F. Eisenberg and D. G. Kleiman), pp. 113–46. The American Society of Mammalogists, Pittsburgh.

Kawai, M. (1958). On the rank system in a natural group of Japanese monkeys. I and II. *Primates*, **1**, 111–48.

Kawamura, S. (1958). Matriarchal social ranks in the Minoo-B troop: A study of the rank system of Japanese monkeys. *Primates*, **1**, 149–56.

Kummer, H. (1967). Tripartite relations in hamadryas baboons. In *Social communication among primates* (ed. S. A. Altmann), pp. 63–72. University of Chicago Press, Chicago.

Kummer, H. (1978). On the value of social relationships to nonhuman primates: a heuristic scheme. *Social Science Information*, **17**, 687–705.

Kummer, H. (1982). Social knowledge in free-ranging primates. In *Animal mind–human mind* (ed. D. R. Griffith), pp. 113–30. Springer-Verlag, New York.

Kummer, H., Dasser, V., and Hoyningen-Huene, P. (1990). Exploring primate social cognition: some critical remarks. *Behaviour*, **112**, 84–98.

Lefebvre, L. and Henderson, D. (1986). Resource defense and priority of access to food by the mate in pigeons. *Canadian Journal of Zoology*, **64**, 1889–92.

Ligon, J. D. (1983). Cooperation and reciprocity in avian social systems. *American Naturalist*, **121**, 366–84.

Lomnicki, A. (1988). *Population ecology of individuals*. Princeton University Press, Princeton.

Macphail, E. M. (1982). *Brain and intelligence in vertebrates*. Clarendon Press, Oxford.

Martin, R. D. (1981). Relative brain size and basal metabolic rate in terrestrial vertebrates. *Nature*, **293**, 57–60.

Martin, R. D. (1984). Body size, brain size and feeding strategies. In *Food acquisition and processing in primates* (eds D. J. Chivers, B. A. Wood, and A. Bilsborough), pp. 73–103. Plenum, London.

Milton, K. (1988). Foraging behaviour and the evolution of primate cognition. In *Machiavellian intelligence: social expertise and the evolution of intellect in monkeys, apes and humans*. (eds R. Byrne and A. Whiten), pp. 285–305. Clarendon Press, Oxford.

Mori, A. (1979). Analysis of population changes by measurement of body weight in the Koshima troop of Japanese monkeys. *Primates*, **20**, 371–97.

Murnighan, J. K. (1978). Models of coalition behavior: game theoretic, social psychological, and political perspectives. *Psychological Bulletin*, **85**, 1130–53.

Noë, R. (1990). A veto game played by baboons: a challenge to the use of the prisoner's dilemma as a paradigm for reciprocity and cooperation. *Animal Behaviour*, **39**, 78–90.

Oki, J. and Maeda, Y. (1973). Grooming as a regulator of behavior in Japanese macaques. In *Behavioral regulators of behavior in primates* (ed. C. R. Carpenter), pp. 149–63, Bucknell University Press, Lewisburg.

Packer, C. (1977). Reciprocal altruism in olive baboons. *Nature*, **265**, 441–3.

Packer, C. (1986). Whatever happened to reciprocal altruism? *Trends in Ecology and Evolution*, **1**, 142–3.

Packer, C. and Pusey, A. (1979). Female aggression and male membership in troops of Japanese macaques and olive baboons. *Folia Primatologica*, **31**, 212–18.

Packer, C. and Pusey, A. (1982). Cooperation and competition within coalitions of male lions: kin selection or game theory? *Nature*, **296**, 740–2.

Passingham, R. E. (1981). Primate specialisation in brain and intelligence. *Symposium of the Zoological Society*, London, **46**, 361–88.

Popp, J. W. (1987). Choice of opponents during competition for food among American goldfinches. *Ethology*, **75**, 31–6.

Pusey, A. E. and Packer, C. (1987). Dispersal and philopatry. In *Primate societies*. (eds B. B. Smuts and D. L. Cheney, R. M. Seyfarth, R. W. Wrangham, and T. T. Struhsaker), pp. 250–66. University of Chicago Press, Chicago.

Ridgway, S. H. (1986). Dolphin brain size. In *Research on dolphins* (eds M. M. Bryden and R. Harrison), pp. 59–70. Clarendon Press, Oxford.

Robinson, J. G. (1988). Group size in wedge-capped capuchin monkeys *Cebus olivaceus* and the reproductive success of males and females. *Behavioral Ecology and Sociobiology*, **23**, 187–97.

Rowell, T. E. (1988). Beyond the one-male group. *Behaviour*, **104**, 189–201.

Sawaguchi, T. (1990). Relative brain size, stratification, and social structure in anthropoids. *Primates*, **31**, 257–72.

Sawaguchi, T. and Kudo, H. (1990). Neocortical development and social structure in primates. *Primates*, **31**, 283–9.

van Schaik, C. P. (1989). The ecology of social relationships amongst female primates. In *Comparative socioecology* (eds V. Standen and R. A. Foley), pp. 195–218. Blackwell, Oxford.

Schaller, G. B. (1972). *The Serengeti lion*. University of Chicago Press, Chicago.

Schilder, M. B. H. (1990). Interventions in a herd of semi-captive plains zebra. *Behaviour*, **112**, 53–83.

Schjelderup-Ebbe, T. (1935). Social behavior of birds. In *Handbook of social psychology* (ed. C. Murchison), pp. 947–72. Clark University Press, Worcester, MA.

Scott, D. K. (1980). Functional aspects of prolonged parental care in Bewick's swans. *Animal Behaviour*, **28**, 938–52.

Seyfarth, R. M. (1977). A model of social grooming among adult female monkeys. *Journal of Theoretical Biology*, **65**, 671–98.

Seyfarth, R. M. (1978). Social relationships among adult male and female baboons. I. Behaviour during sexual consortship. *Behaviour*, **64**, 204–47.

Seyfarth, R. M. (1980). The distribution of grooming and related behaviours among adult female vervet monkeys. *Animal Behaviour*, **28**, 798–813.

Seyfarth, R. M. (1983). Grooming and social competition in primates. In *Primate Social Relationships* (ed. R. A. Hinde), pp. 182–90. Blackwell, Oxford.

Seyfarth, R. M. and Cheney, D. L. (1984). Grooming, alliances and reciprocal altruism in vervet monkeys. *Nature*, **308**, 541–3.

Silk, J. B. (1982). Altruism among female *Macaca radiata*: explanations and analysis of patterns of grooming and coalition formation. *Behaviour*, **79**, 162–88.

Simpson, M. J. A. (1973). The social grooming of male chimpanzees. In *Comparative ecology and behaviour of primates* (eds R. P. Michael, and J. H. Crook), pp. 411–505. Academic Press, London.

Smuts, B. B. (1985). *Sex and friendship in baboons*. Aldine Press, New York.

Smuts, B. B. and Watanabe, J. (1990). Social relationships and ritualized greetings in adult male baboons. (*Papio cynocephalus anubis*). *International Journal of Primatology*, **11**, 147–72.

Stammbach, E. (1988). Group responses to specially skilled individuals in a *Macaca fascicularis* group. *Behaviour*, **107**, 241–66.

Trivers, R. L. (1971). The evolution of reciprocal altruism. *Quarterly Review of Biology*, **46**, 35–57.

Vehrencamp, S. L. (1983). A model for the evolution of despotic versus egalitarian societies. *Animal Behaviour*, **31**, 667–82.

de Waal, F. B. M. (1984). Sex differences in the formation of coalitions among chimpanzees. *Ethology and Sociobiology*, **5**, 239–55.

de Waal, F. B. M. (1989). Dominance 'style' and primate social organization. In *Comparative socioecology* (eds V. Standen and R. A. Foley, eds), pp. 243–63. Blackwell, Oxford.

de Waal, F. B. M. and Luttrell, L. M. (1986). The similarity principle underlying social bonding among female rhesus monkeys. *Folia Primatologica*, **46**, 215–34.

de Waal, F. B. M. and Luttrell, L. M. (1988). Mechanisms of social reciprocity in three primate species: symmetrical relationship characteristics or cognition? *Ethology and Sociobiology*, **9**, 101–18.

Walters, J. (1980). Interventions and the development of dominance relationships in female baboons. *Folia Primatologica*, **34**, 61–89.

Wechsler, B. (1988). Dominance relationships in jackdaws (*Corvus monedula*). *Behaviour*, **106**, 252–64.

Whitten, P. L. (1983). Diet and dominance among female vervet monkeys (*Cercopithecus aethiops*). *American Journal of Primatology*, **5**, 139–59.

Wiggins, D. A. and Morris, R. D. (1986). Criteria for female choice of mates: courtship feeding and parental care in the common tern. *American Naturalist*, **128**, 126–9.

Wilkinson, G. S. (1984). Reciprocal food sharing in the vampire bat. *Nature*, **308**, 181–4.

Wilson, E. O. (1975). *Sociobiology*. Belknap Press, Cambridge, MA.

Wrangham, R. W. (1980). An ecological model of female-bonded primate groups. *Behaviour*, **75**, 262–300.

Wrangham, R. W. (1982). Mutualism, kinship and social evolution. In *Current problems in sociobiology* (ed. King's College Sociobiology Group), pp. 269–89. Cambridge University Press, Cambridge.

Wrangham, R. W. (1983). Social relationships in comparative perspective. In *Primate social relationships* (ed. R. A. Hinde), pp. 325–34. Blackwell, Oxford.

Wrangham, R. W. and Rubenstein, D. I. (1986). Social evolution in birds and mammals. In *Ecological aspects of social evolution* (eds D. I. Rubenstein and R. W. Wrangham), pp. 452–70. Princeton University Press, Princeton.

York, A. D. and Rowell, T. E. (1988). Reconciliation following aggression in patas monkeys, *Erythrocebus patas. Animal Behaviour*, **36**, 502–9.

17

The evolution of reciprocity when conditions vary

ROBERT BOYD

INTRODUCTION

Altruism among unrelated individuals is an important feature of the behaviour of social mammals. Ever since Trivers (1971), such aid has been explained as reciprocal altruism. Hamilton (1964) had previously shown that selection does not favour unconditional altruistic behaviour when individuals interact with unrelated individuals because the costs of performing altruistic acts reduce the fitness of altruists but the benefits increase the fitness of altruists and non-altruists alike. Trivers' insight was that when individuals interact repeatedly, selection may often favour altruism towards individuals who have been altruistic during earlier interactions. Selection can favour such contingent altruism because the benefits of altruistic acts flow disproportionately to other contingent altruists.

More recently, our understanding of reciprocal altruism has been enriched by formal analyses based on the repeated prisoner's dilemma game (Axelrod and Dion 1989; Axelrod and Hamilton 1981). This model assumes that pairs of individuals are sampled from a population and interact t or more times with probability w^t. This assumption implies that the average number of times that a pair of individuals interact over their lifetimes is $1/(1 - w)$, larger values of w meaning more interactions between any pair of individuals. During each interaction, individuals either cooperate (C) or defect (D). Cooperation decreases the fitness the cooperator, but increases the average fitness of the pair of individuals, as shown in Table 17.1. This pattern of payoffs, which defines the prisoner's dilemma game, is a generalization of the 'donor–recipient' model widely used in behavioural ecology to model altruistic social interactions (Boyd 1987). Each cooperative act reduces the fitness of the donor by an amount c and increases the fitness of the recipient by an amount b. Thus, $T = b$, $R = b - c$, $P = 0$, and $S = -c$, and if $b > c > 0$, the donor-recipient model is also a prisoner's dilemma.

To allow the evolution of contingent behaviour, the model assumes that individuals have a *strategy* that determines whether they cooperate (C), or

Table 17.1. Payoff matrix for the prisoner's dilemma game. The left-hand entry in each cell gives the incremental effect of a single interaction on the fitness of individual 1, and the right-hand entry in each cell the effect on individual 2. Thus, two cooperators increase their fitness an amount R, and two defectors P. A cooperator interacting with a defector gets S, while the defector gets T. The matrix is a prisoner's dilemma if defectors do better than cooperators no matter what the other individual does ($T > R$ *and* $P > S$), but cooperation is mutually beneficial ($2R > T + S$).

		Individual 2	
		Cooperate	Defect
Individual 1	Cooperate	R, R	S, T
	Defect	T, S	P, P

defect (D) depending on previous behaviour. Such strategies can be simple unconditional rules such as always defect (ALLD), or they can be contingent rules like tit-for-tat (TFT) which cooperates on the first interaction and then copies the other player's behaviour on the previous interactions. An individual's expected fitness depends on its own strategy and the strategy of the individual with which it interacts. For example, TFT achieves a very high pay-off when interacting with TFT, but a very low pay-off when interacting with ALLD.

The goal of the analysis is to determine what sorts of strategies are favoured by natural selection. This task is complicated by the fact that the average fitness of one strategy depends which other strategies are present in the population. Because there is a combinatorily large number of conceivable strategies, it is not possible to consider all possible social environments. Instead, people have focused on two important special cases which are necessary, but not sufficient, for evolutionary success. First, can a strategy persist once it becomes common in the population? When a common strategy does better against itself than do any rare strategies, the strategy will persist—it is evolutionarily stable (or, for brevity, it is an ESS). Second, are there circumstances which allow the strategy to increase when rare? For a strategy to be an evolutionary success there must be some circumstances in which it can increase. This analysis does not make any assumption about how individuals acquire their strategies: they can be acquired by any combination of genetic inheritance, cultural inheritance, or learning, as long as strategies which lead to higher average fitness tend to increase and strategies which lead to lower average fitness tend to diminish. (See Maynard Smith 1982, 1989 for an extended defence of the ESS approach.) However, the logic of this evolutionary approach is quite different from non-evolutionary equilibrium concepts used in game theory which may yield qualitatively different solutions to the same problem (Binmore 1987).

Axelrod (1984) and Axelrod and Hamilton (1981) argue that if pairs of individuals interact long enough, then evolutionarily successful strategies will be: *nice*, meaning that they are never the first to defect; *provokable*, meaning that they respond to a defection by immediately defecting; and *forgiving*, meaning that they respond to cooperation by cooperating. If

$$wb > c \tag{1}$$

TFT is evolutionarily stable. Furthermore, if the chance that reciprocators interact with other reciprocators is even slightly greater than random, for example due to interaction among relatives, and if interactions go on long enough (w is large), TFT can increase when rare. Finally, computer tournaments suggest that strategies that are nice, provokable, and forgiving do well in a wide range of social environments. Brown *et al.* (1982) have pointed out that there is a close relationship between (1) and Hamilton's rule because w gives the fraction of altruistic acts directed toward other reciprocators.

Subsequent work (reviewed by Axelrod and Dion 1988; see also Boyd 1987) suggests that these conclusions are somewhat sensitive to changes in assumptions.

1. *Group size*. Many social behaviours involve cooperative behaviour among more than two individuals. Several authors have considered the effects of increasing the number of reciprocators (Bendor and Mookerjee 1987; Boyd and Richerson 1988, 1989; Joshi 1987). This work suggests that larger social groups greatly restrict the range of conditions under which reciprocity can be favoured by natural selection. For reciprocal strategies to persist when common, they must not tolerate very many defectors. Otherwise defectors will receive the benefits of cooperation without paying the cost. However, such intolerance of defection makes it difficult for reciprocal strategies to increase when rare because it is unlikely that enough reciprocators will congregate to benefit from cooperation.

2. *Multiple invading strategies*. Axelrod (1984) assumes that populations contain only two strategies. When more than two strategies compete, then no strategy can resist invasion by the right combination of invading strategies if interactions last long enough (Boyd and Lorberbaum 1987; Boyd 1989; Farrell and Ware 1989). For example, TFT can resist invasion by suspicious tit-for-tat (STFT) alone. (STFT defects on the first interaction, and then plays TFT.) However, TFT can be invaded by the combination of STFT and tit-for-two-tats (TF2T), the strategy which is like TFT except that it is provoked by two defections. TF2T achieves the same payoff against TFT as TFT does

against itself, but achieves a higher payoff against STFT. Thus TF2T spreads, and eventually this spread allows STFT to increase as well.

3. *Errors and noise*. Adding errors to the repeated prisoner's dilemma can lead to qualitative changes in the results. It is important to distinguish the mistakes, cases in which individuals know they have made an error, from noise, those cases in which individuals are unsure of the effect of their behaviour on the other individual. When individuals know they have made an error, selection favours 'contrite' TFT, which accepts punishment without retaliating following an error (Boyd 1989; Sugden 1986). This allows reciprocators to maintain a cooperative relationship after an error, and the resulting pattern of behaviour is quite similar to TFT without errors. In contrast, noise seems to favour strategies which are less provokable and less forgiving that TFT. (Bendor 1987; Mueller 1987). Bendor's (1987) model suggests that strategies which require more than one defection before being provoked are favoured because they allow cooperative relationships to persist despite occasional errors. Mueller's (1987) analysis suggests that selection favours strategies that, once provoked, punish defection more severely than does TFT.

Most existing theory assumes that the conditions of the game do not vary. The fitness effects of cooperation do not vary from one individual to the next. Cooperation imposes the same cost on every individual and creates the same benefit for every individual. Every social interaction in an ongoing relationship has the same costs and benefits as every other interaction. Every social relationship persists for the same length of time on the average.

Clearly this assumption is not very realistic. The costs and benefits of cooperative behaviour may vary widely among individuals. A dominant individual may provide extraordinary benefits in a coalition at relatively low cost, while a subordinate may be able to provide only little help, and at a very high cost. The costs and benefits of helping may vary widely from one behaviour to another—for example, it might be that grooming provides a small benefit at an even smaller cost, while aid in agonistic interactions provides a large benefit at a large cost. A long-lasting relationship between two individuals will often consist of a wide range of behaviours. Finally, the expected lifetime of relationships with different individuals may differ. Some individual may provide extraordinary benefits in a coalition at relatively low time, while others may be merely passing through.

Here, I present and analyse models of the repeated prisoner's dilemma in which conditions do vary. First, I consider a model in which costs and benefits of cooperation vary from one individual to another, then one in which costs and benefits vary from one time period to the next, and finally a model in which the expected number of interactions varies from pairing to pairing. In each case, there is a range of parameters in which the basic results derived by Axelrod and others are unchanged—nice, provokable, forgiving strategies like TFT can resist invasion by unconditional defection. However, in each case there is also a range of parameters in which TFT-like strategies are *not* evolutionarily stable, but other contingent strategies which support at least some cooperation *are* stable. These strategies need not be nice or provokable, nor do they necessarily lead to a TFT-like trading of one helping act for another. The fact that some patterns of variation favour TFT, while others favour qualitatively different forms of reciprocity, may provide an avenue for testing this body of theory.

WHEN PAY-OFFS VARY AMONG INDIVIDUALS

In many situations the cost of cooperating and the benefits gained from the cooperation may vary from one individual to the next as a result of differences in sex, age, size, physical condition, reproductive status, or dominance rank. For example, a high-ranked individual may be able to greatly increase the fitness of a low-ranked individual by aiding that individual in a coalition at little cost to itself. In contrast, the aid from a low-ranked individual may produce little benefit at great cost.

To model such variation, we assume that there are two types of individuals, type 1 and type 2. When a type 1 individual helps a type 2 individual, it increases the fitness of that individual an amount b_2 and decreases its own fitness an amount c_1. Similarly, when a type 2 individual helps a type 1 individual, it increases the fitness of that individual by an amount b_1 and decreases its own fitness by an amount c_2. Notice that I label benefits according to the individual who receives the benefit. If, for example, the helping behaviour takes the form of aid in contests, and type 1 individuals are subordinate and type 2 are dominant, it might be that $b_1 > b_2$—the dominant provides a bigger benefit and receives a smaller one, and that $c_1 > c_2$—the subordinate incurs a bigger cost than the dominant in helping. Individuals know their own type, and the type of individuals with whom they interact. Thus, individuals can use different strategies when interacting with different types. Here we are interested in the strategies used by type 1 individuals when they interact with type 2 and vice versa (Table 17.2).

Pairs of individuals, one type 1 and one type 2, are sampled from the population, and interact a number of times. The probability that they interact t or more times is w^t. Each interaction between a type 1 and a type 2 individual has the pay-off matrix shown in table 2. The interaction is a prisoner's dilemma for both individuals if $b_1 > c_1 > 0$, and $b_2 > c_2 > 0$. When these conditions are satisfied both individuals will be better off if they both cooperate, but each individual is better off defecting if the other individual cooperates.

As noted by Axelrod and Hamilton (1981), TFT is not evolutionarily stable unless the long-run benefit of mutual cooperation exceeds the short-term benefits to defection for *both* types. To see this, consider the strategy

Table 17.2. Payoff matrix for prisoner's dilemma game when the costs and benefits of cooperation vary among individuals.

		Individual 2	
		Cooperate	Defect
Individual 1	Cooperate	$b_1 - c_1, b_2 - c_2$	$-c_1, b_2$
	Defect	b_1, c_2	0, 0

TFT1 which plays TFT when it is type 1, but plays ALLD when it is type 2. This strategy can invade TFT whenever $wb_2 < c_2$. Similarly TFT can resist TFT2, the strategy that plays ALLD if type 1 but plays TFT if type 2 if $wb_1 > c_1$. Once either TFT1 or TFT2 becomes common, it can be invaded by ALLD because such strategies derive no benefit from interacting with themselves, and suffer when they interact with ALLD.

There is a generalization of TFT, however, that will allow at least some cooperation even if TFT is not evolutionarily stable. Consider the family of strategies I call 'intermittent tit-for-tat' (denoted ITFT(T)). If an individual using ITFT(T) is type 1, it cooperates on every interaction as long as the other individual cooperates on interactions $1, 1 + T, 1 + 2T, \ldots$ and so on. If an ITFT(T) individual is type 2, it cooperates on the first interaction, and then cooperates every T interactions thereafter as long as the other individual never defects. Notice that if $T = 1$, ITFT(T) is identical to TFT. (A more formal definition is given in the appendix.)

ITFT(T) can be evolutionarily stable if the benefits of intermittent but long-run cooperation exceed the short-term benefits of defection for individuals in both roles. ITFT(T) can resist invasion by ITFT1(T), the strategy which plays ITFT(T) when type 1 and ALLD when type 2, if

$$wB_2 > c_2 \tag{2}$$

where $B_2 = b_2(1 + w + w^2 + \ldots + w^{T-1})$ is the expected benefit to type 2 individuals that results from a type 1 individual cooperating on interactions $1, 2, \ldots, T$. Similarly, ITFT(T) can resist invasion by ITFT2(T), the strategy which plays ITFT(T) when type 2 and ALLD when type 1, if

$$w^T b_1 > C_1 \tag{3}$$

where $C_1 = c_1(1 + w + w^2 + \ldots + w^{T-1})$ is the expected cost incurred by type 1 individuals cooperating for T interactions. If (2) and (3) are both satisfied, ITFT(T) can also resist invasion by ALLD.

These conditions suggest that ITFT(T) is favoured by selection when individuals with lower cost/benefit ratios help less frequently than the individual with the higher cost/benefit ratio. Suppose that $wb_2 < c_2$, so it does not pay type 2 individuals to cooperate every interaction. If, however, the type 2 individual cooperates every T interactions, it still incurs a cost of c_2 per helping act, but now it receives the benefit of T acts by the type 1 individual (discounted by the probability that the interaction persists). Thus, by increasing T the benefit to the type 2 individual can be increased until cooperation does pay, and therefore (2) places a lower bound on the values of T than can be evolutionarily stable. Intermittent cooperation by the type 2 individual means that the type 1 individual incurs the cost of cooperating T times in return for a single benefit. Moreover, the longer lag between cooperation and its reward increases the pay-off to cheaters (represented by the w^I term). Both these factors mean there is a limit to how large T can be before it no longer pays type 1 individuals to cooperate. If the cost/benefit ratio of type 1 individuals is sufficiently large, this upper limit on T will exceed the lower bound imposed by the necessity to sufficiently compensate type 2 individuals, and TFT(I) will be evolutionarily stable. If this is not the case, then there are no reciprocating strategies which can persist.

These results are clearer in the context of an example. Suppose that type 1 individuals have high dominance rank, type 2 individuals have low dominance rank, and the behaviour under consideration is aid in agonistic interactions. For a given cost, it is likely that the benefit that results from aid by a subordinate to a dominant is less than the benefit that results from aid from a dominant to a subordinate. Suppose that TFT is not an ESS because the benefit received by the dominant is not sufficient to compensate for the cost of helping the subordinate. However, it still may be evolutionarily stable for the dominant to cooperate only infrequently, and in return the subordinate cooperates at every opportunity. In this way, the dominant can receive sufficient benefit to compensate it for its cooperation.

These results suggest that reciprocal altruism could be behind observed cooperative behaviour even when aid is unbalanced. Axelrod (1984) argued forcefully that TFT is the likley outcome when individuals interact over a long time in potentially cooperative situations, and TFT leads to balanced reciprocity—each individual helps and is helped by each other individual the same number of times. Some investigators have sought to test for the existence of reciprocity by looking to see whether the observed pattern of aid is balanced. The results just derived suggest that when the costs and benefits vary among individuals selection can favour unbalanced forms of reciprocity—at least when the accounting is in terms of number of helping acts.

This model makes an interesting qualitative prediction that may allow it to be empirically tested. In the example given above it was assumed that the dominant could provide a greater benefit in agonistic encounters than could a subordinate, and therefore selection would favour dominants who cooperate only intermittently. It is also plausible that dominance status might affect the cost of aid to the individual giving aid. If dominants provide the same benefit at lower cost, the model predicts that it will be the subordinates that will help only infrequently compared to dominants. Thus the model makes opposite predictions depending on whether it is the costs or benefits that vary among individuals.

WHEN PAY-OFFS VARY AMONG INTERACTIONS

In the usual model of the iterated prisoner's dilemma, the costs and benefits of cooperation do not vary from interaction to interaction. It seems likely that this assumption is frequently violated in real-world social interactions. Even if attention is limited to a particular behaviour, say grooming or aid during agonistic interactions, the costs and benefits of the behaviour are likely to vary from time to time. The cost of grooming might be low when an individual has little else to do, and high when there are profitable alternatives. The benefit of aid in a coalition is likely to vary with the importance of the contest, the dominance rank of other participants, and the effect of an individual on the outcome. Moreover, there is no reason to suppose that individuals only reciprocate grooming for grooming and aid for aid. It is equally reasonable to view ongoing social relationships as a stream of different behavioural opportunities, some costly and some not.

When costs and benefits vary from interaction to interaction it is no longer obvious that strategies like TFT which demand cooperation on every interaction will be evolutionarily stable. Suppose, for example, that same interactions are very costly, while others are not. It seems plausible that selection might favour individuals who do not help when it is too costly, and do not count it against others when they do not help when it is too costly.

To model the evolution of reciprocity when the fitness effects of behaviour vary among interactions, suppose that there are m mutually exclusive types of interactions, labelled $i = 1, \ldots, m$. During each interaction each individual has the opportunity to help the other individual. During an interaction of type i, helping reduces the fitness of the donor an amount c_i, and increases the fitness of the recipient an amount b_i. Interactions of type i occur with probability p_i. The probabilities from one period to the next are independent. As before we suppose that pairs of individuals are sampled from a population and interact for t or more periods with probability w^t.

The following family of strategies, labelled 'discriminating tit-for-tat', (DTFT) generalizes the concept of TFT to the case in which costs and benefits vary. Let E be a set of types of social interactions. For example, E might be the set $\{3,4,7\}$—perhaps denoting low- and high-cost grooming and aid during coalitions against lower-ranked individuals. Then a DTFT(E) individual cooperates and expects others to cooperate only for behaviours belonging to that subset of all possible behaviours. If the opponent does not go along, it is punished by withholding cooperation until the opponent again cooperates under the appropriate conditions. Strategies in this family range from strategies that are willing to cooperate in all types of interaction, to the strategy DTFT(∅) which never cooperates and thus is the same as ALLD. (A more formal definition is given in the appendix.)

When individuals interact at random, the results of this model are qualitatively similar to Axelrod and Hamilton's results for constant payoffs. Once any member of this family becomes common (1) it can always resist invasion by rare strategies which expand the scope of cooperation; but (2) it can resist invasion by strategies which narrow the scope of cooperation only if interactions between individuals persist a sufficiently long time. Consider a second strategy DTFT(I), where I (for invading) is a set formed by either adding or deleting some elements of E. If E is a subset of I, the invading strategy expands the range of behaviours under which reciprocity will take place, DTFT(E) can always resist invasion by DTFT(I). However, if I is a subset of E, and the invading strategy reduces the scope of reciprocity, DTFT(E) can resist invasion by I only if

$$w > \frac{c_E - c_I}{(p_E - p_I)(b_E - c_E) + c_E - c_I} \tag{4}$$

where c_S and b_S are the average cost and benefit of helping behaviours in the set $S(= E, I)$, and p_S is the probability that an interaction will be of a type contained in S. These quantities are computed as follows:

$$p_S = \sum_{i \varepsilon S} \mathrm{p_i}$$

$$c_S = \sum_{i \varepsilon S} \mathrm{p_i c_i}$$

$$b_S = \sum_{i \varepsilon S} \mathrm{p_i b_i}$$

where the sums are over all elements contained in S. These results are a straightforward generalization of those derived by Axelrod and Hamilton (1981) for the case of constant costs and benefits. Namely, ALLD can always resist invasion by TFT, but TFT can resist invasion by ALLD only if w is sufficiently large. In both cases, a rare strategy that is more cooperative than the population as a whole suffers a short-run cost, but achieves no long-run benefit, and thus can never increase. In contrast, when the rare strategy is less cooperative than the majority, it achieves a short-term benefit at a long-term cost. The more cooperative strategy can only resist invasion if the long-run costs are sufficiently high.

When there is non-random social interaction, however, the results with variable costs and benefits differ strikingly from those that assume constant costs and benefits. Non-random social interaction plays an important role in the theory of the evolution of reciprocity. TFT cannot invade a population in which ALLD is common when interacting pairs are formed at random (assuming that w is constant and there is only a single invading strategy: see Feldman and Thomas 1987, or Boyd and Lorberbaum 1987). However, if interacting pairs are formed assortatively so that like types are even slightly more likely to interact than chance would dictate, then TFT can invade if interactions go on long enough. Interaction among relatives is one mechanism that could lead to such assortment. Here, I model assortative social interaction by assuming that there is a probability r that each strategy interacts with an individual using the same strategy, and a probability $1 - r$ that pairs are formed at random.

When there is assortative social interaction, DTFT(E) can resist invasion by rare DTFT(I) individuals whenever

$$(rb_I - c_I) - (rb_E - c_E) < \frac{(p_E - p_I)w}{1 - w} (r[(b_E - c_E) - (b_I - c_I)] + (1 - r)(b_E - c_E)) \tag{5}$$

efficiency term — coordination term

When there is assortative social interaction, cooperative behaviours can increase even without reciprocity as a result of kin selection. The left-hand side of (5) gives the effect of expanding or reducing the scope of reciprocity on inclusive fitness. Here we are interested in the case in which reciprocity is necessary for expanding the scope cooperation, so this term is assumed to be negative any time E is a subset of I. The right-hand side, which is proportional to the number of interactions, gives the effect of reciprocity. The reciprocity term can be subdivided into two terms—one proportional to r, the efficiency term, and the other proportional to $1 - r$, the coordination term. Remember that r is the probability that like types interact. Thus the efficiency term gives the advantage of the common type relative to the invading type when both interact with individuals like themselves. If $(b_I - c_I) > (b_E - c_E)$, then two DTFT($I$) individuals achieve greater benefits from cooperation, on the average, than two DTFT(E) individuals. The coordination term gives the advantage of the common type relative to the invading type in pairs formed at random. In such pairs the common type almost always interacts with an individual using the same strategy as itself, and thus achieves a long-run benefit proportional to $(b_E - c_E)$. In randomly formed pairs, the rare invading type almost always interacts with a common type. In such pairs cooperation eventually breaks down because the interacting individuals disagree about the contexts in which cooperation is appropriate, and, therefore, such pairs do not achieve lasting cooperation. Thus, the invading strategy can increase only if the additional benefits that are achieved by the invading type when it interacts with itself exceed the disadvantage that result from such disagreements.

This result suggests that when costs and benefits vary among interactions, a wide variety of reciprocal strategies can be evolutionarily stable, even when groups are formed assortatively. Label the set of all interactions for which the benefits exceed the costs (i.e. $b_i > c_i$) M. DTFT(M) yields the maximum long-run benefits to reciprocity. Adding interactions for which c_i exceeds b_i will reduce the expected benefit of reciprocity, as will deleting interactions for which b_i is greater than c_i. If w is large, DTFT(M) is the strongest invading strategy against any common strategy because it maximizes the long-run benefits to cooperation (the efficiency term). If DTFT(M) cannot invade, then no member of the DTFT family can invade. For sufficiently large w, DTFT(E) can resist invasion by DTFT(M) if

$$r < \frac{b_E - c_E}{b_M - c_M} \tag{6}$$

When r is small, it is likely that there will be a wide range of reciprocal strategies that can resist invasion. For example, suppose that $r = 1/16$. Then any set of interactions that produces more than a sixteenth as much net benefit as the most beneficial set, M, can resist invasion by any member of the DTFT family. Such a strategy could specify cooperation only during small subset of M—cooperation would not occur in many situations in which it would pay. More interestingly, it could also specify cooperation in a large superset of M—cooperation would occur in situations in which the cost to the donor exceeds the benefit to the recipient. Both extremes are more likely as the degree of assortment decreases. Among close relatives a narrower range of behaviours can be evolutionarily stable than among more distant relatives.

When costs and benefits vary among interactions, there are a large number of different strategies which can capture at least some of the potential benefits of long-run cooperation. In order to persist when common reciprocating strategies must retaliate against individuals who do not cooperate when it is appropriate. If they did not, they could be exploited by defecting strategies. When a strategy is rare, it will interact mostly with other strategies which cooperate and expect cooperation in a different set of circumstances. Inevitably a rare strategy will retaliate or suffer retaliation, and cooperation will collapse. Thus, a common reciprocating strategy has an advantage relative to rare reciprocating strategies, even if the rare strategy would capture more of the benefits of long-run cooperation once it becomes common.

The most surprising result of this fact is that selection can favour strategies which lead to cooperation when the cost during the particular interaction exceeds the benefit during that interaction as long as cooperation is mutually beneficial in the long run. Rare reciprocating strategies, at least from the family of DTFT, may not be able to invade because of the common type advantage. It is even possible for strategies which lead to cooperation when costs exceed benefits to increase when rare as long as the average benefit captured by the common strategy is even less than their own.

WHEN THE EXPECTED NUMBER OF INTERACTIONS VARIES

Most existing models of the evolution of reciprocity assume that there is a constant probability, w, that a relationship between two individuals will persist from one opportunity for cooperative behaviour to the next opportunity (see Feldman and Thomas 1987 for an exception). This assumption is meant to capture the idea that relationships last a finite but unpredictable length of time. However, this assumption actually implies a great deal more. To see why, consider a pair of individuals just before their first potentially cooperative interaction. On average they can expect to interact $1/(1 - w)$ times. Thus if $w = 0.99$, they will interact 100 times on the average; if it is 0.999, they will interact 1000 times, and so on. In choosing whether to cooperate or defect, individuals must balance the short-run benefits of defection against the long-run benefits that result from sustained cooperation. The expected number of interactions is important because it determines the magnitude of the long-run benefits. After they interact, there is a probability w that their relationship will persist. Suppose that it does, and once again they must weigh the short- and long-term effects of cooperation. Surprisingly, the expected number of interactions is still $1/(1 - w)$. In fact, no matter how long a relationship persists, the expected number of future interactions never changes, and therefore the trade-off between the short-term benefits of defection and the long-term benefits of sustained cooperation does not change either. This phenomenon results from the fact that the probability that a relationship survives from one period to the next is constant and independent of what happened in previous periods. Just as a run of tails tells you nothing about the outcome of the next flip of a coin, the fact that a relationship has survived 1, 10, or 100 periods, tells you nothing about how long the relationship will persist in the future.

There are a wide variety of mechanisms that can cause the expected number of future interactions to change through time. Here I consider only one of them. Most models of the evolution of reciprocity assume that every pair of individuals interacts the same number of times, on the average. In many real-world situations it seems likely that social relationships between different individuals will last different lengths of time on the average. For example, some individuals might be permanent members of social groups while others are merely passing through. Interestingly, such variation will cause the expected number of future interactions to change through time, and, thus affect the kinds of behavioural strategies that can be favoured by natural selection.

To capture the idea that the average lifetime of relationships varies, suppose instead that there are two kinds of relationships—'ephemeral' relationships which persist from one time period to the next with probability w_1, and 'enduring' relationships which persist with probability w_2 where $w_2 > w_1$. A fraction p_1 of relationships are ephemeral and a fraction p_2 are enduring. Individuals do not know whether any particular relationship is ephemeral or enduring. This latter assumption is important; otherwise individuals could simply adopt different strategies in the two contexts. As in previous models we assume that individuals are paired randomly with respect to their strategies, and that payoffs are linear so that $T = b$, $R = b - c$, $P = 0$, and $S = -c$.

First consider a population in which ALLD competes with TFT. As one would expect, ALLD is always evolutionarily stable. TFT is stable if

$$\left(\frac{\bar{n}_1 - 1}{\bar{n}_1}\right) b > c \tag{7}$$

where

$$\bar{n}_1 = \frac{p_1}{1 - w_1} + \frac{p_2}{1 - w_2}$$

The parameter $\bar{n}_1$ is the expected number of interactions at the beginning of the relationship (that is during the first interaction). The ratio $\bar{n}_1/(\bar{n}_1 - 1)$ is a measure of proportion of the benefits created by TFT individuals that are directed non-randomly toward other TFT individuals. Thus, as Brown *et al.* (1982) argue, (7) is closely related to Hamilton's rule. Notice that when all relationships last the same length of time on the average (i.e. $w_1 = w_2 = w$), that (7) reduces simply to $wb > c$.

When (7) is not satisfied, TFT will be invaded by ALLD. However, as long as reciprocity would be favoured in the enduring relationships, that is, whenever $w_2 b > c$, then there is an evolutionarily stable strategy that leads to cooperation. Let T-times wary tit-for-tat (WTFT(T)) be defined as follows. Defect until interaction T, cooperate on interaction $T + 1$, and then play TFT. Thus WTFT(0) is ordinary TFT. As T increases, wary TFT individuals delay longer and longer before attempting to initiate reciprocal cooperation.

WTFT(T) can resist invasion by ALLD whenever

$$\left(\frac{\bar{n}_T - 1}{\bar{n}_T}\right) b > c \tag{8}$$

where

$$\bar{n}_T = \frac{w_1^T p_1}{w_1^T + w^T}\left(\frac{1}{1 - w_1}\right) + \frac{w_2^T p_2}{w_1^T + w_2^T}\left(\frac{1}{1 - w_2}\right)$$

The parameter $\bar{n}_T$ gives the expected number of future interactions given that the relationship has survived T interactions already. Notice that as T increases, the value of $\bar{n}_T$ increases and eventually approaches $1/(1 - w_2)$. The longer a relationship has lasted, the more likely it is that it is of the enduring type, and therefore, the larger the expected number of future interactions. Thus, if $w_2 b > c$, there must be some value of T that is sufficiently large that (8) is satisfied. By delaying the attempt to initiate cooperation, WTFT individuals decrease the chance that they are part of an ephemeral relationship, and are therefore able to risk more to get cooperation started.

When the average length of relationships varies, nice guys may indeed finish last. Axelrod (1984) emphasized that successful strategies are never the first to defect—'nice' strategies finish first. This conclusion makes sense given the assumption that the average length of interactions does not vary. Because the trade-off between the short-term gain of defection and the long-term gain of cooperation does not change, it either pays to cooperate or it does not pay. If it pays to cooperate, it is best to cooperate immediately. However, when some relationships are more enduring than others, there is a benefit to delaying cooperation—the longer the delay, the more likely it is that the relationship is an enduring one. Thus a wary reciprocator can confine cooperation to long-lasting relationships, while nice reciprocators are victimized in ephemeral ones. While the particular model presented here is quite stylized, I believe that this effect will persist in more realistic models.

DISCUSSION

Adding realism to models of reciprocity leads to a rich diversity of outcomes. Most of the essence of kin selection is captured by Hamilton's rule. Different ecological and social contexts lead to different benefits and costs, but the basic logic of cooperation remains the same. This paper, and other recent work, suggest that reciprocity is a different kettle of fish. When potentially cooperative social interactions persist over a long time, selection tends to favour reciprocating strategies. The nature of the strategies that support cooperation may vary widely from one context to another. Even the seemingly innocent introduction of variation in costs and benefits among interactions leads to dizzying variety of outcomes.

This may not be bad news. It is difficult to detect reciprocity empirically. Almost 20 years after the publication of Trivers' (1971) seminal paper, there are only a handful of empirical examples of reciprocity among non-human animals. Part of the problem, as Trivers pointed out, is that reciprocity is expected in the same contexts as kin selection. Perhaps a more important problem is that if a strategy like TFT is supporting cooperation, observable behaviour will look much like unconditional altruism—cooperation is maintained by the threat of retaliation, a threat that may only rarely be carried out.

The fact that distinctive reciprocating strategies are favoured in different circumstances may provide an avenue for empirical detection of reciprocity using the comparative method. The models analysed here suggest two such tests. First, when costs and benfits vary among individuals, the model analysed above predicts that unbalanced reciprocity may result. Thus the observation that unreciprocated aid is more common in situations in which there are significant differences among individuals compared to situations in which individuals are similar would constitute evidence that the observed cooperation results from reciprocity. Second, when there are differences in the expected numbers of interactions among individuals, the model just analysed predicts that selection may favour wary strategies which delay cooperation. Thus, the observation that increasing differences in the duration of social relationships is associated with an initial wariness is also evidence for the action of reciprocity. Other models suggest still more tests. For example, increases in group size have been shown to strongly reduce the possibility for reciprocity (Boyd and Richerson 1988; Joshi 1987). Similarly, Bendor (1987) has shown that as uncertainty about the magnitude of costs and benefits increases, selection tends to favour strategies that are less provokable.

SUMMARY

1. Cooperative relationships between unrelated individuals are an important feature of the behavior of social mammals. Recent theory suggests that that reciprocal cooperation is likely to evolve when pairs of individuals interact over a lengthy period of time.
2. This work makes the unrealistic assumption that the costs and benefits of cooperation do not vary, among individiuals or from time to time.
3. Here, I present and analyze three different models of the evolution of reciprocity in which conditions do vary. In each case, there is a range of parameters in which the basic results of previous theory remain unchanged. However, in each case there is also a range of parameters in which tit-for-tat-like strategies are *not* evolutionarily stable, but other contingent strategies which support at least some cooperation *are* stable.
4. When costs and benefits of cooperation vary from one individual to another, selection may favour strategies that lead to a system of unbalanced reciprocity in which some individuals help more often than others.
5. When the costs and benefits vary from one time period to the next, a very wide variety of strategies may be evolutionarily stable, including strategies

that sometimes cooperate when the cost to the donor exceed the benefit to the recipient.

6. When the expected number of interactions varies from pairing to pairing, selection may favour suspicious strategies which delay the initiation of cooperation.
7. The fact that some patterns of variation favour tit for tat, while others favour qualitatively different forms of reciprocity, may provide an avenue for testing this body of theory.

ACKNOWLEDGEMENTS

I would like to thank Hillard Kaplan, Craig Packer, and Joan Silk for useful discussions of the ideas presented here. Alan Harper pointed out the importance of varying w. The work was partially funded by a fellowship from the John Simon Guggenheim Foundation.

REFERENCES

Axelrod, R. (1984). *The evolution of cooperation*. Basic Books, New York.

Axelrod, R. and Dion, D. (1989). The further evolution of cooperation. *Science*, **242**, 1385–90.

Axelrod, R. and Hamilton, W. D. (1981). The evolution of cooperation. *Science*, **211**, 1390–6.

Bendor, J. (1987). In good times and bad: reciprocity in an uncertain world. *American Journal of Political Science*, **31**, 531–57.

Bendor, J. and Mookerjee, D. (1987). Institutional structure and the logic of ongoing collective action. *American Political Science Review*, **81**, 129–54.

Binmore, K. (1987). Modeling rational players, Part I. *Economics and Philosophy*, **3**, 179–214.

Boyd, R. (1987). Is the repeated prisoner's dilemma a good model of reciprocal altruism? *Ethology and Sociobiology*, **9**, 211–22.

Boyd, R. (1989). Mistakes allow evolutionary stability in the repeated prisoner's dilemma game. *Journal of Theoretical Biology*, **136**, 47–56.

Boyd, R. and Lorberbaum, J. P. (1987). No pure strategy is evolutionarily stable in the repeated prisoner's dilemma game. *Nature*, **327**, 58–9.

Boyd, R. and Richerson, P. J. (1988). The evolution of reciprocity in sizable groups. *Journal of Theoretical Biology*, **132**, 337–56.

Boyd, R. and Richerson, P. J. (1989). The evolution of indirect reciprocity. *Social Networks*, **11**, 213–36.

Brown, J. S., Sanderson, M. J., and Michod, R. E. (1982). Evolution of social behavior by reciprocation. *Journal of Theoretical Biology*, **99**, 319–39.

Farrell, J. and Ware, R. (1989). Evolutionary stability in the repeated prisoner's dilemma. *Theoretical Population Biology*, **36**, 161–6.

Feldman, M. W. and Thomas, E. (1987). Behavior dependent contexts for repeated

plays of the prisoner's dilemma II: Dynamical aspects of the evolution of cooperation. *Journal of Theoretical Biology*, **128**, 297–315.
Hamilton, W. D. (1964). The genetical evolution of social behavior, I & II. *Journal of Theoretical Biology*, **7**, 1–52.
Hirshleifer, J. and Martinez Coll, J. C. (1988). What strategies can support the evolutionary emergence of cooperation? *Journal of Conflict Resolution*, **32**, 367–98.
Joshi, N. V. (1987). Evolution of cooperation by reciprocation within structured demes. *Journal of Genetics*, **1**, 69–84.
Maynard Smith, J. (1982). *Evolution and the theory of games*, Cambridge University Press, London.
Maynard Smith, J. (1989). *Evolutionary genetics*. Oxford University Press, Oxford.
Mueller, U. (1987). Optimal retaliation for optimal cooperation. *Journal of Conflict Resolution*, **31**, 692–724.
Sugden, R. (1986). *The economics of rights, cooperation, and welfare*. Basil Blackwell, Oxford.
Trivers, R. (1971). The evolution of reciprocal altruism. *Quarterly Review of Biology*, **46**, 35–57.

APPENDIX

Definition of intermittent tit-for-tat

Intermittent tit-for-tat (I TFT) is formally defined by the following rule.

If type 1: cooperate on the first interaction. If the other individual cooperates on interaction $1 + jT$, cooperate on interactions $(1 + jT) + 1, (1 + jT) + 2, \ldots (1 + jT) + T$, where $j = 0, 1, 2, \ldots$.

If type 2: cooperate on the first interaction. If the other individual cooperated during interactions $(1 + jT), (1 + jT) + 1, \ldots (1 + jT) + T - 1$, cooperate on interaction $1 + (j + 1)T$, where $j = 0,1,2, \ldots$.

Definition of discriminating tit-for-tat

The strategy DTFT(E) can be defined as follows. During an interaction of type i, cooperate if i is a member of E and your opponent is in good standing. Otherwise defect. An individual is initially in good standing. It loses good standing any time it does not behave according to DTFT(E), but can regain good standing by cooperating during the next interaction in which i belongs to E.

CONCLUSION

18

Cooperation in conflict: from ants to anthropoids

A. H. HARCOURT AND FRANS B. M. DE WAAL

INTRODUCTION

In the concluding chapter, we want to indicate some avenues of future research suggested by the chapters in this book. Before we do this, though, it might be useful to comment on the sometimes difficult dialogue between those who study animals (whom we call zoologists for the purposes of this chapter) and those who study humans (social scientists). One of the goals of this book was to bring together the work of zoologists and social scientists in order to encourage an exchange of data and ideas between the different disciplines. Social scientists tend to be resistant to zoologists' attempts to explain human behaviour. An example is Hallpike's (1988) outright rejection of biological explanations in his characterization of Darwinian thinking about cooperation and competition and the nature of societies. Darwinians cannot explain cooperation, he suggests, because Darwinism is concerned solely with competition (Hallpike 1988, pp. 56, 57). As must be clear from all the chapters of this book, Hallpike completely misses the point that co-operation is usually a means of competing.

Some of the resistance is understandable. Zoologists have rushed in with simplistic explanations of complex human behaviours (see Kitcher's (1985) criticism of sociobiological theory as applied to humans). Also there is a perception among social scientists that the zoologists are trying to explain everything about human behaviour in terms of a few simple biological principles. If some zoologists are trying to do that, they should not: while some aspects of even fairly complex, apparently purely cultural behaviours, like bride prices, can be explained in the biological terms of breeding potential of the bride (Borgerhoff Mulder 1989), many other behaviours certainly cannot be illuminated by such an approach. A third main problem with acceptance of biological explanations of human behaviour is the tendency on the part of social scientists to assume that all biological explanations concern genetic determinism or physiological mechanisms (Alexander 1984). This is not the case.

Explanations of a phenomenon may validly be made at several different levels of analysis, as is evident in Rabbie's review (this volume pp. 175–205) of psychological theories of in-group favouritism (a subject also discussed in this volume by Boehm pp. 137–73, and van Hooff and van Schaik pp. 357–205, but from a different viewpoint). Biologists attempt explanations at four different levels (Tinbergen 1963; see Hinde 1982). Skiers in a group dressed in green might perceive those in a group dressed in blue unfavourably (see Rabbie this volume pp. 175–205) (1) because humans have genes that tend to make us aggressive to those who are different from us (an evolutionary explanation); (2) because we have learnt to be hostile to those who are different from us (a developmental explanation); (3) because the wearing of different colours is perceived as a hostile action (a proximate explanation); or (4) because perceiving the other group unfavourably enhances the probability of success in competition between the groups (a functional explanation). All four levels are valid; all four are to some extent independent; but all four can, with proper integration, benefit from understanding at the other levels (Bateson 1982). It is at the fourth level listed that most of the authors in this book are concerned, the level of consequences of behaviour, specifically the level of the pay-offs of cooperation in conflict.

THE ECONOMICS OF COOPERATION IN CONFLICT

At the simplest level of comparison, the phenomenon of cooperation during competition, far from being primarily a human trait, occurs in a wide array of animal societies. Groups of Amazonian Indians, African lions, and Arizona ants compete with one another, and in all, individuals within groups cooperate in the conflict (Chagnon 1979*b*, Bygott *et al.* 1979, Adams 1990; and see Boehm, van Hooff and van Schaik this volume pp. 137–73, 357–89). Contrary to Alexander (1974, 1984), therefore, lethal intergroup conflict is not uniquely, or even primarily, a characteristic of humans, and humans are not alone in experiencing conspecifics in the form of other groups as 'the principle hostile force of nature'. In addition, in dolphin communities, as well as among humans, separate groups will cooperate in attacks on rival groups (Connor *et al.* this volume pp. 415–43); Meggitt 1978).

Within groups too, humans and non-humans distribute their support in similar ways. Yanomamö males cooperate with kin in fights in villages (Chagnon 1979*a*) as do female rhesus monkeys in fights within social groups (Chapais, Datta, Ehardt and Bernstein this volume pp. 29–59, 61–82, 83–111). Generally, hyaenas, monkeys, children, and nations seem to use similar criteria in choosing whom to side with. For example, often individuals of similar competitive adbility are each other's most frequent coalitionary partner or ally (compare Zabel *et al*, Silk, Grammer, and Falger this volume pp. 113–35, 215–31, 259–83, 323–48). Even in detail, the similarities are apparent. For instance, in human and non-human societies levels of intergroup competition sometimes correlate with the steepness of dominance hierarchies within groups (compare Rabbie with van Hooff and van Schaik). But of course the phenomenological resemblance does not mean that the same causes or processes operate.

Nonetheless, at one level of analysis, the functional level, the level of pay-offs or consequences of action, the processes occurring in animals and humans seem very similar. Thus, the ecologically minded zoologist (Wrangham 1980, van Hooff and van Schaik this volume pp. 357–89), the materialistically minded anthropologist (e.g. Harris 1979, Vayda 1961) and the 'realistically' minded psychologist (Rabbie this volume pp. 175–205) and political scientist (Falger this volume pp. 323–48) all produce essentially the same explanation both for grouping and for intragroup cooperation in the face of intergroup conflict. They all argue that individuals cooperate for the mutual advantages that cooperation can bring in intergroup competition (or perceived competition (Rabbie this volume pp. 175–205). Similarly, the idea that opportunities elsewhere are a determinant of sharing in several human cultures (Bethlehem 1982), and that the threat to form coalitions with others might influence the distribution of pay-offs among partners in coalitions among humans (Murnighan 1978) are ideas very reminiscent of parameters incorporated into mathematical models of cooperative decisions in animals by, for example Emlen (1982), Mangel (1990), Vehrencamp (1983). Put generally, the organism's means of getting resources determine its social structure, a concept familiar to both Karl Marx and Charles Darwin (Harris 1979).

Perhaps it should be repeated here that although we might argue that the function of a coalition, i.e. the beneficial consequence on which selection might act (Hinde 1975), is increased dominance rank, that is not to say that the stimulus to form the coalition, the proximate cause of the coalitionary behaviour, is a desire to rise in rank. Take coalitions against subordinate group members. The proximate cause of these coalitions could simply be a tendency to join in attacks against individuals who are obviously losing a contest, as Zabel *et al.* (this volume pp. 113–35) suggest for their hyaenas. As long as the consequence of the behaviour is that individuals with this tendency increase their dominance status, and as long as individuals of high dominance status reproduce better than do subordinates, then the tendency will be perpetuated in the population. Though linked, proximate cause and ultimate function are logically separate; the consequence can be independent of the immediate mechanism by which it is brought about.

A good example of the similarity between Darwinian functionalist and anthropological materialist explanations is provided by Johnson and Earle (1987), whose presentation of ecological and demographic correlates and causes of territoriality, warfare, and leadership in different levels of society (family, local group, regional polity) could have been lifted out of a zoology book. A change from scattered resources and low population density to clumped resources and high population density correlates with a change from family to group level society, from non-territoriality to territoriality and from egalitarian relationships to a society with leaders. To some extent the similarity is due to the fact that materialist explanations borrow directly from ecological theory. The borrowing is not one-way, however, for ever since Darwin, a number of biological hypotheses about the structure of society have been based on economists' analyses: game theory, now such an important part of biological thinking (Maynard Smith 1982), originated in economics.

So far we have concentrated on just the functional, the 'economic' level of explanation. We suggested at the start that different levels of explanation can, with proper integration, benefit from understanding at the other levels. Rabbie's chapter stands in some contrast to the others in the book by concentrating on the psychology of intragroup cohesion. Nevertheless, his explanations, and those of Tajfel and Turner (e.g. Turner 1982) with whom he contrasts his theories, can be worded in terms of the consequences of behaviour. Might the contrast presented be an instance where integration of functional and proximate levels of explanation could reconcile apparent differences between psychological theories concerning the proximate causes of in-group favouritism? Turner concentrates on the effect of intergroup competetion on favouritism, while Rabbie considers that the important influence is the effect that cooperation with group members will have on resolving the intergroup conflict. The functional cost/benefit analysis would suggest that groups are likely to be in conflict; that cooperation with your own group members is likely to lead to success in conflict between groups; and that cooperation will proceed most smoothly if favourable attitudes prevail about your own group members. Such an analysis sees Turner's and Rabbie's interpretations as different stages in the same process, rather than as competing hypotheses, and therefore implies that both are correct and that humans react both to perceived conflict between groups, and perceived value of cooperation in any conflict that might occur.

Four main issues recur as themes and continuing problems through the chapters of the book. First, how do cooperating individuals reap the benefits of their cooperation. Do they arise as a consequence of the cooperation itself, or through later reciprocation by the partner. Second, given that triadic interactions, those involving more than two animals, are potentially complex, is it in fact the case that complex cognitive abilities are required or being used? Third, under what conditions is it advantageous to establish coalitions and alliances? What environmental and social factors influence the use by individuals of cooperation as a competitive tactic? Finally, how widespread are the ramifying effects of coalitions and alliances on the nature of society?

COALITIONS AND ALLIANCES: RECIPROCATION OR MUTUALISM?

Cooperative behaviour can be explained at the functional level by kin selection, mutualism, or reciprocation. In the first, individuals help relatives because of the chance of increasing the probability that shared genes will be reproduced (Hamilton 1964); in the second, both individuals might benefit from the act of help, for example by obtaining an otherwise unavailable resource; and in the third an individual helps now because by so doing it increases the chances of being later helped by the current recipient (Trivers 1971). (We do not distinguish between 'help' and 'cooperation', for helping is so often merely one side of a cooperative interaction.)

Reciprocity as a reason to cooperate still generates debate and controversy, as a number of chapters in this volume attest. To identify reciprocation as the cause of cooperation, the helpful act needs first to be shown to be costly. Then to confirm reciprocity as the possible benefit of cooperation, it must be demonstrated that the benefit received by the helper comes from the helped animal as a result of the initial cooperative act. It is this requirement that imposes special conditions. In general, reciprocation requires that partners remain together long enough for helpful acts to be exchanged; and that they are capable of recognizing and remembering not only one another, but also previous acts of help as well as refusals to reciprocate help (Axelrod and Hamilton 1981; Trivers 1971; Boyd, Silk, de Waal this volume pp. 473–89, 215–31, 233–57). These special requirements mean that reciprocation is difficult to prove (Seyfarth and Cheney 1988). Consequently, apparent demonstrations of reciprocation are sometimes eagerly received, even when sample sizes are small and possible alternative explanations exist (see Bercovitch 1988 and Noë this volume pp. 285–321 for criticism of Packer 1977; and Lazarus and Metcalfe 1990 and Reboreda and Kacelnik 1990 for criticism of Milinski 1987).

Costs and benefits

Wilkinson's (1984) study of reciprocal feeding of blood meals among vampire bats was so well received, despite the small sample size of apparently reciprocated acts, in part because of his clear demonstration that the cost of giving blood was small and that the benefit to the recipient was great. He could also show how the situation might change such that the next time the same bats met, the current donor might be in severe difficulty, but the recipient in a position to donate blood at little cost to itself. In this volume Silk (pp. 215–31), especially, concentrates on demonstrating the possibility that the act of support in coalitions (among male bonnet macaques) is costly to the supporter but beneficial to the recipient. From a functional viewpoint, benefits and costs should, strictly, be measured in terms of effect on the individual's survival and reproduction, indeed on the composition of the population's gene pool. Receiving a blood meal could save a vampire bat's life, but whether losing a fight, which was Silk's definition of cost, really affects survival or reproduction is unclear. Ornithologists are far ahead of primatologists in their search for and ability to demonstrate lifetime costs and benefits of helping behaviour, in part of course because primates live so long (Harcourt 1987). Nevertheless, perhaps primatologists should be devoting more ingenuity to analysis of costs and benefits of cooperation to individuals' inclusive fitness?

'Cognitive' reciprocation

Where Wilkinson's study perhaps fails to convince is in the amount and quality of his data on reciprocation. de Waal (this volume pp. 233–57) was particularly concerned with the question of whether individuals respond to past favours or not, or indeed past opposition; in other words whether there was any form of record-keeping. Like Wilkinson, but with far more data, he demonstrates that individual chimpanzees and rhesus and stumptail macaques help often those who help them often, independently of consanguinity or time in proximity. Reciprocation of support in contests in these species indeed appears to be calculated.

A major difficulty in identification and analysis of reciprocation concerns the possibility that different types of help might be exchanged: how many minutes of grooming need to be given in exchange for support in a fight (Seyfarth and Cheney 1988)? The calculation becomes especially difficult when, as will usually be the case, reciprocity is delayed, and when it is part of a long-term friendly relationship in which there might be considerable leeway given and taken over any one period of time, resulting in prolonged imbalance in payoffs (Boyd, this volume pp. 473–89).

In human societies, quantifying and measuring favours exchanged can be less of a problem. Nevertheless, so far the analysis applied by anthropologists in this field has not been as quantitative as might have been possible. Lederman (1986), for example, quantified the exchange of goods among the New Guinea Mendi, but not its influence on obligations in coalitions and alliances, although she said that the material exchange influenced social obligations and vice versa. While anthropological studies have produced beautiful qualitative descriptions of how exchange of favours and materials influences alliance formation among traditional peoples, it seems that they have not yet quantitatively tested hypotheses, and certainly have not modelled the processes. Of particular interest would be an analysis of controls on compliance, a subject that is difficult for zoologists to investigate.

Alternative explanations

Sometimes an act of cooperation is not certain to bring benefit, for example in the case of subordinate male baboons cooperating to increase their chances of taking a female from a dominant male (Bercovitch 1988; Noë 1990, this volume pp. 285–321). Sometimes the benefit is delayed, as when male pied kingfishers increase their chance of being accepted by a female in the next breeding season by helping to raise her offspring in this one (Reyer 1984). In such cases, reciprocation can be especially difficult to distinguish from mutualism.

One of the advantages of Boyd's theoretical approach to the question of reciprocation is, he suggests, that his models specify precisely the strategies that would be expected under different conditions were cooperation working through reciprocation. Two specific predictions that Boyd makes are that increasing differences between individuals in competitive ability will be associated with increasing imbalance in reciprocated exchange of favours; and that the greater the range of expected number of interactions between potential partners (e.g. the greater the difference between individuals in time spent in a social group), the more wary will initial cooperators be: individuals should not be nice to those who will emigrate before they can reciprocate. In this second instance, observed differences between the sexes in the use of cooperation as a competitive strategy (Ehardt and Bernstein, Lee and Johnson, de Waal, this volume pp. 83–111, 391–414, 233–57) might be a fruitful place to start testing the model, given sex differences in expected time of residence in the social group (see Lee and Johnson, this volume pp. 391–414).

Noë (1990, this volume pp. 285–321) strongly criticizes the prisoner's dilemma paradigm (iteration of which is the basis of tit-for-tat) as a model not just for cooperation among male baboons in competition over females, but also as a model for most cooperation. Instead he proposes that in the case of baboons, males cooperate because of immediate potential benefits, mutualism. While he is probably correct about male baboons, his general rejection of the prisoner's dilemma paradigm and his reasons for it will be controversial. For instance, Boyd's model (this volume pp. 473–89) appears to resolve one of Noë's objections, namely that cooperation continued in baboon male pairs even though no reciprocation occurred.

If reciprocity is to be pitted against mutualism, we need to ensure that all possible mutual advantages of cooperation have been thought of. Much of the biological analysis of coalitions and alliances concerns their use in achieving and maintaining competitive ability, i.e. power. In political science, a theory which takes account of the relative threat that nations pose has proved of more value than one which studied international alliances in terms of balance of power (Falger this volume pp. 323–48). To generalize, the difference lies in part in incorporating aggressiveness into threat: before

Gorbachev, the USA and the USSR were both powerful, but only the USSR was considered a threat by western Europe. Balance of power explains some of the coalitions and alliances seen by biologists, but perhaps if we took a leaf out of the political scientists' books, and looked at levels of threat posed by other individuals, somehow balancing relative power and relative aggressiveness where they differ, a greater proportion of the observed alliances could be explained (de Waal 1982).

Finally, the influence that changes or differences in the physical, demographic, and social environment can have on the use of coalitions and alliances in competition should be kept in mind in analysis of causes of cooperative behaviour (see Datta 1989, this volume pp. 61–82). Chapais (1983, this volume pp. 29–59) suggested that adult female rhesus macaques supported dominant unrelated adult females for immediate gain; Seyfarth (1980), by contrast, suggested that vervet females did the same for reciprocated benefits. Both arguments could be right. Different conditions might give rise to the same distribution of coalitions and alliances but for different reasons. In the well-fed, rapidly reproducing provisioned population that Chapais studied, it could be very difficult for animals to change their rank, but vital that they maintain their current rank by forming coalitions and alliances to resist rival alliances (mutualism). By contrast, in the wild populations of baboons and vervets that Seyfarth studied, stochastic effects operate (i.e. animals are likely to die soon), and it could be more important for individuals to invest in improving their dominance status by establishing cooperative relationships with currently powerful group members (reciprocation) during the short time they might have available to reproduce.

COALITIONS, ALLIANCES, AND SOCIAL COGNITION

When two animals fight, all they have to assess is their relative competitive ability and their relative readiness to escalate the contest. When a third enters the fight the amount of potentially relevant information immediately increases. In Humphrey's (1976) words, 'it was the arrival of Man Friday on the scene which made things difficult for Robinson Crusoe. If Monday and Tuesday, Wednesday and Thursday had turned up as well then Crusoe would have had every need to keep his wits about him'. The question is whether animals do indeed make use of all the potential information before forming coalitions or establishing alliances. In this volume, Zabel *et al.* (pp. 113–35) suggest that perhaps the decision rules that primatologists have hypothesized to explain primate behaviour are unnecessarily complex (see also Kummer *et al.* 1990).

There are at least three ways of tackling this question. First, is it the case that the size and complexity of animals' information-processing organ, the brain, and in particular the cortex or neocortex, correlates with use of complex social interactions? Second, we can ask what are the simplest rules by which animals could operate to produce the distribution of coalitions and alliances that are seen. Third, if complex social knowledge is being incorporated into tactical cooperative decisions, study of the ontogeny of the use of coalitions and alliances should indicate that apparently more complex behaviours appear later in life. These questions are addressed below.

Brain size and social complexity

Across all taxa of animals, a general correlation between potential complexity of the environment and brain size seems to exist (Jerison 1973). Moreover, experiments on rats demonstrate that increased complexity of both the physical and social environment induces cortical growth and complexity (Diamond 1988). A few studies demonstrate an association between either brain size or neocortical volume across species (corrected for difference in body size) and various measures of social complexity (see Harcourt this volume pp. 445–71). With particular reference to cooperation in conflict, both Connor *et al.* and Harcourt (this volume pp. 415–43, 445–71) suggest that larger-brained taxa show more complex alliances. de Waal (this volume pp. 233–57) also found a difference between primate taxa in the nature of reciprocated behaviour that correlated with a difference in relative brain size. However, Harcourt points out that the correlation could result from lack of reporting for small-brained taxa, rather than absence of complex behaviour; and de Waal warned that the differences in behaviour that he saw might result from the different nature of the species' social systems, specifically the rigidity of their dominance hierarchies, and not from any differences in cognitive ability, a caution that needs to be applied to all cross-taxonomic comparisons (Harcourt this volume pp. 445–71); and Deacon (1990) warns against many brain–body size analyses and conclusions.

If brain size indeed correlates with complexity of social behaviour, the work of Connor *et al.* and Harcourt suggests that the nature of alliances might be a useful place to look for evidence of cognitive constraints on social behaviour. One way forward in this field would be to take advantage of the very different brain to body size ratios in the odontocetes and compare the social behaviour of species with high and low ratios (Connor *et al.* this volume pp. 415–43). Another way would be to extend the work done by Diamond (1988), and compare the size and structure of brains of animals raised in different social environments while keeping the physical environment constant.

Cognitive rules

'If I find a bottle labelled "Chateau Lafite-Rothschild" I am more motivated to ascertain whether that is really true than when the label says: "Fermented fruit juice of undetermined quality"' (Kummer *et al.* 1990). In other words, suspecting complexity in social relationships and the processes of decision-making is heuristically valuable. However, Kummer *et al.* suggest, perhaps workers in the field of non-human primate social cognition are tending too far towards seeing as a *grand cru* what is in fact merely grape juice. For example, Boehm (this volume pp. 137–73) sees detailed similarities between humans and chimpanzees in the decision-making processes during raids into other communities' territories. The implication is that both species are using advanced cognitive assessment of relative competitive ability, yet Adams' (1990) description of raiding between ant colonies provides a comparable account (see the introduction to Part III). Ants are obviously not processing the same information in the same way as are humans or chimpanzees, but the resemblance in behaviour might call into question the complexity of the assessment performed by the larger-brained taxa. Along the same lines of thought, at what point does a monkey's apparent ability to distinguish among partners on the basis of their quality (Harcourt this volume pp. 445–71), or a dolphin's establishment of multiple levels of alliances (Connor *et al.* this volume pp. 415–43) require more sophisticated cognitive abilities, if they do, than choice of a safe but poor resource over a risky but good one by a starling, flycatcher, or shrew (Cheney and Seyfarth 1990; Harcourt 1988)?

One solution is more careful separation of potential influences on behaviour, as demonstrated by de Waal (this volume pp. 233–57). By statistically removing the effects of consanguinity and proximity, he could show that both macaques and chimpanzees were probably remembering and reacting to previous interactions, rather than distributing their behaviour according to more simple general rules (see Zabel *et al.* this volume pp. 113–35). Thus he distinguished between 'symmetry-based reciprocity' (help those with whom you spend a lot of time) and 'calculated', or as we called it earlier in the chapter, 'cognitive reciprocity' (help those who help you). The latter must require a greater ability to remember and process information than the former.

Kummer *et al.* (1990) advocate experimentation. Experimental study of social cognition has yet to reach the level of study of analysis involved in investigation of knowledge about the physical world (see also Essock-Vitale and Seyfarth 1987; Kummer *et al.* 1990), but some recent experiments indicate the potential for sophisticated levels of social cognition. Stammbach (1988) demonstrated that individuals in a macaque group recognized and made use of others' specialized skills; and Keddy Hector *et al.* (1989) cleverly showed by controlled use of one-way mirrors that males altered their behaviour towards infants depending on the perceived presence or absence of the infant's mother, and that females altered their behaviour towards a male according to his behaviour towards their infants.

Determination of the minimum set of cognitive rules necessary to produce the behaviours would presumably benefit from computer simulation models.

Datta's models (1989, this volume pp. 61–82), for example, show that apparently complex social differences can be generated with very similar behavioural rules by simple alteration of a few parameters of the social context. In the process of establishing the cognitive rules by which individuals work, the two questions of what individuals know about their environment, and how they use that knowledge, have to be kept separate. A thorough examination of primates' cognitive abilities based on both naturalistic observations and field and laboratory experiments is provided by Cheney and Seyfarth (1990).

Social experience and social ability

Experiments on the effects of social deprivation in infancy or immaturity on subsequent social behaviour have a long history. The trouble with interpreting the results of many of the experiments is that the animals were not deprived only of social experience. In the present context, a significant study was that of Anderson and Mason (1974). Eleven-month-old rhesus macaques previously reared with their mother and then with peers were more likely to interact as triads (or larger sub-groups) than were deprived individuals (reared on their own from 2 days of age); and only the socially reared immatures apparently recognized, or at least reacted to, differences in status among partners, a crucial ability for the apparent difference in use of allies between primates and non-primates highlighted by Harcourt (this volume pp. 445–71).

While a prediction from the hypothesis that complex coalitionary and alliance behaviour requires advanced cognitive abilities would be that more complex cooperative behaviours appear later in life, it is also the case that the competitive environment and the social situation, and the means of coping with them change through life (Ehardt and Bernstein, Lee and Johnson this volume pp. 83–111, 391–414). Cheney and Seyfarth (1986) showed that juvenile vervet monkeys of over 3 years of age apparently used more information about social companions when redirecting aggression. However, perhaps the younger animals had the information and knew how to use it, but refrained from doing so because it was not advantageous in the current competitive environment? Once more, experimentation seems to be the way forward. The nature of the knowledge available to individuals, and the nature of the situation in which they might use that knowledge advantageously, need to be manipulated in order to separate the two questions of knowledge and use of knowledge (Cheney and Seyfarth 1990).

INFLUENCES ON THE USE OF COOPERATION AS A COMPETITIVE STRATEGY

It is by no means the case that cooperation is always the best way to compete. Work on cooperation in birds (Davies and Houston 1981; Emlen 1984; Vehrencamp 1983), non-human primates (Wrangham 1980; and several authors in this volume) and humans (Turner 1982; Rabbie this volume pp. 175–205) shows clearly that a variety of influences in both the physical and social environment determine when cooperation is advantageous as a competitive tactic. Precisely how these factors operate is another question, and hypotheses differ (compare, for example, van Hooff and van Schaik this volume pp. 357–89) with Rodman (1988) and Wrangham (1980); and Turner (1982) with Rabbie (this volume pp. 175–205). The first group of authors differ on how predation might affect grouping in wild primate populations, and on how intergroup conflict might affect intragroup co-operation; Turner (1982) and Rabbie (this volume pp. 175–205) differ on the extent to which intergroup conflict affects intragroup cohesion in humans. Here we do not detail the differences, but instead suggest that models developed to explain grouping and cooperation in birds might use-fully be applied to analysis of coalitions and alliances in animals and humans.

The zoological work on the ways in which the physical and social environment might affect the use of coalitions and alliances has involved little experimentation (but see Chapais this volume pp. 29–59); and in neither the zoological nor the social scientific studies of coalitions and alliances has mathematical modelling or computer simulation played a large part, although Datta's work (1989, this volume pp. 61–82) provides an exception (see also Dunbar 1988). The lack of modelling is perhaps surprising given its advanced development in analyses of the relation between competition, grouping, and other forms of cooperation (Emlen 1984, Mangel 1990, Vehrencamp 1983). The models are sophisticated enough to consider not merely the existence or size of groups, but also their dominance structure (Mangel 1990, Vehrencamp 1983). The benefit of cooperation in intergroup conflict has not been specifically modelled, but it could easily be incorporated, as could effects of intensity of competition, since equivalent parameters are already in use.

In Wrangham's (1980) and, following him, van Hooff and van Schaik's analyses of primate group structure, divisibility of resources was a key factor. Modelling and associated computer simulation (Mangel 1990) have confirmed the generality of its importance, which has also been demonstrated in experimentation with birds (Elgar 1986). Application of the models to co-operation among male baboons in competition over females (Noë this volume pp. 285–321), which are not divisible, might help to elucidate the potentially special nature of such cooperation, or indeed refine the models.

The ability of current models to incorporate several different parameters is great (Mangel 1990; Mangel and Clark 1988), and it should be possible to develop ones that take into account effects of predation, as well as the conflicting influences of intergroup and intragroup cooperation on individuals' decisions and the structure of the dominance hierarchy within the group. Modelling becomes particularly important in the face of competing hypotheses, for it can lead to quantitative predictions that will separate hypotheses in a way difficult to achieve with verbal argument.

Predation, defence of resources, and foraging efficiency have all been used to explain grouping and variation in group size in birds (Mangel 1990) and primates (Rodman 1988). Currently some confusion might result from contrasting hypotheses that explain different aspects of cooperation in conflict. In the present context, it is necessary to separate hypotheses concerned with the advantages of grouping (e.g. Dunbar 1988 chapter 7; van Schaik 1983) from those concerned with the optimal size of the resultant groups (e.g. Rodman 1988), and both of these from those concerned with the effects of grouping on competition and cooperation between and within groups (van Hooff and van Schaik, this volume pp. 357–89; Wrangham 1980). If predation, intergroup competition, and intragroup competition and cooperation all play some part in determining the existence, size, and structure of groups, as they probably do, then more experimentation aqnd quantitative modelling than has so far been applied will be necessary to sort out their effects.

In primates, the interaction between intergroup competition and intragroup cooperation might be difficult to analyse experimentally, except for the smallest species. However, the results from psychologists' studies on ingroup–out-group differentiation indicate that mere perception of another group can be sufficient to induce cohesion, and sometimes cooperation, within a group (Rabbie this volume pp. 175–205). Such a reaction in humans (especially those of authoritarian personality) indicates that mere proximity of another group, or perhaps even a video presentation of it, could be sufficient to affect cooperation within a social group of an animal species, without the need to induce competition between groups. If so, experimental analysis of the effect of competition and of availability of different types of partners on cooperation within groups should be easier. Datta's (this volume pp. 61–82) and Noë's (this volume pp. 285–321) ideas in this latter area need testing. The sorts of experiments that Chapais (this volume pp. 29–59) conducted to demonstrate the effect of support in contests on acquisition and maintenance of rank would be appropriate. Remembering that much of the work on coalitions and alliances in primates has been done on a very few species, and that species differ in how they solve practical problems (Fragaszy and Mason 1983), generalization of their hypotheses will probably require work on several species.

It should be said also that generalizations will probably have to take into account differences between the sexes in the nature of their competitive and cooperative relationships. In cercopithecines (Lee and Johnson this volume pp. 391–414), chimpanzees and humans (de Waal this volume pp. 238–57), males' interactions seem more determined by current expediency, but females' by the nature of existing relationships: long-term alliances apparently play a greater part in female behaviour. Lee and Johnson (this volume pp. 391–414) argued that in cercopithecine monkeys, at least some of the differences could be related to the sexes' different competitive environments and their contrasting life histories. Females compete over different key resources than do males, and remain in the group in which they were born. Thus in stable populations they are often surrounded by a number of supportive kin, with the result that establishment and maintenance of long-term relationships could be more advantageous to them than to males. However, in chimpanzee and human populations, females, not males, are the emigrant sex (Pusey and Packer 1987). One of the many achievements of behavioural ecology as a discipline over the last 15 years has been the demonstration that animals are decision-makers, reacting to the situation as they find it, not behaving blindly according to rigid genetic 'instructions'. The difference between the sexes among primates (human and non-human) in their co-operative interactions seems to be independent of situation, however.

COALITIONS, ALLIANCES, AND THE NATURE OF SOCIETY

A society is a network of interacting individuals, of relationships. What one individual does affects what others do. This is true whether we are discussing humans or animals. The division that Hallpike (1988, p. 60) would have exist between 'the entirely structureless world of Darwinian theory where every form of organization is seen as merely the sum of a number of traits or elements . . .' and human society does not exist. Contrast Hallpike's statement with the opening of the last paragraph of *The Origin of Species*, 'It is interesting to contemplate an entangled bank, clothed with many [individuals of many species] and to reflect that these elaborately constructed forms, so different from each other, and *dependent on each other in so complex a manner* [my italics] . . .' (Darwin 1859). The Darwinian world is far from structureless. Thus when individuals use coalitions and alliances as a means of competition, the effects on the structure of society can extend way beyond the act itself, as Chapais convincingly demonstrated with his experimental approach, and Datta with her computer simulation (Chapais, Datta this volume pp. 29–59, 61–82). Change simply the number and age of allies, Datta shows, and the nature of the whole dominance hierarchy changes.

Although, Ehardt and Bernstein (this volume pp. 83–111) propose that one of the reasons that adult male macaques intervene in fights is, in effect, to manipulate the structure of their society, in other cases the alliances are probably not formed because of their effects on the stability of the dominance hierarchies or the families. For example, Chapais (this volume pp. 29–59) argues that the necessity for subordinate animals to form alliances with dominant ones in order to control rival alliances has the inevitable result of reinforcing the dominant animals' status; the subordinates maintain their own rank, but at the cost of making it more difficult to improve it. The question of how coalitions and alliances might produce such emergent properties in society (Hinde 1976) is of course closely tied to the question addressed in the previous section of how the nature of resources and competition for them, both between social groups and within them might influence the use of cooperation as a competitive strategy.

CONCLUSION

Like many of the authors throughout the book, in this concluding chapter we have emphasized one class of explanation for the use of coalitions and alliances, the level of analysis of pay-offs. This category of explanation produces one sort of answer to some of the phenomena. It does not explain all the phenomena we see, and is not the only explanation for the ones it does explain. Being zoologists, we have not given enough space in this chapter to cultural, political, psychological, or sociological explanations of cooperative behaviour. Nevertheless, a concentration on consequences of coalitions and alliances has not meant that other explanations have been ignored. Proximate, developmental, and evolutionary explanations are considered in a number of the chapters. An advantage of the functional approach in the context of this book is that its reasoning allows it to be applied in detail to both animal as well as human behaviour. It might therefore provide more potential for integrating the viewpoints of different disciplines than does any of the other three levels of biological analysis. Our hope is that by bringing together studies of coalitions and alliances from different fields of interest we may have stimulated, if not integration, then at least some cross-fertilization.

ACKNOWLEDGEMENTS

We thank Kelly Stewart for comments on this chapter.

REFERENCES

Adams, E. S. (1990). Boundary disputes in the territorial ant *Azteca trigona*: effects of assymetries in colony size. *Animal Behaviour*, **39**, 321–8.

Alexander, R. D. (1974). The evolution of social behavior. *Annual Review of Ecology and Systematics*, **5**, 325–83.

Alexander, R. D. (1984). *The biology of moral systems*. Aldine de Gruyter, New York.

Anderson, C. O. and Mason, W. A. (1974). Early experience and complexity of social organization in groups of young rhesus monkeys (*Macaca mulatta*). *Journal of Comparative and Physiological Psychology*, **87**, 681–90.

Axelrod, R. and Hamilton, W. D. (1981). The evolution of cooperation. *Science*, **211**, 1390–6.

Bateson, P. P. G. (1982). Behavioural development and evolutionary processes. In *Current problems in sociobiology* (ed. King's College Sociobiology Group), pp. 133–51. Cambridge University Press, Cambridge.

Bercovitch, F. B. (1988). Coalitions, cooperation and reproductive tactics among adult male baboons. *Animal Behaviour*, **36**, 1198–209.

Bethlehem, D. W. (1982). Anthropological and cross-cultural perspectives. In *Cooperation and competition in humans and animals* (ed. A. M. Colman), pp. 250–68. Van Nostrand Reinhold, Wokingham.

Borgerhoff-Mulder, M. (1989). Early maturing Kipsigis women have higher reproductive success than late maturing women and cost more to marry. *Behavioral Ecology and Sociobiology*, **24**, 145–53.

Bygott, J. D., Bertram, B. C. R., and Hanby, J. P. (1979). Male lions in large coalitions gain reproductive advantages. *Nature*, **282**, 839–41.

Chagnon, N. A. (1979*a*). Kin selection, and conflict: an analysis of a Yanomamö ax fight. In *Evolutionary biology and human social behavior: an anthropological perspective*. (eds N. A. Chagnon and W. Irons), pp. 213–38. Duxbury Press, Massachusetts.

Chagnon, N. A. (1979*b*). Mate competition, favoring close kin, and village fissioning among the Yanomamö Indians. In *Evolutionary biology and human social behavior: an anthropological perspective*. (eds N. A. Chagnon and W. Irons), pp. 86–132. Duxbury Press, Massachusetts.

Chapais, B. (1983). Adaptive aspects of social relationships among adult rhesus monkeys. In *Primate social relationships* (ed. R. A. Hinde), pp. 286–9. Blackwell, Oxford.

Cheney, D. L. (1983). Extrafamilial alliances among vervet monkeys. In *Primate social relationships* (ed. R. A. Hinde), pp. 278–86. Blackwell, Oxford.

Cheney, D. L. and Seyfarth, R. M. (1986). The recognition of social alliances by vervet monkeys. *Animal Behaviour*, **34**, 1722–31.

Cheney, D. L. and Seyfarth, R. M. (1990). *How monkeys see the world: inside the mind of another species*. University of Chicago Press, Chicago.

Datta, S. B. (1989). Demographic influences on dominance structure among female primates. In *Comparative socioecology* (eds V. Standen and R. A. Foley), pp. 265–84. Blackwell, Oxford.

Darwin, C. (1859). *On the origin of species*. John Murray, London.

Davies, N. B. and Houston, A. I. (1981). Owners and satellites: the economics of territory defence in the pied wagtail, *Motacilla alba*. *Journal of Animal Ecology*, **50**, 157–80.

Deacon, T. W. (1990). Fallacies of progression in theories of brain-size evolution. *International Journal of Primatology*, **11**, 193–236.

Diamond, M. C. (1988). *Enriching heredity: the impact of the environment on the anatomy of the brain*. Free Press, New York.

Dunbar, R. I. M. (1988). *Primate social systems*. Croom Helm, London.

Elgar, M. (1986). House sparrows establish foraging flocks by giving chirrup calls if the resources are divisible. *Animal Behaviour*, **34**, 169–74.

Emlen, S. T. (1982). The evolution of helping. II. The role of behavioral conflict. *American Naturalist*, **119**, 40–53.

Emlen, S. T. (1984). Cooperative breeding in birds and mammals. In *Behavioral ecology* (eds. J. R. Krebs and N. B. Davies), pp. 305–39. Blackwell, Oxford.

Essock-Vitale, S. and Seyfarth, R. M. (1987). Intelligence and social cognition. In *Primate societies* (eds B. B. Smuts, D. L. Cheney, R. M. Seyfarth, R. W. Wrangham, and T. T. Struhsaker), pp. 452–61. University of Chicago Press, Chicago.

Fragaszy, D. M. and Mason, W. A. (1983). Comparative studies of feeding in captive squirrel and titi monkeys. *Journal of Comparative Psychology*, **97**, 310–26.

Hallpike, C. R. (1988). *The principles of social evolution*. Clarendon Press, Oxford.

Hamilton, W. D. (1964). The genetical evolution of social behaviour. *Journal of Theoretical Biology*, **7**, 1–52.

Harcourt, A. H. (1987). Cooperation as a competitive strategy in primates and birds. In *Animal societies: theories and facts* (eds. Y. Ito, J. L. Brown, and J. Kikkawa), pp. 141–57. Japan Scientific Societies Press, Tokyo.

Harcourt, A. H. (1988). Social influences on competitive ability: alliances and their consequences. In *Machiavellian intelligence. Social expertise and the evolution of intellect in monkeys, apes and humans* (eds R. W. Byrne and A. Whiten), pp. 132–52. Clarendon Press, Oxford.

Harris, M. (1979). *Cultural materialism. The struggle for a science of culture*. Random House, New York.

Hinde, R. A. (1975). The concept of function. In *Function and evolution in behaviour* (eds G. P. Baerends, C. de Beer, and A. Manning), pp. 3–15. Clarendon Press, Oxford.

Hinde, R. A. (1976). Interactions, relationships, and social structure. *Man*, **11**, 1–17.

Hinde, R. A. (1982). *Ethology*. Fontana Paperbacks/Oxford University Press, Oxford.

Humphrey, N. K. (1976). The social function of intellect. In *Growing points in ethology* (eds P. P. G. Bateson and R. A. Hinde), pp. 303–17. Cambridge University Press, Cambridge.

Jerison, H. J. (1973). *Evolution of the brain and intelligence*. Academic Press, New York.

Johnson, A. W. and Earle, T. (1987). *The evolution of human societies*. Stanford University Press, Stanford.

Keddy Hector, A. C., Seyfarth, R. M., and Raleigh, M. J. (1989). Male parental care, female choice and the effect of an audience in vervet monkeys. *Animal Behaviour*, **38**, 262–71.

Kitcher, P. (1985). *Vaulting ambition*. MIT Press, Cambridge, MA.

Kummer, H., Dasser, V., and Hoyningen-Huene, P. (1990). Exploring primate social cognition: some critical remarks. *Behaviour*, **112**, 84–98.

Lazarus, J. and Metcalfe, N. B. (1990). Tit-for-tat cooperation in sticklebacks: a critique of Milinski. *Animal Behaviour*, **39**, 987–8.

Lederman, R. (1986). *What gifts engender*. Cambridge University Press, Cambridge.

Mangel, M. (1990). Resource divisibility, predation and group formation. *Animal Behaviour*, **39**, 1163–72.

Mangel, M. and Clark, C. W. (1988). *Dynamic modeling in behavioral ecology*. Princeton University Press, Princeton.

Maynard Smith, J. (1982). *Evolution and the theory of games*. Cambridge University Press, Cambridge.

Meggitt, M. (1977). *Blood is their argument*. Mayfield, Palo Alto.

Milinski, M. (1987). Tit for tat in sticklebacks and the evolution of cooperation. *Nature*, **325**, 433–5.

Murnighan, J. K. (1978). Models of coalition behavior: game theoretic, social psychological, and political perspectives. *Psychological Bulletin*, **85**, 1130–53.

Noë, R. (1990). A veto game played by baboons: a challenge to the use of the prisoner's dilemma as a paradigm for reciprocity and cooperation. *Animal Behaviour*, **39**, 78–90.

Packer, C. (1977). Reciprocal altruism in olive baboons. *Nature*, **265**, 441–3.

Pusey, A. E. and Packer, C. (1987). Dispersal and philopatry. In *Primate societies* (eds B. B. Smuts, D. L. Cheney, R. M. Seyfarth, R. W. Wrangham, and T. T. Struhsaker), pp. 250–66. University of Chicago Press, Chicago.

Reboreda, J. C. and Kacelnik, A. (1990). On cooperation, tit-for-tat and mirrors. *Animal Behaviour*, **40**, 1188–9.

Reyer, H.-U. (1984). Investment and relatedness: a cost/benefit analysis of breeding and helping in the pied kingfisher (*Ceryle rudis*). *Animal Behaviour*, **32**, 1163–78.

Rodman, P. S. (1988). Resources and group sizes of primates. In *The ecology of social behavior* (ed. C. N. Slobodchikoff), pp. 83–108. Academic Press, San Diego.

van Schaik, C. P. (1983). Why are diurnal primates living in groups? *Behaviour*, **87**, 120–44.

Seyfarth, R. M. (1980). The distribution of grooming and related behaviours among adult female vervet monkeys. *Animal Behaviour*, **28**, 798–813.

Seyfarth, R. M. and Cheney, D. L. (1988). Empirical tests of reciprocity theory: problems in assessment. *Ethology and Sociobiology*, **9**, 181–8.

Stammbach, E. (1988). Group responses to specially skilled individuals in a Macaca fascicularis group. *Behaviour*, **107**, 241–66.

Tinbergen, N. (1963). On aims and methods of ethology. *Zeitschrift für Tierpsychologie*, **20**, 410–33.

Trivers, R. L. (1971). The evolution of reciprocal altruism. *Quarterly Review of Biology*, **46**, 35–57.

Turner, J. C. (1982). Intergroup conflict and cooperation. In *Cooperation and competition in humans and animals* (ed. A. M. Colman), pp. 218–49. Van Nostrand Reinhold, Wokingham.

Vayda, A. P. (1961). Expansion and warfare among swidden agriculturalists. *American Anthropologist*, **63**, 346–58.

Vehrencamp, S. L. (1983). A model for the evolution of despotic versus egalitarian societies. *Animal Behaviour*, **31**, 667–82.

de Waal, F. (1982). *Chimpanzee politics*. George Allen and Unwin, London.

Wilkinson, G. S. (1984). Reciprocal food sharing in the vampire bat. *Nature*, **308**, 181–84.

Wrangham, R. W. (1980). An ecological model of female-bonded primate groups. *Behaviour*, **75**, 262–300.

Author index

Taxonomic index

Subject index

The words 'coalition', 'alliance', 'cooperation', etc., are used as sparingly as possible in the index, since most headings refer to them anyway. Thus if you wish to find references to, for instance, the effectiveness of coalitions, or alliances among females, go to 'effectiveness', not 'coalitions', and 'Female–female', not 'Alliances'.